Flow Sensing in Air and

Horst Bleckmann · Joachim Mogdans
Sheryl L. Coombs
Editors

Flow Sensing in Air and Water

Behavioral, Neural and Engineering Principles of Operation

 Springer

Editors
Horst Bleckmann
Joachim Mogdans
Institute for Zoology
University of Bonn
Bonn
Germany

Sheryl L. Coombs
Department of Biological Sciences
Bowling Green State University
Bowling Green, OH
USA

ISBN 978-3-662-50963-0   ISBN 978-3-642-41446-6   (eBook)
DOI 10.1007/978-3-642-41446-6
Springer Heidelberg New York Dordrecht London

Printed on acid-free paper

Springer is part of Springer Science+Business Media (www.springer.com)

Congress Participants

1. J. L. van Hemmen, 2. H. Bleckmann, 3. S. L. Coombs, 4. J. Mogdans, 5. J. Song, 6. B. Fritzsch, 7. M. McHenry, 8. A. Dagamseh, 9. M. Gerson, 10. J. Casas, 11. J. Hellinger, 12. A. Steiner, 13. M. Yoshizawa, 14. P. Pirih, 15. S. van Netten, 16. A. M. Simmons, 17. B. P. Chagnaud, 18. J. F. Webb, 19. A. Chicoli, 20. A.T. Klein, 21. H. Herzog, 22. T. Steinmann, 23. H. Droogendijk, 24. G. Sendin, 25. J. Engelmann, 26. V. Hofmann, 27. J. M. Gardiner, 28. S. Windsor, 29. F. Branoner, 30. N. Wood, 31. T. Kohl, 32. K. Yanase, 33. A. Ghysen, 34. T. Burt de Perera, 35. J. C. Montgomery, 36. G. Krijnen, 37. S. Wieskotten, 38. C. Howard, 39. S. Kranz, 40. D. L. Macmillan, 41. T. Bachmann, 42. F. Oschmann, 43. J. Pujol-Martí, 44. S. Sterbing d'Angelo, 45. H. G. Krapp, 46. Y. Krüger, 47. A. Skorjanc, 48. J. Liao, 49. P. Oteiza, 50. J. Franken, 51. B. Niesterok, 52. G. Dehnhardt, 53. T. Erlinghagen, 54. W. Hanke, 55. M. C. Fiazza, 56. J. B. Coleman, 57. C. M. Harley, 58. R. Zelick, 59. M. Frings, 60. S. Schwarze, 61. F. Clotten, 62. S. Blazek, 63. L. Chambers, 64. F. Rizzi, 65. A. Qualtieri, 66. F. G. Barth, 67. M. Bothe, 68. F. Kaldenbach, 69. J. Winklnkemper, 70. L. Miersch, 71. O. Akanyeti, 72. J. C. Brown, 73. B. Honisch

Preface

Although special flow-sensing abilities are absent in humans, countless aquatic (e.g., fish, cephalopods, crustaceans, some marine mammals), terrestrial (e.g., crickets, spiders and scorpions), and aerial (e.g., bats and perhaps birds) animals have flow sensing abilities that underlie remarkable behavioral feats. These include the ability to (1) identify and localize air or water borne prey signals, (2) follow silent hydrodynamic trails many seconds after the trail blazer has left the scene, (3) form hydrodynamic images of the environment in total darkness, and (4) swim or fly efficiently and effortlessly in the face of destabilizing currents and winds. In recognition of the increasing wealth of information on flow sensing systems in diverse species and the recent surge of engineering interest in biomimetics, an international conference on *Flow Sensing in Air and Water* was convened at the University of Bonn, Germany in July 2011. Leading scientists from all over the world and from different disciplines came together to share information on these fascinating systems so that basic principles of operation might be identified and applied to engineering applications involving autonomous control of underwater or aerial vehicles.

As the published proceedings of this unusual conference, this volume serves as a valuable reference for students and researchers alike in diverse disciplines. Each contribution provides a unique blend of literature review and current research from experts in the fields of sensory biology, neuroethology, computational neuroscience, and engineering. Thus, the entire volume provides a comprehensive survey of flow sensing systems in a variety of different animals over a wide range of topics, including the morphological and functional diversity of flow sensors, spatial and temporal characteristics of air and water flows encountered by animals in their natural environments, mechanosensory transduction mechanisms, processing of flow stimuli by the peripheral and central nervous system, signal analysis, neuronal modeling, and the engineered design of artificial flow sensors and processing algorithms for guiding autonomous vehicles.

Our hope is that this book will serve not only as a reference volume for those interested in flow sensing systems, but also as a source of bioinspiration for engineers and others interested in how behavior is guided by flow.

Acknowledgments

As conference organizers, we would like to thank the Deutsche Forschungsgemeinschaft, the Office of Naval Research, the Company of Biologists, and finally, the Leopoldina for funds to support the conference. Without financial support from these agencies, our conference would have been impossible to convene.

In addition, we would like to thank the University of Bonn, which hosted the conference.

As editors, we thank Mrs. Lindqvist for her excellent editorial assistance.

Finally, as scientists, we would like to extend our gratitude to the conference participants and the authors of the book chapters, who made significant contributions to our understanding of flow sensing systems.

Contents

Part I
Spatio-Temporal Structure of Natural Water and Air Flow Stimuli

Chapter 1
Natural Hydrodynamic Stimuli

Wolf Hanke

Abstract Aquatic animals of all major phyla have developed sensory systems to perceive water movements, so-called hydrodynamic sensory systems. These water movements, or hydrodynamic stimuli, arise from a variety of sources. Some sources of hydrodynamic stimuli are biotic, such as predators, prey, and conspecifics, some are abiotic, such as wind, gravity that induces currents, and others. We have only relatively recently begun to take a closer look at these hydrodynamic stimuli with regard to the question how they may have formed the hydrodynamic sensory systems of aquatic animals during evolution. Hydrodynamic stimuli are measured with several different techniques, some of which are invasive, meaning that a sensor is inserted into the flow, some are noninvasive, such as optical and acoustical techniques. The laser-based technique of particle image velocimetry (PIV) has proven especially helpful. It is an optical technique that measures flow velocities not only in a single point, but simultaneously in hundreds or thousands of points in a selected layer of the fluid, and with more advanced modifications of the technique, even in a limited volume. Furthermore, the laser-based technique of laser Doppler velocimetry has contributed much to our understanding. In this chapter, naturally occurring hydrodynamic stimuli measured with these and other techniques are discussed, along with the advantages and shortcomings of the different experimental approaches.

Keywords Hydrodynamic sensory systems · Hydrodynamic perception · Mechanosensory systems · Lateral line · Vibrissal system · Predator–prey interaction · Sensory ecology · Fish · Pinnipeds · Aquatic mammals · Flow measurement · Particle image velocimetry · Laser Doppler anemometry

W. Hanke (✉)
Institute for Biosciences and Marine Science Center, University of Rostock,
Albert-Einstein-Strasse 3, 18059 Rostock, Germany
e-mail: wolf.hanke@uni-rostock.de; wolf.hanke@gmail.com

H. Bleckmann et al. (eds.), *Flow Sensing in Air and Water*,
DOI: 10.1007/978-3-642-41446-6_1, © Springer-Verlag Berlin Heidelberg 2014

1.1 Introduction

Hydrodynamic stimuli are water movements that can be sensed by animals. This broadest possible definition comprises what we commonly think of as flow, but it also includes acoustic (sound) waves and surface waves. In a narrower sense, the term hydrodynamic stimuli is often used to describe only flow phenomena that are not governed by a wave equation, thus excluding sound waves or surface waves. This chapter briefly reflects on sound waves and surface waves, but focuses on hydrodynamic stimuli in the narrower sense.

Hydrodynamic sensory systems are designed to perceive hydrodynamic stimuli. They are found in nearly all animal phyla that inhabit aquatic environments (Bleckmann 1994). There is a broader and a more narrow definition for hydro-dynamic sensory systems; these are however not related to the broader and the narrower definition of hydrodynamic stimuli.

The broader definition includes the perception of water movements that are so large-scaled that they move the whole animal, leaving little or no relative move-ment between the animal and the water. These hydrodynamic sensory systems are the inner ear of fishes (Fay 1984; Sand 1974), the statocysts of cephalopods (Packard et al. 1990; Mooney et al. 2010), and, depending on where one wants to draw the line between a specialized gravity sensor and a general acceleration sensor, potentially all sensory organs with a similar structural design.

In a narrower sense, hydrodynamic sensory systems measure the flow relative to the animal (Bleckmann 1994; Dijkgraaf 1963). Probably best studied are the sensory hairs of crustaceans (e.g., Wiese 1976), the lateral line systems of fish and amphibians (for reviews, see Bleckmann and Zelick 2009; Bleckmann 2008), and the vibrissal (whisker) system of aquatic mammals (for review, see Hanke et al. 2013). These convergently evolved sensory systems share certain features that reflect their common function, but originate from completely different anatomical structures. In each case, innervated structures that can easily be deflected by water movements emanate from the skin or cuticle. Sensory cells react to the deflection of the structure with hyper- or depolarization, transducing the stimulus into neural information that is transmitted to the central nervous system. In crustaceans, the deflectable structures are sensory hairs of variable shapes, subsets of which are often (not consistently) called setae (Wiese 1976; Yen et al. 1992; Fields et al. 2002). In this context, it is worth mentioning that terrestrial arthropods possess sensory hairs as well, such as the trichobothria of spiders (Barth 2000) or the filiform sensillae of insects (Humphrey and Barth 2007, and references therein); they are used to detect air movements. Fish and aquatic amphibians possess a lateral line system. The basic unit of the lateral line system is the neuromast. The neuromast is a conglomeration of supporting cells and few up to several thousand sensory hair cells whose kinocilia and stereovilli project into a jelly-like cupula, which is tens to a few hundred micrometers in size and is deflected by the flow to be sensed (Coombs et al. 1988). In fish, two lateral line subsystems are present, the superficial lateral line with the superficial neuromasts (SNs) and the (more conspicuous and name-giving) canal lateral line with the canal neuromasts (CNs) (Coombs et al. 1988). The cupulae of the SNs, potentially located on all body parts including the tail fin, project into the water. The cupulae of the CNs are shielded from the free water by being located in canals embedded in the skin or in dermal bones, particularly on the head. The canals contain fluid and open to the environment through canal pores (which can be covered by a thin membrane). The canals are accessory structures that transform pressure gradients along the outside of the canal into fluid velocity inside the canal (Denton and Gray 1988). In mammals, the vibrissae (whiskers) can serve as hydrodynamic sensors. This has been demonstrated in harbor seals (*Phoca vitulina*), California sea lions (*Zalophus californianus*), the American water shrew (*Sorex palustris*) and the Australian water rat (*Hydromys chrysogaster*). Harbor seals have most extensively been studied (Dehnhardt et al., this volume). The vibrissae of aquatic mammals are only a few millimeters or up to several decimeters long and are the largest flow-sensing structures known in animals.

To fully understand the capabilities of a sensory system and the role it plays in the life of an animal, it is important to know the kind of biologically meaningful stimuli that are present in the natural environment, and whether and how these stimuli are masked by noise. This knowledge is not only important for the design of an experiment but also for the correct interpretation of the data obtained in physiological or behavioral experiments (Engelmann et al. 2000, 2002). This is even more important when experiments are designed to uncover higher central nervous processes or behavioral responses that involve sophisticated information processing and feature recognition. This chapter reviews what we know, what we can infer, and what we have reasonable grounds to suspect about natural hydrodynamic stimuli in the animals' habitat.

Hydrodynamic stimuli can be of abiotic origin, self-generated, or externally biotic. Abiotic hydrodynamic stimuli are caused by inanimate objects or events. Self-generated hydrodynamic stimuli are the water movements caused by the sensing animal itself, often in interaction with its inanimate surroundings. External biotic hydrodynamic stimuli are caused by other animals, such as predators, conspecifics, or prey.

After introducing the various flow measurement techniques, this chapter reviews our knowledge about abiotic, self-generated, and external biotic natural hydrodynamic stimuli. In addition, it includes biological experiments that have been conducted to elucidate how and for what purpose these stimuli are sensed.

1.2 Measurement Techniques

1.2.1 Flow Measurement Techniques

A variety of methods have been developed to measure water (and other fluid) flow, some of which are invasive, meaning that a probe is inserted into the water, and some are noninvasive, using optical or acoustical techniques. Overviews are given, e.g., in Nitsche and Brunn (2006) or Durst (2010). Each of these techniques has its advantages and disadvantages; however, none of them matches the sensory abilities of aquatic animals in all respects, rendering all attempts to assess a complete description of the hydrodynamic stimuli at a certain location with only one technique futile. What would be desired is a system that measures the water movements, either in the form of displacement, velocity, or acceleration, (a) in the whole volume that influences the sensory system, and (b) at displacement amplitudes down to the nanometer range, because that is what the lateral line and inner ear of fish may resolve at higher frequencies; (c) with velocity amplitudes well below 1 mm s^{-1}, because that is what the lateral line of fish or the vibrissae of seals can resolve; and (d) at accelerations down to 10^{-5} m s^{-2}, because that is what the inner ear of fish can perceive. These measurements should (e) be based on a sampling rate of at least 200 Hz for studies of those sensory systems that measure water flow relative to the animal, or 600 Hz for studies of those sensory systems that measure whole-body movements (applying the Nyquist criterium to the rule-of-thumb that the former have an operating range of zero to more than 100 Hz, the latter from 0 to 300 Hz). The following section briefly summarizes the characteristic features of different measurement principles, and how they match certain aspects of the abilities of hydrodynamic sensory systems.

1.2.1.1 Invasive Techniques

Wheel flow meters measure flow speed via the rotational speed of a wheel that is driven by the flow. They are a cost-efficient standard in limnological field studies. However, they have limited spatial resolution, and the inertia of the moving mass limits their frequency resolution so that usually mean values over many seconds are reported (e.g., Fulford 2001).

Dynamic pressure probes allow the measurement of flow speed at a single point if the static pressure and the temperature are known. Dynamic and static pressure can be measured simultaneously by combining two probes, e.g., in a Prandtl probe, but cannot be measured at the same point, causing errors especially in turbulent water. Apart from this problem, the measurement principle allows to assess higher frequencies depending on the operating speed of the pressure transducer, but only at moderate to high flow speeds. Dynamic pressure probes depend on the angle of flow impact and thus have a distinct directional characteristic (Nitsche and Brunn 2006).

Hot wire anemometers measure flow speed via the rate at which the flow cools a small heated wire. They have a good velocity and frequency resolution, but require calibration of the probe and the subsequent electronics and are delicate and prone to damage. Velocity measurements are taken at a single point in space (the size of the probe is typically 1 mm). Hot wire anemometers have been used in studies of the fish lateral line (Mogdans and Bleckmann 1998) in the laboratory, but not in the field.

Special-designed and biomimetic measurement devices have been developed to meet special requirements. Artificial hair arrays can measure flow signatures near walls (Brücker, this volume). Artificial lateral line detectors based on the canal lateral line of fish can measure pressure gradients near walls (Klein and Bleckmann, this volume). Their application for the quantification of natural hydrodynamic stimuli has been proposed and seems promising. Kalmijn and Enger developed a direct current (DC) to 300 Hz linear underwater displacement transducer [unpublished, cited after Kalmijn (1988)] and used it for two semi-quantitative measurements of the displacement caused by swimming goldfish (Kalmijn 1989).

1.2.1.2 Noninvasive Techniques

Acoustic Doppler Velocimetry (ADV) Acoustic Doppler velocimeters measure flow speed by emitting sound waves and by measuring the frequency shift (due to the Doppler effect) in the backscattered acoustic wave. They can measure the three-dimensional flow speed with an accuracy of ± 1 mm s^{-1}, averaged across a volume that spans 15 by 5 mm at moderate sampling rates of, e.g., 60–70 Hz.

Laser Doppler Anemometry (LDA) LDA measures flow velocity by the frequency shift (due to the Doppler effect, made accessible via the beat frequency between two laser beams) of back-scattered laser light. LDA can operate in one, two, or three dimensions and measures velocity in a single point. The measurement volume is typically small, e.g., $5 \times 2 \times 2$ mm. The sampling rate of LDAs is potentially high, it does however depend on the seeding of the flow with light-scattering particles. A great advantage is the ability to measure low velocities down to zero with high accuracy (Nitsche and Brunn 2006).

Particle-Image-Velocimetry (PIV) PIV (Adrian 1991; Willert and Gharib 1991; Westerweel 1997) is an imaging technique. A single plane of the fluid is illuminated using a fanned-out or moving laser light beam. The fluid is seeded with neutrally buoyant particles that scatter the laser light. Then the illuminated plane is recorded with a camera. The shift of the particles between two successive camera frames is analyzed using a variety of calculation methods, many of which are based on cross-correlation or Fast Fourier Transformation (FFT). To do so, each camera frame is divided into many subsections called Interrogation Areas (IRs). From each IR, a velocity value is calculated by taking the particle shift and the time difference between the two camera frames into account. An interrogation area must contain at least one particle (in both frames) to yield a meaningful velocity vector, but in order to reduce measurement noise, five or more particles are rec-ommended. Thus, the spatial resolution of PIV depends predominantly on the density of the light-scattering particles. Time resolution depends on the frame rate of the camera, e.g., 25–30 frames per second (fps) with affordable consumer cameras or 1,000, 6,000 and more fps with modern high speed cameras.

PIV is a whole-field technique, meaning that fluid velocities can be measured at many points of the measurement plane simultaneously. This is its specific advantage over all other measurement techniques described here, with the exception of artificial hair arrays. Artificial hair arrays are limited to operating close to walls (where PIV, on the other hand, is difficult to apply) and are not yet readily available outside the laboratories that developed them. One of the disad-vantages of PIV is that a minimal particle shift of at least 0.1 pixels is needed between two camera frames (or more safely 0.5 pixels to 1 pixel in case of imaging noise), meaning that if fluid velocities are low, several frames must be dropped until the minimum shift is reached, which in turn reduces frequency resolution.

The basic PIV design measures two components of the fluid velocity, namely the two components orthogonal to the optical axis of the camera, and it does so in one plane, namely the plane illuminated by the laser light. This design has been modified to measure flow in multiple planes by scanning devices (Brücker 1995; Hanke and Lauder 2009; Hanke and Bleckmann 2004). The flow in three dimensions, i.e., including the velocity component along the optical axis of the camera, has been measured using color-coded light sheets (Brücker 1996), per-pendicular light sheets (Brücker 1997), or the stereoscopic view provided by two or more cameras (Prasad and Adrian 1993). The latter method has been imple-mented in commercially available PIV systems. Special developments to assess three-dimensional flow include holographic PIV (reviewed by Arroyo and Hinsch 2008; Hinsch 2002), which is so far confined to small volumes, and tomographic PIV (reviewed by Arroyo and Hinsch 2008; Kitzhofer et al. 2011), which can be implemented into PIV systems by some commercial suppliers, but has not yet widely replaced easier and more affordable techniques. In short, PIV techniques are a field of engineering that is still developing rapidly, their progress depending not only on new approaches, but also on advances in camera, laser, and computer technology.

1.2.2 Other Measurement Techniques

To a very limited degree and in a limited set of situations that are not commonly encountered in laboratory settings, hydrodynamic stimuli may be inferred from changes in the underwater pressure. Pressure can be measured with hydrophones, i.e., piezoceramic transducers (reviewed in Urick 1996). However, the relation between changes in pressure and the velocity of the water particles can be cal-culated only for freely propagating sound waves far from the sound source. In that case, the velocity amplitude is the ratio of the sound pressure amplitude and the acoustic impedance, where the acoustic impedance depends on the temperature and salinity of the medium and on sound frequency. More often pressure measurements do not allow to calculate the particle velocity. For example, in a standing sound wave that may build up between two reflecting walls, there are locations (knots) where particles do not move in spite of pressure changes. In front of an advancing object, the measured pressure allows the calculation of particle velocity only if the geometry of the object is known, because it determines how easily a particle can escape around the object to its back side. In the vicinity of complex objects with differently moving surfaces, such as animals with body appendages, the relation of pressure and particle velocity is substantially more complex. However, differential measurements of the pressure at multiple points do allow to calculate the average particle acceleration between them. It is clear that pressure measurements at many points are desirable to fully describe the hydrodynamic field, but this is rarely feasible as pressure measurements are always invasive.

Surface waves can be measured with a Wheatstone bridge and a partially immersed conducting wire, since the resistance of the wire-water-system changes with water level (Bleckmann et al. 1994). Also a variety of custom-made optical methods have been used to measure surface waves (e.g., Bleckmann 1980). These devices are mainly suited for low-amplitude surface waves in laboratory settings; in the field, surface waves can be measured using Acoustic Doppler Current Profilers (Shih 2011).

1.3 Abiotic Hydrodynamic Stimuli

Abiotic hydrodynamic stimuli are generated without an organism being involved. Water obviously moves for many reasons such as wind, gravity, the tides at sea, thermal convection, or geological shifting. Not many studies have analyzed abiotic water movements, neither with respect to their role as hydrodynamic stimuli nor as noise that may mask these stimuli.

Particle Image Velocimetry in natural waters Hanke et al. (2000a) used a custom-made PIV device to measure low-velocity water movements in natural waters (Hanke 2001). An array of battery-operated laser diodes was used to illuminate a horizontal plane in natural waters at a depth of up to 30 cm below the surface; the water surface was covered with an acrylic screen to prevent optical distortion by surface waves, and a PAL standard video camera recorded the illuminated plane. Seeding particles were polymer particles that are produced for heat-coating of metal (Vestosint 1101; Degussa-Hüls-AG, Marl, Germany, now Evonic Industries, Essen, Germany). Particle diameter was approximately 100 μm. As expected, the distribution of particles in the illuminated plane was in most of the settings difficult and particle densities much lower than in laboratory settings were achieved, especially in running water. Measurements were taken in two still waters [(1) a garden pond in Bonn, and (2) Holzlarer See near Bonn, Gemany] and two running waters, [(3) River Sieg, Siegburg, and (4) Pleisbach, a 5 m wide creek, Sankt Augustin, Germany]. Characteristic features of the flow (numbers refer to measurement sites) were: (1) at wind speed up to 3 (Beaufort scale) and at a water depth of 12 cm, the water velocity averaged over the camera view of 7 × 9 cm could at times be lower than 1 mm s^{-1}. Maximal water velocities were lower than 3 mm s^{-1} as long as no moving animal appeared. The water movements showed relatively little vorticity. In two measurements of 10 s duration each, maximal vorticity in the camera view was 0.07 or 0.15 s^{-1}, respectively. (2) Measurements were taken at wind speed 0.5 m s^{-1} or less (Huger Wind Speed Meter WSC 888H) in a plane 20 cm below the water surface at a place where water depth was 50 cm. The camera field of view was 9 × 12 cm. During two measurements of 10 s duration each, separated by 5 min, water velocities averaged over the camera view were on average 0.5 mm s^{-1}. Maximum water velocity was 0.9 mm s^{-1}. Flow had a maximum vorticity of 0.05 s^{-1}, even lower than in measurement site (1).

Measurements in the River Sieg (3) were taken below the bridge of Bundesstrasse 56 at a wind speed of less than 0.5 m s^{-1} (Huger Wind Speed Meter WSC 888H) in a plane approximately 15 cm below the water surface at a place where water depth was 40 cm. PIV images were affected by a moderate degree of water turbidity. Seeding particles were inserted approximately 2 m upstream from the measurement site and were partly swept into the camera view; however, this procedure resulted in low particle densities that did not allow PIV evaluation with cross-correlation algorithms. Instead, single particle velocities were measured manually using National Institute of Health Image (NIH, Bethesda, MD). Water velocities were between 60 and 140 mm s^{-1} and fluctuated strongly over both time and space, e.g., changing direction by 90° within 5 s. In the creek Pleisbach (4), three locations were measured: (a) water depth 10 cm, measurement plane 5.5 cm above ground, ground covered with very uneven rocks; (b) water depth 12 cm, measurement plane 4 cm above ground, ground covered with uneven rocks; (c) water depth 35 cm, measurement plane 14 cm above ground, ground covered with rocks and mud. Again, PIV particles were released approximately 2–3 m upstream and then swept through the measurement plane, resulting in low particle densities that allowed only for visualization and manual measurements with NIH Image. Not unexpectedly, water velocities and the maximal deviation of movement direction from the mean flow direction decreased from a to c (a: 254 ± 17 mm s^{-1}, >30°; b: 110 ± 7 mm s^{-1}, <20°; c: 81 ± 16 mm s^{-1}, <11°). In the eddy water behind a stone (water depth 10 cm), velocities less than 10 % of the free stream velocity were found, varying over 360° in direction. In no case was a repeated flow pattern observed that would allow predictions of flow velocity or direction over time, as for example the regular von Kármán vortex street known from laboratory studies (von Kármán 1911).

In summary, these punctual measurements showed, not unexpectedly, highly diverse flow characteristics in limnic waters. In still waters, low velocity and vorticity was even found close to the surface. Thus, calm waters can provide low noise conditions that are ideal for the perception of hydrodynamic stimuli such as water movements produced by predators or prey. However, more measurements are needed to assess how often and where low noise conditions occur, and how they change with wind speed and water depth. Qualitative descriptions from scuba-divers indicate that at greater depth, specifically below the thermocline, extremely low hydrodynamic background noise can prevail, as indicated by nearly immobile suspended particles (L. Miersch, personal communication). In running water, the perception and discrimination of hydrodynamic stimuli probably is very challenging. However, we found that the canal lateral line system of fish can, by means of accessory structures that modify the stimulus before it reaches the CNs, successfully cope with background turbulence (Bleckmann et al. 2004; Engelmann et al. 2000, 2002).

The flow in running waters can not only be viewed as noise, but may carry information that benefits the animal, thus representing a hydrodynamic stimulus in itself. Migrating fish or fish that follow odor plumes use rheotaxis, i.e., orientation toward the flow, to navigate (Engelmann et al. 2001; Gardiner and Atema 2007; Montgomery et al. 2002). Fish may also seek certain positions in the flow that allow them to save locomotor energy, either in the bow-wake in front of an object or behind a stationary object (Liao et al. 2001, 2003; Liao 2007; Przybilla et al. 2010; Taguchi and Liao 2011). However, the fact that vortex streets caused by stationary objects in natural waters are by far less regular than the von Kármán vortex streets used in laboratory studies (own observations) needs further consideration; the performance and the strategies of fish swimming behind irregularly shaped objects such as natural rocks need to be assessed.

1.4 Self-Generated Hydrodynamic Stimuli

Animals may derive information from the water movements they produce while swimming or gliding. Conclusive evidence exists for Mexican blind cave fish of the genus *Astyanax*. [Researchers report to have used *Astyanax jordani*, formerly also known as *Anoptichthys jordani*, also called *Astyanax mexicanus* or *Astyanax fasciatus* as it appears to be a member of the *A. fasciatus* species group. Phylogeny of the genus *Astyanax*, which includes both blind and sighted fish, is still a matter of discussion (Strecker et al. 2012)].

Blind cave fish can detect stationary objects via self-generated hydrodynamic stimuli [von Campenhausen et al. (1981); Windsor, this volume]. von Campenhausen et al. (1981) recorded the water movements on the surface of an artificial fish while it passed a stationary object at short distance. The fish was equipped with a mechano-electric transducer analogue to a SN (in the transducer, a small shield that was moved by the water covered a photodiode to varying degrees). The recordings showed that water movements close to the surface of the fish indicated gaps (several cm wide) in a wall and vertical bars (several mm wide) that the fish passed at close distance.

Hassan (1985, 1992a, b) used potential flow theory to calculate the self-generated hydrodynamic stimuli caused by rigid fish-like objects gliding with a constant speed. The fish-like objects were either infinitely high (two-dimensional calculations, Hassan 1985), which may be a good approximation for a laterally extremely flattened fish, or rotationally symmetrical (Hassan 1992a, b), which approximates a real fish more realistically. It must be noted that potential theory, the theory of vorticity-free flow, does not describe the boundary layer of a moving object correctly, which always contains shear and thus vorticity. However, the boundary layer of a gliding object is thinnest at its front, where most sensors are located. Possible alterations of the flow by surface structures or boundary layer separations are not covered by the theory and were left to flow measurements or computational fluid dynamics (CFD), techniques not available at that time. However, the calculations of Hassan (1985, 1992a, b) certainly approximate some features of the flow. For example, Hassan (1985) found that a cylinder with a fish-like cross section that passes a round cylinder (with their axes aligned and a movement direction orthogonal to the axes, thus representing a two-dimensional model for a fish passing a bar) can at a mid-body position encounter a decrease, an increase, and again a decrease in flow velocity (positive velocity direction defined from head to tail; compare Fig. 3c in Hassan 1985). This is consistent with measurements von Campenhausen et al. (1981) obtained with their flow transducer mounted to a model fish. Hassan (1992a) found that a fish-like three-dimensional object gliding toward a wall encounters significant changes in flow velocity and pressure gradient across the body surface. He also found that the relative changes in pressure gradient at a given distance from the wall always surpassed the relative changes in velocity, which might indicate that the wall is first sensed by the canal lateral line system. However, this result should not be over-interpreted, since too many parameters of the lateral line canal and the canal fluid are not known. For a fish-like object gliding alongside a plane surface, Hassan (1992b), using the same mathematical approach, found significant changes in the flow velocity across the skin in the rostro-caudal and in the dorso-ventral direction. It must be noted that the lack of viscosity in the theoretical approach, meaning the lack of a boundary layer, leads to a more severe error in the case of a fish gliding past a wall than in the case of a fish heading toward the wall. Hassan (1992b) also found significant changes to the pressure gradients, i.e., the stimulus for the canal lateral line, caused by a nearby wall, but hypothesized that the SNs play a bigger role in the resolution of structural details of an object, according to their higher number and spatial resolution. Abdel-Latif et al. (1990), however, found that a structured, grid-like wall was distinguished from a smooth wall by the CNs, but not by the SNs. Montgomery et al. (2001) found that collisions with a grid partition were avoided using the canal system and not the SNs. It is not known if this holds true for differently structured objects as well, or what role the presumably not negligible hydrodynamic background noise may have played. Windsor et al. (2010a, b) performed measurements of the hydrodynamic stimuli caused by swimming *Astyanax* using PIV. PIV measurements were used to validate calculations using CFD, which reached farther into the boundary layer because of the limited spatial resolution of the PIV setup and general limitations of the PIV technique close to solid surfaces. The results are also reviewed by Windsor in this volume, along with behavioral experiments on the sensory abilities of *Astyanax*. The calculations by

Windsor et al. (2010a, b) show strong viscous effects and a different pressure distribution across the surface of the fish than the inviscid models of Hassan (1985, 1992a, b). CFD calculations, which include viscosity, are clearly superior to potential flow theory in this Reynolds number range.

Self-generated hydrodynamic stimuli caused by swimming movements in the free water may also be sensed and may play a so far underestimated role for the control of fish swimming. Yanase et al. (2012) disrupted the trunk SNs in yellowtail kingfish (*Seriola lalandi*) unilaterally and found that swimming performance and energetic efficiency were impaired.

1.5 External Biotic Hydrodynamic Stimuli

1.5.1 Hydrodynamic Stimuli Caused by Predators

Predators in search of prey in dark or turbid waters may benefit from minimizing self-generated water movements. Evidence for this is scarce. Researchers observed predatory fish under infrared illumination, where vision was at least significantly reduced if not completely excluded [note that a recent study indicates near-infrared vision in an African cichlid (Meuthen et al. 2012)]. It is consistently reported that the predator glided calmly with occasional tail flicks. This was found in bluegill *Lepomis macrochirus* (Enger et al. 1989), where it contrasted strongly with the visually guided behavior in light condition, in pike-perch *Sander lucioperca* (Hanke 2001), which under light conditions usually stayed in hiding and attacked prey when it came into close range, and in the peacock chichlid *Aulonocara stuartgranti* (Schwalbe et al. 2012) that performed burst-and-glide swimming in both light and dark conditions, but at approximately half speed in the dark. The calm swimming style under visually deprived conditions could in part be an attempt to reduce perceivable water motions and to avoid being detected, but it may also serve to enhance the sensitivity of the predator's hydrodynamic and/or acoustic sensory systems by reducing self-generated noise. The hypothesis that glide phases serve to increase the probability of prey detection by reducing self-generated noise is supported by Janssen (1997) who reports that under dark conditions, ruffe (*Gymnocephalus cernuus*) detected prey during glide phases. In the same study yellow perch (*Perca flavescens*), which did not show glide phases while swimming, usually detected prey only when they were resting on the ground and the prey was close. Remarkably, none of the researchers who investigated predator–prey interaction in the dark reported a reaction of the prey to the water movements from the predator's calm, searching swimming motions. Specifically designed experiments would be needed to clarify if this lack of reaction is based on a lack of perception. The water movements caused by a calmly approaching predator should mainly consist of a bow wave similar to the water displacement in the local field of a hydrodynamic dipole.

Significantly stronger hydrodynamic stimuli are caused by the predator while attacking. Two cases can be distinguished: lunge attacks and suction feeding. In the first case, the predator lunges forward to get hold of the prey, while in suction feeding, the predator sucks the prey into its mouth. However, both strategies may also be combined (Kane and Higham 2011). The question regarding the kind of hydrodynamic stimuli an attacking predator causes has rarely been investigated in a sensory biological context, but interest in the biomechanics of suction feeding has led to a number of studies that investigated the flow.

The flow field produced by a typical suction feeding predator, the bluegill sunfish *L. macrochirus*, was investigated by Day et al. (2005). They used PIV to measure the flow in front of the fish during suction feeding with high temporal and spatial resolution. Figure 1.1 shows a representative example and compares feeding events of different duration and peak width from one individual. Day et al. (2005) found that significant flow velocities (>5 % of maximum) were limited to a small region of approximately one mouth diameter in front of the mouth. This finding held true for the whole range of mouth gapes and maximum fluid speeds tested. Fluid speed was maximal shortly before gape width was maximal, so that the fish exerted maximum flow on the prey approximately at the time when the zone of high fluid velocity was largest. The authors predict that this temporal pattern may be generally present in suction-feeding fishes, based on functional anatomical studies in fish species as different as *Amia* (Lauder 1980a) and *Lepomis* (Lauder 1980b). They do, however, also show that theoretical models of suction feeding (Muller et al. 1982) have been too simplistic. That means that temporal and spatial patterns of suction feeding in different fish species should be better assessed by direct flow measurements. A finding of great interest with respect to the question of how prey can sense an attack was that steep gradients occur in the velocity field close to the mouth of *Lepomis*, and gradients fall off quickly with decreasing distance. In the velocity graphs in Fig. 1.3 in the article by Day et al. (2005), a typical gradient amounts to about 100 s^{-1} at a distance of 5 mm from the mouth of a 15 cm long bluegill. By contrast, at a distance of 20 mm, the gradient is only about 10 s^{-1}. Strong velocity gradients should be sensed by prey with a hydrodynamic sensory system such as the lateral line.

Lepomis also combines suction feeding with lunge feeding (synonymously called ram feeding in fishes). Higham et al. (2005) quantified fluid speed and flow patterns in 41 events of suction during ram feeding. Ram speed in 41 feeding sequences, measured at the time of maximum gape, ranged between 0 and 25 cm s^{-1}, and the ratio of ram speed to fluid speed ranged from 0.1 to 19.1 %. Fluid speed was not significantly influenced by ram speed, but the volume of ingested water was. For the question of predator detection considered here, it can be stated that hydrodynamic stimuli from suction were not altered in a reproducible way when combined with swimming. Holzman and Wainwright (2009) quantified the flow caused by bluegill sunfish that combined suction and ram feeding and concluded that at the location where the bow wave of the approaching predator and the flow caused by suction overlap, reduced flow speed and deformation rate as compared to suction alone will occur. They propose that similar

Fig. 1.1 Representative example of hydrodynamic stimuli caused by suction feeding. A bluegill sunfish (*Lepomis gibbosus*) fed on a fish larva while the water velocity was measured using PIV in a vertical plane. **a** Water velocity (coded in *gray scale*) and streamlines at the time of peak gape. At this time, water velocity is maximal. The region of significant fluid velocity induced by suction is constrained to a region in close proximity to the mouth. It extends approximately equidistantly in all directions from the mouth. **b** Profiles of water velocity measured along the centerline transects for three different feedings of the same individual. The velocity profile and the velocity gradient (slope of the profile) are affected both by variation in peak gape (*PG*) and time to peak gape (*TTPG*). Adapted from Day et al. (2005), with permission

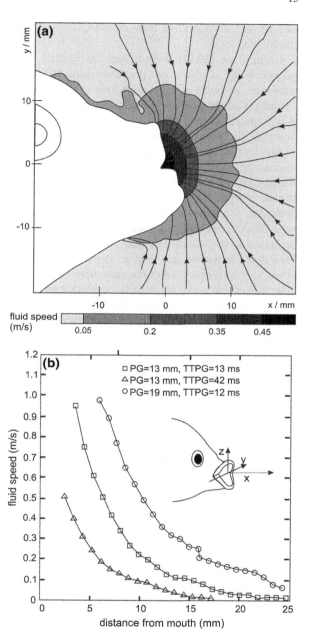

mechanisms are present in many planktivorous fishes. Higham et al. (2005) and Holzman and Wainwright (2009) again used *Lepomis* as a predator; given the diversity in fish functional morphology and the limited use of theoretical studies (see above), comparative flow measurements in feeding fish predators should be conducted.

Colin et al. (2010) investigated the feeding currents of the ctenophore *Mnemiopsis leidyi*, a planktonic predator, using PIV. They found that the deformation rates were an order of magnitude lower than those of predatory fish, and should be virtually undetectable to the ctenophore's prey. This is consistent with the high prey capture rate of *Mnemiopsis*. Other ctenophores may use the same stealth strategy, but have not been investigated quantitatively.

McHenry et al. (2009) showed that larval zebra fish are drastically less likely to respond to a flow stimulus similar to the suction feeding of a predator if their lateral lines are ablated. It should be noted that any similarity between the artificial stimuli in these experiments and the naturally occurring suction feeding flow was limited to acceleration and velocity, while deformation rates were, in contrast with actual suction feeding, minimal. Successful ablation of the lateral line was checked by counting the number of hair cells per neuromast in a separate control group. This does, however, not lead to the conclusion that the lateral line is the only sensory system involved in the detection of suction currents. On the contrary, the inner ear of fish is so sensitive to accelerations in all fish species tested (Sand and Karlsen 1986, 2000; Karlsen 1992a, b) that it must be assumed that it was responsive in the experiments of McHenry et al. (2009) as well. At this point, it can only be speculated on what level of the neural circuitry the responses of the inner ear in the absence of the lateral line are rendered irrelevant in this experimental approach. Karlsen et al. (2004), on the other hand, found that the inner ear was effective in eliciting startle responses. They studied juvenile roach (*Rutilus rutilus*) in a closed water-filled chamber and stimulated the inner ear, but not so much the lateral line, by accelerating the whole chamber. This acceleration elicited startle responses above a threshold of about 0.023 m s^{-2} (r.m.s.) at 6.7 Hz, estimated from the initial half-cycle of the stimulus. The stimulus arrangement made it highly likely, and it was further supported by control experiments where the lateral line was blocked with Co^{2+} in virtually Ca^{2+}-free water (according to Karlsen and Sand 1987), that the lateral line did not play a significant role in eliciting these startle responses. It was further found that startle responses occurred almost exclusively in the trailing half, not in the leading half of the accelerated chamber. In the trailing half, the water is compressed, in the leading half, it is rarefied. Cyprinids are able to discriminate compression from rarefaction of the water (Piddington 1972), and it constitutes a distinguishing feature between the hydrodynamic stimuli caused by a lunge-feeding and a suction feeding predator. Thus, the study of Karlsen et al. (2004) might be interpreted to indicate that roach would rarely flee from a suction feeding-predator. However, not unlike the study of McHenry et al. (2009), it must be noted that the study of Karlsen et al. (2004) lacked the typical velocity gradients and deformation rates found in suction feeding.

1.5.2 Hydrodynamic Stimuli Caused by Conspecifics

There are many reasons why an animal may benefit from knowing the whereabouts and the actions of its conspecifics. Extensive literature exists on benefits and costs of living in groups. An overview is given by Krause and Ruxton (2002). Living in groups can help detecting predators early (Magurran et al. 1985; White and Warner 2007), finding food at a higher rate (Pitcher et al. 1982), and to navigate (Codling et al. 2007) by combining the knowledge of the individual animals into what is known as swarm intelligence (compare Krause et al. 2010).

In dark and/or turbid waters, where vision is limited or impossible, hydrodynamic stimuli are of importance for most animals. Conspecifics that live in groups often stay so close together that vision is still possible even in relatively turbid waters. It has, however, been shown that a blind fish can still school, using hydrodynamic sensory systems alone (Pitcher et al. 1976). ["Schooling" has been defined within the generic term "Shoaling" as a polarized and synchronised swimming behavior (Pitcher and Parrish 1993)]. Experiments by Hanke and Lauder (2009) provided no indication that the structure of a fish school may be determined by hydrodynamic sensing in detail, or that certain positions in a school exist that provide hydrodynamic advantages to school members, as hypothesized earlier (Weihs 1973).

Vision may be useless under water, especially with increased distance to the object of interest. Unfortunately, with increased distance, hydrodynamic perception may also be difficult, as many kinds of hydrodynamic stimuli fall rapidly with the distance. However, this does not hold true for the hydrodynamic trails of swimming animals (see Sect. 1.5.3). Schulte-Pelkum et al. (2007) investigated how a harbour seal (*P. vitulina*) follows the hydrodynamic trail of a conspecific. They trained a harbour seal to swim on one of several paths through the clear water of a pool, approximately 0.5 m below the surface. A second harbor seal was trained to follow the swim path of another harbor seal with a delay of 15–20 s, using only its vibrissal system (Dehnhardt and Kaminski 1995; Dehnhardt et al. 2001; Hanke et al. 2012). Visual and acoustic cues were prevented by an eye mask and headphones; the whole experiment was filmed from above to analyze the swim paths of both the seals. The seal followed the path of its conspecific reliably. We then performed PIV measurements of the hydrodynamic trail of the seal. The experimental setup and a schematic sketch of the structure of a harbour seal's trail are shown in Fig. 1.2. Although safety considerations limited laser power, and particle seeding over the whole width of the trail was not always possible, flow velocities of at least 40 mm s^{-1} after 15 s and 30 mm s^{-1} after 30 s were found after the passage of the seal. Furthermore, the trail spreads into two branches caused by lateral tail beats (Schulte-Pelkum et al. 2007), not unlike the trail of a "typical" fish (Hanke et al. 2000a; Hanke and Bleckmann 2004). It is interesting to note that this kind of hydrodynamic trail was actually utilized by an animal under outdoor conditions, where wind speed and surface waves made it very likely that background flow velocities in the order of centimeters per second must have prevailed.

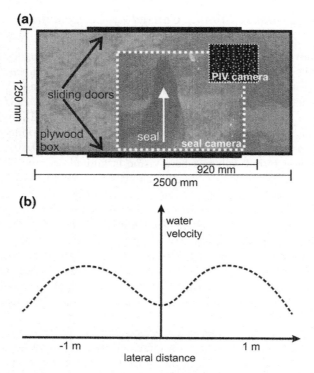

Fig. 1.2 PIV measurements of the hydrodynamic trail behind a harbour seal (*Phoca vitulina*). **a** Experimental setup. The seal was trained to swim through an experimental box (*plywood box*) with two sliding doors. The sliding doors were closed to prepare the measurement area by seeding the water with neutrally buoyant particles. A PIV laser, which was switched off at the beginning of an experiment, illuminated a horizontal plane in the water. After the water had calmed down, the sliding doors were opened and the trained seal was asked to swim through the box. The laser was switched on after the seal's head had exited the box. Two cameras filmed the seal (seal camera) and the PIV particles (PIV camera) from top; the water surface at the position of the PIV camera was covered with an acrylic sheet. **b** Schematic summary of results. PIV measurements with the PIV camera in various positions let conclude that the hydrodynamic trail consisted of two branches of enhanced water movements with a calmer zone in between. After 30 s, the trail was more than 2 m wide and water velocities of about 30 mm s^{-1} were measured in the outer areas of the box. The wake's vertical extent was probably much smaller as our harbour seal swam at a depth of approximately 0.5–1 m, and there was no indication of the wake disturbing the water surface. From Schulte-Pelkum et al. (2007)

Hydrodynamic stimuli in the form of surface waves are used for intraspecific communication in semiaquatic frogs (Walkowiak and Münz 1985) and in fishing spiders (Bleckmann and Bender 1987). Midwater stimuli are used at close range by fish for the synchronization of spawning or in aggressive encounters (reviewed by Bleckmann 1994). For a first example of stimulus quantification using PIV during aggressive behavior, see Hanke et al. (2008).

1.5.3 Hydrodynamic Stimuli Caused by Prey

The detection of prey is probably one of the driving forces in the evolution of hydrodynamic sensory systems. The detection of prey-generated surface waves as well as midwater stimuli has been reviewed by Bleckmann (1994). In a seminal study by Bleckmann et al. (1991), seven species of aquatic animals and the flow that they generated were investigated using LDA (see Sect. 1.2.1). More recently, PIV studies (see Sect. 1.2.1) have begun to complement our knowledge by providing information about the spatial structure and the time course of animal-generated water movements.

Stamhuis and Videler (1995) designed a PIV setup suitable for working with life animals and described three applications: they measured the feeding current of a tethered tetrapod (*Temora longicornis*), the ventilation current of a tube-dwelling thallassinid shrimp (*Callianassa subterranea*), and the vortex street behind a freely swimming juvenile grey mullet (*Chelon* sp.). Detailed results were presented in subsequent studies (Stamhuis and Videler 1998; van Duren et al. 2003; Müller et al. 1997, 2001).

Lauder and coworkers have provided a large amount of PIV data on the flow produced by swimming fishes (Lauder 2011; Lauder and Madden 2008). Starting with the studies by Drucker and Lauder (1999, 2000, 2001) on fluid forces exerted by the sunfish *L. macrochirus*, which has since then become the fish species best studied with PIV, a variety of fish species have been investigated including mackerel *Scomber scombrus* (Nauen and Lauder 2002), salmonids (Standen and Lauder 2007), schools of giant danios *Devario aequipinnatus* (Hanke and Lauder 2009), and sharks (Wilga and Lauder 2000; Flammang et al. 2011). The focus of this work is on biomechanical, functional, and physiological questions of locomotion.

In the context of the perception of hydrodynamic stimuli, Hanke et al. (2000a) and Hanke and Bleckmann (2004) investigated the hydrodynamic trails caused by swimming fish. Hydrodynamic trails are a form of hydrodynamic stimuli that has recently been shown to extend the range of operation of hydrodynamic sensory systems significantly as compared to local hydrodynamic stimuli.

Using a mathematical approach it has been noted early on that hydrodynamic stimuli that do not propagate as a wave fall off steeply with source distance (Harris and van Bergeijk 1962). Figure 1.3a shows the velocity vector field around a dipole, i.e., a sphere oscillating sinusoidally along one axis, calculated using potential flow theory (Milne-Thomson 1968). Hydrodynamic dipoles have been widely used in physiological and behavioral experiments. Figure 1.3b depicts the steep decline of water velocity with the distance from the sphere. This steep decline has led to the well-founded hypothesis that both the lateral line and the inner ear of fish may have evolved to detect the near-field of a hydrodynamic stimulus (Kalmijn 1989). However, it should not be concluded that hydrodynamic stimuli that are not sound waves can only be detected at short distances. For instance, vortices caused by moving animals may outlast their creator by many

Fig. 1.3 Dipole velocity vector field based on potential flow theory. **a** The velocity around a sphere of radius 5 cm, oscillating sinusoidally at 50 Hz at an amplitude of 1 mm, at the time of maximal velocity of the sphere. *Bold arrow* indicates velocity of the sphere. Vectors close to the sphere omitted for clarity. **b** Water velocity along the x axis at $y = 0$ (*lower curve*), and along the y axis at $x = 0$ (*upper curve*). Local flow prevails at these distances. The upper curve contains also water velocities associated with the sound wave; they are, however, of the order of $\mu m\ s^{-1}$

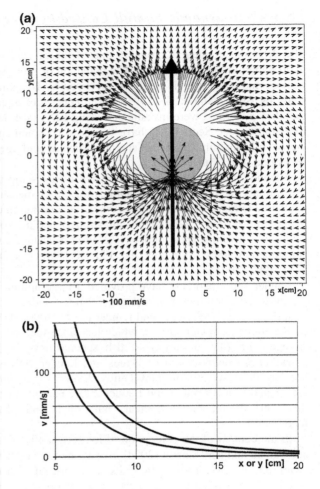

seconds (Bleckmann et al. 1991). Hanke et al. (2000a) investigated the vortex streets left behind by swimming goldfish (*Carassius auratus*) and found that they can be measured for at least 5 min under laboratory conditions, leaving a hydrodynamic trail that offers the potential of being followed by a piscivorous predator (or, for that matter, by a conspecific; compare Sect. 1.5.2). Figure 1.4 (adapted from Hanke and Bleckmann 2004) shows an example of the hydrodynamic trail left behind by a small (86 mm) swimming sunfish (*Lepomis gibbosus*). Figure 1.4a depicts the water velocity vector field measured in a horizontal plane with PIV 60 s after the fish passed. Water velocity is still around 2 mm s^{-1}, and vortex structures are discernible. Figure 1.4b shows the water velocity averaged over the measurement area in three horizontal layers at different heights as a function of time. Along with time, a theoretical distance between the measurement area and the fish is given, based on the assumption that the fish swam away at a speed of 20 cm s^{-1}.

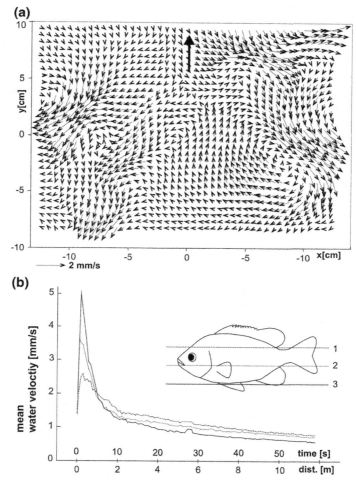

Fig. 1.4 Hydrodynamic trail of a 86 mm long fish (*Lepomis gibbosus*). **a** Water velocity 60 s after the fish passed the area, recorded in a horizontal plane using PIV (see Sect. 1.2.1). *Bold arrow* indicates swimming direction. **b** Water velocity averaged over the recorded area as a function of time in three horizontal layers. Fish passed at t = 0 s; theoretical distance (dist.) to the fish is based on a swimming speed of 20 cm s^{-1}. Adapted from Hanke and Bleckmann (2004)

Average swimming speed in the measurement area was 19 cm s^{-1} (range 2–50 cm s^{-1}). Maximum water velocity after 60 s was still 2 mm s^{-1} (Hanke and Bleckmann 2004). Hydrodynamic trails of this kind would enable an animal that possesses hydrodynamic sensory systems, such as a piscivorous fish, seal, crustacean, or cephalopod, to detect even a small fish at a distance of tens of meters.

Hanke et al. (2000a) found that even the wakes of small, slowly swimming cyprinids (10 cm long goldfish) can still be measured 5 min after the fish swam by.

These values were obtained under laboratory conditions with a background flow of about 0.1 mm s^{-1}. These conditions may be rare in natural habitats, with the exception of the deep zones of limnic waters (compare 1.3), and presumably also with the exception of many waters covered with ice (not yet measured). Niesterok and Hanke (2013) found that water movements caused by the C-starts of teleost fish may last for about 25 min (extrapolated values), even if background noise is present. A more frequent value for how long a trail of a small fish subsists is probably about 1 min. Hanke and Bleckmann (2004) measured the trails of three teleost species, the centrarchid *L. gibbosus* (body length 86 mm), the tetraodontid *Colomesus psittacus* (38 mm), and the chichlid *Thysochromis ansorgii* (86 mm) using scanning PIV. Typical flow velocities measured 60 s after the wake generating fish had passed the measuring plane were 1–2 mm s^{-1}. Trails of the two perciforms typically split up into two or more branches, due to lateral tail movements. Additional branches were caused by movements of the pectoral fins. The trails of the tetraodontid consisted of a single branch, if the tetraodontid swimming style, that is propulsion by undulation of the dorsal and anal fin, was used. Discriminant analysis of the velocity profiles showed that the spatial distribution of the velocity across the width of the trail was sufficient to discriminate the wakes caused by these three species.

While we have reason to believe that the hydrodynamic trails of swimming fish can be detected by predators for up to one minute (provided the background flow is sufficiently low), experimental evidence that predators actually use this strategy is scarce. Dehnhardt et al. (2001) used a miniature submarine to generate hydrodynamic trails that were similar in flow velocity and spatial extent to the wakes of swimming fish, estimated from Hanke et al. (2000a) by linearly scaling up to a fish of 30 cm body length. They trained harbour seals (*P. vitulina*) to follow these trails and to find the submarine using their vibrissal system only (vision and hearing were precluded by an eye mask and headphones, respectively). The harbor seals reliably solved the task even after a delay of 20 s, although background flow in the shallow outdoor facility under varying wind conditions was certainly quite high. Harbor seals could also follow curved paths. Hanke et al. (2000b) and Hanke (2009) observed the piscivorous pike-perch *S. lucioperca* while hunting small prey fish under infrared illumination (i.e., with vision precluded). They found indications of trail following mainly at predator–prey distances up to one predator body length, that is one order of magnitude smaller than in harbor seals. It is conceivable that the good low-frequency hearing abilities of pike-perch as compared to harbor seals makes trail-following, although the physiological prerequisites do exist, less important in the behavioral context in pike-perch than in harbour seals. Pohlmann et al. (2001) demonstrated that the swimming paths of predatory catfish (*Silurus glanis*) and their prey (guppies, *Phoecilia reticulata*) were similar prior to prey capture. However, the three published examples of swimming paths with fish positions (Fig. 2 in Pohlmann et al. 2001; Fig. 1 in Pohlmann et al. 2004) make it appear possible that predator–prey-distances were typically so low that no curved paths were followed.

1.6 Outlook

The current state of research leaves many open questions. Hydrodynamic noise in natural waters should be further quantified, and its influence on hydrodynamic perception should be investigated in specifically designed experiments. Laboratory studies should use more natural hydrodynamic stimuli such as fish wakes. Steps in this direction have been taken by Chagnaud et al. (2006), who used vortex stimuli in physiological experiments on the fish lateral line and by researchers who investigate the vibrissal system of pinnipeds (Wieskotten et al. 2011; Dehnhardt et al., this volume; Wieskotten et al. 2010; Krüger 2011). Behavioral and physiological experiments that used dipole stimuli, i.e., vibrating spheres, and that greatly contributed to our understanding of hydrodynamic sensory systems, should be supplemented by experiments using multipoles. Quadrupole and higher order moment, while negligible in comparison to any dipole moment at large distances form a moving object, must be expected to occur at close range. The role of hydrodynamic trail following and the interaction of hydrodynamic trail perception and hearing in piscivorous predators requires further investigations.

1.7 Conclusions

Natural hydrodynamic stimuli caused by inanimate and animate objects such as submerged rocks or moving animals are highly diverse. After at least six decades of research on hydrodynamic reception, we have only very recently begun to introduce hydrodynamic stimuli that approximate natural hydrodynamic events into experimental settings. It is conceivable that the full capabilities of animals to process, analyze, and recognize hydrodynamic stimuli will only become apparent when natural hydrodynamic stimuli, or close approximations thereof, are applied in physiological and behavioral experiments.

References

Abdel-Latif H, Hassan ES, von Campenhausen C (1990) Sensory performance of blind Mexican cave fish after destruction of the canal lateral neuromasts. Naturwissenschaften 77:237–239. doi:10.1007/bf01138492

Adrian RJ (1991) Particle imaging techniques for experimental fluid mechanics. Annu Rev Fluid Mech 23:261–304

Arroyo MP, Hinsch KD (2008) Recent developments of PIV towards 3D measurements. In: Schroeder A, Willert CE (eds) Particle image velocimetry: new developments and recent applications, vol 112. Topics in Applied Physics, pp 127–154

Barth FG (2000) How to catch the wind: spider hairs specialized for sensing the movement of air. Naturwissenschaften 87:51–58. doi:10.1007/s001140050010

Bleckmann H (1980) Reaction-time and stimulus frequency in prey localization in the surface-feeding fish *Aplocheilus lineatus*. J Comp Physiol A 140:163–172

Bleckmann H (1994) Reception of hydrodynamic stimuli in aquatic and semiaquatic animals. Progress in Zoology 41. Gustav Fischer, Stuttgart, Jena, New York

Bleckmann H (2008) Peripheral and central processing of lateral line information. J Comp Physiol A 194:145–158

Bleckmann H, Bender M (1987) Water surface waves generated by the male pisaurid *Dolomedes triton* (Walckenaer) during courtship behavior. J Arachnol 15:363–369

Bleckmann H, Zelick R (2009) Lateral line system of fish. Integr Zool 4(1):13–25. doi:10.1111/j.1749-4877.2008.00131.x

Bleckmann H, Breithaupt T, Blickhan R, Tautz J (1991) The time course and frequency content of hydrodynamic events caused by moving fish, frogs, and crustaceans. J Comp Physiol A 168:749–757

Bleckmann H, Borchardt M, Horn P, Görner P (1994) Stimulus discrimination and wave source localization in fishing spiders (*Dolomedes triton* and *D. okefinokensis*). J Comp Physiol A 174:305–316

Bleckmann H, Mogdans J, Engelmann J, Kröther S, Hanke W (2004) Das Seitenliniensystem: Wie Fische das Wasser fühlen. Biol unserer Zeit 34:2–9

Brücker C (1995) Digital-Particle-Image-Velocimetry (DPIV) in a scanning light-sheet: 3D starting flow around a short cylinder. Exp Fluids 19:255–263

Brücker C (1996) 3-D PIV via spatial correlation in a color-coded light-sheet. Exp Fluids 21:312–314

Brücker C (1997) Study of the three-dimensional flow in a T-junction using a dual-scanning method for three-dimensional particle-image velocimetry (3D-SPIV). Exp Therm Fluid Sci 14:35–44

Chagnaud BP, Bleckmann H, Engelmann J (2006) Neural responses of goldfish lateral line afferents to vortex motions. J Exp Biol 209(2):327–342

Codling EA, Pitchford JW, Simpson SD (2007) Group navigation and the "many-wrongs principle" in models of animal movement. Ecology 88:1864–1870. doi:10.1890/06-0854.1

Colin SP, Costello JH, Hansson LJ, Titelman J, Dabiri JO (2010) Stealth predation and the predatory success of the invasive ctenophore *Mnemiopsis leidyi*. Proc Natl Acad Sci USA 107:17223–17227. doi:10.1073/pnas.1003170107

Coombs S, Janssen J, Webb JF (1988) Diversity of lateral line systems: evolutionary and functional considerations. In: Atema J, Fay RR, Popper AN, Tavolga WN (eds) Sensory biology of aquatic animals. Springer, New York, Berlin, London, pp 553–593

Day SW, Higham TE, Cheer AY, Wainwright PC (2005) Spatial and temporal patterns of water flow generated by suction-feeding bluegill sunfish *Lepomis macrochirus* resolved by particle image velocimetry. J Exp Biol 208:2661–2671. doi:10.1242/jeb.01708

Dehnhardt G, Kaminski A (1995) Sensitivity of the mystacial vibrissae of harbour seals (*Phoca vitulina*) for size differences of actively touched objects. J Exp Biol 198:2317–2323

Dehnhardt G, Mauck B, Hanke W, Bleckmann H (2001) Hydrodynamic trail following in harbor seals (*Phoca vitulina*). Science 293:102–104. doi:10.1126/science.1060514

Denton EJ, Gray JAB (1988) Mechanical factors in the excitation of the lateral line of fishes. In: Atema J, Fay RR, Popper AN, Tavolga WN (eds) Sensory biology of aquatic animals. Springer, New York, Berlin, London, pp 595–617

Dijkgraaf S (1963) The functioning and significance of the lateral line organs. Biol Rev 38:51–105

Drucker EG, Lauder GV (1999) Locomotor forces on a swimming fish: three-dimensional vortex wake dynamics quantified using digital particle image velocimetry. J Exp Biol 202:2393–2412

Drucker EG, Lauder GV (2000) A hydrodynamic analysis of fish swimming speed: wake structure and locomotor force in slow and fast labriform swimmers. J Exp Biol 203:2379–2393

Drucker EG, Lauder GV (2001) Wake dynamics and fluid forces of turning maneuvers in sunfish. J Exp Biol 204:431–442

Durst F (2010) Fluid mechanics: an introduction to the theory of fluid flows. Springer, Berlin

Engelmann J, Hanke W, Mogdans J, Bleckmann H (2000) Hydrodynamic stimuli and the fish lateral line. Nature 408:51–52

Engelmann J, Hanke W, Bleckmann H (2001) Rheotaxis of still-water fish and running-water fish and the responses of primary lateral line afferents to DC water flow. In: Proceedings of the 28th Göttingen neurobiology conference

Engelmann J, Hanke W, Bleckmann H (2002) Lateral line reception in still- and running water. J Comp Physiol A 188:513–526

Enger PS, Kalmijn AJ, Sand O (1989) Behavioral investigations on the functions of the lateral line and inner ear in predation. In: Coombs S, Görner P, Münz H (eds) The mechanosensory lateral line: neurobiology and evolution. Springer, New York, Berlin, London, Paris, Tokyo, pp 575–587

Fay RR (1984) The goldfish ear codes the axis of acoustic particle motion in three dimensions. Science 225:951–954. doi:10.1126/science.6474161

Fields DM, Shaeffer DS, Weissburg MJ (2002) Mechanical and neural responses from the mechanosensory hairs on the antennule of *Gaussia princeps*. Mar Ecol Prog Ser 227:173–186. doi:10.3354/meps227173

Flammang BE, Lauder GV, Troolin DR, Strand T (2011) Volumetric imaging of shark tail hydrodynamics reveals a three-dimensional dual-ring vortex wake structure. Proc R Soc B 278:3670–3678. doi:10.1098/rspb.2011.0489

Fulford JM (2001) Accuracy and consistency of water-current meters. J Am Water Res Ass 37:1215–1224. doi:10.1111/j.1752-1688.2001.tb03633.x

Gardiner JM, Atema J (2007) Sharks need the lateral line to locate odor sources: rheotaxis and eddy chemotaxis. J Exp Biol 210:1925–1934. doi:10.1242/jeb.000075

Hanke W (2001) Hydrodynamische Spuren schwimmender Fische und ihre mögliche Bedeutung für das Jagdverhalten fischfressender Tiere. PhD thesis, Rheinische Friedrich-Wilhelms-Universität, Bonn

Hanke W (2009) Predation strategy in European pike-perch *Stizostedion lucioperca*: the role of hydrodynamic trail following. Integr Comp Biol 49:E71

Hanke W, Bleckmann H (2004) The hydrodynamic trails of *Lepomis gibbosus* (Centrarchidae), *Colomesus psittacus* (Tetraodontidae) and *Thysochromis ansorgii* (Cichlidae) measured with scanning particle image velocimetry. J Exp Biol 207:1585–1596. doi:10.1242/jeb.00922

Hanke W, Lauder GV (2009) Fish schooling: measurements of flow, school structure, and tail beat frequency. Integr Comp Biol 49:E239

Hanke W, Brücker C, Bleckmann H (2000a) The ageing of the low-frequency water disturbances caused by swimming goldfish and its possible relevance to prey detection. J Exp Biol 203:1193–1200

Hanke W, Meyer E, Bleckmann H (2000b) Behavioural investigations on lateral line function in prey capture in pike-perch (*Stizostedion lucioperca*). Zoology Suppl II I:109

Hanke W, Boyle KS, Tricas TC (2008) Flow measurements during the multimodal communication in Hawaiian butterfly fish. Paper presented at the 16th annual conference of the German Association for Laser Anemometry, Karlsruhe. ISBN: 978-3-9805613-4-1

Hanke W, Wieskotten S, Niesterok B, Miersch L, Witte M, Brede M, Leder A, Dehnhardt G (2012) Hydrodynamic perception in pinnipeds. In: Tropea C, Bleckmann H (eds) Nature-inspired fluid mechanics. Springer, Berlin, pp 225–240

Hanke W, Wieskotten S, Marshall CD, Dehnhardt G (2013) Hydrodynamic perception in true seals (*Phocidae*) and eared seals (*Otariidae*). J Comp Physiol A 199:421–440. doi:10.1007/s00359-012-0778-2

Harris GG, van Bergeijk W (1962) Evidence that the lateral-line organ responds to near-field displacements of sound sources in water. J Acoust Soc Am 34:1831–1841

Hassan E-S (1985) Mathematical analysis of the stimulus for the lateral line organ. Biol Cybern 52:23–36

Hassan ES (1992a) Mathematical description of the stimuli to the lateral line system of fish derived from a 3-dimensional flow field analysis. 1. The cases of moving in open water and of gliding towards a plane surface. Biol Cybern 66:443–452. doi:10.1007/bf00197725

Hassan ES (1992b) Mathematical description of the stimuli to the lateral line system of fish derived from a 3-dimensional flow field analysis. 2. The case of gliding alongside or above a plane surface. Biol Cybern 66:453–461. doi:10.1007/bf00197726

Higham TE, Day SW, Wainwright PC (2005) Sucking while swimming: evaluating the effects of ram speed on suction generation in bluegill sunfish *Lepomis macrochirus* using digital particle image velocimetry. J Exp Biol 208:2653–2660. doi:10.1242/jeb.01682

Hinsch KD (2002) Holographic particle image velocimetry. Meas Sci Technol 13:R61–R72. doi:10.1088/0957-0233/13/7/201

Holzman R, Wainwright PC (2009) How to surprise a copepod: strike kinematics reduce hydrodynamic disturbance and increase stealth of suction-feeding fish. Limnol Oceanogr 54:2201–2212. doi:10.4319/lo.2009.54.6.2201

Humphrey JAC, Barth FG (2007) Medium flow-sensing hairs: biomechanics and models. In: Casas J, Simpson SJ (eds) Advances in insect physiology. Insect mechanics and control, vol 34. Elsevier, Amsterdam, pp 1–80. doi:10.1016/s0065-2806(07)34001-0

Janssen J (1997) Comparison of response distance to prey via the lateral line in the ruffe and yellow perch. J Fish Biol 51:921–930

Kalmijn AJ (1988) Hydrodynamic and acoustic field detection. In: Atema J, Fay RR, Popper AN, Tavolga WN (eds) Sensory biology of aquatic animals. Springer, New York, Berlin, pp 83–130

Kalmijn AJ (1989) Functional evolution of lateral line and inner ear sensory systems. In: Coombs S, Görner P, Münz H (eds) The mechanosensory lateral line: neurobiology and evolution. Springer, New York, Berlin, Heidelberg, pp 187–215

Kane EA, Higham TE (2011) The integration of locomotion and prey capture in divergent cottid fishes: functional disparity despite morphological similarity. J Exp Biol 214:1092–1099. doi:10.1242/jeb.052068

Karlsen HE (1992a) Infrasound sensitivity in the plaice (*Pleuronectes platessa*). J Exp Biol 171:173–187

Karlsen HE (1992b) The inner ear is responsible for detection of infrasound in the perch (*Perca fluviatilis*). J Exp Biol 171:163–172

Karlsen HE, Sand O (1987) Selective and reversible blocking of the lateral line in freshwater fish. J Exp Biol 133:249–262

Karlsen HE, Piddington RW, Enger PS, Sand A (2004) Infrasound initiates directional fast-start escape responses in juvenile roach *Rutilus rutilus*. J Exp Biol 207:4185–4193

Kitzhofer J, Nonn T, Brücker C (2011) Generation and visualization of volumetric PIV data fields. Exp Fluids 51:1471–1492. doi:10.1007/s00348-011-1176-1

Krause J, Ruxton GD (2002) Living in groups. Oxford University Press, Oxford

Krause J, Ruxton GD, Krause S (2010) Swarm intelligence in animals and humans. Trends Ecol Evol 25:28–34. doi:10.1016/j.tree.2009.06.016

Krüger Y (2011) Perception of hydrodynamic stimuli in stationary harbour seals (*Phoca vitulina*). Rostock University, Rostock Diploma thesis

Lauder GV (1980a) Evolution of the feeding mechanism in primitive actinopterygian fishes: a functional anatomical analysis of *Polypterus*, *Lepisosteus*, and *Amia*. J Morphol 163:283–317. doi:10.1002/jmor.1051630305

Lauder GV (1980b) The suction feeding mechanism in sunfishes (*Lepomis*): an experimental analysis. J Exp Biol 88:49–72

Lauder GV (2011) Swimming hydrodynamics: ten questions and the technical approaches needed to resolve them. Exp Fluids 51:23–35. doi:10.1007/s00348-009-0765-8

Lauder GV, Madden PGA (2008) Advances in comparative physiology from high-speed imaging of animal and fluid motion. Annu Rev Physiol 70:143–163. doi:10.1146/annurev.physiol.70.113006.100438

Liao JC (2007) A review of fish swimming mechanics and behaviour in altered flows. Phil Trans R Soc B 362:1973–1993

Liao J, Beal DN, Lauder GV, Triantafyllou M (2001) Novel body kinematics of trout swimming in a von Kármán trail; can fish tune to vortices? Am Zool 41:1505–1506

Liao JC, Beal DN, Lauder GV, Triantafyllou MS (2003) The Kármán gait: novel body kinematics of rainbow trout swimming in a vortex street. J Exp Biol 206:1059–1073

Magurran AE, Oulton WJ, Pitcher TJ (1985) Vigilant behavior and shoal size in minnows. Zeitschrift für Tierpsychologie—J Comp Ethol 67:167–178

McHenry MJ, Feitl KE, Strother JA, Van Trump WJ (2009) Larval zebrafish rapidly sense the water flow of a predator's strike. Biol Lett 5:477–479. doi:10.1098/rsbl.2009.0048

Meuthen D, Rick IP, Thuenken T, Baldauf SA (2012) Visual prey detection by near-infrared cues in a fish. Naturwissenschaften 99:1063–1066. doi:10.1007/s00114-012-0980-7

Milne-Thomson (1968) Theoretical hydrodynamics, 5th edn. Macmillan, London

Mogdans J, Bleckmann H (1998) Responses of the goldfish trunk lateral line to moving objects. J Comp Physiol A 182:659–676

Montgomery JC, Coombs S, Baker CF (2001) The mechanosensory lateral line system of the hypogean form of *Astyanax fasciatus*. Environ Biol Fishes 62:87–96. doi:10.1023/a:1011873111454

Montgomery JC, Macdonald F, Baker CF, Carton AG (2002) Hydrodynamic contributions to multimodal guidance of prey capture behavior in fish. Brain Behav Evol 59:190–198

Mooney TA, Hanlon RT, Christensen-Dalsgaard J, Madsen PT, Ketten DR, Nachtigall PE (2010) Sound detection by the longfin squid (*Loligo pealeii*) studied with auditory evoked potentials: sensitivity to low-frequency particle motion and not pressure. J Exp Biol 213:3748–3759. doi:10.1242/jeb.048348

Muller M, Osse JWM, Verhagen JHG (1982) A quantitative hydrodynamical model of suction feeding in fish. J Theor Biol 95:49–79. doi:10.1016/0022-5193(82)90287-9

Müller UK, van den Heuvel BLE, Stamhuis EJ, Videler JJ (1997) Fish foot prints: morphology and energetics of the wake of a continuously swimming mullet. J Exp Biol 200:2893–2906

Müller UK, Smit J, Stamhuis EJ, Videler JJ (2001) How the body contributes to the wake in undulatory fish swimming: flow fields of a swimming eel. J Exp Biol 204:2751–2762

Nauen JC, Lauder GV (2002) Hydrodynamics of caudal fin locomotion by chub mackerel, *Scomber japonicus* (Scombridae). J Exp Biol 205:1709–1724

Niesterok B, Hanke W (2013) Hydrodynamic patterns from fast-starts in teleost fish and their possible relevance to predator–prey interactions. J Comp Physiol A 199:139–149. doi:10.1007/s00359-012-0775-5

Nitsche W, Brunn A (2006) Strömungsmesstechnik. Springer, Berlin, New York. doi:http://dx.doi.org/10.1007/3-540-32487-9

Packard A, Karlsen HE, Sand O (1990) Low-frequency hearing in cephalopods. J Comp Physiol A 166:501–505

Piddington RW (1972) Auditory discrimination between compressions and rarefactions by goldfish. J Exp Biol 56:403–419

Pitcher TJ, Parrish JK (1993) Functions of shoaling behaviour in teleosts. In: Pitcher TJ (ed) Behaviour of teleost fishes. Chapman & Hall, London

Pitcher TJ, Partridge BL, Wardle CS (1976) A blind fish can school. Science 194:963–965

Pitcher TJ, Magurran AE, Winfield IJ (1982) Fish in larger shoals find food faster. Behav Ecol Sociobiol 10:149–151

Pohlmann K, Grasso FW, Breithaupt T (2001) Tracking wakes: the nocturnal predatory strategy of piscivorous catfish. Proc Natl Acad Sci USA 98:7371–7374

Pohlmann K, Atema J, Breithaupt T (2004) The importance of the lateral line in nocturnal predation of piscivorous catfish. J Exp Biol 207:2971–2978. doi:10.1242/jeb.01129

Prasad AK, Adrian RJ (1993) Stereoscopic particle image velocimetry applied to liquid flows. Exp Fluids 15:49–60

Przybilla A, Kunze S, Rudert A, Bleckmann H, Brücker C (2010) Entraining in trout: a behavioural and hydrodynamic analysis. J Exp Biol 213:2976–2986. doi:10.1242/jeb.041632

Sand O (1974) Recordings of saccular microphonic potentials in the perch. Comp Biochem Physiol 47:387–390

Sand O, Karlsen HE (1986) Detection of infrasound by the Atlantic cod. J Exp Biol 125:197–204

Sand O, Karlsen HE (2000) Detection of infrasound and linear acceleration in fishes. Phil Trans R Soc B 355:1295–1298

Schulte-Pelkum N, Wieskotten S, Hanke W, Dehnhardt G, Mauck B (2007) Tracking of biogenic hydrodynamic trails in a harbor seal (Phoca vitulina). J Exp Biol 210:781–787

Schwalbe MAB, Bassett DK, Webb JF (2012) Feeding in the dark: lateral-line-mediated prey detection in the peacock cichlid Aulonocara stuartgranti. J Exp Biol 215:2060–2071. doi:10.1242/jeb.065920

Shih HH (2011) Real-time current and wave measurements in ports and harbors using ADCP. Oceans 2012 MTS/IEEE Conference, Yeosu, Korea, pp 1–8

Stamhuis EJ, Videler JJ (1995) Quantitative flow analysis around aquatic animals using laser sheet particle image velocimetry. J Exp Biol 198:283–294

Stamhuis EJ, Videler JJ (1998) Burrow ventilation in the tube-dwelling shrimp Callianassa subterranea (Decapoda: Thalassinidea). J Exp Biol 201:2159–2170

Standen EM, Lauder GV (2007) Hydrodynamic function of dorsal and anal fins in brook trout (Salvelinus fontinalis). J Exp Biol 210:325–339. doi:10.1242/jeb.02661

Strecker U, Hausdorf B, Wilkens H (2012) Parallel speciation in Astyanax cave fish (Teleostei) in Northern Mexico. Mol Phylogenet Evol 62:62–70. doi:10.1016/j.ympev.2011.09.005

Taguchi M, Liao JC (2011) Rainbow trout consume less oxygen in turbulence: the energetics of swimming behaviors at different speeds. J Exp Biol 214:1428–1436. doi:10.1242/jeb.052027

Urick RJ (1996) Principles of underwater sound. Peninsula Publishing, Los Altos

van Duren EA, Stamhuis EJ, Videler JJ (2003) Escape from viscosity: kinematics and hydrodynamics of copepod foraging and escape swimming. J Exp Biol 206:269–279

von Campenhausen C, Riess I, Weissert R (1981) Detection of stationary objects in the blind cave fish Anoptichthys jordani (Characidae). J Comp Physiol A 143:369–374

von Kármán T (1911) Über den Mechanismus des Widerstandes, den ein bewegter Körper in einer Flüssigkeit erfährt. Nachrichten von der Wissenschaftlichen Gesellschaft zu Göttingen, Mathematisch-Physikalische Klasse, pp 509–517

Walkowiak W, Münz H (1985) The significance of water surface waves in the communication of fire-bellied toads. Naturwissenschaften 72:49–51. doi:10.1007/bf00405335

Weihs D (1973) Hydrodynamics of fish schooling. Nature 241:290–291

Westerweel J (1997) Fundamentals of digital particle image velocimetry. Meas Sci Technol 8:1379–1392. doi:10.1088/0957-0233/8/12/002

White JW, Warner RR (2007) Safety in numbers and the spatial scaling of density-dependent mortality in a coral reef fish. Ecology 88(12):3044–3054. doi:10.1890/06-1949.1

Wiese K (1976) Mechanoreceptors for near-field water displacements in crayfish. J Neurophysiol 39:816–833

Wieskotten S, Dehnhardt G, Mauck B, Miersch L, Hanke W (2010) Hydrodynamic determination of the moving direction of an artificial fin by a harbour seal (Phoca vitulina). J Exp Biol 213:2194–2200. doi:10.1242/jeb.041699

Wieskotten S, Mauck B, Miersch L, Dehnhardt G, Hanke W (2011) Hydrodynamic discrimination of wakes caused by objects of different size or shape in a harbour seal (Phoca vitulina). J Exp Biol 214:1922–1930. doi:10.1242/jeb.053926

Wilga CD, Lauder GV (2000) Three-dimensional kinematics and wake structure of the pectoral fins during locomotion in leopard sharks Triakis semifasciata. J Exp Biol 203:2261–2278

Willert CE, Gharib M (1991) Digital particle image velocimetry. Exp Fluids 10:181–193

Windsor SP, Norris SE, Cameron SM, Mallinson GD, Montgomery JC (2010a) The flow fields involved in hydrodynamic imaging by blind Mexican cave fish (Astyanax fasciatus). Part I: open water and heading towards a wall. J Exp Biol 213:3819–3831. doi:10.1242/jeb.040741

Windsor SP, Norris SE, Cameron SM, Mallinson GD, Montgomery JC (2010b) The flow fields involved in hydrodynamic imaging by blind Mexican cave fish (*Astyanax fasciatus*). Part II: gliding parallel to a wall. J Exp Biol 213:3832–3842. doi:10.1242/jeb.040790

Yanase K, Herbert NA, Montgomery JC (2012) Disrupted flow sensing impairs hydrodynamic performance and increases the metabolic cost of swimming in the yellowtail kingfish, *Seriola lalandi*. J Exp Biol 215:3944–3954. doi:10.1242/jeb.073437

Yen J, Lenz PH, Gassie DV, Hartline DK (1992) Mechanoreception in marine copepods—ecophysiological studies on the 1st antennae. J Plankton Res 14(4):495–512. doi:10.1093/plankt/14.4.495

Chapter 2
Laser-Based Optical Methods for the Sensory Ecology of Flow Sensing: From Classical PIV to Micro-PIV and Beyond

Thomas Steinmann and Jérôme Casas

Abstract This chapter presents an overview of techniques for laser-based, non-contact fluid flow measurements, and their application to real datasets. Particular consideration is given to particle image velocimetry (PIV)-techniques, from the usual macro-scale PIV, through meso-scale PIV, to micro-PIV, thereby spanning the range from decimeter to micrometer scales. We compare the advantages and limitations of these techniques. The specific requirements of sensory ecology and sensory physiology, as well as the 3D-morphological nature of the organisms studied led us to conclude that the techniques that are used in water are ill-suited for several key tasks when dealing with terrestrial organisms. We therefore propose an innovative mixed technology that exploits the advantages of both standard and micro-PIV techniques while avoiding their main limitations.

Keywords Particle image velocimetry, micro-PIV · Viscous boundary layer · Filiform hairs · Mechanoreceptors · Sensory ecology · Biomimetics · Flow sensing

2.1 Introduction

Flow sensing is used by a vast number of animals (see for example, the reviews of Casas and Dangles (2010) on insects, Engelmann et al. (2002) on fishes, Barth et al. (1993) on arachnids, Denissenko et al. (2007) on crustaceans, Sterbing-D'Angelo et al. (2011) on bats and Hanke et al. (2012) on seals).

T. Steinmann (✉) · J. Casas
Faculté des Sciences et Techniques, Institut de Recherche sur la Biologie de l'Insecte, IRBI
UMR 7261 CNRS, Avenue Monge, Parc Grandmont 37200 Tours, France
e-mail: thomas.steinmann@univ-tours.fr

J. Casas
e-mail: jerome.casas@univ-tours.fr

H. Bleckmann et al. (eds.), *Flow Sensing in Air and Water*,
DOI: 10.1007/978-3-642-41446-6_2, © Springer-Verlag Berlin Heidelberg 2014

The ecological contexts in which flow sensing is relevant are varied: from prey–predator interactions to mate selection, and orientation to flow itself as illustrated in the works cited above. Flow sensing is used in air, water and most likely also in sand and hence in soil (Casas and Dangles 2010; Fertin and Casas 2006, 2007). Flow sensing is one of the several senses used by animals during orientation, and in some cases it is the dominant sense, for example for cave fishes (Windsor et al. 2010a, b). Thus, the study of flow sensing is a vibrant field of research for sensory ecologists and neuroethologists, and also for technologists working on biomimetic sensor design (Casas et al. 2013).

Studies on flow sensing, particularly those at small scale around single sensors, require high spatial precision and nonintrusive measurement methods. Thus, noncontacting measurement methods such as Laser Doppler Anemometry (LDA), Laser Doppler Vibrometry (LDV), and Particle Image Velocimetry (PIV), originally developed by aerodynamics and fluid mechanics engineers, have been used to measure flows of biological relevance. This chapter aims to describe recent technological advances in the measurement of flow around biological and artificial flow sensors in the context of organismal sensory ecology. We start with a description of the state of the art of the various techniques used for flow measurement at the scale of an organism or a sensing organ. We then describe two innovative techniques we have developed for greater spatial resolution, down to the single sensory hair: the first is relatively similar to the standard PIV technique as the particles in the fluid are illuminated by a thin light sheet; the second, micro-PIV, is relatively new to organismal biologists. It is, however, well known to microfluidic engineers, and is based on volumetric illumination. We compare the advantages and limitations of both techniques. In each case, we illustrate our reasoning with a figure or an example. Comprehensive considerations of parameter variations and extended theoretical development can be found in the references provided. The specific requirements of sensory ecology and sensory physiology, combined with the morphological nature of the organisms studied, led us to conclude that both techniques are ill-suited for several key questions. We therefore propose an innovative mixed technology that exploits the advantages of both standard and micro-PIV techniques and avoids their main limitations.

2.2 Classical PIV at the Scale of Bodies and Organs (10^{-2} m–10^{-3} m Resolution)

Due to their improved accessibility, laser-based measurement techniques that were originally restricted to applications in engineering and fluid mechanics are now also used in the fields of locomotion and sensory ecology. In particular, fish physiologists discovered the potential of these techniques to describe complex, unsteady, whole-field flows around an animal's body.

For instance, Bleckmann et al. (1991) used LDA to determine the spectral distribution of hydrodynamic flow fields caused by moving fish, frogs, and crustaceans, and discussed the possible biological relevance of the ability to detect high-frequency hydrodynamic events. Blickhan et al. (1992) used automatic particle tracking and LDA to show that the flow in the wake of a subundulatory swimmer consists of a chain of slightly deformed vortex rings. Two-dimensional PIV was used by Müller et al. (1997) to qualitatively and quantitatively analyze the structure of the wake behind a continuously swimming mullet. More recently, also using PIV, Hanke et al. (2000) found that the wake behind a swimming goldfish can show a vortex structure for at least 30 s. These authors discuss the possible advantage for piscivorous predators of being able to detect and analyze fish-generated wakes. In 2002, Engelmann and colleagues adapted the classical PIV technique for the visualization of the flow around a whole body covered by sensory organs to understand the influence of viscous hydrodynamics on the fish lateral line system (Engelmann et al. 2002).

Chagnaud et al. (2006) analyzed the correlation between neural responses to vortex rings and PIV data. Their artificially generated vortex rings resembled hydrodynamic stimuli that fish might encounter in their natural environments. Denissenko et al. (2007) shed some light on how a hydrodynamic stimulus can be affected by body shape by studying the flow generated by the active olfactory system of the red swamp crayfish. These animals use their anterior fan organs to generate distinct flow patterns that can be used for odor acquisition. The application of PIV to mapping biogenic and biologically relevant flows has been reviewed by Stamhuis et al. (2002).

Catton et al. (2007) quantified the flow fields generated by tethered and free swimming copepods. They report that the area in which the flow velocity was high enough to induce an escape response was 11 times the area of the organism's exoskeletal form. Thus, mechanoreceptive predators may be able to perceive a signal that is spatially extended well beyond the body size (Catton et al. 2007). Windsor et al. (2010a, b), studied the flow field involved in hydrodynamic imaging by blind Mexican cave fishes in open water, both when heading towards a wall and when gliding parallel to a wall. They suggested that the nature of the flow fields surrounding a fish is such that hydrodynamic imaging can be used by fish to detect surfaces at short range.

Wolf spiders live on the surface of leaf litter in forests, where they pursue their cricket prey using two different strategies. Casas et al. (2008) showed, using PIV, that these running spiders are, however, aerodynamically highly conspicuous due to substantial air displacement detectable up to several centimeters in front of them. The airflow in front of running spiders is thus a source of information for escaping prey, such as crickets and cockroaches. More recently, Klopsch et al. (2012) adapted PIV techniques to the study of the flow produced by a fly and sensed by a spider. Their findings led them to propose that the differences in the time of arrival and intensity of the fly signals at their different legs inform spiders about the direction of their prey.

2.2.1 Airflow Over a Cricket Body

The cricket *Nemobius sylvestris* uses hundreds of filiform hairs on two cerci as an early warning system to detect remote potential predators. The direction of the attacking predator can be estimated by arrays of hairs that have different directionalities (Landolfa and Jacobs 1995). However, the correct estimation of the direction of the incoming flow field can be biased by the flow perturbation around the cricket's body as a whole. We used PIV to investigate this flow. This technique is mainly used in experimental fluid mechanics to obtain time-resolved velocity measurements and related properties in fluids. The fluid is seeded with tracer particles which are assumed to follow the flow dynamics, and thus the motion of these seeding particles can be used to calculate the velocity of the flow. We will not describe this technique any further, as many books on this topic are available (Raffel et al. 1998; Adrian and Westerweel 2011).

Living crickets were placed in a glass container, with a single loudspeaker producing an oscillatory flow. The air inside the sealed glass box was seeded with 0.2 μm oil particles (Di-Ethyl-Hexyl-Sebacat) using an aerosol generator. The laser of the PIV system illuminated the airflow within the box. The laser sheet (width = 17 mm, thickness = 1 mm) was operated at low power (3 mJ at 532 nm) to minimize glare. A target area was then imaged onto the CCD array of a digital camera (696 × 512 pixels) using a lens (Macro-Nikkon 60 mm). The use of a macro lens allows us to obtain a ratio of the subject plane on the sensor plane of 1:1. Measurements were conducted at 25 °C, corresponding to an air kinematic viscosity of $v_{air} = 1.59 \times 10^{-5}$ m^2 s^{-1}. The far-field velocity was set to 10 cm/s and the flow frequency was 80 Hz. We used a stroboscopic principle to sample high frequency oscillatory flow signals with a PIV system limited to 20 Hz (Steinmann et al. 2006). We present here the results for an oscillatory flow parallel to the plane defined by the two cerci; the midline along the cricket abdomen was fixed at an angle of 0° to the direction of the flow.

We found that the orientation of the flow on the surface of a cercus differs from the orientation of the free field flow, i.e. the direction of an incoming flow around the two cerci is strongly altered, as illustrated by the streamlines in Fig. 2.1 (left). The directionality of each of the hairs on the two cerci is crucial as the incoming flow from different directions will trigger different patterns of activity across the whole cercal system (Miller et al. 2011). The projection of each hair into the central nervous system forms a functional map of the current air direction that translates the origin of the incoming flow as three-dimensional spatio-temporal activity patterns (Jacobs et al. 2008). Specifically, the flow direction is slightly shifted counter clockwise on the left cercus and clockwise on the right cercus, i.e. in opposite directions. This has important consequences for the overall computation of the direction of the incoming perturbation as these two effects might cancel each other.

We also observed that the flow amplitude was substantially lower close to the surface of the abdomen than further away (Fig. 2.1, right). The presence of the two

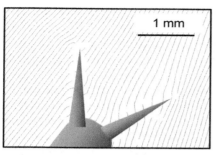
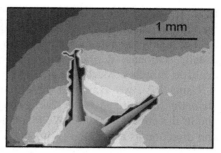

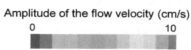

Fig. 2.1 Flow field around the rear of the wood cricket *Nemobius sylvestris*. On the *left*, streamlines represent the direction of airflow around the cricket's abdomen and cerci. The amplitude of the oscillatory flow in a cross section of the cricket's rear is shown on the *right*

cerci tends to decrease the amplitude of the flow field further, such that flow velocity at the bases of the cerci is considerably reduced. We can therefore conclude that there are substantial differences in the amount of energy available to the different hairs along a cercus.

In many cases, the intrinsic mechanical and neurophysiological properties of a sensory system reflect adaptation to ecologically relevant stimuli. We have illustrated how these stimuli can be greatly modified by the presence of the structure supporting the sensory organs. In addition to the cercus, which functions as a complex filter, the boundary layer over the body has significant implications for the ability of sensory organs to sense stimuli. Our findings agree with previous work with fish (Mc Henry et al. 2008) and crickets (Dangles et al. 2008). Thus, a complete understanding of the performance of a sensory system requires not only a knowledge of all the structural properties of the individual sensory elements but also the determination of the physical environment with exact quantification of the forces that drive these individual receptors.

2.2.2 Structure of Acoustic Flow and Inter-Antennal Velocity Differences in Drosophila melanogaster

The fruit fly *Drosophila melanogaster* has bilateral antennae of antisymmetric sensitivity. The arista, as well as the finuculus, two parts of the antenna, rotate about a central axis in response to acoustic stimuli. Acoustic communication is important during courtship, and takes place in the acoustic near field, where the small size of the dipole sound source (the male wing) and the rapid attenuation rate of particle velocity are expected to produce a highly divergent and localized sound field. The small size of the male wing (considered as a dipole sound source) and the rapid attenuation rate of particle velocity produce a spatially divergent sound field of highly variable magnitude. Also, male and female *D. melanogaster* are not usually stationary during courtship, resulting in a variable directionality of the acoustic stimulus (Morley et al. 2012).

Using PIV, we examined the stimulus flow around the head of *Drosophila melanogaster* to identify the true geometry of the acoustic input into the antennae and its directional response. We found that the stimulus changes in both magnitude and direction as a function of its angle of incidence (Fig. 2.2).

Remarkably, directionality is substantial, with inter-antennal velocity differences that were up to 25 dB at 140 Hz [see explanations of the estimation of the intensity of inter-antennal differences in Morley et al. (2012)]. For an organism whose auditory receivers are separated by only 660 $\pm$ 51 μm (mean $\pm$ S.D.), this inter-antennal velocity difference is far greater than differences in intensity observed between tympanal ears for organisms of similar sizes.

Combining these measurements with a laser vibrometry analysis of the vibratory movement of the arista, Morley et al. (2012) demonstrated that the mechanical sensitivity of the antennae changes as a function of the angle of incidence of the acoustic stimulus, with peak responses along axes at 45 and 315° relative to the longitudinal body axis.

This work indicates not only that flies are able to detect a difference in signal intensity according to its direction, but also that the male song structure may not be the sole determinant of mating success; the male spatial position also makes a major contribution to female sound reception and therefore also, perhaps, to her decision making.

2.3 Macro-PIV at the Receptors Scale (10^{-4} m Resolution)

Rigorous understanding of the performance of a sensor generally requires studying the interaction between the input signal and the sensory organs. Airflow sensing by filiform hairs partially immersed in the boundary layer around the body has been extensively studied in arthropods, and especially spiders and crickets (Shimozawa and Kanou 1984; Humphrey et al. 1993; Barth et al. 1993; Bathellier et al. 2012). Filiform hairs can respond to air velocities as small as 0.03 mm s^{-1} (Shimozawa and Kanou 1984), making them one of the most sensitive biological sensors in the animal kingdom (Shimozawa et al. 2003). These outstandingly sensitive structures are used to detect the faintest oscillatory signals produced by the wing beats of prey and predators (Tautz and Markl 1979; Gnatzy and Heusslein 1986; Magal et al. 2006; Steinmann et al. 2006; Casas et al. 2008). The presence of hairs of different lengths allows spiders and crickets to fractionate both the intensity and frequency range of an airflow signal.

Several flow measurement techniques have been exploited to investigate the flow around a sensory hair, or sensory hair deflection, in terms of both phase and

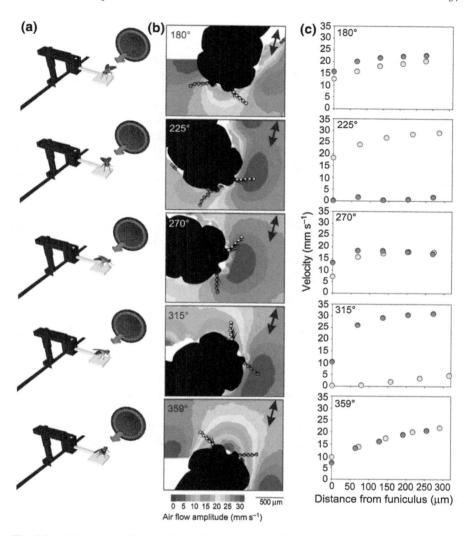

Fig. 2.2   a Orientation of Drosophila melanogaster to a 140 Hz sinusoidal stimulus produced by a loud speaker. **b** PIV velocity maps for five stimulus angles around the head of one individual fly. Points along the aristae correspond to positions at which velocity was determined. Large *black arrows* indicate the direction of the stimulus source. **c** Extracted velocity at five points along each arista for angles corresponding to those in (**a**). *Yellow* points represent the *left arista*, *red* points represent the *right arista*

frequency. In a pioneering study, Barth et al. (1993) adapted LDA to the study of the boundary layer of the flow over a spider leg. Kämper and Kleindienst (1990) and Shimozawa et al. (1998) adapted laser vibrometry to allow measurement of the deflection of cricket hairs.

In water, diverse animals use arrays of hair-like structures for important tasks such as feeding, gas exchange, smelling, and swimming. Koehl used both dynamically scaled physical models and a custom made PIV to study the flow through hairy food-capturing appendages (second maxillae) of calanoid copepods, which are abundant planktonic crustaceans (Koehl 2004). More recently, small-scale PIV has been used to show that antennule morphology and flicking kinematics facilitate odor sampling by the spiny lobster (Reidenbach et al. 2008). PIV has also been used to determine the fine-scale patterns of odor encounter by the antennules of mantis shrimps tracking odor plumes in both wave-affected and unidirectional flow conditions (Mead 2003). Mc Henry et al. (2008) presented direct PIV measurements of mechanical filtering by the boundary layer and fluid-structure interaction in the superficial neuromast of the fish lateral line system.

Following adaptation of PIV to small scales in water, we describe below how we adapted standard PIV protocols for the study of flow in air at a similar small scale. The major differences between macro-PIV and classical PIV are the use of a specialized optical system producing a particularly thin sheet of laser light and of a lens with five-times magnification.

2.3.1 Experimental Instrumentation

Various types of sample, including the cercus of *Nemobius sylvestris*, single 1,000 µm-long micro-electromechanical system (MEMS) hairs, and tandem 1,000 µm-long MEMS hairs on a plate of dimension 10×10 mm^2 (Krijnen et al. 2006) were placed in a glass container (dimensions: $10 \times 10 \times 10$ cm^3), with one loudspeaker (40 W) on one side connected to a signal generator. The seeding of air, the laser sheet generation and the imaging is explained in Sect. 2.2.1. The target area was then imaged onto the CCD array of a digital camera (696×512 pixels) using a binocular lens that allowed observation of a 2×2 mm^2 window around the substrate. The measurement technique is illustrated in Fig. 2.3. Far-field velocities were in the range 10–50 cm/s and flow frequencies 40–320 Hz. We used a stroboscopic principle to sample high frequency oscillatory flow signals with a PIV system limited to 20 Hz (Steinmann et al. 2006).

This technique allowed us to analyze the flow at a relatively high spatial resolution, higher than possible with standard PIV methods. We obtained a whole measurement field of $2,000 \times 2,000$ µm with a 5X magnification binocular objective lens. Given the camera resolution (696×512 pixels), the resolution can be 2.87 µm/pixels. Thus, selecting a 32-pixel correlation area, the spatial resolution is 100 µm. Using a time interval of 500 µs between two images and a sub-pixel interpolation, we were able to follow particle displacements of 0.287–100 µm. This methodology can thus resolve flows between 0.3 and 30 mm/s. The depth of field is given by the thickness of the laser light sheet and is estimated to be almost $\Delta z = 50$ µm at the focus point.

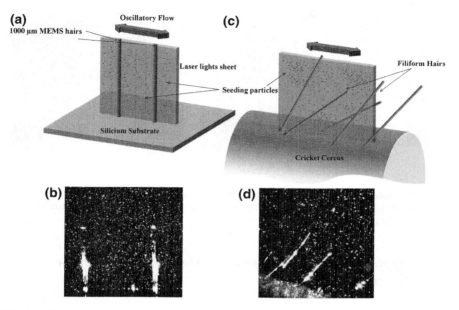

Fig. 2.3 Illustration of the macro-PIV principles. **a** PIV with a silicium substrate covered with MEMS hairs. **b** Raw PIV image of the seeding particles in the plane formed by the two MEMS hairs. **c** The same PIV technique adapted for the measurement of the oscillatory flow over a cricket cercus within the filiform-hair canopy. Filiform hairs are not always in the measurement plane, and sometimes cross it. **d** PIV images of the seeding particles in the light sheet plane. Hairs are in fact longer than can be observed on the image, but parts of their lengths are outside the light sheet

The quality of the measurement is greatly affected by the tracer concentration: if the concentration is too high, the discrimination of individual particles will be nearly impossible; if the concentration is too low, the particle density will not be sufficient for accurate estimation of the velocity at each point in the whole cross section of the flow. Indeed, the concentration must be lower than (Adrian and Westerweel 2011):

$$C_{\max} = \left(\frac{M}{d_i}\right)^2 \frac{1}{\Delta z}$$

where M is the magnification, d_i the particle diameter and Δz the laser light sheet thickness. As we will see in the next chapter on micro-PIV, there is a relationship between the signal-to-noise ratio (SNR) and the thickness of illumination (here referred to as the 'laser light sheet thickness', and also called 'the test section depth').

Fig. 2.4 Amplitude of an oscillatory flow over a canopy of cricket filiform hairs (Frequency = 80 Hz, Amplitude = 100 mm/s). Measurements from two different locations on the cercus are shown

2.3.2 Application of (Conventional) Macro-PIV to Airflow Sensing Organs

2.3.2.1 Cricket Hair Canopy

Figure 2.4 shows the velocity amplitudes of the oscillatory flow over a cricket cercus at two different locations. These findings reveal the complex arrangement of cricket filiform hairs and the corresponding flow patterns that result from the interactions between the boundary layers of different filiform hairs.

The mechanical constraints of the socket at the base of cricket filiform hairs cause a preferential plane of deflection, i.e. directionality. For a given direction of an incoming flow, only some hairs will oscillate in this flow direction, rather than in the plane of the laser sheet. Since it is impossible to determine the exact geometry and the plane of motion of the dense hair canopy in the immediate vicinity of the measurement plane, the interpretation of the flow patterns revealed by the PIV measurement is limited. The "out of plane" hairs probably have a large influence on the flow in the section of measurement. Nevertheless, it is still possible to extract some information about the ability of the hairs to follow the flow. The use of a stroboscopic technique to sample high frequency oscillatory flow allows simultaneous assessment of the dynamics of the flow velocity surrounding the cercus and the angle of displacement of the hairs (see Fig. 2.5).

Hairs of different sizes will be differently out of phase with the far-field flow. In the experiment illustrated, the small hair seems to have a phase advance of 26° with respect to the far-field flow. At a stimulus frequency of 80 Hz, as used here, a phase advance agrees well with the velocity of air within the boundary layer, in which the 260 μm-long hair is totally immersed, being phase advanced by $\pi/4$ (Kumagai et al. 1998; Steinmann et al. 2006). By contrast, the longer hairs (430 and 1,070 μm) were not totally immersed in this boundary layer and were

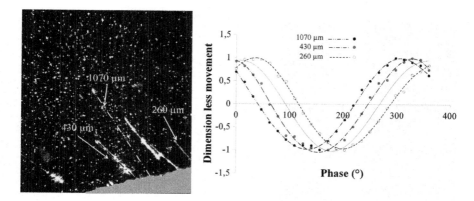

Fig. 2.5 (*Left*) Position and size of 3 filiform hairs, of length of 430, 1,070, 260 μm. (*Right*) Comparison of the phase differences between the far-field flow (*gray line*) and the three hairs. The *gray line* represents the phase of the far-field flow measured 1,500 μm above the cercus surface. The movements of the hairs are dimensionless, their velocity being divided by their maximal velocity. The hair length refers to the visible length and is therefore subject to error

thus less subject to this phase advance: indeed, they showed a phase lag ($-26°$ for the 430 μm-long hair and $-51°$ for the 1,070 μm-long hair). At 80 Hz, such increases of the phase lag with the length of the hair are expected and are coherent with theory. Although they are interesting in themselves, these results cannot be easily generalized. Systematic experimental variation of parameters is not feasible, as it is nearly impossible to find appropriate spatial arrangements of hairs in rows and similarly unlikely that each hair of such rows moves within the plane.

2.3.2.2 Single Cricket Hairs

We applied the macro-PIV technique in combination with a novel compact theoretical framework to describe the cricket hair mechanics (Fig. 2.6). In a systematic fashion, we studied the ability of six hairs of different lengths (410–870 μm) to follow oscillatory flows between 30 and 300 Hz (Bathellier et al. 2012). We found that cricket air-motion sensing hairs (and those in spiders) work close to the physical limit of sensitivity and energy transmission across a broad range of relatively high frequencies. In this range of frequencies, the hairs closely follow the motion of the incoming flow because a minimum of energy is dissipated in their basal articulation. This frequency band depends on hair length, and is between about 40 and 600 Hz, which is beyond the frequency at which the angular displacement of the hair is maximal.

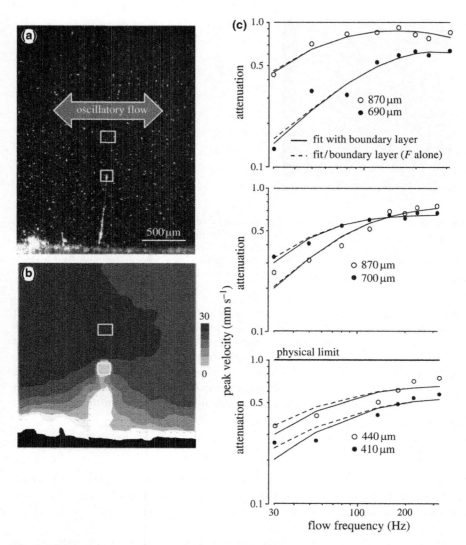

Fig. 2.6 Measurements of filiform hair motion in the cricket *Nemobius sylvestris*. **a** Picture of a cricket cercal hair in macro-PIV recording chamber. **b** Map of instantaneous air particle velocity (grey scale) and direction (*arrows*) for a 80 Hz oscillatory flow. The *white squares* represent the areas selected to compute the far-field flow velocity and the hair-tip velocity, respectively. **c** Deviation from the physical limit calculated by dividing the tip velocity of the hair measured by PIV by its physical limit, for six isolated cricket cercal hairs of different lengths. The solid lines represent the fit of the transfer function of a 2nd order mechanical system. In this model, the effect of the boundary layer was taken into account. It simply translates into a reduction of velocity close to the cercus surface (Bathellier et al. 2012)

Fig. 2.7 Full-field data for flow in the vicinity of 1 mm-long MEMS hairs arranged in tandem, as a function of the frequency and inter-hair distance. Color code as in Fig. 2.6

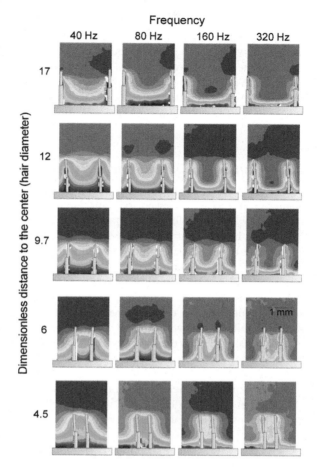

2.3.2.3 The Use of MEMS Hairs as a Surrogate

Investigations of the flow around natural filiforms hairs suffer various limitations: it is difficult to select hairs of the desired lengths and to find tandem hairs with parallel planes of vibration. By using MEMS artificial sensors mimicking biological counterparts, it is however possible to reduce the various uncertainty factors. Indeed, all the MEMS hairs have a same, fixed length, and the use of MEMS technology enables very simple geometrical arrangements to be obtained, such as isolated hairs or tandem hairs with different spacing between the two members of the pair. By using flat plate substrates, all the nonsolvable problems arising from the 3D geometry of the cercus that supports the hairs in vivo can be avoided or minimized.

We used the PIV technique to measure the extent of flow perturbation by single and tandem hairs directly, using MEMS hairs as physical models (Fig. 2.7). Single and tandem MEMS hairs with various inter-hair distances were subjected to oscillatory flows of diverse frequencies. Decreasing hair-to-hair distance markedly reduced flow velocity amplitude and increased the phase shift between the far-field flow and the flow between hairs. These effects were stronger for lower flow frequencies. We therefore predict strong hydrodynamic coupling within natural hair canopies exposed to natural stimuli, and the effects will vary depending on species, hair sizes, and hair densities (Casas et al. 2010).

2.3.3 Limitations of the Macro-PIV Technique

Kähler et al. (2012) reviewed the major sources of uncertainty of PIV near surfaces. They identified four major factors that determine the ability to resolve the flow in the hundreds of micrometers closest to the boundary layer: (1) the use of appropriate tracer particles that follow the fluid motion with sufficient accuracy, (2) the use of fluorescent particles or tangential illumination, such that reflection from the wall can be eliminated or avoided, (3) appropriate imaging of the particles with a lens or a microscope objective such that the particle signal can be suitably captured on a digital camera, and (4) satisfactory estimation of particle image displacement with digital particle imaging analysis methods. In our experiments, we had to deal with two of these four uncertainty factors, the imaging of individual particles and wall reflection. Indeed, one of the most important factors limiting the spatial resolution of the technique is diffraction: the recorded image of a tracer particle is generally greater than its theoretical magnified size (Meinhart et al. 2000). The size of the particle image is mostly determined by the diffraction of the optical system. Let d_i be the particle image diameter given by:

$$d_i = \sqrt{M^2 d_p^2 + d_{\text{Diff}}^2}$$

where M is the linear magnification of the binocular lens, d_p is the particle diameter (200 nm for DEHS oil droplets) and

$$d_{\text{Diff}} = 2.44(1 + M)\lambda \frac{1}{2\text{NA}}$$

where λ is the laser wavelength (532 nm) and NA is the numerical aperture of the optical system (binocular lens) and is 0.1 in our case.

It is also important to note that d_i is the size of the particle as projected onto the camera CCD sensors. Thus, the size of this particle in the measurement field will also depend on the size of the CCD sensor (see Fig. 2.8). As an example, with a 5X magnification, the field of view is 2,000 μm and consequently each pixel corresponds to 2,000/696 = 2.87 μm. At the measurement field scale, the particle image size is 3.24 pixels × 2.87 μm/pixels = 10 μm (Fig. 2.8).

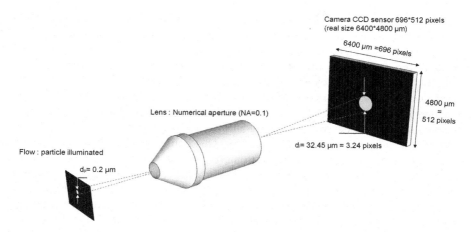

Fig. 2.8 Explanation of the estimation of the particle image size

Another limitation results from the large amount of light re-emitted by the surface of objects present in the flow. Biological sensory organs are generally very small, and therefore the capacity of the technique to resolve flows within a few hundred microns of the surface is crucial.

2.4 Micro-PIV at the Scale of Biological Receptors (10^{-5} m Resolution)

We have shown that the macro-PIV technique has limitations, in terms of spatial resolution, due to the small magnification factors of the optical system used. But even if magnification were to be increased, macro-PIV would still be limited by both the large diffraction of the optics and the substantial reflection from the surfaces of the objects studied. These constraints prevent visualization and quantification of the flows very close to receptors.

We will show in this section that using volume illumination provided by an epifluorescent microscope instead of the conventional optical light sheet allows the analysis of microscopic flows at the required scale (Santiago et al. 1998). In this methodology, the light, provided by a double pulsed monochromatic Nd:YAG laser (532 nm) is reflected by a dichroic mirror, travels through an objective lens that focuses on the point of interest, and illuminates a volume seeded with fluorescent particles. The emission from these particles at a specific wavelength (560 nm), along with the reflected laser light coming from surfaces or interfaces, shines back through the objective, the dichroic mirror and through a band pass filter that blocks the specific wavelength of the laser light (Wereley and Meinhart 2010).

The ability to observe and analyze a plane in macro-PIV is a consequence of the planar nature of the laser sheet that illuminates only a cross section of the flow.

By contrast, micro-PIV is a volume illumination technique, and exploits the ability of the objective lens to focus on a single plane, such that there is a two-dimensional plane in which particles can be viewed. Nevertheless, with this illumination method, the entire depth of the section is illuminated by a cone of light. Consequently, there is often background noise due to the emission from the out-of-focus particles added to the emission of individual particles in the focus plane; this can make discrimination difficult and low seeding concentrations have to be used. In the next section, we describe averaging analysis techniques that must be employed with this technique due to the low seeding. One consequence is that only steady flows can be investigated, in contrast to more conventional PIV techniques which can be used with unsteady or oscillatory flows. Adapted preprocessing techniques must also be used because the images tend to have a zero-displacement bias from background noise and low signal-to-noise ratios. Also, because of the low seeding particle density, high numerical aperture objectives are required to capture as much of the light emission as possible. The choice of optical material is therefore critical.

As far as we know, micro-PIV has rarely, if ever, been applied in the field of sensory ecology. The following section illustrates our attempts to apply this new technique to the visualization and quantification of the flows around single natural and biomimetic hairs with a resolution of 10 μm.

2.4.1 Experimental Instrumentation

The experimental micro-PIV apparatus consisted of a liquid delivery system composed of a syringe pump, a laser system, an epifluorescent microscope, an objective lens and an imaging system, and the cercus sample placed in a test section (a cavity with dimensions: 10 mm × 5 mm × 500 μm in PVC plate). Illumination was provided by a twin Nd:YAG laser system. The laser beams were directed onto the cercus using an objective lens (Olympus X20 or X40). The laser produces pulses of green light (532 nm) re-emitted by the fluorescent tracer particles at 560 nm. The tracer particle images were captured with a PCO Sensicam camera (696 × 512 pixels) and the images were transferred to a computer for processing. The camera and the laser were synchronized via an ILA PIV Synchronizer. The cercus was placed in a small waterproof Plexiglass box (500 μm depth) seeded with fluorescent particles (RhB-labeled beads of $d_p = 360$ nm). The box was open on both sides to allow a continuous flow.

2.4.1.1 Resolution of the Technique

This technique allowed us to acquire high-resolution tracer particle images. We obtained, with a 20X objective lens, a measurement field of 440 × 330 μm. With a 40X objective lens, the measurement field was 220 × 165 μm. The time between two laser impulses was set at 500 μs, to resolve relatively high speed flows. Using the method described in Sect. 2.3.1, we determined that, for a 20X magnification, spatial resolution was 10 μm and flows between 0.13 and 13 mm/s could be studied.

2.4.1.2 Depth of Field and Measurement Depth

For 10 μm resolution in the focus plane, the depth of field and the measurement depth have to be limited to the same scale. However, the generation and alignment of a light sheet thinner than 50 μm is not possible, and therefore the use of volume illumination is the only feasible approach (Meinhart et al. 2000). In macro-PIV techniques, the depth of field is determined by the thickness of the laser light sheet, and the measurement depth by the positioning of the light sheet. By contrast, for volume illumination techniques, the depth of field and the measurement depth must be carefully determined.

The depth of field, which is the distance between the nearest and farthest objects that appear acceptably sharp can be calculated by Inoue and Spring (1997):

$$\delta_z = \frac{n\lambda}{\mathrm{NA}^2} + \frac{ne}{\mathrm{NAM}}$$

where n is the refractive index of the fluid between the objective lens and the test structure, λ is the wavelength of the incident light, NA is the numerical aperture of the objective lens, M is the total magnification of the microscope system, and e is the smallest resolvable distance on the image detector, i.e. the spacing between pixels. In the experimental setup we used, $n = 1$ for air, $\lambda = 532$ nm, NA $= 0.40$, $M = 20$, and $e = 0.64$ μm, resulting in a depth of field of $\delta_z = 3.4$ μm.

The measurement depth is the distance from the center of the object plane beyond which the particle image intensity is too low to contribute significantly to the measurement. It is given by Meinhart et al. (2000):

$$\delta_{z_m} = \frac{3n\lambda}{\mathrm{NA}^2} + \frac{2.16d_p}{\tan\theta} + d_p$$

where d_p is the tracer particle diameter (for RhB-labeled beads, $d_p = 360$ nm) and $\tan\theta = \mathrm{NA}/n$. With the optical settings we used, $\delta_{z_m} = 12.3$ μm.

For our experimental setup, the measurement depth was larger than the expected depth of field. We therefore had to keep the depth estimation criteria that gave the largest numerical result.

2.4.1.3 Air and Water Similitude

Micro-PIV has typically and historically been used in water, whereas cerci and hairs are naturally in air. At constant temperature (27 °C), the kinematic viscosity of air is 20 times greater than that of water. The Stokes boundary layer thickness depends on this kinematic viscosity as follows:

$$\delta_{BL} = \frac{4.64D}{Re^2}$$

where D is the diameter of the object being considered, i.e. the cercus or the hair diameter, and Re is the Reynold's Number defined thus:

$$Re = \frac{UD}{v}$$

where U is the velocity of the fluid and v is the kinematic viscosity of the fluid.

There is a simple relationship between the kinematic viscosity of air, v_{air}, and the kinematic viscosity of water, v_{water}:

$$v_{air} = 20 \ v_{water} \text{ at } 27 \ ^\circ C$$

Consequently, at the same fluid velocity, the boundary layer thickness is 4.5 times smaller in water than in air. During our study, we therefore produced a continuous water flow of 2 mm/s which is equivalent to a 40 mm/s flow in air. The value of 40 mm/s is a biologically relevant value for crickets: it is the flow produced in front of a running spider (Casas et al. 2008).

2.4.2 Application of the Micro-PIV Technique to Airflow Sensors in Crickets

2.4.2.1 Study of Cercus Roughness and the Influence of Setae

The cricket cercus is covered with setae and small spines, up to a few tens of microns long, implanted at high density. The denticles covering shark skin reduce drag by decreasing the vorticity and shear stress in the boundary layer (Bechert and Bartenwerfer 1989; Bechert et al. 2000; Lee and Lee 2001). The drag reduction is proportional to the denticles size, but for larger denticles the proportionality breaks down, and the drag reduction eventually becomes a drag increase (Garcia Mayoral and Jimenez 2011). By analogy, we expected the spines on the cricket cercus to modify the boundary layer in the first hundreds of micron surrounding the cercus (Fig. 2.9).

Indeed, the modification of the boundary layer caused by the setae tends to extend to 150 μm above the cercus surface (Fig. 2.9b), a distance three times longer than the average length of the setae and long enough to be relevant to many sensory hairs. For a meaningful interpretation, these measurements of the biological system should be compared to the corresponding measures for a smooth cylinder surface. The experimental boundary conditions are, however, very difficult to estimate because they involve a superposition of the boundary layers of the

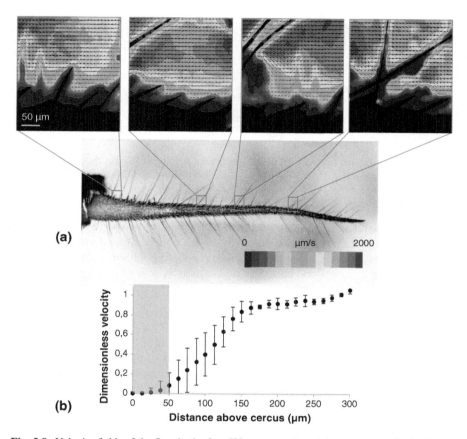

Fig. 2.9 Velocity fields of the flow in the first 500 μm around a cricket cercus acquired with the micro-PIV setup at four positions along a cricket cercus immersed in a water (**a**). Profile of the average flow over a vertical profile at the 4 positions. The distance is estimated from the base of the setae, and the average length of the setae is represented by the *grey box* (**b**). Velocities are normalized to the far-field velocity

channel test chamber with that of the cercus. Finally, the micro-PIV technique can only be used in water and water promotes adhesion of the seeding particles to the cercus surface and to each other (agglomeration).

2.4.2.2 Visualization and Quantification of the Flow Over a Single Hair

The micro-PIV technique was used to visualize and quantify the flow velocity over a single hair in steady flow at three different velocities: 150, 1,000, and 3,000 μm/s. Figure 2.10 shows the velocity field of a continuous flow along a cercus. The hair in the measurement plane had a 10 μm diameter. Two boundary layers can be seen, one around the cercus and the other around the hair. The flow was disturbed around

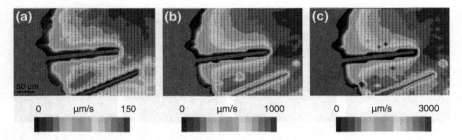

Fig. 2.10 Velocity field around a 10 μm-diameter cricket filiform hair (in the *middle* of each picture) in steady flows at three different velocities corresponding to three different regimes: $V = 150$ μm/s, $Re = 0.002$ (**a**), $V = 1,000$ μm/s, $Re = 0.015$ (**b**), V = 3,000 μm/s, $Re = 0.045$ (**c**)

the hair to a distance of roughly 15 times its diameter. Such strong disturbances over a long distance are comparable to what we observed previously for larger MEMS hairs, suggesting strong hydrodynamic coupling within natural hair canopies, depending on arthropod hair sizes and density (Casas et al. 2010).

2.4.2.3 Steady Flow Over a MEMS Hair

We adapted the micro-PIV technique for investigation of the steady flow along a single MEMS hair (Fig. 2.11). A set of 50 image pairs of the seeding particle flow in a cross section around a hair was obtained. After processing the micro-PIV images for reduction of noise, we used image pairs to compute the correlations to determine the local velocities of the particles. Figure 2.11d is the result of the averaging of the 50 velocity fields.

2.4.3 The Challenges for the Use of Micro-PIV in Sensory Ecology

It is clear that there are numerous problems associated with the adaptation of the micro-PIV technique to applications in sensory ecology, as illustrated above. The use of fluorescent particles and the background noise due to the volume illumination both limit the application of this technique to 2D flow in very simple setups. In most cases, the images are very noisy, such that it was not possible to identify individual particles. The reasons for this large amount of noise are numerous. We present here the three principal problems we had to face and suggest ways to solve them.

2.4.3.1 The Depth of the Micro Channel

Mielnik (2003) provides an estimated relationship between the SNR as function of the particle concentration and the depth of the test section. An increase of the

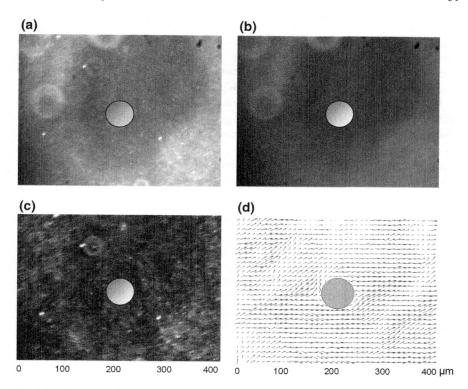

Fig. 2.11 Raw micro-PIV image of the particles around a MEMS hair (**a**), constant background extracted by estimating the minimal *gray* value by a technique of pixel-by-pixel sliding over time (**b**), pre-processed image (**c**) and velocity field around a hair in a viscous steady flow from the *left* (**d**)

micro channel depth or an increase of the particle concentration leads to a decrease of the SNR. The author estimated that test section depths larger than 200 μm would not allow a sufficient SNR, even at very low particle concentration (lower than 0.01 %). We concluded that the cercus was too thick (between 300 and 500 μm, which is thicker than most micro-channels) to provide good measurements because the SNR is very low. A solution to this problem could be to place a small cross section of the cercus in a very shallow micro-fabricated channel.

2.4.3.2 Particle Density

As explained by Mielnik (2003), a reduction of the particle density also leads to an increase of the SNR. Beyond a certain density, seeding particles are too few to provide enough information about the flow velocity at a sufficiently high spatial resolution. Micro-PIV experiment designs have to address this trade-off, lowering the particle density while generating sufficient information. The constraint imposed by the large section depth in our setup made it impossible to find a suitable concentration. Mielnik et al. (2006) developed a novel seeding method, called Selective Seeding. The method involves selective seeding of a thin fluid layer within an otherwise particle-free flow. By analogy to the laser sheet in macro scale PIV, such particle sheets define both the depth and the position of the measurement plane, independent of the details of the optical setup.

It is also possible to make measurements in deep micro channels by confocal micro-PIV (Lima et al. 2007). This technique is based on confocal microscopy, in which pinhole apertures are used for spatial filtering and thus optical sectioning. As a result, it is possible to obtain a series of optical sectioning images at different focal planes, which provides 3D information of the fluid flow and reduces considerably the amount of background noise.

2.4.3.3 Contrast and Particle Fluorescence

The fact that individual particles are often not visible could also be due to insufficient fluorescence of the particles, or to poor transmission through the optical system of a given wavelength. We carried out experiments with three kinds of particles from micro-particles Gmbh: PS-Rhodamine B-particles (Diameter = 439 nm), MF-Rhodamine B-particles (Diameter = 366 nm), and PS-RhB-PEG-particles (Diameter = 448 nm). All these different particles absorb light at 560 nm and re-emit it at 584 nm. The results were similar for the three kinds of particles. We also tried different objective lenses, without improving the results.

2.4.3.4 Particles Agglomeration

We were able to increase the SNR by using a nonionic surfactant, polyvinyl alcohol, and alternatively by using seeding particles with a surface treatment (Polystyrene Rhodamine B particles, treated with Poly Ethylene Glycol). These two solutions greatly reduce both particle agglomeration and the sticking of particles to the walls of channels.

The noise has also been reduced computationally by subtracting a sliding minimum over time, for each pixel independently. The background noise stays the same over all images, while seeding particles move. This operation is executed with a window length of 30 images. For a particular pixel at a given position in image (n), the minimum over all images at that position is calculated and then subtracted from the pixel intensity in image (n) to obtain an enhanced processed image.

2.5 Conclusions: The Way Forward

This survey addresses micro-PIV in particular, as this technique had not previously been used in the field of sensory ecology. Being both novel and having unmatched accuracy, micro-PIV inevitably has great potential. However, our application of this approach to the analysis of cricket sensory hairs has revealed several problems inherent to the technique. Solutions could often, but not always, be found.

Are these or similar problems encountered in other fields of application and how are they solved there? We performed a survey of the many papers published with micro-PIV technology to answer these questions (Appendix Table 2.1). As Wereley and Meinhart (2010) recently estimated, over 100 papers have been published annually reporting the use of this technique. Our survey shows that the problems we encountered are by no means rare. It also reveals that most of these studies were conducted in very small micro channels of thicknesses smaller than 500 µm, usually between 10 and 200 µm. Thus, one has to conclude that micro-PIV is a technique suitable for simple, essentially two-dimensional problems. This can be understood historically, as micro-PIV development was motivated by the need for a methodology to investigate microfluidic viscous flow phenomena. The broader utility of micro-PIV was demonstrated subsequently by applying it to flows in microchannels, micronozzles, BioMEMS, and flows around cells, still essentially 2D structures. Most sensory ecological issues associated with higher organisms are three-dimensional, and consequently micro-PIV is intrinsically not well adapted.

Combining the advantages of classical PIV and micro-PIV (see Appendix Table 2.2 for a comparison) would in many ways be a suitable solution for most of the problems and enable sensory ecologists to work at the receptor scale. There are indeed ways to adapt micro-PIV technology to suit the needs of sensory ecology. One very recent advance is to use confocal techniques (see Lima et al. 2007 and Patrick et al. 2010). However, this development is currently too expensive and sophisticated for most sensory ecology laboratories. Another possibility for channel depths larger than 1 mm is based on traditional laser light sheet illumination combined with long distance microscopic, rather than volume illumination. For an overview of this method see Kähler et al. (2006) and the most recent work by Alharbi and Sick (2010) and Eichler and Sattelmayer (2011). These authors explain that using a telecentric lens instead of a typical microscope solves the problem of the background noise caused by unfocused particles. The depth of focus of the optical system is then larger than the thickness of the laser light sheet, estimated to be 100 µm at a wavelength of 532 nm. More recently an innovative method for determining the three-dimensional location and velocity of particles was developed (Snoeyink and Wereley 2013). The wavefront from a particle is converted through a convex lens into a Bessel beam, the frequency and center of which can be directly related to the three-dimensional position of the particle.

In conclusion, one way forward would be to combine the thin laser sheet of classical and macro-PIV with telecentric lenses and fluorescent markers. Such a setup does not even need to be restricted to water environments, as Burgmann et al. (2011) were able to use fluorescent tracer particles to study small-scale wall-bounded gas flows. None of these ideas have been applied to, or tested in, sensory ecology. The study of flow sensing in organisms therefore has a bright future, but its students must be prepared to design their own tools, as off-the-shelf technology is generally unsuited to the models they study and the problems they face.

Appendix

Table 2.1 Summary of the different techniques, channel flow depth, and applications of micro-PIV from 20 papers on micro-PIV most pertinent to the scope of this review

Author	Year	Channel flow depth	Applications	Techniques	2D/3D	Velocity
Meinhart et al.	1999	30 μm	μ channel flow	NdYag laser	2D	
Cummings	2000	30 μm	Image processing	Argon ion Laser	2D	
Meinhart et al.	2000	90 μm	Inkjet print head	Nd YAG	2D	8 m/s
Kim et al.	2002	7 μm	μ channel, electroosmotic flow	Argon ion laser	2D	260 μm/s
Lee et al.	2002	690 μm	μ channel flow	Nd YAG laser	2D	
Devasenathipathy et al.	2003	50 μm	Electrokinetic flow		2D	
Devasenathipathy and Santiago	2003	107 μm	Silicon channel	Nd YAG laser	2D	20 mm/s
Meinhart and Wereley	2003	100 μm	AC electrokinetic flow	Mercury lamp	2D	110 μm/s
Sato et al.	2003					
Park et al.	2004					
Bitsch et al.	2005	30 μm	Blood flow in a capillary glass	LED	2D	4 mm/s
Liu et al.	2005	225 μm	Microcapillary flow	Nd YAG laser	2D	1 m/s
Boek et al.	2006	40 μm	Wormlike micellor fluid	Nd YAG laser	2D	4 mm/s
Bown et al.	2006	25 μm	DNA concentrator	Nd YAG laser	2D	5 μm/s
Curtin et al.	2006	233 μm	DNA, viscosity estimation	Nd YAG laser	2D	0.4 mm/s
Horiuchi et al.	2006	11 μm	μ channel, electroosmotic flow	Nd YAG laser	2D	100 μm/s
Kähler et al.	2006	Boundary layer of 5–7 mm	Wall-shear-stress and near-wall turbulence	Laser light sheet Nd: Yag laser	2D	5 m/s
Li and Olsen	2006	521 μm	Turbulent flow in μ channel	Nd YAG laser	2D	10 m/s
Lindken et al.	2006	200 μm	μ channel flow	NdYag laser, stereoscopy	3D	0.3 m/s
Mielnik and Saetran	2006	260 μm	Selective seeding	Sheet of seeding particles	2D	2 cm/s
Moghtaderi et al.	2006	300 μm	Baffle plate μ reactor	NdYag laser	2D	0.3 m/s
Sato and Hishida	2006					
Yan et al.	2006					

(continued)

Table 2.1 (continued)

Author	Year	Channel flow depth	Applications	Techniques	2D/3D	Velocity
Yang and Chuang	2005					
Bown and Meinhart	2006	436 μm	μ channel	3D, stereoscopy	3D	16 μm/s
Lima et al.	2007	100 μm	Confocal μPIV	NdYag Laser	2D	0.6 mm/s
Natrajan et al.	2006					
Pereira et al.	2007					
Walsh et al.	2007					
Kim et al.	2008	70 μm	Mixing and pumping	Continuous laser	2D	15 μm/s
Lee et al.	2009	50 μm	Fluid mechanics of blood sucking	Continuous Nd:YAG laser	2D	0.4 cm/s
Mansoor and Stoeber	2010	200 μm	Drying of polymer solutions	HeNe laser	2D	1 μm/s
Raghavan et al.	2009					
Alharbi and Sick	2010	44 mm	Study of boundary layers in internal combustion engines	1 mm laser light sheet	2D	1.5 m/s
Nguyen et al.	2010	200 μm	Improvement of measurement	Continuous DPSS laser	2D	
Poelma et al.	2010	250 μm	Study of outflow tract of embryonic chicken heart	Scanning μPIV	3D	4 cm/s
Wereley and Meinhart	2010		Review of recent advances in μPIV	Confocal imaging, particle image defocusing, stereo imaging…	2D, 3D	
Kloosterman et al.	2010	148 μm	Increasing depth of correlation	Diode-pumped Nd:YLF laser	2D	
Patrick et al.	2010	150 μm	Hemodynamics	Bidirectional scanning micro PIV	2D	45 μm/s
Rossi et al.	2011	Theoretical work	Image preprocessing use of depth of correlation	Particle image defocusing		
Wang et al.	2010	300 μm	Pulsed micro-flows for insulin infusion therapy	Nd Yag laser	2D	30 mm/s
Cierpka et al.	2011	200 μm	Comparative analysis	Astigmatism-μPTV, stereo-μPIV		75 mm/s
Eichler and Sattelmayer	2011	17.5 mm	Measurements of premixed flames	Long-distance micro-PIV	2D	

(continued)

Table 2.1 (continued)

Author	Year	Channel flow depth	Applications	Techniques	2D/3D	Velocity
Jin and Yoo	2011	100–140 μm	Droplet merging	Continuous laser/high speed camera	2D	4 cm/s
Nguyen et al.	2011	500 μm	Improvement of measurement	Volumetric-correlation PIV	3D	
Samarage et al.	2011	200 μm	Optimization of temporal averaging processes	Nd:Yag continuous laser	2D	
Sun et al.	2011	60 μm	Study of a microdiffuser	100 W Halogen lamp	2D	20 mm/s

Table 2.2 Advantages and limitations of both standard macro-PIV and micro-PIV techniques according to seven criteria

	Macro PIV standard thin laser light sheet/non fluorescent particles	Micro PIV Volume illumination/fluorescent particles	Mixed technique: Telecentric lens/thin laser light sheet/ fluorescent particles
1. Glare and body surface reflexion	Standard illumination techniques lead to high reflection close to surfaces, and without fluorescent paint, there is a lack of information in the first 100 μm of the boundary layer	The greatest advantage of the use of fluorescent particles: investigation of the flow in the 100 μm closest to the receptor is possible	Use of fluorescent particles to reduce reflection of laser light close to the surface
2. Signal-to-noise ratio. Capacity of individual particle discrimination	The thin laser sheet thickness tends to reduce the background noise but the noise mainly depends on particle density: $C \ll \frac{1}{\Delta z}\left(\frac{M}{di}\right)^2$	The biggest problem with the technique when applied to biological flows and particularly 3D problems: the SNR increases with the increase of test section depth	Thin laser sheet to reduce undesirable background noise
3. Limit of magnification	Depends on lens magnification, but the upper limit with a standard binocular lens is 5 times, which represents a field of view of 2,000 μm	Theoretically there is no limit of magnification. Practically a 20 times magnification is sufficient for a large number of applications and represents a field of view of 440 μm	Telecentric lens allowing a 10 times magnification leading to a field of view of 880 μm
4. Diffraction of the particles	Diffractive effects and particle image size on CCD sensor are large due to the low numerical aperture (di = 15 μm)	The high numerical aperture of a microscopic lens (0.4) limits the diffraction (di = 3 μm)	Telecentric lens with a numerical aperture of 0.21 leading to an estimated diffraction of 6 μm

(continued)

Table 2.2 (continued)

	Macro PIV standard thin laser light sheet/non fluorescent particles	Micro PIV Volume illumination/fluorescent particles	Mixed technique: Telecentric lens/thin laser light sheet/ fluorescent particles
5. Temporal resolution	At 5 times magnification and given the large SNR, real time and high frequency flow investigation is possible	To maintain a good SNR, the particle density needs to be low, and consequently often not high enough to provide good correlation or velocity fields without time averaging	High SNR expected even at high concentration, real time and high frequency possible
6. Working distance	Depends on lens but typically between 50 and 70 mm	The use of microscopic lens requires small working distances, generally a few mm. This constraint is hard to apply to the investigation of flows around 3D structures	Large working distance, allows the investigation of flows around 3D structures
7. Application fields (in the sensory ecological framework)	Steady and unsteady flows in air and water around 3D structures at a macro scale of a few millimeters	Steady flows in water around 2d structures at the scale of a few hundreds of microns (cells, protozoans, flows in vessels etc.)	The use of fluorescent particles constrains applications to steady and unsteady flows in water around 3D structures at a scale of a few hundreds of microns

The use of gray and white background colors respectively indicates for which criteria standard macro-PIV or micro-PIV perform better. The last column is for the new innovative mixed technique that exploits the advantages of both techniques

References

Adrian RJ, Westerweel J (2011) Particle image velocimetry. Cambridge University Press, Cambridge

Alharbi AY, Sick V (2010) Investigation of boundary layers in internal combustion engines using a hybrid algorithm of high speed micro-PIV and PTV. Exp Fluids 49:949–959. doi:10.1007/s00348-010-0870-8

Barth FG, Wastl U, Humphrey JAC, Devarakonda R (1993) Dynamics of arthropod filiform hairs II Mechanical properties of spider trichobothria (*Cupiennius salei* Keys). Philos Trans R Soc Lond B Biol Sci 340:445–461

Bathellier B, Steinmann T, Barth FG, Casas J (2012) Air motion sensing hairs of arthropods detect high frequencies at near-maximal mechanical efficiency. J R Soc Interface 9:1131–1143. doi:10.1098/rsif-2011-0690

Bechert DW, Bruse M, Hage W (2000) Experiments with three-dimensional riblets as an idealized model of shark skin. Exp Fluids 28:403–412. doi:10.1007/s003480050400

Bechert D, Bartenwerfer M (1989) The viscous flow on surfaces with longitudinal ribs. J Fluid Mech 206:105–129

Bitsch L, Olesen LH, Westergaard CH, Bruus H, Klank H, Kutter JP (2005) Micro particle-image velocimetry of bead suspensions and blood flows. Exp Fluids 39:505–511

Bleckmann H, Breithaupt T, Blickhan R, Tautz J (1991) The time course and frequency content of hydrodynamic events caused by moving fish, frogs, and crustaceans. J Comp Physiol A 168:749–757

Blickhan R, Krick C, Zehren D, Nachtigall W (1992) Generation of a vortex chain in the wake of a subundulatory swimmer. Naturwissenschaften 79:220–221

Boek ES, Padding JT, Anderson VJ, Briels WJ, Crawshaw JP (2006) Flow of entangled wormlike micellar fluids: mesoscopic simulations, rheology and μ-PIV experiments. J Non-Newton Fluid 146:11–21

Bown MR, Meinhart CD (2006) AC electroosmotic flow in a DNA concentrator. Microfluid Nanofluid 2:513–523. doi:10.1007/s10404-006-0097-4

Bown MR, MacInnes JM, RWK Allen (2006) Three-component micro-PIV using the continuity equation and a comparison of the performance with that of stereoscopic measurements. Exp Fluids 42:197–205. doi:10.1007/s00348-006-0229-3

Burgmann S, Van der Schoot N, Asbach C, Wartmann J, Lindken R (2011) Analysis of tracer particle characteristics for micro PIV in wall-bounded gas flows. La Houille Blanche 4:55–61. doi:10.1051/lhb/2011041

Casas J, Dangles O (2010) Physical ecology of fluid flow sensing in arthropods. Annu Rev Entomol 55:505–520. doi:10.1146/annurev-ento-112408-085342

Casas J, Liu C, Krijnen G (2013) Biomimetic flow sensors encyclopedia. Nanotechnology 2013:264–276

Casas J, Steinmann T, Dangles O (2008) The aerodynamic signature of running spiders. PLoS ONE 3:e2116. doi:10.1371/journal-pone-0002116

Casas J, Steinmann T, Krijnen G (2010) Why do insects have such a high density of flow-sensing hairs? Insights from the hydromechanics of biomimetic MEMS sensors. J R Soc Interface 7:1487–1495. doi:10.1098/rsif-2010-0093

Catton KB, Webster DR, Brown J, Yen J (2007) Quantitative analysis of tethered and free-swimming copepodid flow fields. J Exp Biol 210:299–310. doi:10.1242/jeb-02633

Chagnaud BP, Bleckmann H, Engelmann J (2006) Neural responses of goldfish lateral line afferents to vortex motions. J Exp Biol 209:327–342. doi:10.1242/jeb-01982

Cierpka C, Rossi M, Segura R, Mastrangelo F, Kähler CJ (2011) A comparative analysis of the uncertainty of astigmatism-μPTV, stereo-μPIV and μPIV. Exp Fluids 52:605–615. doi:10.1007/s00348-011-1075-5

Cummings EB (2000) An image processing and optimal nonlinear filtering technique for particle image velocimetry in microflows. Exp Fluids 29(1):S42–S50

Curtin DM, Newport DT, Davies MR (2006) Utilising μ-PIV and pressure measurements to determine the viscosity of a DNA solution in a microchannel. Exp Therm Fluid Sci 30:843–852

Dangles O, Steinmann T, Pierre D, Vannier F, Casas J (2008) Relative contributions of organ shape and receptor arrangement to the design of cricket's cercal system. J Comp Physiol A 194:653–663. doi:10.1007/s00359-008-0339-x

Denissenko P, Lukaschuk S, Breithaupt T (2007) The flow generated by an active olfactory system of the red swamp crayfish. J Exp Biol 210:4083–4091. doi:10.1242/jeb-008664

Devasenathipathy S, Santiago JG, Wereley ST, Meinhart CD, Takehara K (2003) Particle imaging techniques for microfabricated fluidic systems. Exp Fluids 34:504–514

Devasenathipathy S, Santiago JG (2003) Electrokinetic flow diagnostics. In: Breuer K (ed) Micro- and Nano-scale diagnostic techniques. Springer, New York, pp 113–144

Eichler C, Sattelmayer T (2011) Premixed flame flashback in wall boundary layers studied by long-distance micro-PIV. Exp Fluids 52:347–360. doi:10.1007/s00348-011-1226-8

Engelmann J, Hanke W, Bleckmann H (2002) Lateral line reception in still- and running water. J Comp Physiol A 188:513–526. doi:10.1007/s00359-002-0326-6

Fertin A, Casas J (2006) Efficiency of antlion trap construction. J Exp Biol 209:3510–3515. doi:10.1242/jeb-02401

Fertin A, Casas J (2007) Orientation towards prey in antlions: efficient use of wave propagation in sand. J Exp Biol 210:3337–3343. doi:10.1242/jeb-004473

García-Mayoral R, Jiménez J (2011) Drag reduction by riblets. Philos T Roy Soc A 369(1940):1412–1427. doi:10.1098/rsta-2010-0359

Gnatzy W, Heusslein R (1986) Digger wasp against crickets: I receptors involved in the antipredator strategies of the prey. Naturwissenschaften 73:212–215

Hanke W, Brücker C, Bleckmann H (2000) The ageing of the low-frequency water disturbances caused by swimming goldfish and its possible relevance to prey detection. J Exp Biol 203:1193–1200

Hanke W, Wieskotten S, Niesterok B, Miersch L, Witte M, Brede M, Leder A et al (2012) Hydrodynamic perception in pinnipeds. Note N Fl Mech Mul D 119:255–270. doi:10.1007/978-3-642-28302-4_16

Horiuchi K, Dutta P, Richards CD (2006) Experiment and simulation of mixed flows in a trapezoidal microchannel. Microfluid Nanofluid 3:347–358. doi:10.1007/s10404-006-0129-0

Humphrey JAC, Devarakonda R, Iglesias I, Barth FG (1993) Dynamics of arthropod filiform hairs: I mathematical modelling of the hair and air motions. Philos Trans R Soc Lond B Biol Sci 340:423–440

Inoué S, Spring KR (1997) Video microscopy. Plenum, Oxford

Jacobs GA, Miller JP, Aldworth Z (2008) Computational mechanisms of mechanosensory processing in the cricket. J Exp Biol 211:1819–1828. doi:10.1242/jeb-016402

Jin BJ, Yoo JY (2011) Visualization of droplet merging in microchannels using micro-PIV. Exp Fluids 52:235–245. doi:10.1007/s00348-011-1221-0

Kähler CJ, Scharnowski S, Cierpka C (2012) On the uncertainty of digital PIV and PTV near walls. Exp Fluids 52:1641–1656. doi:10.1007/s00348-012-1307-3

Kähler CJ, Scholz U, Ortmanns J (2006) Wall-shear-stress and near-wall turbulence measurements up to single pixel resolution by means of long-distance micro-PIV. Exp Fluids 41:327–341. doi:10.1007/s00348-006-0167-0

Kämper G, Kleindienst HU (1990) Oscillation of cricket sensory hairs in a low-frequency sound field. J Comp Physiol A 167:193–200

Kim MJ, Beskok A, Kihm KD (2002) Electro-osmosis-driven micro-channel flows: A comparative study of microscopic particle image velocimetry measurements and numerical simulations. Exp Fluids 33:170–180

Kim BJ, Yoon SY, Lee KH, Sung HJ (2008) Development of a microfluidic device for simultaneous mixing and pumping. Exp Fluids 46:85–95. doi:10.1007/s00348-008-0541-1

Kloosterman A, Poelma C, Westerweel J (2010) Flow rate estimation in large depth-of-field micro-PIV. Exp Fluids 50:1587–1599. doi:10.1007/s00348-010-1015-9

Klopsch C, Kuhlmann HC, Barth FG (2012) Airflow elicits a spider's jump towards airborne prey I Airflow around a flying blowfly. J R Soc Interface 9:2591–2602. doi:10.1098/rsif-2012-0186

Koehl MAR (2004) Biomechanics of microscopic appendages: functional shifts caused by changes in speed. J Biomech 37:789–795. doi:10.1016/j-jbiomech-2003-06-001

Krijnen G, Dijkstra M, van Baar J, Shankar S, Kuipers W, de Boer J, Altpeter D, Lammerink T, Wiegerink R (2006) MEMS based hair flow-sensors as model systems for acoustic perception studies. Nanotechnology 17:84–89. doi:10.1088/0957-4484/17/4/013

Kumagai T, Shimozawa T, Baba Y (1998) The shape of windreceptor hairs of cricket and cockroach. J Comp Physiol A 183:187–192

Landolfa G, Jacobs MA (1995) Direction sensitivity of the filiform hair population of the cricket cercal system. J Comp Physiol A 177:759–766

Lee SY, Wereley ST, Gui L, Qu W, Mudawar I (2002) Microchannel flow measurement using micro Particle Image Velocimetry. In: Proceedings of IMECE2002 ASME international mechanical engineering congress and exposition. New Orleans, Louisiana, 17–22 Nov 2002

Lee SJ, Kim BH, Lee JY (2009) Experimental study on the fluid mechanics of blood sucking in the proboscis of a female mosquito. J Biomech 42:857–864. doi:10.1016/j-jbiomech-2009-01-039

Lee SJ, Lee SH (2001) Flow field analysis of a turbulent boundary layer over a riblet surface. Exp Fluids 30:153–166. doi:10.1007/s003480000150

Li H, Olsen MG (2006) Micro PIV measurements of turbulent flow in square microchannels with hydraulic diameters from 200 μm to 640 μm. Int J Heat Fluid Flow 27:123–134. doi:10.1016/j-ijheatfluidflow-2005-02-003

Lima R, Wada S, Takeda M, Tsubota K, Yamaguchi T (2007) In vitro confocal micro-PIV measurements of blood flow in a square microchannel: the effect of the haematocrit on instantaneous velocity profiles. J Biomech 40:2752–2757. doi:10.1016/j-jbiomech-2007-01-012

Lindken R, Westerweel J, Wieneke B (2006) Stereoscopic micro particle image velocimetry. Exp Fluids 41:161–171. doi:10.1007/s00348-006-0154-5

Liu D, Garimella SV, Wereley ST (2005) Infrared micro-particle image velocimetry in silicon-based microdevices. Exp Fluids 38:385–392

Magal C, Dangles O, Caparroy P, Casas J (2006) Hair canopy of cricket sensory system tuned to predator signals. J Theor Biol 241:459–466. doi:10.1016/j-jtbi-2005-12-009

Mansoor I, Stoeber B (2010) PIV measurements of flow in drying polymer solutions during solvent casting. Exp Fluids 50:1409–1420. doi:10.1007/s00348-010-1000-3

McHenry MJ, Strother JA, van Netten SM (2008) Mechanical filtering by the boundary layer and fluid-structure interaction in the superficial neuromast of the fish lateral line system. J Comp Physiol A 194:795–810. doi:10.1007/s00359-008-0350-2

Mead KS (2003) Fine-scale patterns of odor encounter by the antennules of mantis shrimp tracking turbulent plumes in wave-affected and unidirectional flow. J Exp Biol 206:181–193. doi:10.1242/jeb-00063

Meinhart CD, Wereley ST, Gray MHB (2000) Volume illumination for two-dimensional particle image velocimetry. Meas Sci Technol 11:809–814

Meinhart CD, Wereley ST (2003) The theory of diffraction-limited resolution in microparticle image velocimetry. Meas Sci Technol 14:1047–1053

Meinhart CD, Wereley ST, Santiago JG (1999) PIV measurements of a microchannel flow. Exp Fluids 27:414–419

Mielnik MM (2003) Micro-PIV and its application to some BioMEMS related microfluidic flows. PHD Thesis. ISBN 82-471-6954-1

Mielnik MM, Saetran LR (2006) Selective seeding for micro-PIV. Exp Fluids 41(155–159):1007. doi:10/s00348-005-0103-8

Miller JP, Krueger S, Heys JJ, Gedeon T (2011) Quantitative characterization of the filiform mechanosensory hair array on the cricket cercus. PLoS ONE 6(11):e27873. doi:10.1371/journal-pone-0027873

Moghtaderi B, Shames I, Djenidi L (2006) Microfluidic characteristics of a multi-holed baffle plate micro-reactor. Int J Heat Fluid Fl 27:1069–1077. doi:10.1016/j-ijheatfluidflow-2006-01-008

Morley EL, Steinmann T, Casas J, Robert D (2012) Directional cues in Drosophila melanogaster audition: structure of acoustic flow and inter-antennal velocity differences. J Exp Biol 215:2405–2413. doi:10.1242/jeb-068940

Müller U, Heuvel B, Stamhuis E, Videler J (1997) Fish foot prints: morphology and energetics of the wake behind a continuously swimming mullet (*Chelon labrosus Risso*). J Exp Biol 200:2893–2906

Natrajan VK, Yamaguchi E, Christensen KT (2006) Statistical and structural similarities between micro and macroscale wall turbulence. Microfluid Nanofluid 3:89–100. doi:10.1007/s10404-006-0105-8

Nguyen CV, Carberry J, Fouras A (2011) Volumetric-correlation PIV to measure particle concentration and velocity of microflows. Exp Fluids 52:663–677. doi:10.1007/s00348-011-1087-1

Nguyen CV, Fouras A, Carberry J (2010) Improvement of measurement accuracy in micro PIV by image overlapping. Exp Fluids 49:701–712. doi:10.1007/s00348-010-0837-9

Park J, Choi C, Kihm K (2004) Optically sliced micro-PIV using confocal laser scanning microscopy (CLSM). Exp Fluids 37:105–119. doi:10.1007/s00348-004-0790-6

Patrick MJ, Chen CY, Frakes DH, Dur O, Pekkan K (2010) Cellular-level near-wall unsteadiness of high-hematocrit erythrocyte flow using confocal μPIV. Exp Fluids 50:887–904. doi:10.1007/s00348-010-0943-8

Pereira F, Lu J, Castaño-Graff E, Gharib M (2007) Microscale 3D flow mapping with μDDPIV. Exp Fluids 42:589–599. doi:10.1007/s00348-007-0267-5

Poelma C, Van der Heiden K, Hierck BP, Poelmann RE, Westerweel J (2010) Measurements of the wall shear stress distribution in the outflow tract of an embryonic chicken heart. J R Soc Interface 7:91–103. doi:10.1098/rsif-2009-0063

Raffel M, Willert C, Kompenhans J (1998) Particle image velocimetry, a practical guide. Springer, Berlin

Raghavan RV, Friend JR, Yeo LY (2009) Particle concentration via acoustically driven microcentrifugation: microPIV flow visualization and numerical modelling studies. Microfluidic Nanofluidic 8:73–84. doi:10.1007/s10404-009-0452-3

Reidenbach MA, George N, Koehl MAR (2008) Antennule morphology and flicking kinematics facilitate odor sampling by the spiny lobster, *Panulirus argus*. J Exp Biol 211:2849–2858. doi:10.1242/jeb-016394

Rossi M, Segura R, Cierpka C, Kähler CJ (2011) On the effect of particle image intensity and image preprocessing on the depth of correlation in micro-PIV. Exp Fluids 52(4):1063–1075. doi:10.1007/s00348-011-1194-z

Samarage CR, Carberry J, Hourigan K, Fouras A (2011) Optimisation of temporal averaging processes in PIV. Exp Fluids 52:617–631. doi:10.1007/s00348-011-1080-8

Santiago JG, Wereley ST, Meinhart CD, Beebe DJ, Adrian RJ (1998) A particle image velocimetry system for microfluidics. Exp Fluids 25:316–319. doi:10.1007/s003480050235

Sato Y, Hishida K (2006) Electrokinetic effects on motion of submicron particles in microchannel. Fluid Dyn Res 38:787–802. doi:10.1016/j-fluiddyn-2006-04-003

Sato Y, Inaba S, Hishida K, Maeda M (2003) Spatially averaged time-resolved particle-tracking velocimetry in microspace considering Brownian motion of submicron fluorescent particles. Exp Fluids 35:167–177. doi:10.1007/s00348-003-0643-8

Shimozawa T, Kanou M (1984) The aerodynamics and sensory physiology of range fractionation in the cercal filiform sensilla of the cricket *Gryllus bimaculatus*. J Comp Physiol A 155:495–505

Shimozawa T, Kumagai T, Baba Y (1998) Structural scaling and functional design of the cercal wind-receptor hairs of cricket. J Comp Physiol A 183:171–186

Shimozawa T, Murakami J, Kumagai T (2003) Cricket wind receptors: thermal noise for the highest sensitivity known. In: Barth FB, Humphrey JAC, Secomb T (eds) Sensors and sensing in biology and engineering. Springer, Berlin, pp 145–157

Snoeyink C, Wereley S (2013) A novel 3D3C particle tracking method suitable for microfluidic flow measurements. Exp Fluids 54:1453. doi:10.1007/s00348-012-1453-7

Stamhuis EJ, Videler JJ, Duren LAV, Mu UK (2002) Applying digital particle image velocimetry to animal-generated flows: traps, hurdles and cures in mapping steady and unsteady flows in Re regimes between 10–2 and 105. Exp Fluids 33:801–813. doi:10.1007/s00348-002-0520-x

Steinmann T, Casas J, Krijnen G, Dangles O (2006) Air-flow sensitive hairs: boundary layers in oscillatory flows around arthropod appendages. J Exp Biol 209:4398–4408. doi:10.1242/jeb-02506

Sterbing-D'Angelo S, Chadha M, Chiu C, Falk B, Xian W, Barcelo J, Zook JM et al (2011) Bat wing sensors support flight control. PNAS 108:11291–11296. doi:10.1073/pnas-1018740108

Sun C, Lee HC, Kao RX (2011) Diagnosis of oscillating pressure-driven flow in a microdiffuser using micro-PIV. Exp Fluids 52:23–35. doi:10.1007/s00348-011-1204-1

Tautz J, Markl H (1979) Caterpillars detect flying wasps by hairs sensitive to airborne vibration. Behav Ecol Sociobiol 4:101–110

Walsh PA, Walsh EJ, Davies MRD (2007) On the out-of-plane divergence of streamtubes in planar mini-scale flow focusing devices. Int J Heat Fluid Fl 28:44–53. doi:10.1016/j-ijheatfluidflow-2006-05-006

Wang B, Demuren A, Gyuricsko E, Hu H (2010) An experimental study of pulsed micro-flows pertinent to continuous subcutaneous insulin infusion therapy. Exp Fluids 51:65–74. doi:10.1007/s00348-010-1033-7

Wereley ST, Meinhart CD (2010) Recent advances in micro-particle image velocimetry. Annu Rev Fluid Mech 42:557–576. doi:10.1146/annurev-fluid-121108-145427

Windsor SP, Norris SE, Cameron SM, Mallinson GD, Montgomery JC (2010a) The flow fields involved in hydrodynamic imaging by blind Mexican cave fish (*Astyanax fasciatus*) Part I: open water and heading towards a wall. J Exp Biol 213:3819–3831. doi:10.1242/jeb-040741

Windsor SP, Norris SE, Cameron SM, Mallinson GD, Montgomery JC (2010b) The flow fields involved in hydrodynamic imaging by blind Mexican cave fish (*Astyanax fasciatus*), Part II: gliding parallel to a wall. J Exp Biol 213:3832–3842. doi:10.1242/jeb-040790

Yan DG, Yang C, Huang XY (2006) Effect of finite reservoir size on electroosmotic flow in microchannels. Microfluid Nanofluid 3:333–340. doi:10.1007/s10404-006-0135-2

Yang CT, Chuang HS (2005) Measurement of a microchamber flow by using a hybrid multiplexing holographic velocimetry. Exp Fluids 39:385–396. doi:10.1007/s00348-005-1022-4

Part II
Flow Sensing and Animal Behavior

Chapter 3
The Role of Flow and the Lateral Line in the Multisensory Guidance of Orienting Behaviors

Sheryl Coombs and John Montgomery

Abstract This chapter summarizes what is known about the role of lateral line flow sensors in different types of fish orienting behaviors and the spatiotemporal characteristics of flow patterns that guide fish. Where possible, fundamental and shared principles of flow guidance are identified for behaviors as diverse as rheotaxis, prey orientation, and predator avoidance. Multisensory, egocentric direction maps in the midbrain optic tectum are thought to underlie oriented movements toward discrete targets of interest (e.g., prey), whereas differential biasing of bilaterally paired Mauthner cells by multisensory inputs are involved in fast-start oriented behaviors such as predator avoidance. Wide-field integration of global flow fields inspired by optic flow studies in flies is suggested as a central mechanism by which cells are tuned to expected flow patterns over large body regions during common motion states, such as forward translational movement. Disruptions in those patterns caused by, e.g., the destabilizing influences of currents could then be used as part of closed-loop feedback control for corrective actions. Flow guidance of behaviors is likely to be best understood when viewed from a multisensory, evolutionary perspective in which nonvisual and visual systems work in tandem to guide behavior.

S. Coombs (✉)
Department of Biological Sciences and JP Scott Center for Neuroscience, Mind and Behavior, Bowling Green State University, Bowling Green, OH 43403, USA
e-mail: scoombs@bgnet.bgsu.edu

J. Montgomery
School of Biological Sciences and Department of Marine Science,
University of Auckland, Auckland 1142, New Zealand
e-mail: j.montgomery@auckland.ac.nz

H. Bleckmann et al. (eds.), *Flow Sensing in Air and Water*,
DOI: 10.1007/978-3-642-41446-6_3, © Springer-Verlag Berlin Heidelberg 2014

3.1 Introduction

It is claimed that animals spend more time orienting than in any other type of behavior. The list of orienting behaviors is thus very long and covers a wide range of behavioral and ecological contexts. Based on the landmark work of Fraenkel and Gunn (1961) and Jander (1975), orienting behaviors can be grouped into four broad categories: (1) orienting toward (or away from) other animals to, e.g., locate prey or avoid predators, (2) orienting toward topographical features/geographical destinations for finding resources (e.g., food, shelter, breeding grounds) or away from obstacles that block access to resources and finally, maintaining a particular body orientation with respect to (3) others in a social group (e.g., schooling fish), or (4) persistent global stimuli such as light (phototaxis) or currents (rheotaxis).

Orienting behaviors can be brief and short-range (e.g., orienting movements of the head or eyes toward a nearby sound source of interest) or long-lasting and long-range (e.g., long-distance bird or monarch butterfly migrations over thousands of kilometers). They can rely on one or more sensory cues and a myriad of different sensory mechanisms and sensorimotor strategies for guiding the behavior, such as following a scent trail, moving along an intensity gradient or instantaneously computing the location of a sound source from binaural time and intensity cues.

In this chapter, we summarize and evaluate the role that lateral line flow sensors play in different types of orienting behaviors and the flow characteristics that guide them (Sects. 3.2, 3.3, 3.4, 3.5). To the extent possible, we examine how lateral line contributions dovetail with those from other senses for a given behavior and how the relative contributions depend on a number of factors, including the sequence of motor acts in the overall behavior, sensory conditions (e.g., day vs. night), sensory specializations, and life stage (e.g., larval vs. adult fish). Most importantly, we consider the multisensory and sensorimotor integration sites in the brain that underlie orienting behaviors (Sect. 3.6) and in this context, compare and contrast the sensory mechanisms and sensorimotor strategies for flow-based orientation (Sect. 3.7). Our overriding goal is to identify fundamental and shared principles by which flow information is merged with other sensory information to provide multisensory guidance of orienting behaviors under different behavioral contexts. For excellent, earlier reviews on flow-based orientation, see Montgomery and Walker (2001) and Sand and Bleckmann (2008).

3.2 Orienting Toward or Away from Other Animals: Finding Prey and Avoiding Predators

Animals adopt many different solutions to the problems of how to find food and avoid predators, not all of which involve orientation. Those involving a flow-guided orienting component can be grouped into several categories (Table 3.1). Behaviors involving the pursuit of prey include: orienting responses to a nearby

Table 3.1 Orienting behaviors and their characteristics

Behavior	Duration and range	Stimulus source	Proximal stimulus to lateral line	Proposed CNS structures
Prey-orienting (initial reaction)	Brief, short-range	Motile prey (submerged or surface)	Incompressible flow patterns or surface waves	OT
Prey-orienting (trail-following)	Long-lasting, long-range	Motile prey	Hydrodynamic wake	OT
Prey-orienting (flow-dispersed odors)	Long-lasting, long-range	Motile or dead odorous prey	Hydrodynamic wakes and/or ambient flows	OT
Fast-start escape	Brief, short-range	Ram or suction-feeding predators	Incompressible flow	M cell
Obstacle avoidance	Brief, short-range	Submerged obstacles	Perturbations in self-generated or ambient flows	OT and/or M cell
Obstacle entrainment (wall following or station holding)	Long-lasting, short-range	Submerged obstacles	Perturbations in self-generated or ambient flow	OT
Schooling	Long-lasting, short-range	Schooling conspecifics	Incompressible flow; vortex wakes	OT and/or M cell
Propagated c-starts (waves of agitation)	Brief, short-range	Schooling conspecifics	Incompressible flow (water jets)	OT and/or M cell
Rheotaxis	Variable, NA	Ambient currents; stationary features in the environment	Ambient flows	OT

prey that attracts immediate attention via local incompressible flows (Sect. 3.2.1) or surface capillary waves (Sect. 3.2.2); following the hydrodynamic wake (trail) of a swimming prey that is no longer in the immediate vicinity (Sect. 3.2.3), and (3) following flow-dispersed odors (Sect. 3.2.4).

Although there are similarly many different types of predator-avoidance behaviors, the C-shaped escape response, which rapidly propels fish away from a predator when the threat of being eaten is imminent, is the best-studied behavior known to have a lateral line component (Sect. 3.2.4). In solitary fish, these escape responses can be evoked by two fundamentally different predator actions (suction and ram), resulting in accelerating flows of opposite directions. In schooling fish, startle responses may occur as a direct response to a predator or, alternatively, as an indirect response to nearby startled fish, in which case startle responses can be rapidly propagated as a *wave of agitation* (Radakov 1973) that may serve as a form of intra-school transmission of information about an external threat (see Sect. 3.4).

3.2.1 Prey-Orienting Responses to Incompressible Flow Sources

Lateral line mediated prey-orienting responses have been reported for several benthic and mid-water fishes, which detect the local incompressible flows produced by appendage and/or whole-body movements of their small invertebrate or vertebrate prey. Although the flow patterns generated by prey vary in detail, the dipole component tends to dominate (Kalmijn 1988, 1989) and so is a common feature of flow patterns around prey as diverse as tiny copepods (Fig. 3.1b) and fish (Fig. 3.1a, c). Some fish (e.g., nocturnal predators like the Lake Michigan mottled sculpin) rely heavily if not exclusively on the lateral line (Hoekstra and Janssen 1985; Coombs et al. 2001). In these cases, prey-orienting responses are distance-limited by the nature of incompressible flow, which attenuates steeply with distance from the source (1/distance^3 for a dipole source, Kalmijn 1988, 1989). The general rule-of-thumb is that the detection range is less than one body length away (Kalmijn 1988, 1989). However, the precise distance depends on many factors (e.g., the amplitude of prey movements) and thus, the distance at which mottled sculpin can detect the minute water movements of a small water flea (*Daphnia magna*) is only a fraction of a body length ($< \sim 1/10$th of a body length or 1 cm) (Hoekstra and Janssen 1986).

At the other end of the multisensory continuum are fish that rely more heavily on vision for the initial detection of prey, which can occur from much longer distances and beyond the normal working range of the lateral line. In these cases, the lateral line has still been shown to play a role in guiding prey capture behaviors, but only when from a close distance. For example, bluegill sunfish, *Lepomis macrochirus* (Enger et al. 1989), largemouth bass (Gardiner and Motta 2012) and blacktip sharks (Gardiner 2012) are all capable of detecting and striking at prey without vision, but

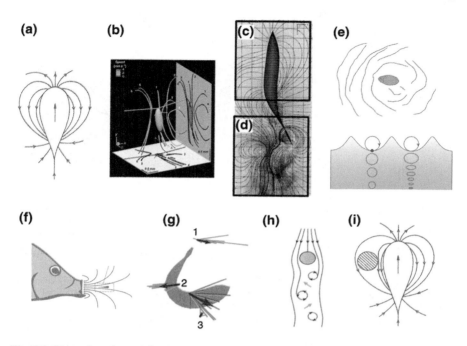

Fig. 3.1 Examples of water flow patterns that direct various orienting behaviors. **a** Mathematically modeled dipole field flow around an advancing body (from Hassan 1985), which can be characteristic of both invertebrate (copepod) (**b**) (copied with permission from Malkiel et al. 2003) and **c** vertebrate (fish) (adapted with permission from Wolfgang et al. 1999) prey that elicit short-range prey-orienting responses. Hydrodynamic wakes left behind by a swimming fish (**d**) (adapted with permission from Wolfgang et al. 1999) provide a long-range trail for predators to follow. **e** Surface waves produced by insects guide surface-feeding fish to the central source of the outwardly propagating wave; each wave consists of *circular* or *elliptical* orbits of local flow patterns just below the water surface. Ram predators produce a local incompressible flow in the bow wake as they move forward (**a** and **c**) and suction feeders additionally produce a suction flow created by the rapid expansion of the jaws (**f**) (copied with permission from Stewart et al. 2013). C-start escape responses elicited from predators (**g**) produce three distinct local water jets (*blue vectors*) that may generate a cascade of escape responses from neighboring fish to transmit information about external threats (copied with permission from Tytell and Lauder 2008; *red vectors* indicate the fish's momentum). Fish often orient or entrain to (**h**) bluff bodies in a stream, which generate vortex wakes and region of reduced flow behind the body to create steep velocity gradients (from Drucker and Lauder 2002). Blind cavefish can orient away from stationary obstacles by sensing distortions in their own self-generated flow field as they swim by the obstacle (**i**) (from Hassan 1985)

only if the prey are less than 0.5 body lengths away. In bluegill sunfish, lateral line deprivation causes a total abolishment of prey capture abilities, but in largemouth bass and blacktip sharks, it causes a decline in strike accuracy and a failure in their normal braking behavior immediately before capture. Similarly, muskellunge, *Esox masquinongy* rely on both vision and the lateral line for the final strike behavior, but predominantly on vision for the initial orienting and slow (stalk) approach behavior

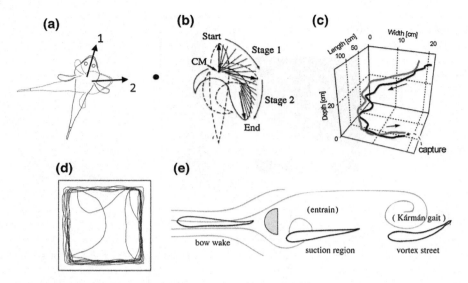

Fig. 3.2 Examples of different orienting behaviors involving the lateral line: **a** initial prey-orienting response of Lake Michigan mottled sculpin to a dipole source (*black dot*), **b** two-stage, C-start escape response of goldfish (adapted with permission from Eaton and Emberly 1991), **c** hydrodynamic trail following by a predatory European catfish (*black*) following the trail of its goldfish prey (*grey*) [copied with permission from Pohlman et al. (2001). Copyright (2001) National Academy of Sciences, USA], **d** wall-following trajectories of a blind cavefish around a square arena, and **e** three different kinds of station-holding behaviors of trout (see text) (from Liao 2007, copied with permission from the Royal Society)

that bring them within striking distance of their prey (New et al. 2001). Interestingly, sharks and likely some bony fishes (e.g., catfish) require odor cues to initiate prey tracking and capture behaviors (Gardiner and Atema, chapter in this volume).

The prey-orienting behavior of Lake Michigan mottled sculpin has been studied extensively with a combination of behavioral, neurophysiological, and modeling approaches. In the absence of visual cues, these fish exhibit a naturally occurring and unconditioned orienting response (Fig. 3.2a) that can be triggered by both live (e.g., *Daphnia*) and artificial prey (a small sphere that vibrates intermittently) (Hoekstra and Janssen 1985, 1986). Under these conditions, the lateral line system and in particular, canal but not superficial neuromasts (SNs), are required for the initial orienting response (Coombs et al. 2001). In addition, unilateral, but not bilateral comparisons of outputs from neuromasts are necessary for accurate orienting responses to ipsilateral sources (Conley and Coombs 1998). Finally, orientation does not require closed-loop guidance, as these fish usually launch the orienting response from a stationary position within 500 ms of stimulus onset (Coombs and Conley 1997a). Together, all of these findings suggest that fish use spatial patterns of activation along the lateral line for determining prey direction.

Various models of dipole flow fields have shown that pressure-gradient patterns (the stimulus to canal neuromasts) change in predictable ways as a function of

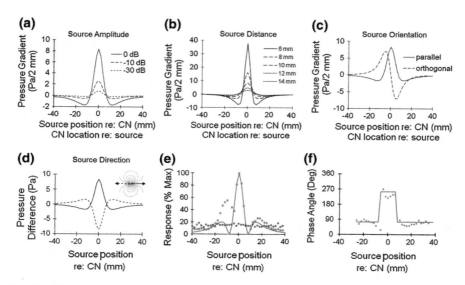

Fig. 3.3 Illustration of the *sequential pattern of activation* of a single CN (*x*-axis title = Source position re: CN) when the source changes its location along a linear transect with respect to CN location (*X* = 0). The same figure also illustrates the *instantaneous spatial pattern of activation* across an array of CNs when the location of the source is fixed (at *X* = 0) (*x*-axis title = CN location re: source). Both sequential and instantaneous activation patterns vary with different source characteristics, including **a** source amplitude, **b** source distance, **c** source orientation—i.e., the axis of source vibration is parallel or perpendicular to the long axis of the sensory array, and **d** source direction—i.e., whether the source is moving forward or backward along its axis of vibration. In (**a–c**), the pressure gradient across the surrounding two canal pores at each source position is plotted for the same phase of the sinusoidal cycle of source oscillation. In (**d**), the pressure gradient is plotted for two opposite phases of the sinusoidal cycle to show how the pressure-gradient direction (fore or aft along the canal axis as indicated by positive and negative values on the *Y*-axis) oscillates over time at any given position. In **e** and **f**, the magnitude (**e**) and phase angle (**f**) response of a PLLN fiber (*red closed symbols*) are plotted relative to spontaneous response rates (*black open symbols*) and modelled predictions for the pressure-gradient stimulus in *blue* (data from Coombs et al. 1996). The slight mismatch between neuronal and modelled predictions are due to the fact that the fish's body was tilted slightly downward with respect to the axis of source translocation in these neurophysiological studies

source amplitude, distance, and orientation (Coombs et al. 1996; Ćurčić-Blake and van Netten 2006; Goulet et al. 2008) (Fig. 3.3a–d) and moreover, are faithfully represented by the activity of lateral line nerve fibers (Fig. 3.3e, f) (Coombs et al. 1996; Coombs and Conley 1997b). Furthermore, modeled predictions about how the pressure gradient pattern should affect behavioral performance have been experimentally verified in a few cases. For example, the model predicts that the spatial region of maximum excitation (the 'hot spot' in the spatial excitation pattern) will broaden as source distance (Fig. 3.3b), but not amplitude (Fig. 3.3a) increases. If fish were using the somatotopic location of the hot spot to localize the source, then their localization precision should decline with increasing distance, even when source amplitude is adjusted to keep the received amplitude at the skin surface the same. Indeed, the orienting accuracy of mottled sculpin declines significantly for a mere doubling of source distance from 3 to 6 cm (Coombs and Patton 2009). Nevertheless, the hot spot or other features that may signify the location of the source along the body surface is insufficient for informing fish where the prey is with respect to the eyes and mouth and must be combined with information about source distance before source direction can be computed (see Sect. 3.7.1).

3.2.2 Prey-Orienting Responses to Surface Wave Sources

Surface-feeding fish and amphibians exhibit prey-orienting responses to struggling insects on the water surface (reviewed in Bleckmann et al. 1989; Sand and Bleckmann 2008). However, the hydrodynamic stimuli and the directional cues provided by prey-generated surface waves are quite different from those provided by local flows around submerged prey. Unlike local flows, the amplitude and duration of surface-wave signals are strongly modified in a frequency-dependent way as they propagate outward from the source. For all but the lowest frequencies ($<\sim 13$ Hz), which are dominated by gravity-wave properties, high-frequency components (dominated by capillary-wave properties) travel faster than and are more strongly attenuated than low-frequency components over the frequency range of interest (13–150 Hz). Furthermore, submerged prey create broad and spatially complex flow patterns that may be axisymmetric, but not radially symmetric about the source, whereas surface-wave sources generate radially symmetric wave fronts (concentric waves that spread outward from the source), each of which consists of vertically oriented circular or elliptical flow patterns just below the water surface (Fig. 3.1d).

As estimated with broadband puffs of air at the water surface, the orienting responses of surface feeding fish like the striped panchax, *Aplocheilus lineatus*, and African butterflyfish, *Pantodon buchholzi* are remarkably accurate over a wide range of target angles (nearly 360°) and distances (up to 20 cm) (Bleckmann et al. 1989). Furthermore, orientation remains accurate if lateral line inputs are restricted to only two cephalic neuromasts ipsilateral to the source, even when the two neuromasts have the same axis of hair cell orientation and hence, directional tuning (Tittel et al. 1984). Based on these and other experimental manipulations, Bleckmann and colleagues have concluded that surface feeding fish use time-of-arrival differences among neuromasts to compute source direction (Tittel et al. 1984; Tittel 1985, 1991; Bleckmann et al. 1989). For slowly propagating surface waves (~ 20–50 cm/s in the relevant frequency range of 10–150 Hz) (reviewed in Bleckmann et al. 1989), time-of-arrival differences between neuromasts separated by as little as 1 mm (~ 2–5×10^{-3} s) are well above binaural time differences (~ 20–160×10^{-6} s) effectively used by barn owls to localize sound pressure waves in air (Moiseff and Konishi 1981).

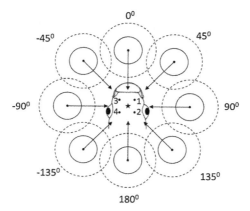

Fig. 3.4 Time-of-arrival differences between unilateral (1:2 and 3:4) and bilateral (1:3 and 2:4) pairs of neuromasts for surface waves (*concentric circles*) emanating from sources at different locations. The time-of-arrival difference of a wave front at each neuromast in the pair depends on the source direction re: a central reference point on the head (*asterisk*). For a source direction of 0°, neuromast 1 leads neuromast 2; at 180°, neuromast 1 lags neuromast 2 and at 90° neuromast 1 and 2 are stimulated simultaneously. However, 'mirror image' contralateral sources produce the same time differences on their respective ipsilateral neuromast pair (3:4) to create *left/right* ambiguities. For example, source angles of 90° and −90° both produce a 0 time difference, but yet their directions are opposite. Likewise, source angles of 45° and −45° produce the same time-of-arrival differences, but their directions differ by 90°. If both bilateral (*left/right*) (1:3 and/or 2:4) and unilateral (rostrocaudal) (1:2 and/or 3:4) comparisons are made, these *left/right* ambiguities would be resolved. That is, if neuromast 1 leads neuromast 2 but lags neuromast 3, there is sufficient information to determine that this stimulus comes from the *upper right* (e.g., 45°) instead of the *upper left* (−45°). With this same reasoning, it can be shown that similar rostro/caudal ambiguities arise if fish use only bilateral comparisons (e.g., comparing time-of-arrival between neuromasts 1 and 3). Thus, *left/right* and rostro/caudal ambiguities can only be resolved by making both bilateral and unilateral comparisons

The ability of surface-feeding fish to localize surface-wave sources using only two neuromasts on the same side of the head suggest that bilateral comparisons of time-of-arrival cues may be unnecessary. However, accurate localization is possible only when sources are ipsilateral but not contralateral to the intact neuromasts; contralateral sources always produce inaccurate turning responses toward the unblocked side (Schwartz 1965; Tittel et al. 1984). Thus, both unilateral (on the same side of the body but at different rostrocaudal locations) and bilateral (on opposite sides of the body but at the same rostrocaudal location) comparisons of time-of-arrival cues may be necessary for accurate localization of sources in all directions. This necessity can be understood in terms of the mirror-image ambiguities that would arise if source direction were based solely on time-of-arrival differences computed in either a unilateral or bilateral fashion, as illustrated in Fig. 3.4. For example, fish making only unilateral comparisons between neuromasts separated in a rostrocaudal direction (e.g., neuromast pair 1:2 in Fig. 3.4) would be unable able to distinguish a +45° source on the right from a −45° source on the left; likewise, fish making only bilateral comparisons (e.g., neuromast pair 1:3 in Fig. 3.4) would be unable to distinguish a +45° source in front from a +135° source from the rear. In theory, these ambiguities could be resolved by making both unilateral (rostrocaudal) and bilateral (left–right) comparisons.

3.2.3 Hydrodynamic Trail Following

Following the hydrodynamic wake left behind by a prey no longer in the immediate vicinity is a third way that fish can orient to their prey (Fig. 3.2c). Wake-following behaviors have been observed in seals, which use their flow-sensing vibrissae (reviewed in Dehnhardt, chapter in this volume) and the European catfish, *Silurus glanis,* which rely on their lateral line (Pohlmann et al. 2001, 2004). European catfish are able to follow the persistent (up to 10 s old) wake of prey fish (a small guppy) in total darkness and this ability is lost in individuals with blocked lateral lines (Pohlmann et al. 2001, 2004).

Wake-following behavior is very different from the brief and open-loop prey-orienting behaviors discussed previously (Sects. 3.2.1 and 3.2.2); it is indirect, longer lasting and requires continuous sensory feedback to guide the predator to its prey. The wake stimulus at the trailing edge of a swimming fish (Fig. 3.1d), which is quite distinct from the local incompressible (nearfield) flows at the leading edge (Fig. 3.1c), may be long and winding (Pohlmann et al. 2001) and detectable over longer distances $(20 > x < 100 \text{ cm})$ (Bleckmann 1993). Like local incompressible flows, wakes provide a hydrodynamic signature of prey characteristics. For example, the vortex shedding frequency depends on tail beat frequency, the vertical height of the wake depends on caudal fin size, and the width of the wake on tail beat amplitude and elapsed time/distance to the wake-generating source (Hanke et al. 2000; Hanke and Bleckmann 2004).

Little is known about the actual cues that fish use to follow wakes. However, recent potential flow models indicate that information about vortex magnitude, orientation, translational speed, spacing and distance from the fish's body is encoded by pressure-gradient patterns, which are available to the lateral line (Franosch et al. 2009; Ren and Mohseni 2012). Although a significant role of gustation in wake following behavior has been experimentally ruled out for the European catfish (Pohlmann et al. 2001), the idea that fish (in general) might use a combination of both chemosensory and lateral line cues to find the source of a wake or other forms of current-dispersed odor, as described in the next section, must be considered.

3.2.4 Odor-Gated Rheotaxis and Eddy Chemotaxis

Odor-gated rheotaxis and eddy chemotaxis rely on the hydrodynamic dispersal of odors to find prey. In still waters, odors from a stationary source are dispersed by a non-directional diffusional process, but in fluvial or large bodies of water with currents, odor dispersal is by advection and is dominated by fluid dynamics, as are odors that are dispersed in the wake of a swimming prey (Weissburg 2000). Odor-gated rheotaxis is part of a general phenomenon in which odors elicit search behaviors, with rheotaxis being one component of the overall search strategy. Given that odors are always distributed in a downstream direction, positive rheotaxis is an important, directional component of the search strategy. Both sharks and many different species of bony fishes use odor-gated rheotactic strategies to navigate upstream until they are in the vicinity of the odor source, after which other strategies take over (reviewed in Baker et al. 2002; Gardiner and Atema 2007, chapter in this volume).

In blind cavefish, the threshold flow speed for rheotaxis is lowered in the presence of food odors—an effect that is dependent on lateral line SNs, which are important to rheotaxis even in the absence of food odors (Baker and Montgomery 1999b) (see also Sect. 3.5 on rheotaxis). Similarly, SNs have been shown to play a crucial role in the rheotactic component of olfactory search behavior in banded kokoupou, *Galaxias fasciatus*, (Baker et al. 2002). As might be expected, the removal of SNs has no impact on search behavior in still water. In slow (2 cm/s) flows, however, rheotaxis is seriously impaired, resulting in more convoluted search paths and an increase in the amount of time that it takes fish to find the odor source.

Whereas rheotactic strategies to search for and/or track odorous prey involves upstream orientation to large-scale flow (Sect. 3.5), eddy chemotaxis refers to the process by which fish track small-scale, 'packets' of odors that are dispersed by vortex eddies (Gardner and Atema 2007). In any but the slowest of flows, where laminar conditions prevail (Re $\leq$ 10), chemical distribution will be spatially and temporally 'patchy', being governed by turbulent structures like the wakes behind swimming fish (Fig. 3.1d) and bluff bodies (Fig. 3.1h) (Moore and Atema 1991). Some sharks use eddy chemotaxis to track the odor-plume left in the wake of a swimming prey and this ability requires the lateral line (Gardiner and Atema 2007).

3.2.5 Fast-Start Escape Responses

Escape behaviors classified as *fast-start* (rapid tail-flip) or *startle* behaviors can be evoked by both ram and suction-feeding predators. The low pressure created by buccal expansion of a suction-feeding predator creates a negative pressure gradient that drives a local, incompressible flow of water into the mouth, with flow velocity increasing toward the mouth (Fig. 3.1f) (Day et al. 2005; Higham et al. 2005). In contrast, the high pressure created by the acceleration of a ram predator, creates a positive pressure gradient that accelerates water out and away from the leading edge of the predator (the bow-wave) and around and back toward the trailing edge (Fig. 3.1a, c). Pressure plays a particularly important role in the initiation of the startle responses in otophysan species like goldfish (*Carassius auratus*) (Canfield and Eaton 1990), which have specialized pressure-sensitive ears (Fay and Popper 1974). In fish without these auditory specializations, the local flow, which can be sensed by the inertial ear as the flow-induced acceleration of the fish's body and/or by the lateral line as differential movement between fish and the surrounding water motions (Kalmijn 1988, 1989), plays a more prominent role (Canfield and Rose 1996). In both cases, however, the local incompressible flow plays an important role in the directional guidance of the response (Casagrand et al. 1999; Mirjany et al. 2011).

The most widely studied and best understood startle response is the C-start, which is mediated by the Mauthner (M) cell/reticulospinal system (Sect. 3.6.2) within 10 ms or less. The C-start consists of two major stages—(1) the initial rapid turn or C-shaped body bend and (2) a lesser body bend in the opposite direction, which launches the final escape trajectory (Eaton et al. 2001; Domenici and Blake 1997) (Fig. 3.2b). The C-start generates its own hydrodynamic signature (Fig. 3.1g), which can direct a whole cascade of escape responses from neighboring fish in a school (see Sect. 3.4).

The precise role of the lateral line in escape responses has been difficult to assess and likely depends on a number of factors, including (1) the nature of the stimulus (ram or suction), (2) the form of behavior that is elicited (e.g., C or S-start), (3) the stage of the behavior (e.g., stage 1 or 2 of the C-start), (4) the developmental stage of the fish (e.g., larval or adult) and (5) the relative importance of other senses, including the auditory system, which may or may not have gas-filled structures (e.g., the swim bladder) and other associated auditory structures (e.g., weberian ossicles of otophysan fish) for sensing the pressure component of the stimulus. For example, comparative studies by Canfield and Rose (1996) have shown that blocking the lateral line of a species without pressure-sensing specializations (the cichlid, *Haplochromis burtoni*) reduces the probability of a C-start to a compressive sound pulse (similar to what a ram predator might produce), whereas blocking the lateral line of the pressure-sensitive goldfish does not. The implication of this finding is that cichlids rely primarily on the lateral line to detect the incompressible flow component of the compressive sound pulse, whereas goldfish rely primarily on the auditory system to detect the pressure component.

More recent studies on goldfish escape behavior have revealed that the lateral line may play a bigger role in modulating both response probability and direction than previously thought and moreover, that different body regions of the lateral line (head vs. trunk) may provide directional guidance at different stages of the behavior (Mirjany et al. 2011). Transection of the posterior lateral line nerve (PLLN) supplying the trunk region of the fish had little effect on response latency, probability or initial direction of goldfish escape responses, but pharmacological blocking of the entire system significantly increased escape latency and orientation errors (Mirjany et al. 2011). Interestingly, PLLN transection nevertheless increased the rate at which goldfish collided with the walls of the tank during the escape, suggesting that PLLN input can still play a role in the subsequent guidance of the behavior after it is initiated.

When considering the role that the lateral line and other senses might play in escape responses, it is instructive to remember that both the timing (not too soon, but not too late) and execution (e.g., speed and direction) of the response are important to its success (Webb 1982). Whereas longer range systems (vision and audition) might be expected to provide advance notice of the predator, the short-range lateral line system (one body lengths away) might kick in at the last moment as a final deciding factor to initiate the escape maneuver. Furthermore, for otophysan fishes like the goldfish, which rely heavily on non-directional pressure information from the ear, the lateral line is well-suited to provide the necessary directional guidance. Indeed, the findings of Mirjany et al. (2011)—that goldfish deprived of lateral line inputs have significantly longer response latencies and increased directional errors—are supportive of both a timing and directional role for the lateral line.

Response direction is important not only for avoiding the oncoming predator, but also for avoiding collision with nearby obstacles or other fish while escaping the predator. In this respect, it is interesting that both PLLN transection and pharmacological blocking of the entire lateral line interferes with the ability of goldfish to avoid collisions with surrounding walls, suggesting that inputs from both head and trunk regions are important (Mirjany et al. 2011).

More recent studies on larval zebrafish have established clear evidence that the lateral line also plays a major role in the escape responses of larval fish (McHenry et al. 2009; Liao and Haehnel 2012; Olszewksi et al. 2012), as proposed earlier by Blaxter and Fuiman (1990). In these experiments, three types of stimuli were used: broad-scale, impulse flows across the width of the test arena (McHenry et al. 2009), localized water jets delivered to different body regions (Liao and Haehnel 2012) and a centrally located, small suction source in the floor of the tank from which the flow speed decreased in an outward, radial direction from the source (Olszewiski et al. 2012). Both impulse flows and local water jets produced characteristic C-start escape behaviors, whereas the continuous suction source elicited fast-start swimming bursts. Experiments by Liao and colleagues indicate that in contrast to findings with adult goldfish, escape behaviors in larval zebrafish are associated with stimulation of more caudally located neuromasts near the tail (Liao 2010; Liao and Haehnel 2012; Olszewski et al. 2012). The innervating fibers of these caudal neuromasts represent some of the first-born and largest cells in the developing lateral line system and there is strong evidence that they project to the region of the brain that orchestrates escape behaviors (Liao and Haehnel 2012) (see also Sect. 3.6.2). Considering that the visual fields of fish typically have a blind spot in the rear (reviewed in Collin et al. 2003), lateral line-mediated escape responses to stimuli from the rear of larval fish would seem to be a good way to compensate for this blind spot.

Like goldfish, zebrafish are otophysans and thus, the question arises as to why the auditory system does not play as big a role in larval zebrafish as it does in adult goldfish. One possibility is that the auditory system of larval zebrafish is not yet developmentally mature enough to respond proficiently (or at all) to the pressure component of sound. In support of this hypothesis is the finding that deflation of the swimbladder (the pressure transducer in the otophysan auditory system) in adult, but not larval zebrafish leads to a significant elevation in hearing thresholds (Zeddies and Fay 2005). An additional explanation is that the flow signal to the lateral line is reduced in adult fish compared to larval fish because adult fish have a larger swimbladder volume and hence, a specific gravity that is closer to that of water. A fish with the same specific gravity as water will tend to be accelerated at the same phase and amplitude as the surrounding water, meaning that the stimulus to the lateral line (the net difference between the motions of the fish and surrounding water) will be minimized if not nullified. Indeed, recent modeling studies using flow signals similar to those generated by suction-feeding predators, indicate that the initial inflation of the zebrafish swimbladder causes a 5% reduction in specific gravity and an 80% reduction in the magnitude of the flow signal to the lateral line (Stewart and McHenry 2010). Swimbladder volume, whether via its effects on specific gravity and/or pressure-sensing abilities of the auditory system, would thus seem to be an important factor in determining the relative importance of the lateral line to both larval and adult fish in sensing predators.

3.3 Orienting Toward or Away from Topographical Features: Obstacle Entrainment and Avoidance

Any flow will be disturbed by a stationary object or obstacle, regardless of whether the flow is generated by abiotic (e.g., gravity-driven currents) or biotic (e.g., a swimming fish) means. Thus, fish are capable of using flow information in different ways to detect and respond to stationary features in their environment. The best studied examples are (1) obstacle avoidance and wall-entrainment (following) behaviors of blind cavefish (*Astyanax mexicanus*), both of which involve self-generated flows for *active* flow sensing, and (2) station entrainment (holding) behaviors of stream-dwelling trout (*Oncorhynchus* sp.), which rely predominantly on *passive* sensing of object-generated disturbances in the flow. These behaviors have been well-reviewed in recent papers (Liao 2007; Windsor et al. 2008; chapter by Windsor, this volume; Bleckmann et al. 2012).

3.3.1 Obstacle Avoidance and Wall-Following Behaviors of Blind Cavefish

Blind cavefish are well known for their use of active flow sensing to avoid obstacles (reviewed in Windsor et al. 2008; Windsor chapter in this volume) and follow walls (reviewed in Sharma et al. 2009; Patton et al. 2010). It is likely that other species make use of active flow sensing too, especially when visual abilities are limited. Blind cavefish swim in a burst (caudal body and fin propulsion) and coast fashion (Windsor et al. 2008, chapter by Windsor in this volume). Relative to the burst phase, the flow field during the coast phase is steady and predictable and has some of the same features as a dipole flow field (Hassan 1989; Windsor et al. 2008) (Fig. 3.1a). That blind cavefish rely on the coast, rather than burst phase for active flow sensing is demonstrated by the fact that they avoid collisions with obstacles more frequently during the coast than burst phase (Teyke 1985; Windsor et al. 2008). That they rely on their lateral line is demonstrated by the fact that the collision rate goes up when the lateral line is blocked (Windsor et al. 2008). Because the reaction distance for obstacle avoidance is short ($< \sim 10$ mm) (Windsor et al. 2008), evasive maneuvers to avoid collisions involve rapid turning maneuvers. Whether or not these involve some of the same circuitries as fast-start escape behaviors is unknown, but likely.

Blind cavefish also exhibit a strong, unconditioned tendency to follow the perimeter of a novel, enclosed environment in which they maintain a fixed distance and nearly parallel orientation to the wall's surface for long periods of time (Fig. 3.2d) (Sharma et al. 2009). Unconditioned, wall-following (or thigmotactic) behaviors are widespread in the animal kingdom, although their functional utility is debated (reviewed in Sharma et al. 2009). Wall-following may serve a protective purpose for finding shelter or escape routes and/or it may be part of an overall exploratory behavior to acquire spatial knowledge. In most terrestrial animals, this ability is mediated primarily by tactile senses, but blind cavefish may make use of both tactile and lateral line senses, depending on whether walls are straight, concave, or convex (Patton et al. 2010). Even so, the lateral line appears to dominate, as blind cavefish swim significantly closer to and collide more frequently with hydrodynamically transparent (mesh) walls than solid walls (Windsor et al. 2011). The relative role of superficial versus canal neuromasts in wall-following and other active flow sensing applications is not well understood, but blocking of canal neuromasts alone alters the swim velocity of these fish, as well as their ability to follow walls and discriminate between a solid wall and a gridded wall (Abel-Latif et al. 1990).

3.3.2 Station Holding by Trout

Trout and probably other rheophilic species are capable of holding station in close proximity to a nearby bluff body, such as a rock or boulder in a stream. This can be done in strong currents with little apparent effort. This ability is closely associated with the more general rheotactic abilities of fish to maintain an upstream direction (Sect. 3.5). Both station-holding and rheotactic behaviors are thought to have a number of benefits, including reduced energetic costs and improved interception of downstream planktonic drift (reviewed in Arnold 1974; Montgomery et al. 1997). Station-holding behaviors can be subdivided into three categories: bow wake swimming, entrainment and Kármán gaiting (reviewed by Liao 2007; Bleckmann et al. 2012) (Fig. 3.2e), each of which are associated with one of three distinct regions of the disturbed flow around a bluff body (Figs. 3.1h, 3.2e). These are: (1) a stagnation zone in the bow wake, in which pressure is elevated above ambient levels, but flow velocity is reduced (zero at the stagnation point), (2) a suction region just behind the body in which pressure is reduced and the current direction is opposite that of the overall current, and (3) a vortex wake region some downstream distance from the body. For Reynolds numbers in the 40–200,000 range, a Kármán vortex street is generated in which vortices are periodically and alternately shed from the left and right sides of the bluff body (Vogel 1994). When swimming in the bow wake, fish tend to orient directly upstream, but when entraining in the suction region, they tend to position their bodies and angle their heads to the left or right side of the body (Przybilla et al. 2010) (Fig. 3.2e). When Kármán gaiting in the vortex street, fish slalom in and between the shed vortices, presumably to capture energy from the upstream-directed edge of the shed vortices (Liao et al. 2003) (Fig. 3.2e).

Sutterlin and Waddy (1975) were the first to report an involvement of the lateral line in the station-holding abilities of brook trout (*Salvelinus fontinalis*) under variable flow speed and bluff body conditions well within the vortex street range. Bilateral denervation of the PLLN had little effect on station-holding behaviors as long as fish were tested in the light and had access to visual cues. When fish were tested in the dark, however, bilateral denervation greatly reduced, but did not completely abolish station-holding behavior, which fish exhibited as long as their snout contacted the bluff body. Unilateral removal of the PLLN introduced a lateral bias such that fish deprived of lateral line information on, e.g., the left side spent more time with the bluff body on their right side, even when tested in the light. More recent studies on rainbow trout (*Oncorhynchus myskiss*) have confirmed a role of the lateral line in station holding behaviors (Montgomery et al. 2003; Liao 2006; Pryzbilla et al. 2010), which appear to depend on both canal and superficial neuromasts (Montgomery et al. 2003). These and other studies also show that the interplay between vision and the lateral line during station holding is complex. For example, lateral line information seems to be important for the kinematics of Kármán gaiting in the vortex street region, whereas vision plays an apparent role in the decision (and/or preference) of fish to hold station in the vortex street region (Liao 2006). Interestingly, rainbow trout have the lateral line dependent ability to track (follow) a moving bluff body (a cylinder), regardless of whether they are swimming in the bow wake, entraining in the suction region, or Kármán gaiting in the vortex street region (Bleckmann et al. 2012).

3.4 Maintaining Body Orientation to Others in a Social Group: Schooling Behavior

Schooling fish are able to maintain a fixed (on average) distance, position and orientation with respect their schooling neighbors and this ability has long been thought to involve both visual and lateral line senses (Pitcher et al. 1976; Partridge and Pitcher 1980). Pivotal experiments by Pitcher and colleagues in the late 1970s and early 1980s on saithe (*Pollachius virens*) revealed that blind fish can school (Pitcher et al. 1976), but also that sighted fish exhibit subtle alterations in school structure and dynamics when deprived of lateral line information (Partridge and Pitcher 1980). The fact that nearest-neighbor distances decreased when fish were deprived of lateral line information and increased when fish were deprived of vision, led to the suggestion that nearest-neighbor distance was maintained by opposing forces of visually mediated attraction and lateral line-mediated repulsion. Vision was furthermore suggested to play a primary role in maintaining angular orientation and rostrocaudal position with respect to neighboring fish, whereas the lateral line was implicated in the regulation of swim speeds.

One of the difficulties in interpreting the results of earlier studies, however, is that the lateral line deprivation technique—cutting the PLLN—deprived fish of information from neuromasts on the trunk, but not the head. More recent experiments on strongly schooling firehead tetra (*Hemigrammus bleheri*), in which the entire lateral line system was blocked with aminoglycosides, indicate a greater disruption to schooling structure and dynamics than previously reported (Faucher et al. 2010). Moreover, complete elimination of the lateral line produced some effects that were actually opposite of those obtained with partial elimination. For example, nearest-neighbor distances increased (rather than decreased) and the school radius therefore expanded by 25%. In addition, fewer fish were observed swimming in the same direction and collisions with other fish increased, thus implicating the lateral line (rather than vision alone) as having a role in maintaining body orientation and swimming direction.

Surprisingly little is known about the specific hydrodynamic cues that fish use for maintaining distance, position and orientation with respect to neighboring fish. As has already been described, swimming fish produce a bow wave of local incompressible flow, which spreads out in both forward and lateral directions to move out in front of the fish and around to its rear (Fig. 3.1a, c). In addition, swimming fish generate hydrodynamic wakes (Fig. 3.1c). Just as nearby stationary objects alter the flow field of a swimming fish (Fig. 3.1i) (Sect. 3.3.1), adjacent fish in a school will alter each other's flow fields in ways not yet well-characterized or understood. In addition, the escape responses of schooling fish to an approaching predator produces a different type of hydrodynamic signal that could be used in the social transmission of information about external threats. That is, each C-start produces three distinct water jets pointing more or less toward or away from the startled fish: two during the initial (stage 1) C-bend and a third during the stage 3 rebound (Fig 3.1g) (Tytell and Lauder 2008).

A spectacular example of information transfer observed in some schooling fish is the so-called *wave of agitation* (Radakov 1973). Rapid body movements associated with startle or fright reactions propagate in a wave-like fashion across the school at speeds that can be greater than the swimming speed of an individual fish and even that of the approaching predator (Radakov 1973; Godin and Morgan 1985). The rapid schooling maneuvers of clupeid fishes (e.g., herring, anchovy, sprat) are also known to produce fast pressure pulses, detectable by highly specialized, pressure-sensitive ears and perhaps also by a cephalic portion of the lateral line system that is in close association with the same air bubble that renders the ear sensitive to pressure (Gray and Denton 1991). In addition, it is reasonable to expect that clupeid and other schooling species use their lateral line to detect the local water jets in a directionally appropriate manner to transmit information and avoid collisions with each other. Indeed, schooling saithe deprived of trunk lateral line input collide more frequently with each other than normal control fish when startled with a looming visual stimulus (Partridge and Pitcher 1980). In addition, the startle response latency of lateral line-deprived, but not normal control saithe varied as a function of stimulus azimuth and distance. Thus, these studies indicate that vision alone is insufficient for the rapid transmission of information and that the lateral line plays a critical role.

3.5 Maintaining Body Orientation to Global Stimuli: Rheotaxis

Fish and other aquatic organisms exhibit unconditioned orienting responses (rheotaxis) to water currents (Lyon 1904; Arnold 1974), which are typically in either an upstream or downstream direction. Rheotaxis confers many potential benefits (reviewed in Arnold 1974; Montgomery et al. 1997), including improved interception of downstream-drifting prey, energetic cost savings, and directional guidance for long-range migrations or short-range olfactory search strategies (Sect. 3.2.4).

The sensory cues that control rheotactic behavior arise not from the water motions alone, but from the relative motions between the fish's body and the surrounding water and visual field. Optic flow cues (movement of the visual surround with respect to the fish; see also chapter by Krapp) have long been thought to play a dominant role in rheotactic behavior when flow speed is high enough to displace fish downstream (Lyons 1904). Indeed, optic flow cues appear to be sufficient for many species (Lyons 1904), but certainly not necessary, as rheotaxis persists in the absence of visual cues in species like the blind cavefish (Montgomery et al. 1997; Baker and Montgomery 1999b). Nonvisual cues implicated in rheotaxis include tactile (Lyon 1904; Baker and Montomery 1999a, b), vestibular (Pavlov and Tyuryukov 1993) and the lateral line (Montgomery et al. 1997; Baker and Montgomery 1999a, b; Olszewski et al. 2012; Suli et al. 2012). Additionally, odor cues can lower the flow speed (rheotactic threshold) that elicits rheotactic behavior for fish that use both odor and hydrodynamic cues to find upstream odor sources (Baker and Montgomery 1999b; Gardiner and Atema 2007) (see also Sect. 3.2.4).

A role of the lateral line in rheotaxis has now been demonstrated for several species in both adult and larval fish. Initial studies on adult blind cavefish, torrent fish (*Cheimarrichthys fosteri*) and the antarctic fish (*Pagothenia borchgrevinki*) revealed that the rheotactic performance of fish deprived of superficial, but not canal neuromasts, dropped to chance levels at low ($< \sim 5$ cm/s or ~ 1 fish body length/s), but not higher flow speeds (Montgomery et al. 1997; Baker and Montgomery 1999a, b). More recently, the use of SNs in rheotaxis by the Port Jackson shark (*Heterodontus portusjacksoni*) has also been demonstrated (Peach 2002).

Larval zebrafish also exhibit rheotactic behavior to extremely slow flows (<0.2 cm/s) and this behavior is significantly impaired in the dark (but not in the light) when the lateral line is pharmacologically blocked (Suli et al. 2012). In a separate study, larval zebrafish were shown to exhibit orienting behaviors to the radial currents generated by a small suction source in the center of a circular arena (Olszewiski et al. 2012). At the perimeter of the arena, where flow velocity was lowest (~ 0.1 cm/s) and the spatial velocity gradient was shallowest, fish responded with steady, upstream swimming (i.e., rheotaxis). Fish sucked into the inner regions of the tank, where the spatial velocity gradients were steepest, responded with rapid swimming bursts away from the suction source and in an upstream direction. While the latter behavior might arguably be characterized as escape behavior instead of rheotaxis, rapid swimming bursts could theoretically be elicited by both biotic (e.g., a suction-feeding predator) and abiotic (e.g., a narrow constriction in a stream bed) sources that create steep velocity gradients.

Surprisingly, a more recent study on the giant danio, *Devario aequipinnatus*, has failed to show any effects of lateral line deprivation on rheotactic performance at either high (7 cm/s) or low (3 cm/s) flow speeds, although subtle effects on the form of the rheotactic behavior were observed (Bak-Coleman et al. 2013). This study furthermore demonstrated no significant decrement in performance when fish were deprived of both visual and lateral line information together, confirming the robust nature of rheotactic behavior and the ability of fish to compensate for multiple sensory losses. The absence of any evidence for a substantial role of the lateral line in this study is perplexing, especially in light of the growing body of evidence for such a role. However, this study differed from previous studies in a number of important ways, including that fish were tested individually, rather than in groups, and in a flow tank designed specifically to minimize spatial heterogeneities in the flow field. Further studies will be needed before these differences can be understood.

When considering the relative contributions of the lateral line to rheotaxis, it is important to note that as flow speed increases, so do the passive lift and drag forces on the fish's body and the speed at which fish are swept downstream. As a result, the magnitude of optic flow cues to the visual system and body motion cues to the vestibular system will be likewise increased. Furthermore, when optic flow cues are maximal (i.e., when fish are moving at the same phase and amplitude as the surrounding water), hydrodynamic cues to the lateral line will theoretically be minimized. Conversely, when fish move forward against the current to stabilize the image of their visual surround on the retina (the so-called optomotor response that underlies visually mediated rheotaxis) (reviewed in Arnold 1974), the lateral line stimulus will theoretically be maximized. The same applies to fish that are able to stabilize their position using tactile or vestibular cues. In general, theoretical considerations predict that lateral line stimuli are likely to be most effective as rheotactic cues at low flow speeds, a prediction supported by a number of different studies (Montgomery et al. 1997; Baker and Montgomery 1999a, b; Suli et al. 2012).

3.6 Central Control of Orienting Movements: Major Sites of Multisensory and Sensorimotor Integration

Although there may be a few examples of orienting behaviors that are open-loop, many if not most behaviors involve closed-loop control systems that provide continuous sensory feedback on targets (e.g., prey, predators, obstacles) of interest, as well as the consequences of the animal's own movements through the environment. Nevertheless, both open and closed-loop control systems for guiding orienting behaviors share three essential elements (Fig. 3.5a): (1) sensory inputs that provide information about the external world, (2) information processing and control systems and, (3) motor activation/instruction systems that initiate and play out the behavior. The corresponding (but still greatly oversimplified) control elements in a 'generic' fish brain are illustrated schematically in Fig. 3.7b (see McCormick 1989; Wulliman and Grothe 2013 for more precise details on the central pathways of the lateral line system in different fish taxa).

Evidence from multiple vertebrate groups indicates that the deep layers of the midbrain optic tectum (OT_d) of fish (the *multisensory direction map* of Fig. 3.5b) and the homologous structure in mammals (the superior colliculus) play a central role in the orienting behaviors of the eyes, head and body toward or away from individual targets of interest (e.g., prey) (reviewed in Stein and Meredith 1993 and Gandhi and Katnani 2011). The OT has also been implicated in body orientation to optic flow (Springer et al. 1977), one of the visual cues guiding rheotactic behaviors (Sect. 3.5). Rapid escape responses to a rapidly approaching and threatening target (i.e., predator escape responses) rely instead on a bilateral pair of large M cells (the *escape actuator* in Fig. 3.5b) and an associated network of hindbrain cells (Eaton et al. 2001). Paradoxically, the M cell is also thought to be involved in other types of fast-start behaviors, including what might arguably be classified in the prey-orienting category—i.e., strike and prey capture behavior (see Sect. 3.6.2). In any event, the midbrain OT and hindbrain M cell-reticulospinal network are two of the most highly conserved and well-studied sites of multisensory and sensorimotor integration in the vertebrate brain (Nissanov and Eaton 1989; Stein 1984). Recent work on larval zebrafish suggest that lateral line inputs to the

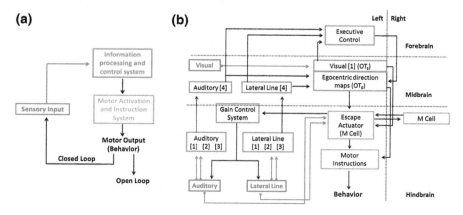

Fig. 3.5 Simplified overview of closed and open-loop control systems (**a**) and the corresponding regions of the fish brain that are part of the control process (**b**) (see key for anatomical structures). *Red arrows* represent primary afferent inputs from the sensory organs (*red boxes*), whereas *black arrows* represent the flow of information to and from various processing regions in the hindbrain, midbrain and forebrain (*green squares*) in relation to motor control regions in the hindbrain (*blue squares*). Flow of information involves two key multisensory and sensorimotor integration sites: (1) the deep layers of the optic tectum, OT_d in the midbrain, which orchestrate egocentric orienting behaviors and (2) the Mauthner (*M*) cells in the hindbrain, which actuate fast-start escape behaviors. Many details on the complex interconnections between the various processing regions and the extent of bilateral and contralateral projections are not included. Key for corresponding anatomical structures: *Auditory [1]* 1st order processing regions in the hindbrain, the anterior and descending octavolateralis nuclei, *Auditory [2]* and *[3]* secondary and tertiary hindbrain and isthmal nuclei, *Auditory [4]* torus semicircularis in the midbrain, *Egocentric Direction Maps*: multisensory processing regions in the deep layers of the midbrain optic tectum, OT_d, *Escape Actuator* Mauthner cells and associated reticulospinal networks, *Executive Control* processing regions in the forebrain that receive sensory input, *Lateral line [1]* 1st order processing region in the hindbrain, the medial octavolateralis nucleus, *Lateral line [2]* and *[3]* secondary and tertiary hindbrain and isthmal nuclei, *Lateral Line [4]* torus semicircularis, *Motor Instructions* reticulospinal network, *Visual [1]* 1st order processing region in the superficial layers of the midbrain optic tectum, OT_s

brain may be subdivided into fast-conducting pathways to the M cell/reticulospinal network and slower-conducting pathways to the first-order hindbrain region (Liao and Haehnel 2012), which relays information to the midbrain OT.

Motor instructions for executing both prey-orienting and predator escape responses reside in the reticulospinal motor network (the *motor instruction* center in Fig. 3.5b). This descending pathway relays motor commands from higher centers in the forebrain and midbrain to spinal motor neurons, which ultimately play out the motor actions (Zotolli et al. 2007). In fishes, there is no direct connection between the optic tectum and the spinal cord (Meek 1990), so motor commands from the midbrain orienting center are relayed through the reticulospinal network. Indeed, lesioning of the reticulospinal pathway in larval zebrafish abolishes prey-orienting responses (Gahtan et al. 2005).

In addition to brain areas that process sensory information to guide behavior, the octavolateralis efferent nucleus in the hindbrain (reviewed in Roberts and Meredith 1989; Köppl 2011) functions as a gain control system to modulate the sensitivity of lateral line (and auditory) hair cells in context-dependent ways. Both sensory-evoked states of arousal (Highstein and Baker 1986; Tricas and Highstein 1990), as well as vigorous swimming movements (Roberts and Russell 1972; Ayali et al. 2009) are known to activate the efferent system, but in opposite ways. Whereas vigorous swimming movements result in a reduction of lateral line hair cell sensitivity, arousal results in an increase in sensitivity. In terms of orienting behaviors, the efferent system might thus be expected to enhance sensitivity during odor-gated rheotaxis when fish are first alerted to a prey by its odor (Sect. 3.2.4), but to inhibit lateral line sensitivity during vigorous swimming movements, as happens during a rapid escape maneuver, during the burst phase of a blind cavefish's swim cycle and during steady upstream swimming against a strong current.

3.6.1 Midbrain Orienting Center: Deep Layers of the Optic Tectum

Electrical stimulation of the optic tectum generates saccadic eye, head and whole-body orienting movements in fish (Al-Akel et al. 1986; Herrero et al. 1998) and other vertebrates (reviewed in Ghandi and Katnani 2011). Since the direction and amplitude of the electrically evoked orienting response varies systematically with the location of electrical stimulation (Stein and Meredith 1993), the optic tectum can be regarded as an egocentric map of source location in motor coordinates (Fig. 3.5a). The origin of the motor coordinate system corresponds to a point in space directly in front of the head and thus, the map (often referred to as a *motor error map*) essentially provides information on the amplitude and direction of motor movements required to bring the front of the head and eyes in line with the source, thus reducing the motor error to zero.

In cases where midbrain multisensory integration has been studied in detail, it is clear that other nonvisual modalities, such as auditory and electrosensory systems, form maps of space that are in register with visual motor error maps in a wide range of vertebrates, including fish (e.g., Wubbels et al. 1995). Lateral line information is also integrated with visual and other nonvisual modalities in multisensory midbrain areas (reviewed in Schellart and Kroese 1989; Schellart 1992; Fay and Edds-Walton 2001).

3.6.2 Mauthner Cells and the Hindbrain Escape Network

Rapid escape responses involve a well-studied hindbrain escape network: the M cell and associated reticulospinal cells (Eaton et al. 2001; Korn and Faber 2005).

The M cell is the largest cell in the network and a primary site of multisensory integration, receiving spatially segregated inputs from multiple senses (visual, auditory, vestibular, and lateral line) (Zottoli et al. 1995). The visual input is indirectly relayed from a processing region of the brain, the superficial layers of the optic tectum (Visual [1] in Fig. 3.5b, black arrow connection), whereas lateral line and auditory inputs to the M cell are direct (red arrow connections in Fig. 3.5b). For this and other reasons (e.g., processing that takes place in the retina itself), transmission of information from the visual system is slower and delayed relative to that from mechanosensory systems. Although C-starts do not require the M cell, they rarely occur without M cell activity and the M cell is always the first to fire among cells in the network (Casagrande et al. 1999; Eaton et al. 2001). Thus, the M cell is regarded as a key, decision-making element that actuates the escape response (Korn and Faber 2005).

M cells not only actuate the escape response, but they also play a large role in determining the final escape trajectory, since the initial C-bend generally has a larger influence on the final escape trajectory than stage 2 (Eaton et al. 1988; Domenici and Blake 1991). However, stage 2 can also influence the final escape trajectory. Moreover, the fact that fish can alter the direction of the final trajectory to avoid collisions with obstacles (see Sect. 3.2) and other fish (see Sect. 3.2.5) indicates that the escape response represents a true act of sensorimotor coordination that includes closed-loop control.

From the work of Mirjany et al. (2011) on goldfish escape responses (summarized in Sect. 3.2.5), both the anterior and posterior lateral lines appear to be involved in the closed-loop guidance of stage 2 behavior to avoid collisions with walls. In contrast, inputs from the head alone, which arrive via a recently discovered fast-conducting pathway to the M cell (Mirjany et al. 2011), will influence the excitability of the M cell, which in turn, determines which M cell fires and thus, the initial direction of the stage 1 C-bend. Interestingly, unlike sound pressure or a visual looming stimulus, lateral line input alone has not yet been shown to be sufficient for eliciting goldfish startle responses (Eaton et al. 1977; Preuss et al. 2006). However, lateral line inputs alone may very well be sufficient for evoking startle responses in larval zebrafish, as well as adult fish that do not possess pressure-sensitive auditory systems (see Sect. 3.2.5).

As a final note, fast-start strike behaviors, which are more closely associated with prey-orienting than predator escape behaviors, may involve elements of the escape system. For example, some species (e.g., northern pike, *Esox lucius*) exhibit S-start strike behaviors that are very similar to S-start escape behaviors, suggesting that they share elements of the same brainstem escape circuitry (Hale et al. 2002). Although it is now clear that these two behaviors can be differentiated kinematically and otherwise (Schriefer and Hale 2004), their underlying circuitries can arguably be considered to have more in common with each other than with those underlying prey-orienting responses. Indeed, Canfield and Rose (1993) have shown with chronically implanted electrodes that goldfish M cells are active during both predator escape and prey capture (strike) behaviors. The M cell may thus be recruited in different fast-start orienting behaviors besides escape.

3.7 Sensory Mechanisms and Sensorimotor Strategies for Flow-based Orientation

Given that orienting behaviors are as varied as the flow signals that guide them (Table 3.1; Fig. 3.2), what, if anything, can be concluded about the sensory mechanisms and sensorimotor strategies that guide fish? Are there separate mechanisms and strategies for each major category of behavior or stimulus? Are there shared principles of operation and if so, what are they? Key to an understanding of shared principles is the functional organization and operation of multisensory and sensorimotor integration sites in the brain that control orienting behaviors (Sect. 3.6). Knowledge of how fish use vision and other sensory cues also provides a multisensory context within which the role of the lateral line system can be better understood.

In consideration of these key factors, we have identified three principles by which central processing of the lateral line information is likely to guide orienting behaviors: brain maps (Sect. 3.7.1), multisensory biasing of M cell actuators (Sect. 3.7.2) and wide-field integration (WFI) and matched filters (Sect. 3.7.3). The first two principles are involved in the guidance of behavior toward or away from discrete hydrodynamic targets (e.g., prey or predators), whereas the third involves guidance by global currents.

3.7.1 Brain Maps: Determining the Direction of Discrete Hydrodynamic Sources

There are at least two different kinds of brain maps that could provide fish with information about target location, even under open-loop conditions. *Topographic* maps preserve the spatial order of the sensory surface by mapping neighboring points on the sensory surface as neighboring points in the brain (Fig. 3.5b). In contrast, *egocentric direction* maps preserve the spatial order of target direction in space with respect to an eye-centered point of reference on the head (Fig. 3.5a). For all practical purposes, topographic maps of the retinal surface (retinotopic maps), observed in the optic tectum and elsewhere, preserve the spatial order of both the sensory surface and the target direction in visual space (Fig. 3.5d). However, topographic maps of the lateral line sensory surface preserve only the spatial order of lateral line inputs along the body surface (a *somatotopic* map). Both anatomical and physiological evidence (reviewed in Bleckmann and Mogdans 2013; Chagnaud and Coombs 2013) indicate that the rostrocaudal order of lateral line inputs is mapped in the medial octavolateralis nucleus (MON) (lateral line [1] in Fig. 3.5b) and perhaps also in the midbrain torus semicircularis (TS) (lateral line [4] in Fig. 3.5b). However, the only clear evidence for an egocentric map of source direction in the OT (midbrain direction map in Fig. 3.5b) comes from physiological studies on the response of OT cells in surface-feeding clawed frogs to surface wave sources (Zittlau et al. 1986; Claas et al. 1989).

If the optic tectum contains a map of the turning magnitudes and directions needed to orient to a hydrodynamic source, such a map would likely underlie the brief, open-loop prey-orienting responses described for a wide variety of fishes, including benthic (Sect. 3.2.1) and surface-feeding fishes (Sect. 3.2.2). This raises the question of how egocentric maps might be constructed from such seemingly disparate types of hydrodynamic cues—flow patterns along the body surface created by dipole-like sources, on the one hand, and time-of-arrival cues from a slowly propagating surface wave, on the other. The answer is simply that we do not know. The elegant body of work on how barn owls use delay line and coincident detector circuits for constructing auditory space maps from binaural time differences (e.g., Carr and Konishi 1990) may prove to be instructive. Theoretical considerations suggest that both bilateral and rostrocaudal comparisons of arrival times are necessary in order to resolve mirror-image ambiguities about wave source directions (Fig. 3.4). Theoretical considerations also suggest that information about source distance is unnecessary when using time-of-arrival cues to determine the direction of a surface-wave source, but essential for determining the direction of dipole-like sources when relying on spatial activation patterns, as explained in the following paragraph.

For fish stimulated by dipole-like sources, spatial activation patterns provide information about the somatotopic location of the source, as well as its distance (Fig. 3.3). However, somatotopic information about source location alone is insufficient for determining egocentric direction. This is because lateral line stimulus sources at the same somatotopic location (Fig. 3.6c), but at different distances away from the body (distance 1 and 2), lie at different visual angles (α_1 and α_2), and thus require different orienting movements (Coombs and Patton 2009). Therefore, information about both source distance and its somatotopic location must be combined in order to compute the egocentric direction of the source. Exactly how or even if this is done is presently unknown, but information about both of these parameters could, in theory, be encoded by population codes—i.e., patterns of activity across populations of neurons that are somatopically mapped in the brain.

3.7.2 Differential Biasing of Bilaterally Paired Actuators

As a key-decision making element, the 'M cell actuator' (Fig. 3.5b) relies on the sum of all of its inputs, which includes not only ipsilateral inputs from multiple sensory systems, but also contralateral inputs from the other M cell (Korn and Faber 2005) (Sect. 3.6.2). As such, M cells behave as classic integrate and fire neurons; if the sum of excitatory and inhibitory influences exceed threshold at the axon hillock, the M cell fires. A single action potential is sufficient for activating a fast body bend (C-start) on the opposite side of the body. Reciprocal inhibitory

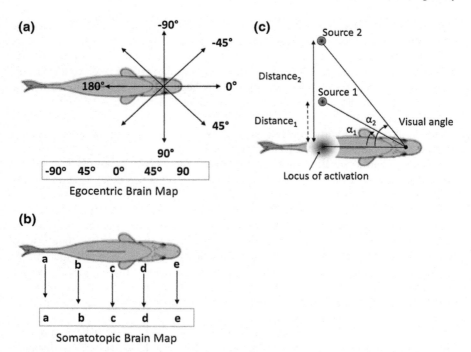

Fig. 3.6 Schematic illustration of the difference between egocentric (**a**) and somatotopic (**b**) brain maps. In (**c**), sources at different distances, but at the same somatotopic location are shown to lie at different visual angles (α_1 and α_2) and thus, different egocentric directions. Therefore, information about both source distance and somatotopic location is needed to construct an egocentric map

connections between left and right M cells ensure that (1) the M cell that fires first suppresses the ability of the opposite M cell to fire and (2) muscles on one side of the body contract, whereas those on the opposite side relax. Thus, the first of the two M cells to fire determines the direction of the initial bend (Stage 1) (Fig. 3.2b), which in turn has a heavy influence on the final escape trajectory. The final escape trajectory is likely determined by the relative magnitude and timing of activity in two groups of cells (Eaton et al. 2001); the first includes the M cell and other network cells to control the timing and direction of the initial turn (C-bend) and the second consists of a different network of cells to control the second (rebound) stage of the behavior.

Different senses in different species have different relative weights in biasing the M cell toward its threshold level. In goldfish, for example, stimulation of either the visual or pressure-sensitive auditory system is sufficient, by itself, to excite the M cell to threshold, whereas stimulation of the lateral line appears to influence the probability of firing, but only when combined with inputs from other senses. In larval zebrafish and other species without a pressure-sensitive auditory system, however, lateral line input alone may be sufficient for bringing the M cell to threshold and thus, is likely to play a more dominant role in actuating the response. In all cases that have been investigated, the lateral line seems to play an important role in influencing the direction of the escape by preferentially biasing one M cell relative to the other. In this respect, ipsilateral inputs to each M cell from both the anterior and posterior lateral line play a role. The lateral line system also guides corrective actions during the escape that are necessary to avoid collisions with walls (Eaton and Emberley 1991; Mirjany et al. 2011) or other fish in a school (Partridge and Pitcher 1980).

3.7.3   Wide-Field Integration: Inspirations from Optic Flow

An elegant body of work on the use of optic flow by flying insects such as flies (Krapp and Hengstenberg 1996; reviewed in chapter by Krapp in this volume) suggests a basic principle (WFI neurons to encode global flow patterns) that could very well be used by the lateral line system in a number of behavioral contexts. This work reveals that the fly's visual system contains WFI neurons, so-called lobula plate tangential cells (LPTCs), whose receptive fields appear to match the global flow fields expected for various rotational and translational movements of the body (Fig. 7a). That is, some WFI neurons respond best to the global flow field associated with a roll, whereas others respond best to that associated with a lift and so on. Based on these and other findings, workers in this field have proposed that WFI neurons function as matched filters for sensing different body states of motions (Krapp and Hengstenberg 1996; Krapp et al. 1998; Franz and Krapp 2000, reviewed in Krapp, chapter in this volume). Body motions could be either self-induced or exogenously induced (e.g., by wind, in the case of flies, or currents, in the case of fish), but in either case, sensory feedback could be used to generate corrective or compensatory actions. Indeed, LPTC cells in the fly's visual system have been shown to be involved in compensatory optomotor responses (Huston and Krapp 2008). Since it is well known that optic flow cues play a dominant role in the rheotactic abilities of fish (Lyons 1904; Arnold 1974), WFI processing of optic flow may very well be used. If so, it is reasonable to suggest that the lateral line might complement the visual system in its task to inform fish about body movements relative to their stationary surroundings.

The lateral line system has several of the features predicted to be necessary for the processing of global flow fields. One such feature is the local, directionally sensitive elementary movement detectors (EMD), which are the proposed building blocks of WFI neurons and which determine the receptive field organization and filter properties of these neuron (Fig. 3.7a). Applied to the lateral line, the idea is that the preferred directions of somatotopically arranged EMDs will be matched to the orientation of local velocity vectors of a specific hydrodynamic flow field, which in turn is associated with a particular state—e.g., heading directly upstream. In the fly's visual system, EMDs must be constructed from local circuits that combine information from neighboring ommatidia in localized regions of the compound eye. In the lateral line system, however, EMD properties already exist at the level of individual sensory hair cells (Fig. 3.7b). That is, each hair cell is anatomically and physiologically polarized to respond best (maximal excitation) to a single direction of motion (Flock 1965). Hair cells don't function in isolation, however, but are grouped into many sense organs (neuromasts) that are distributed all over the head and body (Fig. 3.7c). Each neuromast consists of two, oppositely oriented populations of hair cells, each of which are contacted by a separate nerve that integrates information from hair cells of the same orientation (Fig. 3.7b). As a result, each neuromast has a single, bidirectional axis of best sensitivity; water motion in one direction along this axis causes one population to be maximally stimulated (firing rate of the innervating fiber increases) and the other to be inhibited (firing rate of the innervating fiber decreases) and this information is transmitted to the brain in separate channels.

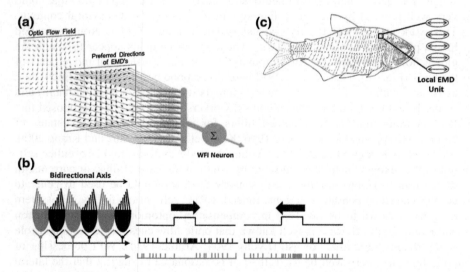

Fig. 3.7 a Schematic illustration of a wide-field integration (*WFI*) neuron that is tuned to a particular global flow pattern (adapted from Krapp, chapter in this volume). WFI neurons receive inputs from many directionally sensitive, EMDs responding to visual motion in small areas of space. **b** Proposed LL-EMDs in the lateral line system consist of two populations of oppositely oriented hair cells, each of which are anatomically and physiologically polarized to respond best to flow in one direction. When apical hairs (stereovillae) are bent in the direction of the longest hair (kinocilium), the hair cell is depolarized and the firing rate of the innervating fiber increases. When bent in the opposite direction, the hair cell is hyperpolarized and the firing rate of the innervating fiber decreases. **c** Spatial distribution of superficial neuromasts (*SNs*) (each small dot = 1 neuromast) in the blind cavefish illustrating the wide-distribution of lateral line sensors on the body surface; SNs form a single vertically oriented row on each scale (*boxed region*), but the bidirectional axis of sensitivity is in the rostrocaudal direction. Hair cells of the same orientation in any given row are typically innervated by the same fiber, so each row of SNs is likely to function as a single unit

As with the EMDs proposed for the fly visual system, each neuromast is situated to respond to a small region of the entire lateral line 'space,' which for the purpose of this hypothetical comparison, may be regarded as the total (combined) receptive fields of neuromasts on, e.g., the entire left or right side of the fish's body. In fish like the goldfish and blind cavefish, there are literally thousands of SNs all over the body surface (Schemmel 1967; Schmitz et al. 2008) and these are particularly well suited for WFI applications. In goldfish, the bidirectional axis of sensitivity for SNs on the head varies, but that for SNs on the trunk is predominantly in the rostrocaudal axis (Schmitz et al. 2008). Such an arrangement would seem to be a good match to the expected flow patterns produced by the forward movements of a fish in still water (Fig. 3.1a) or, alternatively, by the downstream flow of water past a fish exhibiting positive rheotaxis (pointing upstream).

There is fairly good evidence that spatially segregated inputs from lateral line neuromasts on different body regions are integrated centrally, even at the earliest level of processing—the MON in the hindbrain (lateral line [1] in Fig. 3.5b) (reviewed in Bleckmann and Mogdans, chapter in this volume). The evidence comes from studies in which the receptive fields of CNS cells in the goldfish were mapped with a dipole source (vibrating sphere) that changed its location along the length of the fish. Using this approach, a subpopulation of cells in both the MON (Künzel et al. 2011) and in the midbrain TS (lateral line [4] in Fig. 3.5b) (Engelmann and Bleckmann 2004; Meyer et al. 2012) were identified to have very broad receptive fields, some greater than or equal to the length of the fish. A separate study (Wojtenek et al. 1998) provided evidence for two of the remaining properties required for the proposed WFI function—i.e., motion sensitivity and directional sensitivity (Krapp, chapter in this volume). In this study, cells in the midbrain TS of goldfish responded to the movement of an object along the length of the fish in a directionally selective manner—i.e., some cells responded to a rostrocaudal direction of movement, whereas others responded to a caudal-rostral direction of movement.

Wide-field encoding of hydrodynamic flow patterns could be useful in other orienting behaviors besides just rheotaxis. In fact, WFI techniques have been applied by engineers to both optic (Humbert et al. 2009; Hyslop and Humbert 2010) and hydrodynamic (Ranganathan et al. 2013) flow patterns, in order to produce information about relative speed and proximity to obstacles. Rather than using WFI outputs to estimate self-motion states (e.g., forward translation), however, Humbert and colleagues used departures from the desired 'state' patterns as feedback signals. This WFI application would seem to be particularly useful in the context of active flow sensing, in which fish need to detect perturbations in their steady-state flow field during the coast phase of their swim cycle (Sect. 3.3.1). As proof-of-concept, potential flow theory, in combination with WFI analytical methods, have been used to derive feedback signals to simulate rheotactic, as well as wall-avoidance and wall-following behaviors of fish (Ranganathan et al. 2013).

3.8 Summary and Conclusions

The relative contributions of the lateral line and flow cues to orienting behaviors depends on a number of factors, including life stage, temporal stage of the behavior, and the nocturnal/diurnal habits of fish. Auditory specializations for detecting the pressure changes produced by rapid (accelerating) bodies also impact the relative contributions of the lateral line and the auditory system. The interdependence of these two mechanosensory systems follows from the fact that moving sources are capable of stimulating both systems, but in different ways. Thus, although flow cues may dominate the guidance of some orienting behaviors in some species under some circumstances, the vast majority of orienting behaviors are likely guided by information from multiple senses. For this reason, a complete understanding of how flow guides fish behavior must include a multisensory perspective. With rare exceptions (e.g., hydrodynamic trail following), the lateral line operates as a close-range system and thus, contributes in a range-fractionated manner to complement longer-range systems like vision or audition in the context of prey-orienting or predator avoidance behaviors. The contribution of the lateral line to rheotactic behavior is also range-limited in that its contributions are largely restricted to the very low end of the flow speed range.

At least two major multisensory and sensorimotor integration sites in the brain—the deep, multisensory layers of the midbrain optic tectum and the hindbrain M cell-reticulospinal network—are involved in orienting behaviors. The optic tectum is likely to be involved in a wide variety of orienting behaviors, ranging from brief prey-orienting behaviors in response to discrete targets to sustained orienting behaviors like rheotaxis, whereas M cell reticulospinal networks are more likely to be associated with fast-start orienting behaviors (Table 3.1). In some cases, circuits in both brain regions may interact or be recruited for different stages of an orienting behavior.

Egocentric direction maps in the multisensory regions of the OT are key to understanding the flow-based guidance of prey-orienting behaviors. An important gap in our knowledge is how or even if various hydrodynamic cues are used by the nervous system to compute these egocentric space maps. The proposed use of spatial activation patterns to localize dipole-like sources and time-of-arrival cues to localize surface wave sources raises the less-than satisfying specter that lateral line direction maps are computed in entirely different ways under these two circumstances. Future studies will have to determine if OT direction maps do indeed underlie both behaviors and if so, whether or not there is a unifying principle for constructing these maps from hydrodynamic information.

In contrast to the proposed guidance of prey-orienting behaviors by egocentric direction maps in the OT, directional guidance of escape responses by M-cell circuits arises in a fundamentally different way that involves the differential biasing of the bilaterally paired M cells by multisensory inputs. In this case, lateral line inputs exert directional influences by biasing the probability of an M cell response. The weight of lateral line inputs in bringing the M cell to threshold depends on a number of factors, including species, life stage and whether or not the auditory system is adapted for pressure-sensitivity. The lateral line may provide later-stage directional guidance as well, in which case M cell biasing is not involved.

Finally, WFI of global flow fields, inspired by optic flow studies in flies, is proposed for the lateral line. This hypothetical construct involves WFI neurons in the central nervous system that are tuned to different global flow patterns corresponding to different motion states (e.g., forward translation). Perturbations in these motion states (and thus, optic flow patterns) can, in principal, be used as sensory feedback for corrective actions under different behavioral contexts, including station holding, rheotaxis, obstacle avoidance and wall-following, as demonstrated by engineered applications. Although lateral line cells in the midbrain have some of the required properties of WFI neurons, further experiments will be needed to determine whether this hypothetical construct is actually implemented by the nervous system for these or other behaviors. In this respect, recent technological advances in the ability to observe brain activity in awake and behaving larval zebrafish under conditions which provide for the adaptive control of locomotion (e.g., Ahrens et al. 2012), holds real promise for a better understanding of flow guidance of orienting behaviors.

Acknowledgments We thank Holger Krapp and Badri Ranganathan for their comments on earlier drafts of this chapter.

References

Abdel-Latif H, Hassan E, Campenhausen C (1990) Sensory performance of blind Mexican cave fish after destruction of the canal neuromasts. Naturwissenschaften 77:237–239

Ahrens MB, Li JM, Orger MB, Robson DN, Schier AF, Engert F, Portugues R (2012) Brain-wide neuronal dynamics during motor adaptation in zebrafish. Nature 485:471–477

Al-Akel AS, Guthrie DM, Banks JR (1986) Motor responses to localized electrical stimulation of the optic tectum in the freshwater perch (*Perca fluviatilis*). Neuroscience 19:1381–1391

Arnold GP (1974) Rheotropism in fishes. Biol Rev Cam Phil Soc 49:515–576

Ayali A, Gelman S, Tytell E, Cohen A (2009) Lateral-line activity during undulatory body motions suggests a feedback link in closed-loop control of sea lamprey swimming. Can J Zool 87:671–683

Bak-Coleman J, Court A, Paley DA, Coombs S (2013) The spatiotemporal dynamics of rheotactic behavior depends on flow speed and available sensory information. J Exp Biol 216:4011–4024

Baker CF, Montgomery JC (1999a) Lateral line mediated rheotaxis in the Antarctic fish *Pagothenia borchgrevinki*. Polar Biol 21:305–309

Baker CF, Montgomery JC (1999b) The sensory basis of rheotaxis in the blind Mexican cavefish, Asytanax fasciatus. J Comp Physiol A 184:519–527

Baker CF, Montgomery JC, Dennis TE (2002) The sensory basis of olfactory search behavior in banded kokopu (*Galaxias fasciatus*). J Comp Physiol A 188:560

Blaxter J, Fuiman L (1990) The role of the sensory systems of herring larvae in evading predatory fishes. J Mar Biol Assoc UK 70:413–427

Bleckmann H (1993) Role of the lateral line in fish behaviour. In: Pitcher TJ (ed) Behavior of teleost fishes, 2nd edn. Chapman & Hall, New York, pp 201–246

Bleckmann H, Mogdans J (2013) Central processing of lateral line information. In: Coombs S, Bleckmann H (eds) The Lateral Line System. Springer, New York, pp 253–280

Bleckmann H, Tittel G, Blübaum-Gronau E (1989) Lateral line system of surface-feeding fish: anatomy, physiology and behavior. In: Coombs S, Görner P, Münz H (eds) The mechanosensory lateral line: neurobiology and evolution. Springer, New York; Berlin, pp 501–526

Bleckmann H, Przybilla A, Klein A, Schmitz A, Kunze S, Brücker C (2012) Station holding of trout: behavior, physiology and hydrodynamics. In: Tropea C, Bleckmann H (eds) Nature-inspired fluid mechanics. Springer, New York, pp 161–177

Canfield JG, Eaton RC (1990) Swimbladder acoustic pressure transduction initiates Mauthner-mediated escape. Nature 347:760–762

Canfield JG, Rose GJ (1993) Activation of Mauthner neurons during prey capture. J Comp Physiol A 172:611–618

Canfield JG, Rose GJ (1996) Hierarchical sensory guidance of Mauthner-mediated escape responses in goldfish (Carassius auratus) and cichlids (Haplochromis burtoni). Brain Behav Evol 48:137–156

Carr C, Konishi M (1990) A circuit for detection of interaural time differences in the brain stem of the barn owl. J Neurosci 10:3227

Casagrand JL, Guzik AL, Eaton RC (1999) Mauthner and reticulospinal responses to the onset of acoustic pressure and acceleration stimuli. J Neurophysiol 82:1422–1437

Chagnaud BP, Coombs S (2013) Information encoding and processing by the peripheral lateral line system. In: Coombs S, Bleckmann H, Popper AN, Fay RR (eds) The lateral line, springer handbook of auditory research, vol 48. Springer, New York

Claas B, Münz H, Zittlau KE (1989) Direction coding in central parts of the lateral line system. In: Coombs S, Görner P, Münz H (eds) The mechanosensory lateral line: neurobiology and evolution. Springer, New York; Berlin, pp 409–419

Collin SP, Shand J, Collin S, Marshall J (2003) Retinal sampling and the visual field in fishes. Sens Process Aquat Environ 139–169

Coombs S, Braun CB, Donovan B (2001) The orienting response of Lake Michigan mottled sculpin is mediated by canal neuromasts. J Exp Biol 204:337–348

Coombs S, Patton P (2009) Lateral line stimulation patterns and prey orienting behavior in the Lake Michigan mottled sculpin (Cottus bairdi). J Comp Pysiol A 195:279–297

Conley RA, Coombs S (1998) Dipole source localization by mottled sculpin. III. Orientation after site-specific, unilateral blockage of the lateral line system. J Comp Physiol A 183:335–344

Coombs S, Conley RA (1997a) Dipole source localization by mottled sculpin. I. Approach strategies. J. Comp. Physiol 180:387–399

Coombs S, Conley RA (1997b) Dipole source localization by mottled sculpin. II. The role of lateral line excitation patterns. J Comp Physiol 180:401–415

Coombs S, Hastings MC, Finneran JJ (1996) Modeling and measuring lateral line excitation patterns to changing dipole source locations. J Comp Physiol 178:359–371

Ćurčić-Blake B, van Netten SM (2006) Source location encoding in the fish lateral line canal. J Exp Biol 209:1548–1559

Day SW, Higham TE, Cheer AY, Wainwright PC (2005) Spatial and temporal patterns of water flow generated by suction-feeding bluegill sunfish Lepomis macrochirus resolved by particle image velocimetry. J Exp Biol 208:2661–2671

Domenici P, Blake RW (1991) The kinematics and performance of the escape response in the angelfish (Pterophyllum eimekei). J Exp Biol 156:187–205

Domenici P, Blake R (1997) The kinematics and performance of fish fast-start swimming. J Exp Biol 200:1165–1178

Drucker EG, Lauder GV (2002) Experimental hydrodynamics of fish locomotion: Functional insights from wake visualization. Integr Comp Biol 42:243–257

Eaton RC, Bombardieri RA, Meyer DL (1977) The mauthner-initiated startle response in teleost fish. J Exp Biol 66:65–81

Eaton RC, Emberley DS (1991) How stimulus direction determines the trajectory of the Mauthner-initiated escape response in a teleost fish. J Exp Biol 161:469–487

Eaton RC, DiDomenico R, Nissanov J (1988) Flexible body dynamics of the goldfish C-start: implications for reticulospinal command mechanisms. J Neurosci 8:2758–2768

Eaton R, Lee R, Foreman M (2001) The Mauthner cell and other identified neurons of the brainstem escape network of fish. Prog Neurobiol 63:467–485

Engelmann J, Bleckmann H (2004) Coding of lateral line stimuli in the goldfish midbrain in still and running water. Zoology (Jena) 107:135–151

Enger PS, Kalmijn AJ, Sand O (1989) Behavioral investigations on the functions of the lateral line and inner ear in predation. In: Coombs S, Görner P, Münz H (eds) The mechanosensory lateral line: neurobiology and evolution. Springer, New York; Berlin, pp 575–587

Ranganathan BN, Dimble KD, Faddy, JM, Humbert JS (2013) Underwater navigation behaviors using wide-field Integration methods. In: IEEE international conference on robotics and automation, Karlsruhe Germany, pp 4132–4137

Faucher K, Parmentier E, Becco C, Vandewalle N, Vandewalle P (2010) Fish lateral system is required for accurate control of shoaling behaviour. Anim Behav 79:679–687

Fay RR, Edds-Walton PL (2001) Bimodal units in the torus semicircularis of the toadfish (*Opsanus tau*). Biol Bull 201:280–281

Fay RR, Popper AN (1974) Acoustic stimulation of the ear of the goldfish (*Carassius auratus*). J Exp Biol 61:243–260

Flock Å (1965) Electron microscopic and electrophysiological studies on the lateral line canal organ. Acta Otolaryngol 199:1–90

Fraenkel GS, Gunn DL (1961) The orientation of animals. Kineses, taxes and compass reactions. Oxford University Press, Oxford and New York, p 352

Franosch JMP, Hagedorn HJA, Goulet J, Engelmann J, van Hemmen JL (2009) Wake tracking and the detection of vortex rings by the canal lateral line of fish. Phys Rev Lett 103:78102

Franz MO, Krapp HG (2000) Wide-field, motion-sensitive neurons and matched filters for optic flow fields. Biol Cybern 83:185–197

Gahtan E, Tanger P, Baier H (2005) Visual prey capture in larval zebrafish is controlled by identified reticulospinal neurons downstream of the tectum. J Neurosci 25:9294–9303

Gandhi NJ, Katnani HA (2011) Motor functions of the superior colliculus. Annu Rev Neurosci 34:205

Gardiner JM (2012) Multisensory integration in shark feeding behavior

Gardiner JM, Atema J (2007) Sharks need the lateral line to locate odor sources: rheotaxis and eddy chemotaxis. J Exp Biol 210:1925–1934

Gardiner JM, Motta PJ (2012) Largemouth bass (*Micropterus salmoides*) switch feeding modalities in response to sensory deprivation. Zoology 115:78–83

Godin JGJ, Morgan MJ (1985) Predator avoidance and school size in a cyprinodontid fish, the banded killifish (*Fundulus diaphanus* Lesueur). Behav Ecol Sociobiol 16:105–110

Goulet J, Engelmann J, Chagnaud B, Franosch JP, Suttner MD, van Hemmen JL (2008) Object localization through the lateral line system of fish: theory and experiment. J Comp Physiol A 194:1–17

Gray JAB, Denton EJ (1991) Fast pressure pulses and communication between fish. J Mar Biol Assoc U K 71:63–106

Hale ME, Long JH Jr, McHenry MJ, Westneat MW (2002) Evolution of behavior and neural control of the fast-start escape response. Evolution 56:993–1007

Hanke W, Bleckmann H (2004) The hydrodynamic trails of *Lepomis gibbosus* (Centrarchidae), *Colomesus psittacus* (Tetraodontidae) and *Thysochromis ansorgii* (Cichlidae) investigated with scanning particle image velocimetry. J Exp Biol 207:1585–1596

Hanke W, Brucker C, Bleckmann H (2000) The ageing of the low-frequency water disturbances caused by swimming goldfish and its possible relevance to prey detection. J Exp Biol 203:1193–1200

Hassan E (1985) Mathematical analysis of the stimulus for the lateral line organ. Biol Cybern 52:23–36

Hassan E (1989) Hydrodynamic imaging of the surroundings by the lateral line of the blind cave fish *Anoptichthys jordani*. In: Coombs S, Görner P, Münz H (eds) The mechanosensory lateral line: neurobiology and evolution. Springer, New York; Berlin, pp 217–227

Herrero L, Rodriguez F, Salas C, Torres B (1998) Tail and eye movements evoked by electrical microstimulation of the optic tectum in goldfish. Exp Brain Res 120:291–305

Higham TE, Day SW, Wainwright PC (2005) Sucking while swimming: evaluating the effects of ram speed on suction generation in bluegill sunfish *Lepomis macrochirus* using digital particle image velocimetry. J Exp Biol 208:2653–2660

Highstein SM, Baker R (1986) Organization of the efferent vestibular nuclei and nerves of the toadfish, *Opsanus tau*. J Comp Neurol 243:309–325

Hoekstra D, Janssen J (1985) Non-visual feeding behavior of the mottled sculpin, *Cottus bairdi*, in Lake Michigan. Envir Biol Fish 12:111–117

Hoekstra D, Janssen J (1986) Lateral line receptivity in the mottled sculpin (*Cottus bairdi*). Copeia 1986:91–96

Humbert JS, Conroy JK, Neely CW, Barrows G (2009) Wide-field integration methods for visuomotor control. Flying Insects Robots 63–71

Huston SJ, Krapp HG (2008) Visuomotor transformation in the fly gaze stabilization system. PLoS Biol 6:e173

Hyslop AM, Humbert JS (2010) Autonomous navigation in three-dimensional urban environments using wide-field integration of optic flow. J Guidance, Control, Dyn 33:147–159

Jander R (1975) Ecological aspects of spatial orientation. Ann Rev Ecol Syst 6:171–188

Kalmijn AJ (1988) Hydrodynamic and acoustic field detection. In: Atema J, Fay RR, Popper AN, Tavolga WN (eds) Sensory biology of aquatic animals. Springer, New York, pp 83–130

Kalmijn AJ (1989) Functional evolution of lateral line and inner-ear sensory systems. In: Coombs S, Görner P, Münz H (eds) The mechanosensory lateral line: neurobiology and evolution. Springer, New York; Berlin, pp 187–215

Köppl C (2011) Evolution of the octavolateral efferent system. In: Ryugo D, Fay RR and Popper AN (eds) Auditory and vestibular efferents, vol 38. The springer handbook of auditory research. Springer, New York, pp 217-259

Korn H, Faber DS (2005) The Mauthner cell half a century later: a neurobiological model for decision-making? Neuron 47:13–28

Krapp HG, Hengstenberg R (1996) Estimation of self-motion by optic flow processing in single visual interneurons. Nature 384:463–466

Krapp HG, Hengstenberg B, Hengstenberg R (1998) Dendritic structure and receptive-field organization of optic flow processing interneurons in the fly. J Neurophysiol 79:1902–1917

JC (2006) The role of the lateral line and vision on body kinematics and hydrodynamic preference of rainbow trout in turbulent flow. J Exp Biol 209:4077–4090

Liao JC (2007) A review of fish swimming mechanics and behaviour in altered flows. Philos Trans R Soc Lond B 362:1973

Liao JC (2010) Organization and physiology of posterior lateral line afferent neurons in larval zebrafish. Biol Lett 6:402–405

Liao JC, Haehnel M (2012) Physiology of afferent neurons in larval zebrafish provides a functional framework for lateral line somatotopy. J Neurophysiol 107:2615–2623

Liao JC, Beal DN, Lauder GV, Triantafyllou GS (2003) The Karman gait: novel body kinematics of rainbow trout swimming in a vortex street. J Exp Biol 206:1059–1073

Lyon EP (1904) On rheotropism. I. Rheotropism in fishes. Am J Physiol 149–161

Malkiel E, Sheng J, Katz J, Strickler JR (2003) The three-dimensional flow field generated by a feeding calanoid copepod measured using digital holography. J Exp Biol 206:3657–3666

McCormick CA (1989) Central lateral line mechanosensory pathways in bony fish. In: Coombs S, Görner P, Münz H (eds) The mechanosensory lateral line: neurobiology and evolution. Springer, New York; Berlin, pp 341–364

McHenry M, Feitl K, Strother J, Van Trump W (2009) Larval zebrafish rapidly sense the water flow of a predator's strike. Biol Lett 5:477–479

Meek HJ (1990) Tectal morphology: connections, neurones and synapses. In: Anonymous The visual system of fish. Springer, New York, pp 239–277

Meyer G, Klein A, Mogdans J, Bleckmann H (2012) Toral lateral line units of goldfish, *Carassius auratus*, are sensitive to the position and vibration direction of a vibrating sphere. J Comp Physiol A 198:639–653

Mirjany M, Preuss T, Faber DS (2011) Role of the lateral line mechanosensory system in directionality of goldfish auditory evoked escape response. J Exp Biol 214:3358–3367

Moiseff A, Konishi M (1981) The owl's interaural pathway is not involved in sound localization. J Comp Physiol 144:299–304

Montgomery JC, Baker CF, Carton AG (1997) The lateral line can mediate rheotaxis in fish. Nature 389:960–963

Montgomery JC, McDonald F, Baker CF, Carton AG, Ling N (2003) Sensory integration in the hydrodynamic world of rainbow trout. Proc R Soc Lond B 270:S195

Montgomery JC, Walker MM (2001) Orientation and navigation in elasmobranchs: which way forward? Environ Biol Fishes 60:109–116

Moore PA, Atema J (1991) Spatial information in the three-dimensional fine structure of an aquatic odor plume. Biol Bull 181:408–418

New JG, Fewkes LA, Khan AN (2001) Strike feeding behavior in the muskellunge, *Esox masquinongy*: contributions of the lateral line and visual sensory systems. J Exp Biol 204:1207–1221

Nissanov JN, Eaton RC (1989) Reticulospinal control of rapid escape turning maneuvers in fishes. Am Zool 29:103–121

Olszewski J, Haehnel M, Taguchi M, Liao JC (2012) Zebrafish larvae exhibit rheotaxis and can escape a continuous suction source using their lateral line. PLoS ONE 7:e36661

Partridge BL, Pitcher TJ (1980) The sensory basis of fish schools: relative roles of lateral line and vision. J Comp Physiol 135:315–325

Patton P, Windsor SP, Coombs S (2010) Active wall following by Mexican blind cavefish (*Astyanax mexicanus*). J Comp Physiol A 1–15

Pavlov DS, Tyuryukov SN (1993) The role of lateral-line organs and equilibrium in the behavior and orientation of the dace, *Leuciscus leuciscus*, in a turbulent flow. J Ichthyol 33:45–55

Peach MB (2002) Rheotaxis by epaulette sharks, *Hemiscyllium ocellatum* (Chondrichthyes: Hemiscylliidae), on a coral reef flat. Aust J Zool 50:407–414

Pitcher TJ, Partridge BL, Wardle CS (1976) A blind fish can school. Science 194:963–965

Pohlmann K, Grasso FW, Breithaupt T (2001) Tracking wakes: The nocturnal predatory strategy of piscivorous catfish. Proc Natl Acad Sci USA 98:7371–7374

Pohlmann K, Atema J, Breithaupt T (2004) The importance of the lateral line in nocturnal predation of piscivorous catfish. J Exp Biol 207:2971–2978

Preuss T, Osei-Bonsu PE, Weiss SA, Wang C, Faber DS (2006) Neural representation of object approach in a decision-making motor circuit. J Neurosci 26:3454–3464

Przybilla A, Kunze S, Rudert A, Bleckmann H, Brücker C (2010) Entraining in trout: a behavioural and hydrodynamic analysis. J Exp Biol 213:2976

Radakov DV (1973) Schooling in the ecology of fish. J. Wiley

Ren Z, Mohseni K (2012) A model of the lateral line of fish for vortex sensing. Bioinspiration Biomimetics 7:036016

Roberts BL, Meredith GE (1989) The efferent system. In: Coombs S, Görner P, Münz H (eds) The mechanosensory lateral line: neurobiology and evolution. Springer, New York; Berlin, pp 445–459

Roberts BL, Russell IJ (1972) The activity of lateral-line efferent neurones in stationary and swimming dogfish. J Exp Biol 57:435–448

Sand O, Bleckmann H (2008) Orientation to auditory and lateral line stimuli. Fish Bioacoust 183–231

Schellart NAM (1992) Interrelations between the auditory, the visual and the lateral line systems of teleosts; a mini-review of modelling sensory capabilities. Neth J Zool 42:459–477

Schellart NAM, Kroese ABA (1989) Interrelationship of acousticolateral and visual systems in the teleost midbrain. In: Coombs S, Görner P, Münz H (eds) The mechanosensory lateral line: neurobiology and evolution. Springer, New York; Berlin, pp 421–443

Schemmel C (1967) Comparative studies of the cutaneous sense organs in epigean and hypogean forms of *Astyanax* with regard to the evolution of Cavernicoles. Z Morphol Tiere 61:255–316

Schmitz A, Bleckmann H, Mogdans J (2008) Organization of the superficial neuromast system in goldfish, *Carassius auratus*. J Morphol 269:751–761

Schriefer JE, Hale ME (2004) Strikes and startles of northern pike (*Esox lucius*): a comparison of muscle activity and kinematics between S-start behaviors. J Exp Biol 207:535–544

Schwartz E (1965) Bau und Funktion der Seitenlinie des Streifenhechtlings (*Aplocheilus lineatus* Cuv. u. Val.). Z Vergl Physiol 50:55–87

Sharma S, Coombs S, Patton P, de Perera TB (2009) The function of wall-following behaviors in the Mexican blind cavefish and a sighted relative, the Mexican tetra (*Astyanax*). J Comp Physiol A 195:225–240

Springer AD, Easter SS, Jr., Agranoff BW (1977) The role of the optic tectum in various visually mediated behaviors of goldfish. Brain Res 128:393–404

Stein BE (1984) Multimodal representation in the superior colliculus and optic tectum. In: Vanegas H (ed) Comparative neurology of the optic tectum. Plenum, New York, pp 819–841

Stein BE, Meredith MA (1993) The merging of the senses. MIT Press, Cambridge

Stewart WJ, McHenry MJ (2010) Sensing the strike of a predator fish depends on the specific gravity of a prey fish. J Exp Biol 213:3769–3777

Stewart WJ, Cardenas GS, McHenry MJ (2013) Zebrafish larvae evade predators by sensing water flow. J Exp Biol 216:388–398

Suli A, Watson GM, Rubel EW, Raible DW (2012) Rheotaxis in larval zebrafish is mediated by lateral line mechanosensory hair cells. PLoS ONE 7:e29727

Sutterlin AM, Waddy S (1975) Possible role of the posterior lateral line in obstacle entrainment by brook trout (*Salvelinus fontinalis*). J Fish Res Board Can 32:2441–2446

Teyke T (1985) Collision with and avoidance of obstacles by blind cave fish *Anoptichthys jordani* (Characidae). J Comp Physiol A 157:837–843

Tittel G (1985) Determination of stimulus direction by the topminnow, *Aplocheilus lineatus*. A model of two dimensional orientation with the lateral line system. In: Barth FG (ed) Verhandlungen der Deutschen Zoologischen Gesellschaft, vol 78, 1st edn. Gustav Fischer, Stuttgart, p 242

Tittel G (1991) Verhaltensphysiologische, ultrastrukturelle und ontogenetische Studien am Seitenliniensystem von *Aplocheilus lineatus*. Ein vorläufiges Modell zur Richtungsdetermination von Beuteobjekten. Doktorarbeit, pp 1–445

Tittel G, Muller U, Schwartz E (1984) Determination of stimulus direction by the topminnow *Aplocheilus lineatus*. In: Varju D, Schnitzler HU (eds) Localization and orientation in biology and engineering. Springer, Berlin; New York, pp 69–72

Tricas TC, Highstein SM (1990) Visually mediated inhibition of lateral line primary afferent activity by the octavolateralis efferent system during predation in the free-swimming toadfish, *Opsanus tau*. Exp Brain Res 83:233–236

Tytell ED, Lauder GV (2008) Hydrodynamics of the escape response in bluegill sunfish, *Lepomis macrochirus*. J Exp Biol 211:3359–3369

Vogel S (1994) Life in moving fluids: the physical biology of flow. Princeton University Press, Princeton

Webb PW (1982) Avoidance responses of fathead minnow to strikes by four teleost predators. J Comp Physiol A 147:371–378

Weissburg MJ (2000) The fluid dynamical context of chemosensory behavior. Biol Bull 198:188–202

Windsor SP, Tan D, Montgomery JC (2008) Swimming kinematics and hydrodynamic imaging in the blind Mexican cavefish (*Astyanax fasciatus*). J Exp Biol 211:2950–2959

Windsor S, Paris J, de Perera TB (2011) No role for direct touch using the pectoral fins, as an information gathering strategy in a blind fish. J Comp Physiol A 197:321–327

Wojtenek W, Mogdans J, Bleckmann H (1998) The responses of midbrain lateral line units of goldfish, *Carassius auratus*, to objects moving in the water. Zoology 101:69–82

Wolfgang M, Anderson J, Grosenbaugh M, Yue D, Triantafyllou M (1999) Near-body flow dynamics in swimming fish. J Exp Biol 202:2303–2327

Wubbels RJ, Schellart NAM, Goossens JHHLM (1995) Mapping of sound direction in the trout lower midbrain. Neurosci Lett 199:179–182

Wullimann MF, Grothe B (2013) The central nervous organization of the lateral line system. In: Coombs S, Bleckmann H, Popper AN, Fay RR (eds) The lateral line system. Springer, NY

Zeddies DG, Fay RR (2005) Development of the acoustically evoked behavioral response in zebrafish to pure tones. J Exp Biol 208:1363–1372

Zittlau KE, Claas B, Münz H (1986) Directional sensitivity of lateral line units in the clawed toad *Xenopus laevis* Daudin. J Comp Physiol A 158:469–477

Zottoli S, Bentley A, Prendergast B, Rieff H (1995) Comparative studies on the Mauthner cell of teleost fish in relation to sensory input. Brain Behav Evol 46:151–164

Zottoli SJ, Cioni C, Seyfarth E (2007) Reticulospinal neurons in anamniotic vertebrates: a celebration of Alberto Stefanelli's contributions to comparative neuroscience. Brain Res Bull 74:295–306

Chapter 4
Hydrodynamic Imaging by Blind Mexican Cavefish

Shane P. Windsor

Abstract Blind Mexican cavefish (*Astyanax mexicanus*) live in complete darkness in underground streams and pools. These eyeless fish use hydrodynamic imaging to sense their surroundings. Hydrodynamic imaging involves fish using their mechanosensory lateral line system to sense changes in the water flows around their body caused by the presence of nearby objects. This allows them to sense detailed information about their surroundings as they move through complex environments. The fluid dynamics associated with this remarkable ability have been revealed using experimental flow measurements and computational modelling. Measurements of the fish's behavior and of the flow fields around the fish show that hydrodynamic imaging has a short range, of the order of 10 % of the fish's body length, and that fish need fast reactions in order to use it for collision avoidance. Due to the fluid dynamics of the flow fields involved, this sensory range is not increased when fish swim faster, contrary to previous expectations. This chapter summarises the behaviors and fluid dynamics involved with hydrodynamic imaging as used by blind cavefish.

Keywords Blind Mexican cavefish · Hydrodynamic imaging · Astyanax mexicanus · Lateral line · Fluid dynamics · Behavior

4.1 Introduction: What is Hydrodynamic Imaging?

Hydrodynamic imaging involves fish using their mechanosensory lateral line system to sense changes in the water flows around their body caused by the presence of nearby objects (Fig. 4.1). The hypogean (cave dwelling) form of

S. P. Windsor (✉)
Department of Aerospace Engineering, University of Bristol, Bristol, UK
e-mail: shane.windsor@bristol.ac.uk

H. Bleckmann et al. (eds.), *Flow Sensing in Air and Water*,
DOI: 10.1007/978-3-642-41446-6_4, © Springer-Verlag Berlin Heidelberg 2014

Fig. 4.1 Streamlines representing the flow field around a fish and how it is altered by the presence of an object. (Based on Hassan 1985)

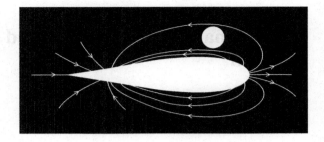

Astyanax mexicanus, commonly known as the blind Mexican cavefish, is the best known example of hydrodynamic imaging. These fish do not have eyes, yet by using hydrodynamic imaging they are capable of swimming through complex environments without collision. Here we will concentrate on hydrodynamic imaging as observed in blind Mexican cavefish, but this ability is also used by other fish species. Other eyeless cave dwelling fish species also use hydrodynamic imaging to avoid unfamiliar obstacles without touching them (Poulson 1963). In addition it has been found that eyed *A. mexicanus,* which live in above ground streams and rivers, behave in the same manner as cave *A. mexicanus* when experimentally blinded (Breder 1943; John 1957; Hassan et al. 1992). Like their eyeless counterparts, the blinded fish swam continuously and did not collide with obstacles. There are also many instances in the scientific literature where experimentally blinded fish have been observed to avoid obstacles without touching them. Experimentally blinded saithe were able to avoid collision with tanks walls and were also capable of schooling with other fish (Pitcher et al. 1976), indicating just how much navigational information is sensed by the lateral line system. Blinded goldfish also show the ability to avoid obstacles without collision (Yasuda 1973; Teyke 1988) and Dijkgraaf (1962) cites literature for 13 different species that show similar behavior when blinded.

Given that when many species of fish are deprived of visual information they are capable of using hydrodynamic imaging suggests that the use of hydrodynamic imaging may be common place in many fish species. Indeed, there are many situations in aquatic environments when vision is of limited use, such as in turbid waters, in the deep sea, or simply at night. In these situations fish need a non-visual way of getting information about their surroundings and hydrodynamic imaging appears to be well suited for this task.

4.2 Blind Mexican Cavefish

Blind Mexican cavefish are found in subterranean fresh water pools and streams in cave systems in the Huastecan Province area of east central Mexico (Mitchell et al. 1977). These small characid fish grow to approximately 90 mm in length and are currently widely available as aquarium fish from commercial suppliers. The limestone caves the fish naturally inhabit were formed by the flow of water from subterranean springs and it is thought that at various times these underground water ways have had connections with surface rivers. Many of the caves the fish are found in are subject to regular flooding during the wet season, but during the rest of the year the caves contain streams and pools. The fauna of the cave systems is generally very limited and there are no known predators of A. *mexicanus* present. Their diet is not well known, but it is likely that they feed on material washed into caves and bought in by bats through their droppings.

4.2.1 Comparison with Epigean A. mexicanus

The epigean (surface dwelling) form of A. *mexicanus*, which has well developed eyes and skin pigmentation, is found in rivers in the same area of Mexico and can freely interbreed with the hypogean form. Thus the evolutionary relationship between the two forms and the evolutionary mechanisms responsible for the loss of eyes and pigmentation in the cave dwelling populations is an area of great scientific interest (reviewed in Jeffery 2005). There are more than 30 caves which support subterranean populations of A. *mexicanus* and the different cave populations show different degrees of eye and pigmentation reduction. Some populations are completely eyeless and totalling lacking pigment, while others have fully developed eyes and pigmentation, while still other caves are inhabited by phenotypically intermediate forms (Romero 1985). It is thought that some of these populations originated at different times and that some may have evolved eye and pigment reduction independently (Jeffery and Martasian 1998).

The hypogean form of A. *mexicanus* shows a number of morphological and behavioral differences from the epigean form of A. *mexicanus*. These include:

- lack of pigmentation
- reduction or loss of eyes
- minor skeletal changes (Jeffery 2001)
- increased numbers of external taste buds on jaw (Schemmel 1967)
- larger superficial neuromasts (Teyke 1990)
- no schooling behavior (Breder 1943; Gregson and Burt de Perera 2007)
- greatly reduced aggression (Parzefall 1983)
- different bottom feeding behavior (Schemmel 1967)

The currently accepted scientific name for what is commonly known as the blind Mexican cavefish is A. *mexicanus*, however there have been a number of other species names used in the scientific literature which can cause some confusion (see Jeffery et al. 2003, for discussion). When the blind cavefish were first discovered they were originally named *Anoptichtys jordani* (Hubbs and Innes 1936). Later another cavefish population was discovered and named *Astyanax hubbsi* with the epigean population known as A. *mexicanus*. Following this it came

Fig. 4.2 Diagram of the lateral line of *A. mexicanus*. Large *black circles* with *white outlines* represent canal pores; canals are shown with *white lines*. A canal neuromast is located approximately midway between each pair of pores. The approximate locations of the superficial neuromasts are shown with *small black dots*. *io* infraorbital canal, *mc* mandibular canal, *po* preopercular canal, *so* supraorbital canal, *st* supratemporal canal, *tc* trunk canal. The scale bar represents one body length. (From Windsor et al. 2010a)

to be recognized that the two forms were in fact a single species, for which many authors, including myself, used the name *Astyanax fasciatus* (Romero 1985). Currently both *A. mexicanus* and *A. fasciatus* are commonly used in the scientific literature. Unless otherwise indicated here "blind cavefish" or "cavefish" refers to the hypogean form of *A. mexicanus*.

4.2.2 Lateral Line Morphology

The general morphology of the canal lateral line system found in *A. mexicanus* is typical of many teleost fish (Fig. 4.2) and there is little difference in canal morphology between epigean and hypogean populations of *A. mexicanus* (Schemmel 1967). Both populations show similarly large numbers of superficial neuromasts distributed densely over the entire body of the fish, with 4–8 superficial neuromasts per scale (Fig. 4.2). Other characid fish show similar numbers of superficial neuromasts, so the high density of superficial neuromasts in *A. mexicanus* is not unexpected, even though it is much higher than the number of superficial neuromasts seen in the majority of other fish species (Schemmel 1967). Where the lateral lines of the epigean and hypogean populations of *A. mexicanus* do show a difference is in the morphology of the superficial neuromasts (Teyke 1990). The cupula of the superficial neuromasts of the hypogean form of *A. mexicanus* are longer and wider than those of the epigean fish. In the hypogean fish, the base of the superficial neuromasts is about 80×50 μm, with the length of the cupula averaging 180 μm at the head of the fish reducing down the length of the fish to 100 μm at the caudal fin. In comparison, in the epigean fish the base of the superficial neuromasts is about 50×30 μm, with an average cupula length of only 42 μm. It is hypothesized that longer cupula will be more sensitive, as they extend further out of the boundary layer and are exposed to higher flow velocities. In both populations the superficial neuromasts are unusual as the long axis of the cupula is aligned in the dorsal–ventral axis, yet the axis of sensitivity of the haircells is aligned along the rostral-caudal direction. Normally the long axis of the cupula corresponds with the axis of sensitivity of the haircells in a neuromast. The arrangement found in *A. mexicanus* may serve to make the superficial neuromasts more sensitive. The morphology of the canal neuromasts has not been studied, so it is not known if similar differences exist between hypogean and epigean populations.

4.3 Behavior and Sensing

Hydrodynamic imaging is an active sensory modality as fish must move in order to generate the flow that is used to probe their surroundings. As such, the movement and behavior of blind cavefish can tell us a lot about what they are able to sense through the use of hydrodynamic imaging.

4.3.1 Sensing Their Surroundings

Blind cavefish are able to use hydrodynamic imaging to get detailed information about their surroundings. They can discriminate between openings of different shapes (von Campenhausen et al. 1981) and dimensions (Weissert and von Campenhausen 1981) and the resolution of hydrodynamic imaging appears to be relatively high, with fish being able to discriminate differences in the spacing of objects of less than 1.5 mm (Hassan 1986). Hydrodynamic imaging appears to provide more information to the fish about lateral rather than frontal objects. When fish are prevented from being able to glide across openings they are no longer able to discriminate between them (Weissert and von Campenhausen 1981). This increased lateral information is likely due to the way that an object is sequentially sampled by the entire array of neuromasts down the side of a fish, as the fish passes beside it.

Blind cavefish seem to navigate through familiar environments using both hydrodynamic imaging and memory. When unfamiliar obstacles are placed in their environment they are successfully avoided, and similarly, when obstacles are removed, the fish immediately swim through the past location of the obstacles (John 1957). Both of these behaviors show that the fish are using instantaneous sensory information rather than memory for navigation. However, when an obstacle that cannot be detected by hydrodynamic imaging, such as wall of pins, is placed in their environment the fish quickly learn to avoid this area of the tank and continue to do so for a period after the wall has been removed (John 1957), indicating that fish can memorize some aspects of their surroundings.

It is believed that blind cavefish memorize a spatial map of their surroundings. Teyke conducted a series of experiments (Teyke 1985, 1988, 1989) examining the behavior of blind cavefish when they were released into a new environment. It was found that blind cavefish exploring a novel environment increased their swimming velocity and then, gradually over the course of the following hours, their swimming velocity decreased to a slower constant value. It was hypothesized that the increase in swimming velocity optimized the lateral line stimulus. Then once the fish built up a spatial map of their surroundings, they could reduce their swimming speed. This hypothesis was supported by the finding that the fish did not show the same increase in swimming velocity when they were removed from a tank for less than two days and then returned. If removed for a longer period the fish showed the same increased speed. This indicated that the fish remembered the layout of the tank for up to two days. Teyke also found that the fish would decrease their swimming velocity more quickly in tanks which were not symmetrical, indicating that the fish had to compensate for the increased ambiguity in symmetrical environments. It has also been shown that when the sensitivity of the lateral line system is reduced by altering the ionic make-up of the water that blind cavefish increase their swimming velocity in proportion to the change in the ionic components of the water (Hassan et al. 1992). From this it was inferred that the fish changed their swimming velocity to optimize the hydrodynamic information available to them.

A prediction of the hypothesis that fish increase their swimming speed to optimize hydrodynamic imaging is that fish swimming at higher velocities should react to objects at greater distances. However, when the distances at which fish reacted to an obstacle were compared to the swimming velocity of the fish there was no significant correlation found (Windsor et al. 2008). This result also held when estimates of reaction time were taken into consideration. Instead, it was found that fish reacted to avoid collision with a wall directly in front of them (Fig. 4.3a) at a relatively constant distance of 4.0 ± 0.2 mm (0.09 ± 0.01 body lengths (BL)) irrespective of their swimming velocity. The working distance of hydrodynamic imaging appears to be similar in front of and beside the fish. When swimming alongside a wall the fish kept a relatively constant mean distance of 4.7 ± 0.5 mm (0.10 ± 0.01 BL) from the wall, again irrespective of swimming velocity. These findings prompted studies looking at how the flow fields around blind cavefish changed with swimming velocity and the implications of these changes to the information present in the flow field (Windsor et al. 2010a, b). These studies will be discussed in Sect. 4.4.

The relatively short range of hydrodynamic imaging may at first seem surprising given that the lateral line is normally thought to enable fish to detect prey movements from 1 to 2 body lengths away (Coombs and Montgomery 1999). The order of magnitude decrease in the effective distance of hydrodynamic imaging relative to prey sensing is due to the differences in the hydrodynamics of the signals involved. In the case of sensing the water movements generated by prey, the fish is detecting the presence of a signal with properties that normally differ strongly from any background signal. In hydrodynamic imaging the fish

(a)

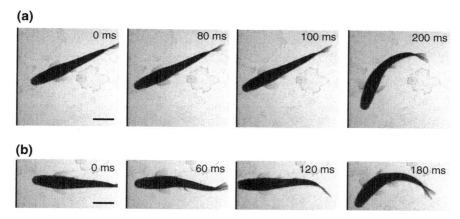

(b)

Fig. 4.3 Series of images of blind cavefish approaching a wall head-on. The wall is indicated by the *black line* to the *left* of each image. **a** The fish glides towards the wall, then at 100 ms extends the pectoral fins away from the body; at this point the nose is 2.7 mm away from the wall. The fish then *curves* its body to the *left*, turning to follow along the wall, without making any contact. **b** The fish starts a tail beat as it approaches the wall and shows no sign of detecting the wall before it collides. After colliding the fish turns to the *left* and swims along the wall. The length of both scale bars is 10 mm. (From Windsor et al. 2008)

need to detect much more subtle changes in the strong signal generated by their own movement.

Blind cavefish are not always successful at detecting the changes in the flow field around them and sometimes collide with obstacles (Fig. 4.3b). It has been found that when approaching a wall head-on that fish collided approximately 25 % of the time (Windsor et al. 2008). Whether a fish collided with an obstacle was highly correlated with whether or not a fish was beating its tail as it approached the wall (Teyke 1985; Windsor et al. 2008). Fish were much more likely to collide with the wall if tail beating rather than gliding when approaching the wall. This could be due to a number of factors. Firstly, when the fish is tail beating, it is changing the flow field around its body, generating noise which makes the task of detecting the changes created by the wall more difficult. Secondarily, the motor activity of the fish may activate the efferent system of the lateral line reducing its sensitivity (Russell and Roberts 1974). In addition to these effects, the act of tail beating may itself make it more difficult for the fish to turn, as its tail may not be in a suitable position for initiating a sharp turn.

4.3.2 Wall Following

When released into a new environment blind cavefish show a clear preference for staying near the boundaries of the environment (Weissert and von Campenhausen 1981; Teyke 1985, 1989; Patton et al. 2010). This wall following behavior

Fig. 4.4 Typical swimming sequence of a blind cavefish swimming beside a wall with a netted section (the *darker* region in the *centre*) that could be sensed with tactile, but not hydrodynamic sensors. The fish was initially swimming parallel to the solid wall with pectoral fins extended but not touching the wall (*1*) following a tail beat. The fish then glided parallel to the wall (*2*) until coming alongside the netted region. Once alongside the netted region the fish turned into the netted wall and collided with it (*3*). The fish then turned away from the wall (*4*) and once alongside the solid wall again (*5, 6*) proceeded to swim parallel to the wall without making contact. (From Windsor et al. 2011)

appears to be advantageous for exploration using short range senses. By swimming along boundaries the fish sequentially brings its sensors within range of features in the environment. There is evidence to suggest that this allows fish to form a map of their environment based on the sequence of features encountered (Burt de Perera 2004b) and also possibly the distances between them (Burt de Perera 2004a). When the swimming behavior of both blind and sighted *Astyanax* was compared in darkness it was found that the blind fish spent longer periods swimming in a constant direction relative to walls and kept a more parallel orientation to the wall (Sharma et al. 2009). It was thought that these behavioral changes are beneficial for exploration based on hydrodynamic imaging. There is also evidence that blind cavefish change their swimming kinematics when swimming beside walls in comparison to when swimming away from any boundaries. Away from boundaries fish were found to swim more slowly, with longer gliding periods between tail beats (Windsor et al. 2008). By increasing the proportion of time spent gliding the fish would increase their chances of detecting obstacles in front of them for the reasons already discussed.

4.3.3 Sensory Modalities

When blind cavefish swim alongside walls they frequently make contact with the wall with their pectoral fins (Windsor et al. 2008). This could provide additional tactile sensory information to the fish and indeed it has been shown that blind cavefish can follow along walls with a disabled lateral line system (Patton et al. 2010), presumably relying solely on tactile information. However, when tactile information is placed in conflict with hydrodynamic information, blind cavefish appear to respond to hydrodynamic information in preference to tactile information. When fish swam alongside a wall with a section that could be sensed with tactile, but not hydrodynamic sensors (Fig. 4.4), they collided more frequently and swam closer to that section even though it provided the same tactile feedback (Windsor et al. 2011). It therefore appears that hydrodynamic imaging is the primary sensory modality used in wall following by blind cavefish.

In the case of approaching an obstacle head-on it is clear that the lateral line is responsible for the fish being able to detect and avoid the obstacle. With a disabled lateral line system (Windsor et al. 2008), or with hydrodynamically invisible objects (Windsor et al. 2011), blind cavefish collide with obstacles in front of them. Even with the sensitivity of the lateral line slightly reduced, the ability of fish to avoid collisions is greatly reduced (Tan 2007). The relative roles of the two sub-systems of the lateral line, the canal and superficial neuromasts, in hydrodynamic imaging is not as clear. The rate of collisions of blind cavefish with a barrier was found to increase when the canal neuromasts were selectively disabled using the antibiotic gentamicin (Montgomery et al. 2001). In the same experiment when the superficial neuromasts were physically ablated the rate of collision remained low. This suggests that hydrodynamic imaging is mediated by the canal system. However, recent studies looking at the effects of gentamicin have shown that it disables both populations of neuromasts (Van Trump et al. 2010), bringing the conclusions of earlier studies using this method into question.

4.4 Fluid Dynamics

In order to study the mechanics of hydrodynamic imaging, it is necessary to understand the fluid flows involved. When fish move through the water they create a flow field around their body. It is by sensing changes in this flow field, caused by the presence of an object, that the fish are able to get information about their surroundings. Therefore, it is important to measure the flow fields around fish in open water, when there are no objects nearby, and then look at how the flow field changes as the fish approach an object. The next step is to then look at how these changes would be encoded by the lateral line in order to estimate what information hydrodynamic imaging can give fish about their surroundings.

Blind cavefish swim in an intermittent manner, normally making a single tail beat and then gliding for a period with their body held straight. Blind cavefish spend approximately 70 % of their time gliding (Windsor et al. 2008), and as already discussed it appears that this is when hydrodynamic imaging is most effective. As such, it is the flow field around gliding fish that is of most interest in terms of hydrodynamic imaging. As well as having benefits for hydrodynamic imaging, intermittent swimming is thought to be more energetically efficient than continuous swimming (Weihs 1974; Wu et al. 2007). This would be a major advantage for blind cavefish, which swim almost constantly.

There is a considerable amount known about the flow fields created by fish when they are actively swimming (reviewed in Lauder and Tytell 2005), but surprisingly little work has been done looking at the flow fields around gliding fish. Early experimental work measured the pressure at points along the bodies of swimming bluefish (*Pomatomus saltatrix*) using Pitot pressure tubes (Dubois et al. 1974) and the pressure distribution around a fish-shaped model has also been investigated (Kuiper 1967). Following this, a series of studies (Hassan 1985,

1992a, b) looked at hydrodynamic imaging using mathematical potential flow models. These models explored the flows generated by the interaction of a three-dimensional (3D) fish shape and a flat surface. With the development of new technologies it is now possible to experimentally measure the flows involved with hydrodynamic imaging and to model these flows in detail using computational methods. The remainder of this chapter will focus on what has been learnt about the fluid flows involved in hydrodynamic imaging using these approaches.

4.4.1 Flow Fields in Open Water

The flow field around a blind cavefish gliding in open water can be thought of as the field that is used by the fish to probe its surroundings. The flows around the fish in this situation were experimentally measured using particle image velocimetry (PIV) (Windsor et al. 2010a). In this technique, the water is seeded with small neutrally buoyant particles and their motion in the plane of a laser light sheet is recorded with a high speed video camera. The velocity of the flow field is then calculated by looking at the movement of particles between consecutive video frames. With this information it is also possible to calculate the pressure field around the fish (see Windsor 2008, for details). These experimental measurements can then used to validate computational fluid dynamic (CFD) models of the same situation.

The velocity and pressure distributions of the flow around a fish in open water show a number of distinct features (Fig. 4.5). At the nose of the fish there is a stagnation point where the velocity of the flow is zero relative to the fish. This corresponds to a region of high pressure around the nose of the fish. As the water flows around either side of the fish it accelerates, creating a region of low pressure around the widest part of the body. The flow then gradually decelerates as it moves towards the tail. Along the body of the fish there is a clear boundary layer, where the fluid is being entrained and moves along with the fish. This boundary layer is created by the viscous interaction of the water with the body, and the thickness of this layer grows as the flow moves further down the body. The boundary layer cannot be seen in the pressure distribution as pressure is constant across a boundary layer. All of the flows measured around blind cavefish were laminar with no indications of turbulence.

It is informative to compare the flows around blind cavefish to flows around other streamlined bodies, such as aerofoils. However, to make this comparison the effects of Reynolds number need to be taken into account. The Reynolds number (*Re*) represents the relative importance of viscosity compared to inertia and is defined as:

$$Re = \frac{UL}{v} \tag{4.1}$$

Where U is velocity of the body, L is the characteristic length (normally body length for a fish), and v is the kinematic viscosity of the fluid. Flows at high

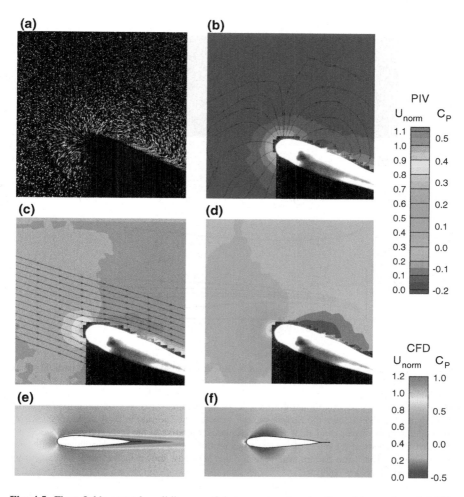

Fig. 4.5 Flow fields around a gliding cavefish in open water at a Reynolds number of 6,000. **a** PIV particle streak image created by adding together sequential video frames. The laser sheet was projected from top of the frame casting a *shadow* beside the fish. **b** Normalized velocity (U_{norm}) contours measured using PIV, presented in the frame of reference of a stationary observer. The *black lines* represent instantaneous streamlines. **c** Normalized velocity contours measured using PIV, presented in a frame of reference moving with the fish. The *black lines* represent instantaneous streamlines. The velocity field can be transformed to that shown in the previous panel by subtracting the velocity of the fish. **d** Normalized pressure field (C_P) contours calculated from the PIV velocity data. **e** Normalized velocity field in the frame of reference of the moving fish calculated from a 2D CFD model. **f** Normalized pressure field calculated from a 2D CFD model

Reynolds numbers ($Re > 100,000$) are dominated by inertial effects; this is the regime in which most aircraft operate. At the other end of the scale, at very low Reynolds numbers ($Re < 1$) viscous effects dominate; this is the regime in which

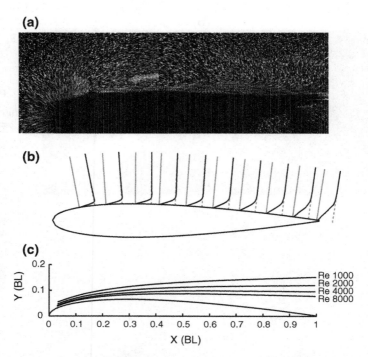

Fig. 4.6 Boundary layer over the surface of a gliding blind cavefish. **a** Synthetic particle streak image from a PIV video sequence of a fish (length 60 mm) viewed from a dorsal perspective moving at a Reynolds number of 6,000. **b** CFD boundary layer velocity profiles of flow around a revolved 3D body based on an aerofoil at a Reynolds number of 6,000, showing the tangential velocity profile (*black line*), the zero tangential velocity axes normal to body surface (*solid grey line*), and the free-stream tangential velocity component relative to the body surface (*dashed grey line*). **c** CFD results of the effect of the Reynolds number on the 99 % boundary layer profiles for a 2D aerofoil. The same pattern was seen for the 3D body shown in the panel above. (From Windsor and McHenry 2009)

bacteria operate. In between there is an intermediate range of Reynolds numbers where both inertial and viscous effects are important; this is the regime in which the flows around blind cavefish fall.

Almost all the work that has been done looking at flows around aerofoils has been done at high Reynolds numbers. The general form of the flow field around a symmetric aerofoil in a uniform flow is similar to that found around a gliding blind cavefish. There is a stagnation point at the nose, accelerated flow around the widest part of the body and corresponding high and low pressure regions. However, at high Reynolds number the pressure increases over the rear of the aerofoil to a positive value at the trailing edge, while for Reynolds numbers appropriate for cavefish the pressure is still slightly negative at the trailing edge. This is due to the formation of a thick laminar boundary layer at low Reynolds numbers (Fig. 4.6), whereas at high Reynolds numbers the boundary layer is turbulent and much thinner.

Over the range of Reynolds numbers experienced by blind cavefish (1,000–8,000) the boundary layer remains laminar but decreases in thickness with increasing Reynolds number (Fig. 4.6). The pressure gradients along the body effect the rate of growth of the boundary layer, with the decreasing pressure from the nose of the fish to the widest part of the body reducing the rate of growth and the increasing pressure past this point causing the boundary layer to grow rapidly. This means that the boundary layer over the body of the fish is not well approximated by the model sometimes used of a flat plate with a zero pressure gradient.

Having measured the flow field in open water the next question is to estimate what stimulus this is going to give the two sub-modalities of the lateral line. The response of a superficial neuromast is generally considered to be proportional to the velocity of the flow around the cupula of the neuromast. However, as the cupula lies within the boundary layer, where velocity changes rapidly with distance from the surface, there is no obvious height at which to measure the velocity of the flow. This means that detailed models are needed to realistically estimate the details of the response of superficial neuromasts (McHenry et al. 2008). However, a good approximation of the magnitude of the stimulus to the superficial neuromasts is provided by the wall shear stress (τ_w) on the surface of the skin.

$$\tau_w = \mu \left(\frac{\delta u_t}{\delta y} \right) \bigg|_{y=0} \tag{4.2}$$

where u_t is the tangential velocity, y is the direction normal to the surface and μ is dynamic viscosity. In regions where the velocity of the flow is high close to the body, the wall shear stress will be high, while in regions where the flow is slow close to the body the wall shear stress will be low. The normalized version of wall shear stress is the skin friction coefficient (C_f).

$$C_f = \frac{\tau_w}{0.5\rho U^2} \tag{4.3}$$

where ρ is the density of the fluid. Blind cavefish have a large number of superficial neuromasts distributed relatively uniformly over their entire body (Fig. 4.2) leading to the assumption that the fish can sense flow over their entire body surface.

The response of canal neuromasts is proportional to the pressure difference across the canal pores (ΔP) between which the neuromasts are located (Denton and Gray 1983; Kalmijn 1988). For blind cavefish, the canal pores are located approximately 2 % of their body length apart. This gives fish a measure of the pressure gradient along each of their lateral line canals (Fig. 4.2). The normalized version of the pressure field is the coefficient of pressure (C_P).

$$C_P = \frac{P}{0.5\rho U^2} \tag{4.4}$$

Using these approximations the stimulus to the two sub-modalities of the lateral line system can be estimated for a given flow field. In the open water case there is a

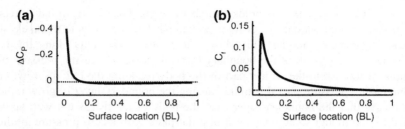

Fig. 4.7 Open water lateral line stimulus distribution based on a CFD model of a 2D aerofoil at a Reynolds number of 6,000. **a** Normalized pressure differences across canal pores at 2 % BL spacing. Note that the Y axis is inverted in the aerodynamic convention, so a negative pressure difference represents a more positive pressure at the rostral canal pore. **b** Normalized wall shear stress distribution representing the presumed stimulus to the superficial neuromasts

large peak in the wall shear stress close to the nose of the fish and then a rapid decrease to lower values down the body of the fish (Fig. 4.7a). For the stimulus to the canal lateral line system there is a strong negative pressure gradient at the nose of the fish, then a rapid decrease to smaller magnitude positive values around the widest part of the body, before a gradual decrease to a zero pressure gradient at the tail (Fig. 4.7b).

It is again informative to compare these results to those for higher Reynolds numbers. The potential flow models of Hassan (Hassan 1985; Hassan et al. 1992; Hassan 1992a) do not include the effects of viscosity, modelling very high Reynolds number flows. These models predict smaller magnitude stimuli to the lateral line and smaller relative changes in the stimuli due to the presence of objects. These differences are due to the way potential flow models neglect the effects of viscosity. At the intermediate Reynolds numbers at which blind cavefish swim, viscosity has a large effect on the form of the flow field around the body of the fish and therefore on the stimulus to the lateral line.

4.4.2 Head-on Approaches

As a blind cavefish approaches a wall head-on, the flow field around the head of the fish changes as it gets closer to the wall (Windsor et al. 2010a). From the point of view of a stationary observer, flow is pushed forward and away from the nose when the fish is far from the wall, but increasingly to either side of the nose as the fish gets closer to the wall (Fig. 4.8). As this happens the pressure at the stagnation point at the nose of the fish increases. The changes in the flow field are concentrated around the head of the fish, with the flow around the rest of the body remaining relatively constant.

The stimulus to both sub-modalities of the lateral line system increase around the nose of the fish as it approaches a wall (Fig. 4.9). In line with the changes in the flow field, the stimulus to both the superficial and canal neuromasts on the head increase rapidly as the fish gets closer to the wall, while the stimulus over the rest of the body remains relatively constant.

Fig. 4.8 Flow fields as measured using PIV around a 60 mm blind cavefish approaching a wall at the *top* of the frame at a Reynolds number of 3,500.
a–d Normalized velocity contours with streamlines in the frame of reference of a stationary observer. The streamlines were drawn from the same points on the body of the fish for each frame.
e–h Normalized pressure contours. **a, e** 0.28 BL from the wall (0 s). **b, f** 0.21 BL from the wall (0.07 s). **c, g** 0.13 BL from the wall (0.15 s).
d, h 0.07 BL from the wall (0.22 s). The fish then started to turn *left* away from the wall, avoiding making any contact with the wall. (From Windsor et al. 2010a)

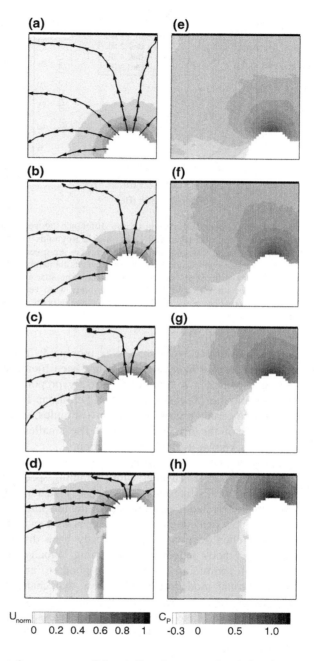

In order for a fish to sense the presence of the wall as it approaches it head-on, the fish needs to detect the changes the wall causes in the stimulus to its lateral line. This requires the fish to detect the difference in its lateral line stimulus from

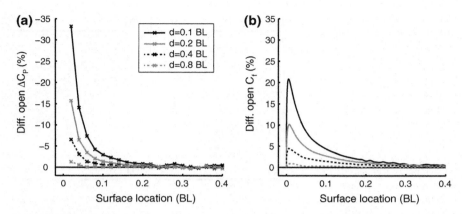

Fig. 4.9 CFD modelling results of the difference between the lateral line stimuli for a 2D aerofoil as it approached a wall head-on at a Reynolds number of 6,000 and the open water stimuli at the same Reynolds number. Differences are given as a percentage of the maximum C_P and C_f values in open water. Each line represents the stimuli at a certain distance from the wall (d). Only the differences in the stimuli along the first 0.4 BL are shown for clarity as the differences in stimuli down the rest of the body were very small. **a** Difference in normalized pressure differences across canal pores from when in open water. **b** Difference in normalized wall shear stress from when in open water. (From Windsor et al. 2010a)

the stimulus it receives when it is moving in open water. For many sensory modalities, across many species, it has been found that the smallest change that can be detected in a stimulus is directly proportional to the magnitude of the original stimulus (Teghtsoonian 1971; Schiffman 1996). This means that an animal can sense a certain relative change in a stimulus, such as 10 %, irrespective of the magnitude of the original stimulus. The smallest level of change that can be detected is known as the Weber fraction.

In the case of blind cavefish, we are interested in how the distance at which they can detect an object changes with swimming speed. If we assume that the lateral line system works like most other sensory systems, and that the smallest change that can be detected is proportional to the magnitude of the stimulus, then we can calculate how detection distance changes with swimming velocity if we assume a certain Weber fraction. As has already been discussed, Reynolds number is a function of body length and swimming velocity. So for an individual fish an increase in swimming velocity corresponds to an increase in Reynolds number. Computational modelling results show that changes in detection distance with Reynolds number are small and that in fact by swimming faster the distance at which a fish will detect a wall actually decreases slightly (Fig. 4.10). This is the opposite of what we would expect based on the hypothesis that swimming faster improves hydrodynamic imaging. In reality, the change in detection distance with Reynolds number over the behaviorally relevant range is so small that we would expect to see the fish detect the wall at a relatively constant distance irrespective of

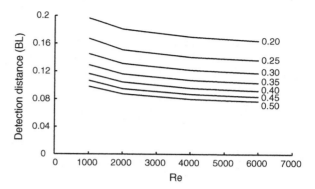

Fig. 4.10 Effect of Reynolds number on the distance that a fish would detect a wall, assuming the fish responded to a certain relative change in canal lateral line stimuli. Sensitivity thresholds in terms of Weber fractions are labelled. Results from a fish shaped 2D CFD model. (From Windsor et al. 2010a)

swimming speed. This matches closely with what has been found in behavioral experiments (Windsor et al. 2008).

Detecting when the properties of a signal change is made more difficult when the signal is noisy. The greater the magnitude of the noise relative to the magnitude of the signal the more difficult it is to measure the properties of the signal. For blind cavefish, one of the advantages of swimming faster is that the signal-to-noise ratio would increase since only the active flow signal (but not ambient noise) levels would increase (Fig. 4.11). This could possibly explain why blind cavefish swim faster in novel environments.

All of the analysis and experiments done with blind cavefish mentioned so far have been done in conditions with no background noise. In an effort to test whether swimming faster increased the distance at which blind cavefish detected objects in the presence of background noise we conducted a preliminary study. This involved repeating the head-on behavioral experiments (Windsor et al. 2008) already discussed with varying levels of background noise. No significant change in swimming velocity, detection distance or collision rate was found with levels of background noise up to that which started to disrupt the swimming trajectories of the fish during periods of gliding. These experiments showed that blind cavefish are capable of discriminating the signal generated by the wall from these levels of background noise at the speeds they swim in novel environments. This result however does not provide evidence to either support or refute the hypothesis that the fish swim faster in new environments to improve the signal to noise ratio of hydrodynamic imaging. An alternative explanation to the behavior of blind cavefish when introduced into novel environments may simply be that the fish swim faster in order to explore their new surroundings more quickly.

4.4.3 Beside a Wall

For blind cavefish, sensing changes in the flow field down the sides of their body is important for following walls and for distinguishing different objects or landmarks.

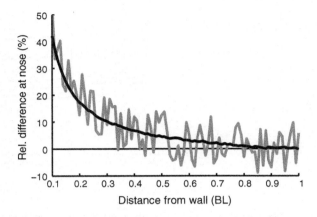

Fig. 4.11 CFD model results of the effect of a set level of environmental noise on the relative difference in pressure across the canal pores closest to nose of a 2D aerofoil as it approaches a wall head-on at different Reynolds numbers. Random white noise with a maximum magnitude of 10 % of the open water pressure difference at a Reynolds number of 1,000 was added to both sets of CFD results. The *gray line* is at a Reynolds number of 1,000, the *black line* is at a Reynolds number of 2,000. (From Windsor et al. 2010a)

The flows around a fish swimming parallel to a vertical wall have been studied using experimental PIV measurements combined with CFD modelling (Windsor et al. 2010b). It was found that when the fish were more that 0.25 BL from the wall there was very little noticeable change in the flow field from that seen in open water (Fig. 4.12). As the distance between the fish and the wall was reduced, the velocity of the accelerated flow in the region between the fish and the wall increased. At the same time, the stagnation point moved progressively from the tip of the nose around to the side of the fish adjacent to the wall. Corresponding with the changes in the velocity field, the high pressure region around the stagnation point expanded towards the wall, and the low pressure region between the fish and the wall increased in magnitude and moved down the body.

As the fish drew closer to the wall, the flow, accelerated by the movement of the fish, caused a boundary layer to form on the wall. The maximum velocity of the flow between the fish and the wall increased as the distance between the two decreased down to 0.05 BL. When the distance between the fish and the wall was decreased to 0.02 BL there was a change in the flow field as the boundary layer around the fish and on the wall merged; the size of the region of accelerated flow decreased and the wake behind the fish shifted towards the wall.

As the fish swam closer to the wall the stimulus to the lateral line increased as the magnitude of the changes in the flow field increased. In behavioral trials, blind cavefish swam a mean distance of 0.10 BL from the wall at an average Reynolds number of 6,000 (Windsor et al. 2008). Under these conditions, the CFD modelling indicated that the stimulus to the different canals of the lateral line varied depending which side of the fish they were on, their location and their orientation

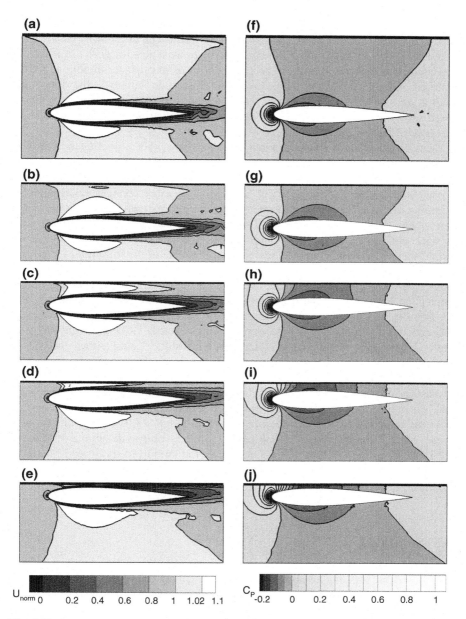

Fig. 4.12 CFD results of the effect of the distance from the wall (*d*) on the flow field about a 3D body of revolution at a Reynolds number of 6,000. **a–e** Normalized velocity distribution. *Shading* highlights the flow regions where the velocity was different from the inlet velocity. Velocity line contours are spaced at 0.2 intervals. **f–j** Normalized pressure distribution. *Shading* highlights the flow regions where the pressure was different from zero. C_P contour lines are spaced at 0.05 intervals. **a, f** *d* = 0.50 BL. **b, g** *d* = 0.25 BL. **c, h** *d* = 0.10 BL. **d, i** *d* = 0.05 BL. **e–j** *d* = 0.02 BL. (From Windsor et al. 2010b)

(Windsor et al. 2010b). The stimuli to the supraorbital, infraorbital and mandibular canals was very similar on the same side of the body. The differences between the same canals on different sides of the body were greatest near the nose of the fish. In the preopercular canal, the pressure gradients were in opposite directions on the different sides of the body, with the largest magnitude stimuli at the dorsal and ventral ends of the canal. In regard to the stimulus to the superficial neuromasts, there were small but significant differences in the distribution of the wall shear stress between the two sides of the body.

In order to gain information about a wall or any other object using hydrodynamic imaging, fish need to measure changes in the stimulus to their lateral line. This raises the question as to what the change is measured relative to. The two most obvious options would be for the fish to measure changes relative to the stimulus that they experience in open water, or alternatively to measure the differences in the stimuli to their lateral line on either side of their body. If fish were to sense changes in lateral line stimuli relative to when they are swimming in open water, they would need to have some form of template of the stimuli they expect at their current swimming velocity. Sensing changes relative to the other side of their body would not require a template, but would have disadvantages when there were objects on both sides of the fish. Given the limits to our current understanding of the sensory processing of lateral line stimuli we can not say which, if either of these methods is most likely to be used by blind cavefish. CFD modelling suggests both options are viable and produce similar results. Using either method of comparison, the stimulus to the lateral line undergoes the largest change around the head of the fish on the side of the fish closest to the wall. These changes increased in magnitude as the distance between the wall and the fish decreased. The magnitude of the relative changes in stimulus were approximately constant over the behaviorally relevant Reynolds number range. At 0.10 BL from the wall these changes appear to be sufficient for the fish to be able to detect the wall, but at 0.25 BL from the wall the relative change in the stimuli to the lateral line is so small it is unlikely that blind cavefish will be able to detect a wall at this distance (Windsor et al. 2010b).

4.5 Summary

Hydrodynamic imaging involves fish using their lateral line system to sense changes in the water flows around their body caused by the presence of nearby objects. Blind Mexican cavefish are the best known example of hydrodynamic imaging, but it is likely that many other species of fish also use it in environments where visibility is limited. Blind cavefish use hydrodynamic imaging to sense detailed information about their surroundings and they make use of this information to move through complex environments. Their lateral line has a number of adaptations which appear to increase its sensitivity. Despite these adaptations, blind cavefish regularly collide with obstacles if they are beating their tail as they approach.

Blind cavefish exhibit a wall following behavior which appears to be beneficial for exploration using hydrodynamic imaging. While following walls, blind cavefish use hydrodynamic imaging to learn about their surroundings. Measurements of the fish's behavior and of the flow fields involved show that hydrodynamic imaging has a short range, of the order of 0.10 BL, and that fish need fast reactions in order to use it for collision avoidance. This sensory range is not increased when fish swim faster due to the fluid dynamics of the flow fields involved. As such, the finding that blind cavefish increase their swimming velocity as they explore novel environments cannot simply be explained as fish changing their behavior to increase the sensory range of hydrodynamic imaging. This behavior may increase the signal to noise ratio of the stimulus to the lateral line in noisy environments, but this remains to be shown.

The fluid dynamics involved in hydrodynamic imaging fall in an intermediate Reynolds number range, where both viscous and inertial effects shape the form of the flow field around the fish. Modern flow quantification and modelling techniques have allowed us to explore the properties of these flow fields and estimate the stimulus they give to a fish's lateral line system. The next step in understanding more about hydrodynamic imaging is to look at how these stimuli are encoded by the lateral line and how this sensory information is then processed in the central nervous system.

References

Breder CM (1943) Problems in the behavior and evolution of a species of blind cave fish. Trans N Y Acad Sci 168–176

von Campenhausen C, Riess I, Weissert R (1981) Detection of stationary objects by the blind cave fish *Anoptichthys jordani* (Characidae). J Comp Physiol (A) 143(3):369–374

Coombs S, Montgomery JC (1999) The enigmatic lateral line system. In: Fay RR, Popper AN (eds) Comparative hearing: fish and amphibians. Springer, New York, pp 319–362

Denton EJ, Gray J (1983) Mechanical factors in the excitation of clupeid lateral lines. Proc R Soc Lond Ser B 218(1210):1–26

Dijkgraaf S (1962) Functioning and significance of lateral-line organs. Biol Rev Camb Philos Soc 38(1):51–105

Dubois AB, Cavagna GA, Fox RS (1974) Pressure distribution on body surface of swimming fish. J Exp Biol 60(3):581–591

Gregson JNS, Burt de Perera T (2007) Shoaling in eyed and blind morphs of the characin *Astyanax fasciatus* under light and dark conditions. J Fish Biol 70(5):1615–1619

Hassan ES (1985) Mathematical-analysis of the stimulus for the lateral line organ. Biol Cybern 52(1):23–36

Hassan ES (1986) On the discrimination of spatial intervals by the blind cave fish (*Anoptichthys jordani*). J Comp Physiol (A) 159(5):701–710

Hassan ES (1992a) Mathematical-description of the stimuli to the lateral line system of fish derived from a 3-dimensional flow field analysis: I the cases of moving in open water and of gliding towards a plane surface. Biol Cybern 66(5):443–452

Hassan ES (1992b) Mathematical-description of the stimuli to the lateral line system of fish derived from a 3-dimensional flow field analysis: II the case of gliding alongside or above a plane surface. Biol Cybern 66(5):453–461

Hassan ES, Abdel-Latif H, Biebricher R (1992) Studies on the effects of Ca++ and Co++ on the swimming behavior of the blind Mexican cave fish. J Comp Physiol (A) 171(3):413–419

Hubbs CL, Innes WT (1936) The first known blind fish of the family Characidae: a new genus from Mexico. Occas Paper Mus Zool Univ Mich 342:1–9

Jeffery WR (2001) Cavefish as a model system in evolutionary developmental biology. Dev Biol 231(1):1–12

Jeffery WR (2005) Adaptive evolution of eye degeneration in the Mexican blind cavefish. J Hered 96(3):185–196

Jeffery WR, Martasian DP (1998) Evolution of eye regression in the cavefish *Astyanax:* apoptosis and the pax-6 gene. Am Zool 38(4):685–696

Jeffery WR, Strickler AG, Yamamoto Y (2003) To see or not to see: evolution of eye degeneration in Mexican blind cavefish. Integr Comp Biol 43(4):531–541

John KR (1957) Observations on the behavior of blind and blinded fishes. Copeia 2:123–132

Kalmijn AJ (1988) Hydrodynamic and acoustic field detection. In: Atema J, Fay RR, Popper AN, Tavolga WN (eds) Sensory biology of aquatic animals. Springer, New York, pp 83–130

Kuiper JW (1967) Frequency characteristics and functional significance of the lateral line organ. In: Cahn PH (ed) Lateral line detectors. Indiana University Press, Bloomington, pp 105–121

Lauder GV, Tytell ED (2005) Hydrodynamics of undulatory propulsion. In: Shadwick RE, Lauder GV (eds) Fish biomechanics, vol 23. Academic Press, USA, pp 425–468

McHenry MJ, Strother JA, van Netten SM (2008) Mechanical filtering by the boundary layer and fluid-structure interaction in the superficial neuromast of the fish lateral line system. J Comp Physiol (A) 194(9):795–810

Mitchell RW, Russell WH, Elliot W (1977) Mexican eyeless characin fishes, genus *Astyanax* : Environment, distribution, and evolution. Spec Publ Mus Texas Tech Univ 12:1–89

Montgomery JC, Coombs S, Baker CF (2001) The mechanosensory lateral line system of the hypogean form of *Astyanax fasciatus*. Environ Biol Fishes 62(1–3):87–96

Parzefall J (1983) Field observation in epigean and cave populations of the Mexican characid, *Astyanax mexicanus* (Pisces, Characidae). Mem Biospeol 10:171–176

Patton P, Windsor S, Coombs S (2010) Active wall following by Mexican blind cavefish (*Astyanax mexicanus*). J Comp Physiol (A) 196(11):853–867

Burt de Perera T (2004a) Fish can encode order in their spatial map. Proc R Soc Lond Ser B 271(1553):2131–2134

Burt de Perera T (2004b) Spatial parameters encoded in the spatial map of the blind Mexican cave fish, *Astyanax fasciatus*. Anim Behav 68:291–295

Pitcher TJ, Partridge BL, Wardle CS (1976) Blind fish can school. Science 194(4268):963–965

Poulson TL (1963) Cave adaptation in amblyopsid fishes. Am Midl Nat 70(2):257–290

Romero A (1985) Ontogenetic change in phototactic responses of surface and cave populations of *Astyanax fasciatus* (Pisces, Characidae). Copeia 4:1004–1011

Russell IJ, Roberts BL (1974) Active reduction of lateral-line sensitivity in swimming dogfish. J Comp Physiol 94(1):7–15

Schemmel C (1967) Vergleichende untersuchungen an den hautsinnesorganen ober und unterirdisch lebender *Astyanax-Formen*. Z Morph Tiere 61:255–316

Schiffman HR (1996) Sensation and perception: an integrated approach, 4th edn. Wiley, New York

Sharma S, Coombs S, Patton P, Burt de Perera T (2009) The function of wall- following behaviors in the Mexican blind cavefish and a sighted relative, the Mexican tetra (*Astyanax*). J Comp Physiol (A) 195(3):225–240

Tan D (2007) Can blind cave fish compensate for decreased lateral line sensitivity when approaching an obstacle head-on. BSc honors, University of Auckland

Teghtsoonian R (1971) On the exponents in Stevens' law and the constant in Ekman's law. Psychol Rev 78(1):71–80

Teyke T (1985) Collision with and avoidance of obstacles by blind cave fish *Anoptichthys jordani* (Characidae). J Comp Physiol (A) 157(6):837–843

Teyke T (1988) Flow field, swimming velocity and boundary layer: Parameters which affect the stimulus for the lateral line organ in blind fish. J Comp Physiol (A) 163(1):53–61

Teyke T (1989) Learning and remembering the environment in the blind cave fish *Anoptichthys jordani*. J Comp Physiol (A) 164(5):655–662

Teyke T (1990) Morphological differences in neuromasts of the blind cave fish *Astyanax hubbsi* and the sighted river fish *Astyanax mexicanus*. Brain Behav Evol 35(1):23–30

Van Trump WJ, Coombs S, Duncan K, McHenry MJ (2010) Gentamicin is ototoxic to all hair cells in the fish lateral line system. Hear Res 261(1–2):42–50

Weihs D (1974) Energetic advantages of burst swimming of fish. J Theor Biol 48(1):215–229

Weissert R, von Campenhausen C (1981) Discrimination between stationary objects by the blind cave fish *Anoptichthys jordani* (Characidae). J Comp Physiol (A) 143(3):375–381

Windsor SP (2008) Hydrodynamic imaging by blind Mexican cave fish. PhD thesis, University of Auckland

Windsor SP, McHenry MJ (2009) The influence of viscous hydrodynamics on the fish lateral-line system. Integr Comp Biol 49(6):691–701

Windsor SP, Tan D, Montgomery JC (2008) Swimming kinematics and hydrodynamic imaging in the blind Mexican cave fish (*Astyanax fasciatus*). J Exp Biol 211(18):2950–2959

Windsor SP, Norris SE, Cameron SM, Mallinson GD, Montgomery JC (2010a) The flow fields involved in hydrodynamic imaging by blind Mexican cave fish (*Astyanax fasciatus*). Part I: open water and heading towards a wall. J Exp Biol 213(22):3819–3831

Windsor SP, Norris SE, Cameron SM, Mallinson GD, Montgomery JC (2010b) The flow fields involved in hydrodynamic imaging by blind Mexican cave fish (*Astyanax fasciatus*). Part II: gliding parallel to a wall. J Exp Biol 213(22):3832–3842

Windsor SP, Paris J, Burt de Perera T (2011) No role for direct touch using the pectoral fins, as an information gathering strategy in a blind fish. J Comp Physiol (A) 197(4):321–327

Wu G, Yang Y, Zeng L (2007) Kinematics, hydrodynamics and energetic advantages of burst-and-coast swimming of koi carps (*Cyprinus carpio koi*). J Exp Biol 210(12):2181–2191

Yasuda K (1973) Comparative studies on swimming behavior of blind cave fish and goldfish. Comp Biochem Physiol 45(2A):515–527

Chapter 5
Flow Sensing in Sharks: Lateral Line Contributions to Navigation and Prey Capture

Jayne M. Gardiner and Jelle Atema

Abstract Elasmobranchs (sharks, skates, and rays), like other fishes, possess a mechanosensory lateral line system that detects weak water motions. The anatomy of the lateral line system of elasmobranchs is subtly different from that of bony fishes. Found along the head and body in species-specific patterns, it is composed of neuromasts that lie in grooves or pits on the surface of the skin, as well as neuromasts that line fluid-filled subepidermal canals which may be open to the environment via pores or may be closed (nonpored). While there is a growing wealth of knowledge on lateral line function in bony fishes, comparatively less is known about the behavioral role of this sensory system in elasmobranchs. Recent research suggests that in sharks, as in bony fishes, the lateral line functions in navigation and obstacle avoidance, orientation to currents, and feeding behavior, where it contributes to prey tracking, prey localization, and capture precision.

5.1 Introduction

Like other fishes and aquatic amphibians, elasmobranchs possess a mechanosensory lateral line system that functions as a hydrodynamic detector. The lateral line system of bony fishes has been shown to function in several behaviors, such as schooling (Faucher et al. 2010; Pitcher et al. 1980; Partridge and Pitcher 1980),

J. M. Gardiner (✉)
Department of Integrative Biology, University of South Florida, 4202 E. Fowler Ave, Tampa, FL 33620, USA
e-mail: jayne@mote.org

J. M. Gardiner
Center for Shark Research, Mote Marine Laboratory, 1600 Ken Thompson Parkway, Sarasota, FL 34236, USA

J. Atema
Boston University Marine Program, 5 Cummington Mall, Boston, MA 02215, USA

H. Bleckmann et al. (eds.), *Flow Sensing in Air and Water*,
DOI: 10.1007/978-3-642-41446-6_5, © Springer-Verlag Berlin Heidelberg 2014

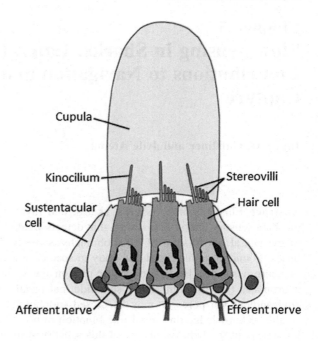

Fig. 5.1 An elasmobranch
neuromast organ (Drawn
after Liem et al. 2001;
Roberts and Ryan 1971)

predator avoidance (Blaxter and Fuiman 1989), rheotaxis (Montgomery et al.
1997), obstacle avoidance (Windsor et al. 2010a, b; Blaxter and Batty 1985), and
prey capture (Janssen et al. 1995; Schwalbe et al. 2012). In contrast, much less is
known about the behavioral functions of the lateral line system in elasmobranchs
(Gardiner et al. 2012).

5.2 Anatomy of the Shark Lateral Line

5.2.1 The Elasmobranch Neuromast

As in bony fishes, the basic unit of the elasmobranch lateral line system is the
neuromast which consists of sensory hair cells that have an eccentrically located
kinocilium and a bundle of directionally sensitive stereovilli (often erroneously
referred to as stereocilia) in a staircase arrangement at their apices (Peach and
Rouse 2000), topped by an acellular gelatinous cupula (Tester and Kendall 1968;
Fig. 5.1). The receptor cells of the neuromasts in the cephalic region are innervated by afferents and efferents from the anterior lateral line nerve (VIII) while
those of the neuromasts found along the body are innervated by the posterior
lateral line nerve (Northcutt 1978, 1989).

5.2.1.1 Superficial Neuromasts

In sharks, superficial (or free) neuromasts are found distributed along the head and body on the surface of the skin in shallow pits and are called pit organs (Figs. 5.2a, 5.3). In most sharks, they sit beneath a pair of modified scales with enlarged, thickened bases. While the anterior scale of each pair bears a ridge pattern similar to that of the adjacent scales, the posterior scale often bears a single, thickened, and rounded ridge that is more elevated than that of the adjacent scales (Tester and Nelson 1967; Fig. 5.4). In a few shark species, superficial neuromasts are situated in grooves on raised papillae, like those of skates and rays. Elasmobranch pit organs are on the order of 40–60 μm wide, 80–120 μm long, and 70–100 μm tall, from basement membrane to apex (Peach and Rouse 2000). The number of superficial neuromasts varies by species, from only a few per side in the horn shark (*Heterodontus* spp.) to over 600 per side in the scalloped hammerhead (*Sphyrna lewini*) (Tester and Nelson 1967; Peach and Marshall 2000; Peach 2003b; Fig. 5.5). In general, benthic elasmobranchs have fewer surface neuromasts than pelagic elasmobranchs (Peach and Rouse 2004). The distribution pattern of superficial neuromasts is also species-specific. In general, they are found on the dorsolateral and lateral surface of the trunk and caudal fin (dorsolateral neuromasts, Fig. 5.5a–d) and in two groups on the ventral surface, posterior to the mouth (mandibular row, Fig. 5.5e), and between the pectoral fins (umbilical row, Fig. 5.5e). The umbilical row may disappear during ontogeny in some species (Johnson 1917), while in others, it is retained in older juveniles or adults (Tester and Nelson 1967). Some species also possess one or two pairs anterior to the endolymphatic ducts on the head (supratemporal pit organs, Fig. 5.5f) and a few demersal species possess a group in the region of the eye and/or spiracle (spiracular pit organs; reviewed in Peach 2003b; Fig. 5.5g).

5.2.1.2 Canal Neuromasts

Neuromasts are also found within subepidermal fluid-filled canals (canal neuromasts, Figs. 5.2b, and 5.6). Supraorbital, infraorbital, hyomandibular, and mandibular canals are located on the head (Fig. 5.7a, b). The posterior lateral line canal extends caudally from the endolymphatic pores on the dorsal side of the head, along the flanks to the tip of the tail (Tester and Kendall 1969; Boord and Campbell 1977; Roberts 1978; Maruska 2001; Chu and Wen 1979; Fig. 5.7c). These canals may be pored and nonpored, both contain neuromasts. Pored canals are open to the exterior via neuromast-free tubules that extend to the surface of the skin and terminate in pores. In elasmobranchs, the canals are 0.3 × 0.5 mm in diameter and the canal neuromasts within are 30–50 μm wide, 150–300 μm long, and 30–50 μm tall, from basement membrane to apex (Peach and Rouse 2000). The canal neuromasts are situated adjacent to one another and the gap between them is so small that they form a nearly continuous sensory epithelium, with

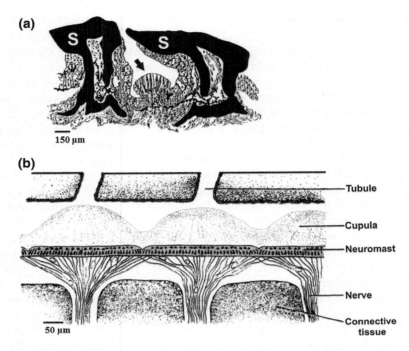

Fig. 5.2 Shark neuromasts **a** Diagram of a pit organ in the nurse shark, *Ginglymostoma cirratum*. *Arrow* indicates the neuromast, located between modified scales (*S*). Cupulae not shown (modified from Budker 1958) **b** Diagram of the neuromasts of the canal lateral line system in a silky shark, *Carcharhinus falciformis* (Modified from Tester and Kendall 1969)

multiple neuromasts between adjacent pores (Ewart and Mitchell 1892; Johnson 1917; Hama and Yamada 1977; Figs. 5.3b, 5.6a). This is in contrast to bony fish that has large gaps, on the order of 2–5 mm between canal neuromasts, and only a single neuromast between each pore (Webb and Northcutt 1997). The functional significance of this difference is as of yet unknown.

Nonpored canals are found on the head of many shark species (Maruska 2001; Chu and Wen 1979; Fig. 5.7a, b). They are isolated from the environment and therefore cannot respond to water motion but have been suggested to function as tactile receptors by responding to the velocity of skin movements that result from contact with objects in the environment such as the substrate, prey, or conspecifics during social interactions (Maruska and Tricas 2004; Maruska 2001). Subtle water pressure patterns caused by turbulence could also cause skin motion that would result in local fluid accelerations in the nonpored canals. Electrophysiological studies have shown them to be far more sensitive than cutaneous tactile receptors and responsive to low-frequency ($\leq$10 Hz) stimuli (Maruska and Tricas 2004). The number and distribution pattern of pored and nonpored canals vary among species.

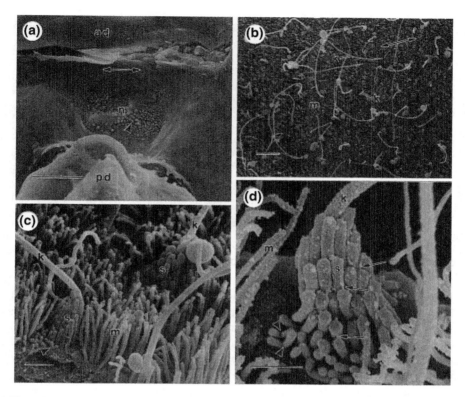

Fig. 5.3 Scanning electron micrograph (SEM) of the pit organs of the gummy shark, *Mustelus antarcticus*. **a** Surface view of a pit organ, cupula is absent and scales have been trimmed, **b** Hair cells and supporting cells, **c** Small and large stereovilli bundles, **d** Stereovilli bundle. *Double arrow* indicates directional sensitivity, *single arrow* indicate possible tip links, *arrowheads* indicate branching of stereovilli, *ad* anterior denticle, *k* kinocilium, *m* microvilli, *n* neuromast, *pd* posterior denticle, *s* stereovilli. Scale bars: **a** 50 µm; **b** 5 µm; **c, d** 1 µm (Peach and Rouse 2000)

5.2.2 Spiracular Organ

Elasmobranchs also possess a spiracular organ, a specialized organ which is not found in teleosts and is associated with the first (spiracular) gill slit. This blind tube or pouch is lined with patches of sensory epithelium, consisting of neuromasts topped with cupulae and innervated by a branch of the anterior lateral line nerve (Barry and Boord 1984; Barry et al. 1988a). It is believed to be isolated from the motion of water through the spiracle; it is relatively insensitive to electrical and vibrational stimuli, but stimulated by flexion of the hyomandibular-cranial joint. It is believed to function as a proprioceptor, though its biological role is unknown (Barry et al. 1988b).

Fig. 5.4 Modified scales associated with a pit organ. Scanning electron micrograph (SEM) of a pair of modified scales covering a pit organ in the bonnethead, *Sphyrna tiburo*. *a* anterior modified scale, *p* posterior modified scale (Reproduced from Gardiner 2012, with permission)

5.3 Lateral Line-Mediated Behaviors in Sharks

In contrast to bony fish, much less is known about the biological roles of the lateral line system in elasmobranchs. Though it has been known for some time that the shark lateral line responds to hydrodynamic motion (Sand 1937), until recently, direct evidence of its behavioral functions was lacking. In the last decade, several studies have specifically examined the role of this sensory modality, most notably in navigation and prey detection.

5.3.1 Navigation and Obstacle Avoidance

Sharks generally navigate well through both captive and natural environments. In the captive environment, healthy animals rarely collide with the tank walls, plumbing, or décor. Obstacle avoidance is mediated by vision and the lateral line system. In three species of sharks, blacktip sharks, *Carcharhinus limbatus*, bonnetheads, *Sphyrna tiburo*, and nurse sharks, *Ginglymostoma cirratum*, a significant increase in the frequency of wall collisions was observed when vision and the lateral line system were simultaneously blocked (Gardiner 2012; Fig. 5.8). A similar effect was observed in captive scalloped hammerheads, *Sphyrna lewini*, after damage to the lateral line system from a fungal infection (Crow et al. 1995). This suggests that in sharks, as in bony fishes, lateral line-mediated hydrodynamic imaging is used to detect obstacles in the environment, particularly when visual information is lacking (Windsor et al. 2008, 2010a, b, 2011; Abdel-Latif et al. 1990; Hassan 1989; Weissert and Von Campenhausen 1981; Dijkgraaf 1962). Hydrodynamic imaging involves using the flow field that is generated around the body of

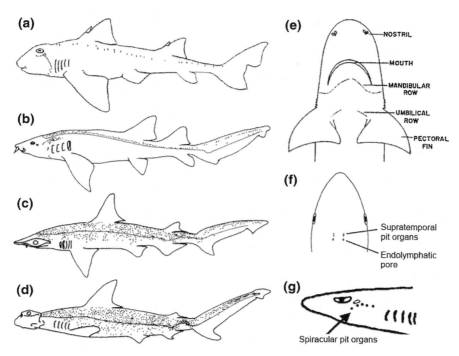

Fig. 5.5 Distribution of the superficial neuromasts (pit organs) in sharks. Each dot represents a single neuromast. Line along the flank represents the posterior lateral line canal. Superficial neuromasts on the dorsolateral surface of **a** the Port Jackson shark, *Heterodontus portusjacksoni*, **b** the nurse shark, *Ginglymostoma cirratum*, **c** the bonnethead, *Sphyrna tiburo*, and **d** the scalloped hammerhead, *Sphyrna lewini* **e** Superficial neuromasts on the ventral surface of the lemon shark, *Negaprion brevirostris* **f** Supratemporal pit organs of the smooth dogfish, *Mustelus canis*, **g** Spiracular pit organs of the Australian spotted catshark, *Asymbolus analis* (**a** modified after Peach 2001; **b-e** modified after Tester and Nelson 1967; **f** drawn after Johnson 1917; **g** modified after Peach and Rouse 2004)

a moving fish as it displaces water in front of the head (Hassan 1989). As the fish approaches an obstacle, this self-generated flow field becomes distorted and by detecting these changes using the lateral line system, the fish senses the presence of the obstacle and thereby avoids colliding with it (Hassan 1989; Windsor et al. 2010a, b). Walls exist not only in the captive environment; sharks must also successfully navigate around various natural and manmade obstacles in the natural environment. Hydrodynamic imaging would presumably be of particular importance for obstacle avoidance in species which are obligate ram-ventilators, such as bonnetheads and blacktip sharks, and therefore cannot stop swimming, even at night when visual cues may be diminished (Carlson et al. 2004).

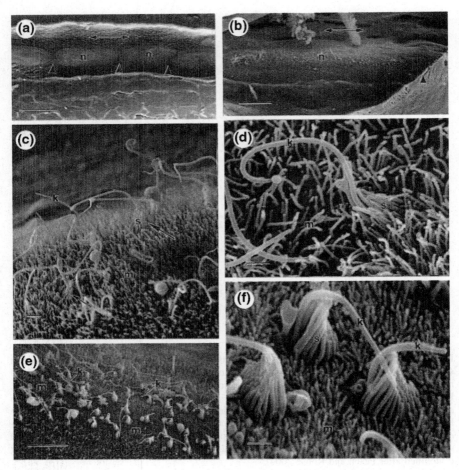

Fig. 5.6 Scanning electron micrograph (SEM) of the canal neuromasts of the gummy shark, *Mustelus antarcticus*. Neuromasts from the trunk canal (**a, c**) and the mandibular canal (**b, d, e, f**); cupulae are absent. *Double arrows* indicate directional sensitivity, *arrowheads* indicate junctions between neuromasts, *small* and *large single arrows* indicate small and large bundles of stereovilli, respectively. *k* kinocilium, *m* microvilli, *n* neuromast, *s* stereovilli. Scale bars: **a** 100 μm; **b** 50 μm; **c** 2 μm; **d** 1 μm; **e** 10 μm; **f** 1 μm (Reproduced from Peach and Rouse 2000, with permission)

5.3.2 Prey Capture

Feeding behavior in sharks can be broken down into five phases: detection, tracking, orienting, striking, and capture (Gardiner 2012). Recently, it has been shown that the lateral line plays a role in all but one phase of feeding.

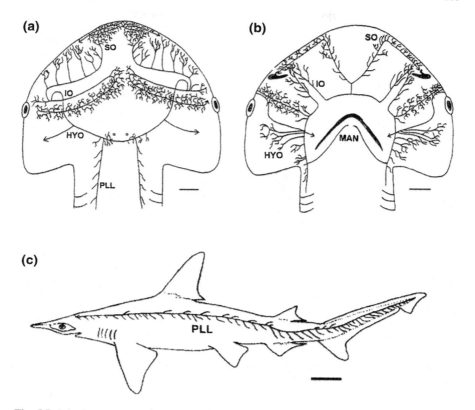

Fig. 5.7 Distribution of the lateral line canals in sharks. Canals are found on the (**a**) dorsal and (**b**) ventral sides of the head and along (**c**) the flank of the bonnethead, *Sphyrna tiburo*. *HYO* hyomandibular canal, *IO* infraorbital canal, *MAN* mandibular canal, *PLL* posterior lateral line canal, *SO* supraorbital canal (Modified after Maruska 2001)

5.3.2.1 Prey Detection

In the species examined to date, lateral line cues alone are not sufficient to alert the animals to the presence of food. Smooth dogfish, *Mustelus canis*, do not show any interest in turbulent wakes, unless they are flavored with food odor (Gardiner and Atema 2007; Fig. 5.9a, b). Blacktip sharks, *Carcharhinus limbatus*, and bonnetheads, *Sphyrna tiburo*, detect live, moving prey visually or by smell, but lateral line cues alone are not sufficient to induce feeding behavior (tracking or striking, see below) in these species (Gardiner 2012). This is in contrast to a number of species of bony fishes that can detect prey with the lateral line alone (Hoekstra and Janssen 1985; Müller and Schwartz 1982; Janssen 1997; Schwalbe et al. 2012), even orienting to and striking at an odorless source of hydrodynamic motion, such as a vibrating sphere (Hoekstra and Janssen 1986; Abdel-Latif et al. 1990; Coombs and Conley 1997; Abboud and Coombs 2000; Janssen 1990). A reliance on the lateral line for prey detection in bony fish is often found in nocturnally active fish

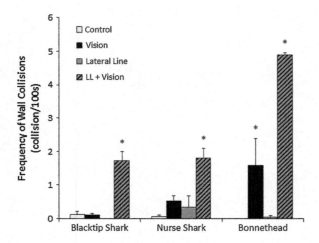

Fig. 5.8 Frequency of wall collisions after sensory blocks. The frequency of collisions with the wall, expressed as collisions/100 s, in three species of sharks, the blacktip shark, *Carcharhinus limbatus*, the nurse shark, *Ginglymostoma cirratum*, and the bonnethead, *Sphyrna tiburo*. Data are shown for animals with all senses intact (control) and following blocks of the sensory systems as indicated (*LL* lateral line). *Error bars* are ±s.e.m. The * denotes treatments that are significantly different from control at α = 0.05 (Tukey pairwise post hoc comparisons following ANOVA and Benjamini-Hochberg corrections). (Modified after Gardiner 2012, with permission)

or fish from low-light environments, such as cave fish (Yoshizawa et al. 2010), but even the nocturnal nurse shark, *Ginglymostoma cirratum*, will not strike at live, moving prey if olfactory cues are blocked (Gardiner 2012; Fig. 5.10). This suggests that sharks rely primarily on odor cues for prey detection in the dark, rather than the lateral line system. The planktonic prey consumed by the teleosts in question produces a potent, localized lateral line stimulus that matches the peak frequency sensitivity of the teleost lateral line system (Montgomery and Mac-Donald 1987, 1988). The larger crustaceans and teleosts consumed by the sharks examined in these studies generally produce wakes, predominantly composed frequencies that are slightly below the best frequency of the elasmobranch lateral line (Bleckmann et al. 1991; Maruska and Tricas 2004), that persist in the environment for several minutes after the prey has passed (Hanke and Bleckmann 2004; Hanke et al. 2000). These wakes provide good directional information, but since both biotic and abiotic wakes are everywhere in the natural environment, for sharks, odor is likely more relevant for identifying the nature of the source (Atema 1985; Gardiner and Atema 2007).

5.3.2.2 Prey Tracking

Beyond detection, however, the lateral line plays an important role in shark feeding behavior. Many species of sharks approach their prey from downstream, using tight circles and Fig. 5.8 patterns (Mathewson and Hodgson 1972; Gardiner

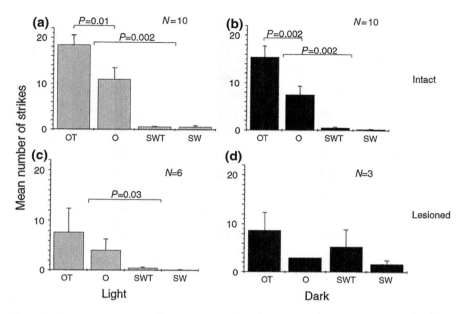

Fig. 5.9 Mean number of strikes by smooth dogfish, *Mustelus canis*, on four targets: odor with turbulence (*OT*), odor alone (*O*), seawater with turbulence (*SWT*) and seawater alone (*SW*). **a** Intact, in the light, all animals (10 out of 10, see Fig. 5.11) successfully located the targets and the animals preferred the odor side (OT + O) over the seawater side (SWT + SW; WSR = 27.5, $P = 0.002$, $N = 10$) and the source of odor with turbulence over the odor alone (WSR = 23.5, $P = 0.01$, $N = 10$). **b** Intact, in the dark, all animals (10 out of 10, see Fig. 5.11) successfully located the targets and the animals preferred the odor side over the seawater side (WSR = 27.5, $P = 0.002$, $N = 10$) and the source of odor with turbulence over the source of odor alone (WSR = 27.5, $P = 0.002$, $N = 10$). **c** Lateral line lesioned, in the light, most animals (6 out of 8, see Fig. 5.11) located the targets and the animals preferred the odor side over the seawater side (WSR = 10.5, $P = 0.03$, $N = 6$) but did not discriminate between odor/turbulence and odor alone (WSR = 5.5, $P = 0.1$, $N = 6$). **d** Lateral line lesioned, in the dark, most animals could not locate the targets and those that did (3 out of 8, see Fig. 5.11) did not display a preference for any of the targets (WSR = 3.0, $P = 0.2$, $N = 3$). *Gray bars* fluorescent light (*light*), *black bars* infrared light (*dark*) (Reproduced from Gardiner and Atema 2007, with permission)

and Atema 2007; Kleerekoper 1978, 1982; Hobson 1963; Hodgson and Math-ewson 1971; Gardiner 2012). Tracking behavior is, therefore, characterized by frequent, rapid turns, as the animal moves upstream. This behavior requires the simultaneous use of olfactory and hydrodynamic cues (Gardiner and Atema 2007; Gardiner 2012; Fig. 5.11). Odor motivates tracking behavior; when odor cues are lacking, turns are slow and infrequent (Figs. 5.12, 5.13). Upstream movement is accomplished in one of two ways. Sharks can detect the bulk flow and navigate upstream (rheotaxis) to the vicinity of the source, at which point other sensory cues will prompt them to strike (see Orientation, Striking, and Capture, below), or they can detect and precisely track the odor-flavored eddies of a source-directed wake (odor plume) to its source through eddy chemotaxis (Gardiner and Atema 2007).

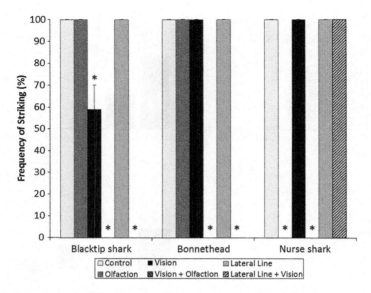

Fig. 5.10 Frequency of striking behavior. The frequency of striking behavior in three species of sharks the blacktip shark, *Carcharhinus limbatus*, the bonnethead, *Sphyrna tiburo*, and the nurse shark, *Ginglymostoma cirratum*, feeding on live prey with all senses intact (control) and following blocks of the sensory systems indicated in the figure legend. *Error bars* are ±s.e.m. * denotes treatments that are significantly different from control at $\alpha = 0.05$ (Modified after Gardiner 2012, with permission)

As in teleosts (Montgomery et al. 1997; Lyon 1904; Chagnaud et al. 2008), sharks can determine the direction of the bulk flow using the lateral line or vision (visual flow field) (Gardiner and Atema 2007; Peach 2001). In epibenthic shark species, when vision and the lateral line are simultaneously blocked, turns are rapid, but they are infrequent, and the animals are unable to reach the source (Figs. 5.11, 5.12, 5.13). Some benthic species, such as the epaulette shark, *Hemiscyllium ocellatum*, have been suggested to be capable of orienting to the bulk flow using touch (Peach 2003a), as has been described for bony fishes (Arnold 1974; Baker and Montgomery 1999). The nurse shark, *Ginglymostoma cirratum*, maintains contact with the bottom while tracking, occasionally using its pectoral fins to propel itself (Moss 1972; Limbaugh 1963), and appears to be capable of tracking using olfaction in combination with touch (Figs. 5.12, 5.13). It will successfully reach the prey in the absence of vision and the lateral line (Fig. 5.10), but the process is much slower (Fig. 5.14). Precisely tracking the flavored eddies contained within an odor plume (eddy chemotaxis; Atema 1996), such as the wakes left behind by swimming prey (Hanke and Bleckmann 2004; Hanke et al. 2000), to the source requires the lateral line (Gardiner and Atema 2007; Gardiner 2012; Figs. 5.9, 5.11).

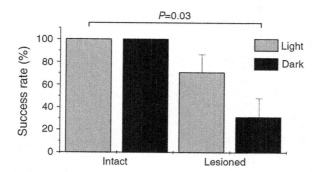

Fig. 5.11 Success rate (%) of smooth dogfish, *Mustelus canis*, in locating the source of turbulent odor plumes under four experimental conditions: intact and lateral line lesioned with streptomycin in the light and in the dark. Lateral line lesion reduced success rate: nonsignificant in the light (WSR = 3.0, $N = 8$, $P = 0.2$) and significant in the dark (WSR = 10.5, $P = 0.03$, $N = 8$). Lighting alone did not affect success rate (WSR = 0, $N = 10$, $P = 1$). *Gray bars* fluorescent light (*light*), *black bars* infrared light (*dark*) (Reproduced from Gardiner and Atema 2007, with permission)

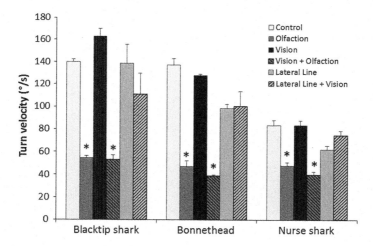

Fig. 5.12 Turn velocity during tracking. Turn velocity (°/s) during tracking of live prey in three species of sharks, the blacktip shark, *Carcharhinus limbatus*, the bonnethead, *Sphyrna tiburo*, and the nurse shark, *Ginglymostoma cirratum*, in animals with all senses intact (control) and after blocking the sensory systems indicated in the figure legend. *Error bars* ±s.e.m. * denotes treatments that are significantly different from control at $\alpha = 0.05$ (Modified after Gardiner 2012, with permission)

5.3.2.3 Orientation, Striking, and Capture

Orienting and striking are visually mediated in many sharks (Fouts and Nelson 1999; Gilbert 1963; Gardiner 2012). The long-distance (2 body lengths), rapid strikes (2 body lengths/s) performed by blacktip sharks, *Carcharhinus limbatus*, as

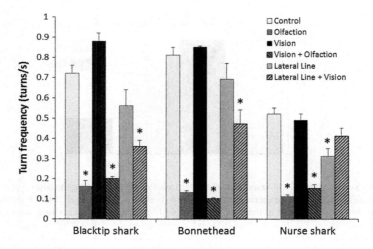

Fig. 5.13 Turn frequency during tracking. Turn frequency (turns/s) during tracking of live prey in three species of sharks, the blacktip shark, *Carcharhinus limbatus*, the bonnethead, *Sphyrna tiburo*, and the nurse shark, *Ginglymostoma cirratum*, in animals with all senses intact (control) and after blocking the sensory systems indicated in the figure legend. *Error bars* ±s.e.m. * denotes treatments that are significantly different from control at $\alpha = 0.05$ (Modified after Gardiner 2012, with permission)

Fig. 5.14 Search time. Time (s) spent tracking live prey by nurse sharks, *Ginglymostoma cirratum*, feeding on live prey with all senses intact (control) and after blocking the sensory systems indicated in the figure legend. *Error bars* ±s.e.m. * denotes treatments that are significantly different from control at $\alpha = 0.05$ (Modified after Gardiner 2012, with permission)

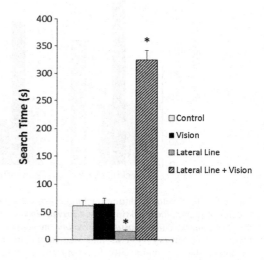

they chase down elusive midwater teleost prey are visually mediated (Gardiner 2012), as they are in other ram-feeding fishes, such as largemouth bass, *Micropterus salmoides* (Gardiner and Motta 2012). The ram-feeding strategy requires a predator to precisely locate their prey from a distance of several body lengths, in order to have sufficient room to accelerate. However, in the absence of vision, the blacktip shark can use lateral line cues to orient and strike, though from a much closer

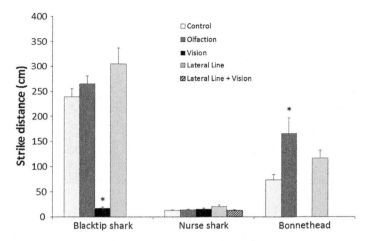

Fig. 5.15 Strike distance. Strike distance (cm) of three species of sharks feeding on live prey, the blacktip shark, *Carcharhinus limbatus*, the bonnethead, *Sphyrna tiburo*, and the nurse shark, *Ginglymostoma cirratum*, with all senses intact (control) and after blocking the sensory systems indicated in the figure legend. *Error bars* ±s.e.m. * denotes treatments that are significantly different from control at $\alpha = 0.05$ (Modified after Gardiner 2012, with permission)

proximity, less than half a body length, similar to the largemouth bass (Gardiner 2012); Gardiner and Motta 2012; Fig. 5.15). Ram-feeding fishes typically decelerate just prior to beginning to move the jaws to capture prey, which has been suggested to increase capture accuracy (Higham et al. 2005, 2006). When the lateral line system is disabled, both largemouth bass and blacktip sharks fail to brake prior to capture, and at the start of capture, they position their mouth at greater angles to prey (Gardiner 2012; Gardiner and Motta 2012), which decreases capture efficiency (Higham et al. 2005). In the blacktip shark, high velocity strikes tend to result in the predator missing the prey; if however, the strikes begin with a lower velocity, captures are typically successful, despite the failure to brake (Gardiner 2012; Fig. 5.16). With laterally positioned eyes, sharks possess a frontal binocular overlap, but also a blind area directly in front of the head (McComb et al. 2009). Therefore, the final moments just prior to capture are blind. The lateral line and electrosensory systems function to fill this void. In the blacktip shark, the lateral line functions to fine-tune and provide precision to the final moments of their very rapid strikes (Gardiner 2012). This would be particularly useful to counter any prey escape responses. In slower-striking ram-feeding species, such as the bonnethead (0.75 body lengths/s) (Gardiner 2012), strikes are still successful without lateral line input. This may be because the bow wave, generated in front of the head, which displaces the prey and alerts it to the approach of the predator, prompting an escape response is smaller with slower strikes (Ferry-Graham et al. 2003). In teleosts, this has been suggested to decrease the reaction distance of the prey (Viitasalo et al. 1998). Lateral line-mediated strike adjustment may, therefore, not be as critical to strike precision in slower-striking species.

Fig. 5.16 Strike velocity of successful captures versus misses in lateral line-blocked blacktip sharks, *Carcharhinus* limbatus. *Error bars* are ±s.e.m. * denotes significant differences at α = 0.05 (Modified after Gardiner 2012, with permission)

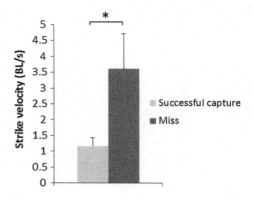

In suction-feeding species, such as the nurse shark, *Ginglymostoma cirratum*, strikes occur from a closer proximity (Fig. 5.15), as suction feeding is only effective over a distance of a few centimeters (Nauwelaerts et al. 2007; Motta et al. 2002, 2008; Wilga and Sanford 2008; Lowry and Motta 2007). As a result of the very close proximity of the strikes, this species can also use electroreception to orient their strike and elimination of the lateral line system does not impact this behavior (Gardiner 2012). Since suction-feeding species use little to no forward motion (Motta et al. 2008; Lowry and Motta 2007), the prey is primarily alerted to the presence of the predator by the suction flow (Holzman and Wainwright 2009), which is often too late for a successful escape response.

5.4 Summary and Conclusions

Based on the research conducted to date, the behavioral role of the lateral line system in sharks appears to be similar to that of in teleosts: functioning in hydrodynamic imaging and obstacle avoidance, playing an alternating role with vision in rheotaxis, mediating wake tracking, contributing to orientation and striking, particularly in the dark, and in ram-feeding species, mediating braking just prior to capture and thereby contributing to capture precision. Though recently we have made much progress in understanding the behavioral functions of the lateral line system in sharks, many questions still remain. In particular, the functional significance of the anatomical differences between the lateral line system of elasmobranchs, especially sharks, and that of bony fishes remains unknown, and this remains an area that is ripe for future research.

Acknowledgments We thank Robert Hueter and Philip Motta for helpful comments on the manuscript. Aspects of this work were supported by a University of South Florida Presidential Doctoral Fellowship, a Lerner-Gray Grant for Marine Research, an American Elasmobranch Society Donald R. Nelson Behavioral Research Award, and an American Society of

Ichthyologists and Herpetologists Raney Fund Award to JMG, a collaborative National Science Foundation grant (IOS-0843440, IOS-0841478, and IOS-081502), and support from the Porter Family Foundation. This manuscript was prepared during JMG's tenure as a Mote Postdoctoral Fellow.

References

Abboud JL, Coombs S (2000) Mechanosensory-based orientation to elevated prey by a benthic fish. Mar Freshw Behav Physiol 33:261–279

Abdel-Latif H, Hassan ES, von Campenhausen C (1990) Sensory performance of blind Mexican cave fish after destruction of the canal neuromasts. Naturwissenschaften 77:237–239

Arnold GP (1974) Rheotropism in fishes. Biol Rev 49:515–576

Atema J (1985) Chemoreception in the sea: adaptations of chemoreceptors and behaviour to aquatic stimulus conditions. Symp Soc Exp Biol 39:386–423

Atema J (1996) Eddy chemotaxis and odor landscapes: exploration of nature with animal sensors. Biol Bull 191:129–138

Baker CF, Montgomery JC (1999) The sensory basis of rheotaxis in the blind Mexican cave fish, *Astyanax fasciatus*. J Comp Physiol A 184:519–527

Barry MA, Boord RL (1984) The spiracular organ of sharks and skates: anatomical evidence indicating a mechanoreceptive role. Science 226:990–992

Barry MA, White RL, Bennett MVL (1988a) The elasmobranch spiracular organ I. Morphological studies. J Comp Physiol A 163:85–92

Barry MA, White RL, Bennett MVL (1988b) The elasmobranch spiracular organ II. Physiological studies. J Comp Physiol A 163:93–98

Blaxter JHS, Batty RS (1985) Herring behaviour in the dark: responses to stationary and continuously vibrating obstacles. J Mar Biol Assoc UK 65:1031–1049

Blaxter JHS, Fuiman LA (1989) Function of the free neuromasts of marine teleost larvae. In: Coombs S, Gorner P, Munz H (eds) The mechanosensory lateral line: neurobiology and evolution. Springer, New York, pp 481–499

Bleckmann H, Breithaupt T, Blickhan R, Tautz J (1991) The time course and frequency content of hydrodynamic events caused by moving fish, frogs and crustaceans. J Comp Physiol A 168:749–757

Boord RL, Campbell CBG (1977) Structural and functional organization of the lateral line system of sharks. Am Zool 17:431–441

Budker P (1958) Les organes sensoriels cutanés des sélaciens. In: Grassé PP (ed) Traité de Zoologie. Anatomie, Systémique, Biologie, vol Tome XIII. Agnathes et Poissons. Masson et Cie, Paris, pp 1033–1062

Carlson JK, Goldman KJ, Lowe CG (2004) Metabolism, energetic demand, and endothermy. In: Carrier JC, Musick JA, Heithaus MR (eds) Biology of sharks and their relatives. CRC Press, Boca Raton

Chagnaud BP, Brücker C, Hofmann MH, Bleckmann H (2008) Measuring flow velocity and flow direction by spatial and temporal analysis of flow fluctuations. J Neurosci 28:4479–4487

Chu YT, Wen MC (1979) Monograph of fishes of China (No. 2): a study of the lateral-line canals system and that of Lorenzini ampullae and tubules of elasmobranchiate fishes of China. Science and Technology Press, Shanghai

Coombs S, Conley RA (1997) Dipole source localization by mottled sculpin. I. Approach strategies. J Comp Physiol A 180:387–399

Crow GL, Brock JA, Kaiser S (1995) *Fusarium solani* infection of the lateral line canal system in captive scalloped hammerhead sharks (*Sphyrna lewini*) in Hawaii. J Wildl Dis 31:562–565

Dijkgraaf S (1962) The functioning and significance of the lateral-line organs. Biol Rev 28:51–105

Ewart JC, Mitchell HC (1892) On the lateral sense organs of elasmobranchs. II. The sensory canals of the common skate (*Raja batis*). Trans R Soc Edinb 37:87–105

Faucher K, Parmentier E, Becco C, Vandewalle N, Vandewalle P (2010) Fish lateral system is required for accurate control of shoaling behaviour. Anim Behav 79:679–687

Ferry-Graham LA, Wainwright PC, Lauder GV (2003) Quantification of flow during suction feeding in bluegill sunfishes. Zoology 106:159–168

Fouts WR, Nelson DR (1999) Prey capture by the Pacific angle shark, *Squatina californica*: visually mediated strikes and ambush-site characteristics. Copeia 1999:304–312

Gardiner JM (2012) Multisensory Integration in Shark Feeding Behavior. Dissertation, University of South Florida, Tampa

Gardiner JM, Atema J (2007) Sharks need the lateral line to locate odor sources: rheotaxis and eddy chemotaxis. J Exp Biol 210:1925–1934

Gardiner JM, Hueter RE, Maruska KP, Sisneros JA, Casper BM, Mann DA, Demski LS (2012) Sensory physiology and behavior of elasmobranchs. In: Carrier JC, Musick JA, Heithaus MR (eds) Biology of sharks and their relatives, vol. I, 2nd edn. CRC Press, Boca Raton, pp 349–401

Gardiner JM, Motta PJ (2012) Largemouth bass (*Micropterus salmoides*) switch feeding modalities in response to sensory deprivation. Zoology 115:78–83

Gilbert PW (1963) The visual apparatus of sharks. In: Gilbert PW (ed) Sharks and survival. DC Heath & Co, Boston, pp 283–326

Hama K, Yamada Y (1977) Fine structure of the ordinary lateral line organ. 2. The lateral line canal organ of spotted shark, Mustelus manazo. Cell Tiss Res 176:23–36

Hanke W, Bleckmann H (2004) The hydrodynamic trails of *Lepomis gibbosus* (Centrarchidae), *Colomesus psittacus* (Tetraodontidae) and *Thysochromis ansorgii* (Cichlidae) investigated with scanning particle image velocimetry. J Exp Biol 207:1585–1596

Hanke W, Brücker C, Bleckmann H (2000) The ageing of the low frequency water disturbances caused by swimming goldfish and its possible relevance to prey detection. J Exp Biol 203:1193–2000

Hassan ES (1989) Hydrodynamic imaging of the surroundings by the lateral line of the blind cave fish *Anoptichthys jordani*. In: Coombs S, Gorner P, Munz H (eds) The mechanosensory lateral line: neurobiology and evolution. Springer, New York, pp 217–228

Higham TE, Day SW, Wainwright PC (2005) Sucking while swimming: evaluating the effects of ram speed on suction generation in bluegill sunfish *Lepomis macrochirus* using digital particle image velocimetry. J Exp Biol 208:2653–2660

Higham TE, Day SW, Wainwright PC (2006) Multidimensional analysis of suction feeding performance in fishes: fluid speed, acceleration, strike accuracy, and the ingested volume of water. J Exp Biol 209:2713–2725. doi:10.1242/jeb.02315

Hobson ES (1963) Feeding behavior in three species of sharks. Pac Sci 17:171–194

Hodgson ES, Mathewson RF (1971) Chemosensory orientation in sharks. Ann N Y Acad Sci 188:175–182

Hoekstra D, Janssen J (1985) Non-visual feeding behaviour of the mottled sculpin, *Cottus bairdi*, in Lake Michigan. Environ Biol Fishes 12:111–117

Hoekstra D, Janssen J (1986) Lateral line receptivity in the mottled sculpin (*Cottus bairdi*). Copeia 1986:91–96

Holzman R, Wainwright PC (2009) How to surprise a copepod: strike kinematics reduce hydrodynamic disturbance and increase stealth of suction-feeding fish. Limnol Oceanogr 54:2201–2212

Janssen J (1990) Localization of substrate vibrations by the mottled sculpin (*Cottus bairdi*). Copeia 1990:349–355

Janssen J (1997) Comparison of response distance to prey via the lateral line in the ruffe and yellow perch. J Fish Biol 51:921–930

Janssen J, Jones WR, Whang A, Oshel PE (1995) Use of the lateral line in particulate feeding in the dark by juvenile alewife (*Alosa pseudoharengus*). Can J Fish Aqua Sci 52:358–363

Johnson SE (1917) Structure and development of the sense organs of the lateral canal system of selachians (*Mustelus canis* and *Squalus acanthias*). J Comp Neurobiol 28:1–74

Kleerekoper H (1978) Chemoreception and its interaction with flow and light perception in the locomotion and orientation of some elasmobranchs. In: Hodgson ES, Mathewson RF (eds) Sensory biology of sharks, skates, and rays. U.S. Office of Naval Research, Arlington

Kleerekoper H (1982) The role of olfaction in the orientation of fishes. In: Hara TJ (ed) Chemoreception in fishes: developments in aquaculture and fisheries science. Elsevier, Amsterdam, pp 201–225

Liem KF, Bemis WE, Walker J, W.F., Grande L (2001) Functional Anatomy of the Vertebrates: an Evolutionary Perspective. Harcourt College Publishers, New York

Limbaugh C (1963) Field Notes on Sharks. In: Gilbert PW (ed) Sharks and Survival. DC Heath and Company, Lexington, pp 63–94

Lowry D, Motta PJ (2007) Ontogeny of feeding and cranial morphology in the whitespotted bambooshark *Chiloscyllium plagiosum*. Mar Biol 151:2013–2023

Lyon EP (1904) On rheotropism. I. Rheotropism in fishes. Am J Physiol 12:149–161

Maruska KP (2001) Morphology of the mechanosensory lateral line system in elasmobranch fishes: ecological and behavioral considerations. Environ Biol Fishes 60:47–75

Maruska KP, Tricas TC (2004) Test of the mechanotactile hypothesis: neuromast morphology and response dynamics of mechanosensory lateral line primary afferents in the stingray. J Exp Biol 207:3463–3476

Mathewson RF, Hodgson ES (1972) Klinotaxis and rheotaxis in orientation of sharks toward chemical stimuli. Comp Biochem Physiol 42:79–84

McComb DM, Tricas TC, Kajiura SM (2009) Enhanced visual fields in hammerhead sharks. J Exp Biol 212:4010–4018

Montgomery JC, Baker CF, Carton AG (1997) The lateral line can mediate rheotaxis in fish. Nature 389:960–963

Montgomery JC, MacDonald JA (1987) Sensory tuning of lateral line receptors in antarctic fish to the movements of planktonic prey. Science 235:195–196

Montgomery JC, MacDonald JA (1988) Lateral line function in antarctic fish related to the signals produced by planktonic prey. J Comp Physiol A 163:827–833

Moss SA (1972) Nurse shark pectoral fins: an unusual use. Am Midl Nat 88:496–497

Motta PJ, Hueter RE, Tricas TC, Summers AP (2002) Kinematic analysis of suction feeding in the nurse shark *Ginglymostoma cirratum* (Orectolobiformes, Ginglymostomidae). Copeia 2002:24–38

Motta PJ, Hueter RE, Tricas TC, Summers AP, Huber DR, Lowry D, Mara KR, Matott MP, Whitenack LB, Wintzer AP (2008) Functional morphology of the feeding apparatus, feeding constraint and suction performance in the nurse shark *Ginglymostoma cirratum*. J Morphol 269:1041–1055

Müller U, Schwartz E (1982) Influence of single neuromasts on the prey localizing behavior of the surface feeding fish *Aplocheilus lineatus*. J Comp Physiol A 149:399–408

Nauwelaerts S, Wilga CD, Sanford CP, Lauder GV (2007) Hydrodynamics of prey capture in sharks: effects of substrate. J R Soc Interface 4:341–345

Northcutt RG (1978) Brain organization in the cartilaginous fishes. In: Hodgson ES, Mathewson RF (eds) Sensory biology of sharks, skates, and rays. Office of Naval Research Department of the Navy, Arlington, pp 117–193

Northcutt RG (1989) The phylogenetic distribution and innervation of craniate mechanoreceptive lateral lines. In: Coombs S, Görner P, Münz H (eds) The mechanosensory lateral line: neurobiology and evolution. Springer, New York, pp 17–78

Partridge BL, Pitcher TJ (1980) The sensory basis of fish schools: relative roles of lateral line and vision. J Comp Physiol A 135:315–325

Peach MB (2001) The dorso-lateral pit organs of the Port Jackson shark contribute sensory information for rheotaxis. J Fish Biol 59:696–704

Peach MB (2003a) The behavioral role of pit organs in the epaulette shark. J Fish Biol 62:793–802

Peach MB (2003b) Inter- and intraspecific variation in the distribution and number of pit organs (free neuromasts) of sharks and rays. J Morphol 256:89–102

Peach MB, Marshall NJ (2000) The pit organs of elasmobranchs: a review. Philos Trans R Soc Lond B Biol Sci 355:1131–1134

Peach MB, Rouse GW (2000) The morphology of the pit organs and lateral line canal neuromasts of *Mustelus antarcticus* (Chondrichthyes: Triakidae). J Mar Biol Assoc UK 80:155–162

Peach MB, Rouse GW (2004) Phylogenetic trends in the abundance and distribution of pit organs of elasmobranchs. Acta Zool 85:233–244

Pitcher TJ, Partridge BL, Wardle CS (1980) A blind fish can school. Science 194:963–965

Roberts BL (1978) Mechanoreceptors and the behavior of elasmobranch fishes with special reference to the acoustico-lateralis system. In: Hodgson ES, Mathewson RF (eds) Sensory biology of sharks, skates, and rays. U.S. Office of Naval Research, Arlington, pp 331–390

Roberts BL, Ryan KP (1971) The fine structure of the lateral-line sense organs of dogfish. Proc R Soc Lond B Biol Sci 179:157–169

Sand A (1937) The mechanism of the lateral sense organs of fishes. Proc R Soc Lond B Biol Sci 23:472–495

Schwalbe MAB, Bassett DK, Webb JF (2012) Feeding in the dark: lateral-line-mediated prey detection in the peacock cichlid *Aulonocara stuartgranti*. J Exp Biol 215:2060–2071

Tester AL, Kendall JI (1968) Cupulae in shark neuromasts: composition, origin, generation. Science 160:772–774

Tester AL, Kendall JI (1969) Morphology of the lateralis canal system in shark genus *Carcharhinus*. Pac Sci 23:1–16

Tester AL, Nelson GJ (1967) Free neuromasts (pit organs) in sharks. In: Gilbert PW, Mathewson RF, Rall DP (eds) Sharks, skates, and rays. John Hopkins Press, Baltimore, pp 503–531

Viitasalo M, Kiørboe T, Flinkman J, Pedersen LW, Visser AW (1998) Predation vulnerability of planktonic copepods: consequences of predator foraging strategies and prey sensory abilities. Mar Ecol Prog Ser 175:129–142

Webb JF, Northcutt RG (1997) Morphology and distribution of pit organs and canal neuromasts in non-teleost bony fishes. Brain Behav Evol 50:139–151

Weissert R, Von Campenhausen C (1981) Discrimination between stationary objects by the blind cave fish *Anoptichthys jordani* (Characidae). J Comp Physiol A 143:375–381

Wilga CD, Sanford CP (2008) Suction generation in white-spotted bamboo sharks *Chiloscyllium plagiosum*. J Exp Biol 211:3128–3138

Windsor SP, Norris SE, Cameron SM, Mallinson GD, Montgomery JC (2010a) The flow fields involved in hydrodynamic imaging by blind Mexican cave fish (*Astyanax fasciatus*). Part I: open water and heading towards a wall. J Exp Biol 213:3819–3831

Windsor SP, Norris SE, Cameron SM, Mallinson GD, Montgomery JC (2010b) The flow fields involved in hydrodynamic imaging by blind Mexican cave fish (*Astyanax fasciatus*). Part II: gliding parallel to a wall. J Exp Biol 213:3832–3842

Windsor SP, Paris J, De Parera TB (2011) No role for direct touch using the pectoral fins, as an information gathering strategy in a blind fish. J Comp Physiol A 197:321–327

Windsor SP, Tan D, Montgomery JC (2008) Swimming kinematics and hydrodynamic imaging in the blind Mexican cave fish (*Astyanax fasciatus*). J Exp Biol 211:2950–2959

Yoshizawa M, Goricki S, Soares D, Jeffery WR (2010) Evolution of a behavioral shift mediated by superficial neuromasts helps cavefish find food in darkness. Curr Biol 20:1631–1636

Chapter 6
Hydrodynamic Perception in Seals and Sea Lions

Guido Dehnhardt, Wolf Hanke, Sven Wieskotten, Yvonne Krüger and Lars Miersch

Abstract Marine mammals often forage in dark and turbid waters. While dolphins use echolocation under such conditions, pinnipeds seem to lack this sensory system. Instead, species of the families Phocidae (true seals) and Otariidae (eared seals) both possess richly innervated whiskers (synonymously "vibrissae") representing highly sensitive hydrodynamic receptors that enable these animals to detect fish-generated water movements. The third family of pinnipeds, the Odobenidae (walruses), is less well studied. As water movements in the wake of fishes persist for several minutes, they constitute hydrodynamic trails that should be trackable by piscivorous predators. Hydrodynamic trail following has indeed been shown for the harbor seal (*Phoca vitulina*) and the California sea lion (*Zalophus californianus*). However, in experiments with a sea lion aging of the trails resulted in an earlier decrease in performance. This difference in tracking performance most likely is due to differences in the structure of the respective vibrissal hair shaft. In the harbor seal the high sensitivity and excellent tracking performance is ascribed to the specialized undulated structure of the whiskers that largely suppresses self-generated noise in the actively moving animal. In contrast, the whiskers of a swimming California sea lion, which are smooth in outline, are substantially affected by self-generated noise. However, in the sea lion such self-generated noise contains a characteristic carrier frequency that might allow hydrodynamic reception by being modulated in response to hydrodynamic stimuli impinging on the hair. Thus, in the course of pinniped evolution at least two types of whiskers evolved that realized different mechanisms for the reception of external hydrodynamic information.

G. Dehnhardt (✉) · W. Hanke · S. Wieskotten · Y. Krüger · L. Miersch
University of Rostock, Biosciences, Sensory and Cognitive Ecology,
Albert-Einstein-Strasse 3, 18059 Rostock, Germany
e-mail: guido.dehnhardt@uni-rostock.de

H. Bleckmann et al. (eds.), *Flow Sensing in Air and Water*,
DOI: 10.1007/978-3-642-41446-6_6, © Springer-Verlag Berlin Heidelberg 2014

Abbreviations

F-SC   Follicle sinus complex
PIV   Particle image velocimetry
VRGs   Vortex ring generators
HMC   Head-mounted camera
CCD   Charge-coupled device
VIVs   Vortex-induced vibrations
SNR   Signal-to-noise ratio

6.1 Introduction

Most marine mammals must search for food under highly variable conditions requiring corresponding sensory specializations. For the detection of distant objects in dark and/or murky waters odontocetes have developed an echolocation system accompanied by exceptional hearing skills at high frequencies (Au 1993). In pinnipeds no sonar system like that of toothed whales has been demonstrated so far, although a sensory modality allowing the localization of prey in the dark has long been debated for these marine mammals.

For decades vision has been suggested to be the predominant source of sensory information in foraging pinnipeds (Walls 1942; Hobson 1966; Lavigne et al. 1977; Levenson and Schusterman 1999). Indeed, the eyes of the pinniped species studied so far are highly adapted for under water vision and to low light intensities encountered so often in the aquatic environment (Hanke et al. 2006a, b, 2008a, b, 2009a, b, 2011; Scholtyssek et al. 2008). A marine mammal relying primarily on vision strongly depends on the transparency of the water. Therefore turbidity is an important ecological factor in aquatic habitats, especially for visual underwater object detection (Aksnes and Giske 1993; Aksnes and Utne 1997). Accordingly, psychophysical experiments indicated a dramatic loss of visual acuity in harbor seals (*Phoca vitulina*) even at moderate levels of turbidity (Fig. 6.1) so that visual object localization is hardly possible under such conditions (Weiffen et al. 2006).

As many fish species produce sounds (see, e.g., Popper and Fay 1993; Akamatsu et al. 2002; Wahlberg and Westerberg 2003; Wilson et al. 2004) pinnipeds searching for pelagic fish may benefit from this acoustic information for prey localization. The high sound localization acuity, particularly with regard to frequencies below 1 kHz, as determined in the California sea lion (*Zalophus californianus*) and the harbor seal (*P. vitulina*), may facilitate the localization of a soniferous fish from a distance (Møhl 1964, 1968; Gentry 1967; Terhune 1974; Moore and Au 1975; Kastak and Schusterman 1998; Bodson et al. 2006, 2007). However, as such signals are generally not continuously produced by a fish, this source of information should not reliably allow a predator to pinpoint its prey.

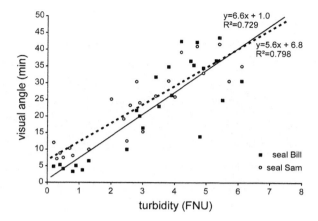

Fig. 6.1 Visual acuity angles of two harbor seals as a function of water turbidity. Sam *dashed line*, Bill *solid line*. Adapted from Weiffen et al. (2006)

In this paper we provide evidence that pinniped whiskers provide the sensory information for hydrodynamic perception and therewith allow the detection of a swimming fish at a considerable distance as well as its pursuit.

6.2 Morphology and Innervation of Pinniped Whiskers

Pinniped whiskers are generally well developed. They are large in size, have a conspicuous structure, and the follicle sinus complex (F-SC) of each whisker is highly innervated. The term F-SC is derived from the complex sinus system (whiskers are also called "sinus hairs"), a characteristic feature of vibrissal follicles (a third term for these tactile hairs: "vibrissae") not found in normal body hairs (Stephens et al. 1973; Hyvärinen and Katajisto 1984; Hyvärinen 1989; Marshall et al. 2006).

Although the structure of the sinus system varies considerably across mammalian species, F-SCs of many terrestrial mammals show a ring sinus and a cavernous sinus below it (see, e.g., Ebara et al. 2002). In contrast, pinnipeds seem to be the only group of mammals possessing an additional cavernous sinus situated above the ring sinus (tripartite sinus system, Hyvärinen 1989). In adaptation to the high thermal conductivity and the large potential cooling power of the aquatic environment, this additional and extraordinarily long upper cavernous sinus (60 % of the total follicle length in phocid seals Fig. 6.2) serves the thermoregulation of the system (Dehnhardt et al. 1998a; Mauck et al. 2000) that is essential for the reception of mechanosensory information. Although negative effects of cooling have been described at the receptor level (Bolanowski et al. 1988; Gescheider et al. 1997), increasing stiffness of the F-SC surrounding tissue might be even more important by affecting the transduction of mechanical stimulation via the hair shaft to the main receptor level at and below the ring sinus. However, while seals have been shown to cool down their outermost tissue layers close to ambient

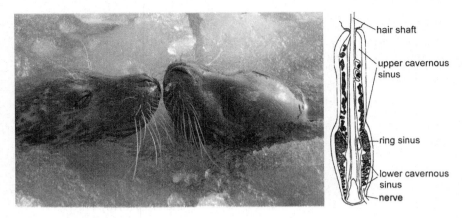

Fig. 6.2 Harbor seals use their mystacial vibrissae in a cold environment. The drawing on the *right* shows a longitudinal section of a vibrissal follicle of *Phoca hispida* (adapted from Dehnhardt et al. 1998a)

temperature even when exposed to very cold conditions (Irving 1969; Folkow and Blix 1989; Hokkanen 1990; Worthy 1991; Watts et al. 1993; Ryg et al. 1993; Kvadsheim et al. 1997) infrared-thermography demonstrated that the respective pads of mystacial and supraorbital whiskers can show thermal emissions substantially higher than in adjacent skin areas (Fig. 6.3). In phocid seals this regional thermoregulation of vibrissal follicles is accompanied by an excess of low-melting-point monoenoic fatty acids in the adipose tissue surrounding the follicles so that an optimal tissue flexibility and thus mobility of the whiskers inside the follicle should be realized under cold conditions (Käkelä and Hyvärinen 1993, 1996). In accordance with this sophisticated morphological organization of the vibrissal system, psychophysical experiments with harbor seals have shown that their high haptic vibrissal sensitivity for texture differences at water temperatures of about 20 °C remained unaltered even at temperatures close to 0 °C (Dehnhardt et al. 1998a). However, the follicle adaptations that allow active touch to work in a thermally hostile environment may also facilitate hydrodynamic reception.

Compared to terrestrial mammals the degree of innervation found in pinniped F-SCs is also outstanding. Each follicle is innervated by a single "deep vibrissal nerve" of the infraorbital branch of the trigeminal nerve that enters the capsule surrounding the follicle at its base. At the point where the nerve enters the capsule, 1,000–1,600 myelinated axons have been counted in the ringed seal (*Phoca hispida*, Hyvärinen and Katajisto 1984; Hyvärinen 1989) and the bearded seal (*Erignathus barbatus*, Marshall et al. 2006). Thus, the innervation density of vibrissal follicles of these pinniped species exceeds that calculated for richly innervated terrestrial species (Rice et al. 1986) by a factor of 10. Calculated for the bearded seal endowed with 240 single mystacial whiskers, the mystacial vibrissal system is innervated by about 320,000 myelinated nerve fibers.

Fig. 6.3 Infrared thermogram showing the typical distribution of temperatures measured on the surface of a seal's face immediately after the animal had left the water

Mechanoreceptors found in a follicle correspond to those typical for the mammalian skin. With about 15,000 per F-SC the Merkel cell–neurite complex is by far the most dominating sensory element (Hyvärinen 1995), although it is not clear yet whether this receptor type can be further differentiated based on ultrastructure and function (Baumann et al. 2003). In addition there are 1,000–4,000 lanceolate endings and 100–400 lamellated endings per F-SC, as well as numerous small free nerve endings at the level of the ring sinus and the lower cavernous sinus (Dehnhardt et al. 2003). As demonstrated for the Northern fur seal (*Callorhinus ursinus*) this high degree of innervation of F-SCs corresponds to a strong and somatotopic representation of the vibrissal system in the somatosensory cortex (Ladygina et al. 1985). It remains to be investigated however, whether the somatosensory cortex shows a barrel-like organization similar to that known from, e.g., rodent species (Woolsey and Van Der Loos 1970).

While we do not understand the system at the neural level yet, it became evident during the last few years that the geometry of the hairs is most crucial for their function as a hydrodynamic sensor. In contrast to the round cross-section of the whiskers of terrestrial mammals, whiskers of sea lions and fur seals (Otariidae) are oval in cross-section, which also applies to the whiskers of the walrus (Odobenidae) and those of some phocid species (*E. barbatus* and *Monachus spp.*, see Fig. 6.4 bottom). However, like a typical hair shaft found in terrestrial mammals, whiskers of these pinniped species are smooth in outline. Whiskers of all other phocid species differ considerably from those of the pinnipeds described above and represent a unique hair type among mammals by being extremely flattened and showing a distinct sinusoidal beaded profile (Watkins and Wartzok 1985; Hyvärinen 1989; Dehnhardt and Kaminski 1995; Ginter and Marshall 2010; Ginter et al. 2012; see Fig. 6.4 top). According to our observations on mystacial whiskers in harbor seals, this beaded profile is present from the shortest to the longest hairs, but diminishes to some degree during the course of a year due to mechanical abrasion. However, different from descriptions by Ling (1966, 1977), harbor seals shed their whiskers during the annual molting season so that most of the year the beaded profile is sufficiently present.

Fig. 6.4 Vibrissae of a
harbor seal (*Phoca vitulina,*
top) and a California sea lion
(*Zalophus californianus,*
bottom). Note the undulated
outline of the harbor seal
vibrissa. *Scale bar* 1 mm.
Adapted from Hanke et al.
(2010)

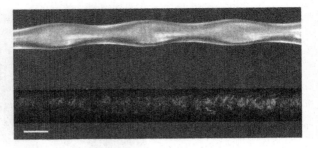

6.3 Hydrodynamic Sensitivity and Trail Following in Harbor Seals and Sea Lions

Based on single unit recordings from the infraorbital branch of the trigeminal nerve of harbor seals, gray seals, and domestic cats, Dykes (1975, p. 650) concluded that these experiments "...did not reveal any way in which the seal's vibrissae are better adapted to an aquatic environment than the cat's" and "...their thresholds are too high for most airborne or waterborne vibrations." Instead, he suggested that whiskers are primarily designed for mechanosensory information obtained by active touch, such as the recognition of surface texture and the shape and size of an object. A number of psychophysical studies supported this hypothesis by showing that the whiskers of the walrus (Kastelein and van Gaalen 1988), the California sea lion (Dehnhardt 1990, 1994; Dehnhardt and Dücker 1996) and the harbor seal (Dehnhardt and Kaminski 1995; Dehnhardt et al. 1997, 1998a; Grant et al. 2013) indeed represent very efficient haptic systems that work in air and under water. However, contradictory to Dykes' hypothesis a combination of behavioral studies and fluid mechanics has since demonstrated that seal whiskers are highly adapted to the reception of hydrodynamic stimuli that appear as mechanosensory information in the aquatic environment.

After Renouf (1979) and Mills and Renouf (1986) showed that harbor seals respond to a vibrating rod directly contacting the whiskers, Dehnhardt et al. (1998b) used dipole water movements to demonstrate that the whiskers of harbor seals represent a hydrodynamic receptor system analogous to the lateral line of fish (Bleckmann 1994). Dipole water movements were generated at variable distances (5–50 cm) from the whiskers of a stationary harbor seal by means of a constant-volume oscillating sphere. This way, detection thresholds for water movements in the range of 10–100 Hz were determined (Fig. 6.5). Detection thresholds of the harbor seal varied across frequencies with the highest velocity sensitivity of 245 μms^{-1} at 50 Hz. The same experimental approach was used to test a California sea lion (*Z. californianus*) which revealed even lower thresholds at 20 and 30 Hz (Dehnhardt and Mauck 2008). These experiments established the sensory modality of hydrodynamic perception in seals and sea lions. As has been described for other hydrodynamic sensory systems (Bleckmann 1994), the tuning curve of the harbor seal suggests that it responded to particle acceleration at frequencies

Fig. 6.5 Detection of hydrodynamic dipole stimuli by a harbor seal. *Top* experimental setup. A harbor seal was trained to station in a hoop and jaw station. Hydrodynamic stimuli were generated with a sinusoidally oscillating sphere in front of the animal. The animal was trained to leave the station if it detected a hydrodynamic stimulus, and to remain in station otherwise. *Bottom* results. The detection threshold, here in terms of water velocity, is shown as a function of the oscillation frequency of the sphere. Adapted from Dehnhardt et al. (1998b)

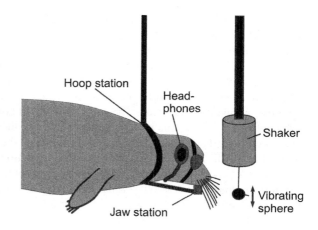

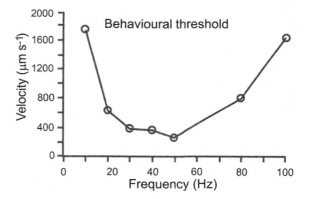

<50 Hz and water displacement at frequencies >50 Hz. Further experiments revealed that harbor seals can not only detect but also discriminate the amplitudes of sinusoidal water movements (Dehnhardt and Mauck 2008).

In dipole experiments a seal remains stationary until it receives hydrodynamic stimulation. With respect to natural situations, this resembles the sit-and-wait strategy known, e.g., from antarctic leopard seals (*Hydrurga leptonyx*) sitting in ambush at the ice edge for a penguin that leaves the water or at the breathing hole of a seal. However, the majority of pinniped species forage on pelagic fish (Hauksson and Bogason 1997; Andersen et al. 2004). This requires active predation, including searching for and pursuing prey. As known from canids, active predation in terrestrial predator–prey interactions is often based on olfactory trail following (Hepper and Wells 2005). A similar concept has been established by Hanke et al. (2000) for the aquatic environment which shows that even a small fish generates a wake showing a distinct vortex structure for at least 30 s while water velocities significantly higher than background noise can still be measured 3 min after a small fish passed by. Thus, a swimming fish leaves a hydrodynamic trail that could be detected and tracked by a piscivorous predator equipped with a hydrodynamic receptor system. Furthermore, the persistence of a fish generated wake may allow the pursuit from a considerable distance. Hydrodynamic trail following was first demonstrated in harbor seals by Dehnhardt et al. (2001). In behavioral experiments hydrodynamic trails were generated with an autonomously operating miniature submarine. The submarine's trail comprised a narrow street of turbulent water movements with velocities similar to those calculated for the wake of a 30 cm long fish. To exclude vision during tests the experimental animal was blindfolded with a stocking mask. After trail detection the harbor seals meticulously tracked the hydrodynamic trails left by the submarine and arrived at its final position in almost 90 % of the trials. However, seals always failed to detect the trail when wearing an additional stocking mask that also covered the mystacial vibrissae. Even delays between the start of the submarine and the start of a seal's search of up to 20 s did not affect a seal's performance. While these delays already simulated trails as long as 40 m, a seal could, hypothetically, continue to follow a hydrodynamic trail indefinitely as long as the traced fish continues swimming and other hydrodynamic events do not disturb the trail. It remains to be shown over which distance a seal can track a hydrodynamic trail under natural conditions.

As mentioned above, the geometry of vibrissal hair shafts of sea lions and fur seals differs from that of most phocid species, like the harbor seal, raising the question whether, e.g., a sea lion is also capable of hydrodynamic trail following. Using the methods applied by Dehnhardt et al. (2001) recent experiments conducted with a California sea lion (Z. californianus) demonstrated that this eared seal is, indeed, capable of hydrodynamic trail following (Gläser et al. 2011). The performance was high and comparable to that of harbor seals, provided a linear trail was followed. However, unlike in harbor seals it decreased substantially with trails older than 5–7 s. Particle image velocimetry (PIV) analysis demonstrated that 7 s old trails still contained water velocities higher than 50 mms^{-1}, while water velocities in the 20 s old trails reliably tracked by harbor seals were in the range of 20 mms^{-1} (Wieskotten et al. 2010a). Additionally, although the sea lion showed a significant performance when following curved trails (Gläser et al. 2011) the performance was much lower than that of the harbor seals confronted with the same task (Dehnhardt et al. 2001). In contrast to the finding that stationary animals of both species are similarly sensitive to hydrodynamic dipole stimuli, these results suggest that during forward swimming the structural differences of vibrissal hair shafts of sea lions and harbor seals account for the difference in their capability to use this sensory system for hydrodynamic trail following.

In the wild, fish-generated hydrodynamic trails will not always result from continuous swimming movements, but can be affected by cyclic burst-and-glide swimming, associated with various modifications regarding hydrodynamic parameters of the trail (compare Fig. 3k of Hanke et al. 2000). To test for effects of burst-and-glide swimming, Wieskotten et al. (2010a) used a remote-controlled miniature submarine to investigate the impact of glide phases on the trackability of differently aged hydrodynamic trails in a harbor seal. It was shown that gliding phases during the generation of a hydrodynamic trail had a negative impact on trackability when trails were ≥15 s old. The seal lost a trail more often within the transition zones, where the submarine switched from a burst to a glide phase. PIV revealed that during the gliding phase the smaller dimensions and faster decay of hydrodynamic parameters were responsible for the decrease in performance. Another hydrodynamic parameter that might have affected the seal's ability to track the trail at the gliding section was the change from a rearwards directed stream in the burst phase to a water flow passively dragged behind the submarine during gliding, which might cause a weaker deflection of the vibrissae. Burst-and-glide swimming has been reported for a variety of fish species (Videler and Weihs 1982; Blake 1983; Hinch et al. 2002; Standen et al. 2004), some of which belong to the prey spectrum of harbor seals. These results suggest that gliding in fish not only plays a role in energy saving during locomotion (as discussed in Blake 1983) but might also constitute an anti-predator strategy by interrupting the continuous traceability of hydrodynamic trails. However, as hydrodynamic parameters resulting from this movement style differ considerably between the wakes of fishes and that of the submarine used in the experiment by Wieskotten et al. (2010a), experiments with live fish are needed to arrive at a better understanding of this potential hydrodynamic camouflage against predators capable of hydrodynamic trail following and how in return seals counteract this strategy under natural conditions.

Hydrodynamic trail following provides a conclusive explanation for how pinnipeds may successfully hunt in dark and murky waters. In addition to this predator–prey scenario it could be demonstrated that harbor seals can also detect and follow hydrodynamic trails of conspecifics (Schulte-Pelkum et al. 2007). This may play an important role in, e.g., mate detection and mother–pup cohesion. In this experiment two harbor seals were used, one as the trail generator and the other as the trail follower. In principle hydrodynamic trails generated by the sub-carangiform or thunniform swimming style of phocid seals (Williams and Kooyman 1985; Fish et al. 1988) compare well in structure to those left by many fish species. However, as trail width is also a function of the body size of the trail generator seal trails were about 2 m wide (Schulte-Pelkum et al. 2007). The trail following seal tracked these biogenic hydrodynamic trails with high accuracy. Video analysis revealed that it precisely followed even small deviations from a straight course by sticking exactly to the center of the trail, thus indicating that it was able to analyze the inner structure of the trail. This behavior while following a biogenic hydrodynamic trail as well as the finding that harbor seal whiskers are quite sensitive to amplitude differences of hydrodynamic dipole stimuli suggests that a seal reading a trail obtains qualitative information about the trail generator. Indeed, Hanke and Bleckmann (2004) showed that the velocity profiles of the wakes of different fishes can differ between species. Trails should thus allow a predator to extract information regarding, e.g., the fish species, its size, swimming style, and swimming speed. The characteristic three-dimensional structure of a hydrodynamic trail could be recognized by a seal due to the spatial arrangement of protracted mystacial whiskers that allows simultaneous multiple-point measurements in the wake.

6.4 Qualitative Information Deducible from Hydrodynamic Trails

That a harbor seal can discern information beyond the mere presence of a wake has been shown in a series of psychophysical experiments conducted under PIV-control, this way not only providing information about the discrimination performance of the tested animal, but also about the hydrodynamic parameters it might have used for stimulus discrimination. This was possible because the eyes of the experimental animal were protected against the laser light by the stocking mask while the white and reflecting vibrissae were illuminated by the laser in the same way as the neutrally buoyant particles used for PIV.

After Dehnhardt et al. (2001) provided first indication that a harbor seal can use hydrodynamic cues to determine the swimming direction of a miniature submarine, Wieskotten et al. (2010b) studied how the aging of a hydrodynamic trail affects the directional sensitivity of seals. In the calm water of a closed experimental box situated in the seal's holding pool hydrodynamic trails were generated using a fin-like paddle, moving from left to right or from right to left. PIV measurements revealed that the structure of the hydrodynamic trail generated by the fin-like paddle corresponded well to that found in fish wakes (see Blickhan et al. 1992; Hanke et al. 2000). The blindfolded seal could enter the box through a circular gate up to its pectoral flippers, where it approached the trail with its mystacial vibrissae protracted to the most forward position. Maximum contact time to a trail was less than 0.5 s. In most trials the seal performed a minute head movement toward the moving direction of the fin-like paddle, thus already indicating its correct detection. Then the seal left the box for touching one of two response targets (left target: moving direction of the fin-like paddle, from right to left, right target: moving direction of the fin-like paddle, from left to right). The seal reliably recognized the direction of the paddle movement when the hydrodynamic trail was up to 35 s old. Based on PIV analysis, the contact points of the mystacial whiskers to the hydrodynamic trail suggest that the seal used the rotation direction of the vortex rings and/or the direction of the jet flow between two counter-rotating vortex rings to determine the moving direction of the fin-like paddle (Wieskotten et al. 2010b).

As has been shown in various studies the structure of a fish generated hydrodynamic trail correlates with the size, shape, and swimming style of the species (Hanke et al. 2000; Drucker and Lauder 2002; Nauen and Lauder 2002; Hanke and Bleckmann 2004; Standen and Lauder 2007; Tytell et al. 2008). Consequently, one can assume that a piscivorous predator like a seal is able to distinguish between different trail generators. Based on this assumption, Wieskotten et al. (2011) conducted experiments that elucidated to what extent a seal can obtain information about the size and shape of a moving object from its hydrodynamic trail. For these experiments the same experimental setup and response paradigm was used as for tests on the perception of directional parameters inherent in a hydrodynamic trail. Trails and the whisker contacts to different trail structures were also visualized and measured by PIV.

For trials on shape discrimination trail generating objects consisted of vertically oriented paddles (flat rectangular paddle and flat paddle with undulated edges) and rods (round and triangular in cross section). First, these objects only differed in cross-sectional shape, while the length of all objects was 30 cm and the width 4 cm. However, PIV-data revealed that the lateral extension of the trail strongly correlated with the object shape. Thus, to eliminate the relevance of this cue, object width was randomized from 2–7 cm so that the seal had to rely on the spatial arrangement of the trail for its decision. As long as paddles of equal width were presented, the harbor seal was able to discriminate the flat rectangular paddle from the cylindrical, triangular, and the undulated paddle, as well as the cylindrical paddle from the undulated paddle. In trials where the width of the paddles was randomized the seal could still discriminate rectangular from triangular paddles. PIV measurements showed that vortices detached asymmetrically from the rectangular paddle, while from the triangular paddle they detached symmetrically. This suggests that especially for these two objects discrimination was based on the salient spatial arrangement of vortices in the respective trail.

Size difference thresholds were determined for three rectangular standard paddles that were equal in length (30 cm) and 2, 6 and 8 cm wide, respectively. These were presented against a series of test paddles varying in width from 2–8 cm. Paddles were moved on a horizontal circular path in clockwise direction first at a constant speed of 0.55 ms^{-1} and afterwards at speeds differing from trial to trial between 0.31 and 0.85 ms^{-1} (Fig. 6.6). Speed variation trials were conducted because at constant speed the animal could associate the larger paddles with the greatest water velocities and the widest extension of the area of high velocities within the wake. By varying the speed of the moving paddle the seal could not use these prevailing hydrodynamic parameters but was required to decide on object size on the basis of the mean spatial extension of the trail and/or the diameter of single vortices within the wake. When objects were moved at the same speed, the harbor seal was able to discriminate size differences of trail generating objects of 2.8–3.2 cm. In contrast, if speed was randomized the performance slightly dropped to size differences of 4.1–4.3 cm (Fig. 6.6). As in speed randomization trials the mean spatial extension of the trail and the diameter of single vortices correlated with paddle size these hydrodynamic parameters were identified to be the prime candidates for the seal's trail discrimination. However, as the lateral spread of the trails showed a higher variation between trials than did the size of single vortices, the latter can be expected to be the decisive factor.

All experiments conducted so far on qualitative information deducible from hydrodynamic trails suggest that single vortex structures are most relevant for the ability of a seal to identify the swimming direction, size, or shape of the trail generator. Consequently, recent psychophysical experiments addressed the question whether and to what extent harbor seals can detect and analyze single vortex rings similar to those found in trails generated by swimming fish (Krüger et al. unpublished). Vortex rings of predefined size, velocity, and acceleration were generated by custom made vortex ring generators (VRGs). VRGs consisted of $20 \times 20 \times 20$ cm cubes with a 2 cm circular aperture on one side. On top of the

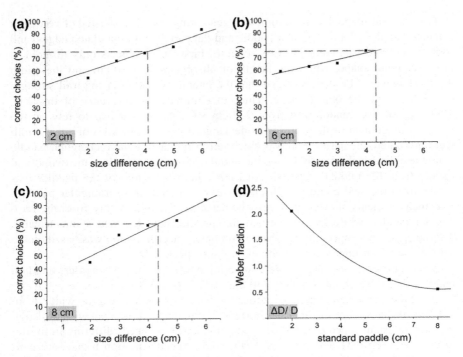

Fig. 6.6 Psychometric functions of a harbor seal's performance on discriminating different sized objects moving with different speeds ranging from 31 up to 85 cms^{-1}. **a–c** Percentages of correct choices are plotted against differences in size between standard objects ranging from 2 to 8 cm and different sized comparison objects. The *dashed lines* at 75 % correct choices mark difference thresholds. Each data point represents the result of 80 single trials. **d** Weber function showing the relationship between Weber fractions (the ratio of the size difference at threshold ΔD to the size of the comparison object D) and comparison object size (D)

cube, a 10 cm diameter vertical pipe was mounted in which a plunger was moved by a linear motor. A vortex ring was generated by the plunger moving down, this way expelling water through the aperture (see Fig. 6.7). PIV was used to quantify the hydrodynamic parameters of vortex rings generated this way.

In a first set of experiments the ability of blindfolded stationary harbor seals to perceive and detect the direction of single vortex rings was investigated. Two VRGs were fixed on a semicircular horizontal profile on either side of a point directly ahead of the animal's snout. For every trial, one of the two VRGs was oriented toward the animal in order to produce a vortex ring travelling to the seal's mystacial vibrissal pad (see Fig. 6.7). The second vortex ring generator simultaneously produced a vortex ring that travelled away from the seal and was operated in the same way as the first VRG to avoid acoustical directional cues. Seals were trained to indicate the direction from which a vortex ring impinged on their vibrissae by touching a left or right response target.

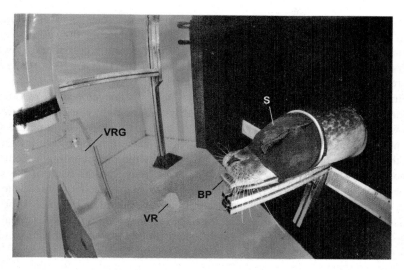

Fig. 6.7 Experimental setup for investigating the ability of harbor seals to detect and analyze single vortex rings. For visualization purpose the water in the VRGs was dyed. *VRG* vortex ring generator, *VR* single vortex ring, *BP* bite plate, *S* stocking mask. Picture: E. Krumpholz (http://www.photos-subjektiv.de)

The two test animals spontaneously detected the direction of a vortex ring, either coming from the right or the left side at an angle of 60° to the midline of the animal. The minimum angle from which the direction of a vortex ring is still detectable by the vibrissal system was determined by gradually decreasing the angle between the VRGs. The harbor seals could still detect the direction of a vortex ring coming from an angle of ±5.7° to the body midline. Although the seals performance was highly significant at this angle, the distance of the VRGs could not be reduced further so that a threshold could not be determined yet. However, in these still ongoing experiments many further questions concerning the detectability and discrimination of single vortex rings will be addressed.

6.5 Mechanisms of Hydrodynamic Reception in Seals and Sea Lions

When in search of hydrodynamic information, a seal usually keeps its moveable mystacial vibrissae in an abducted position largely perpendicular to its swimming direction. This means that while the seal is swimming with a velocity of about $1~\text{ms}^{-1}$ the vibrissae are maximally exposed to the resulting flow effects. Although the vibrissal hair shafts of seals and sea lions are much stiffer than the hair shafts of normal body hair, they are quite flexible structures. This is especially true for phocid vibrissae and here in particular while these hygroscopic hairs are immersed in water, although the degree of soaking has not been quantified for this hair type yet. These observations raise the question of how harbor seals and California sea lions, both capable of hydrodynamic trail following, counter the flow effects acting on their vibrissae like bending and vibrations resulting from vortex shedding while the hair is dragged through the water. For the harbor seal this has been studied by different experimental approaches by Hanke et al. (2010).

First, a head-mounted camera (HMC) system was developed to record the posture and potential movements of vibrissae within the right vibrissal pad while a seal was following the hydrodynamic trail of a remote controlled miniature submarine. An analog charge-coupled device (CCD) camera module was integrated into the stocking mask used to blindfold the seal, while the recording system and its power supply was mounted in a waterproof housing on the back of the seal. Immediately after blindfolding, the still stationary seal abducted its vibrissae to the most forward position. The video recordings revealed that the seal kept the vibrissae in that position while tracking the trail at swimming speeds of about 1 ms^{-1}. Despite their flexibility the vibrissae neither showed significant bending nor did they vibrate to a degree that was resolved by the recording system. These first data obtained on the organismic level already demonstrated that the vibrissal hair shafts of a harbor seal do not respond to laminar flow. It was suggested that the special hair structure characterized by the flat elliptical cross-section and the sinusoidal beaded longitudinal profile is responsible for this effect. To elucidate this effect the wake flow of harbor seal vibrissae was studied by a combination of micro-stereo-PIV and 3D direct numerical simulation (Hanke et al. 2010; Witte et al. 2012). These studies revealed that the primary vortex structure in the wake flow of an isolated whisker differs considerably from that of a circular cylinder. In comparison to the typical Kármán-vortex-street in the wake of a cylinder, two counter rotating vortex-segments shed simultaneously from the whisker, but they are shifted spanwise (Fig. 6.8), reducing the fluid–structure interaction. These vortex rings shedding from a whisker are more unstable so that no side alternating forces act on the structure. In fact, the fluctuating lift and drag forces acting on the hair are reduced by more than 90 % as compared to a cylinder (Fig. 6.9). This effective suppression of the periodic forces acting on the whisker provokes that vortex induced vibrations (VIVs) are almost eliminated while the hair is dragged through the water. Thus, as already indicated by data obtained by the HMC-system, the harbor seal's vibrissae move almost quiescent through the water, e.g., while searching for relevant hydrodynamic information.

The specialized whisker type of harbor seals explains how these pinnipeds cope with the flow effects affecting their vibrissae while extracting hydrodynamic information from their environment at considerable swimming speed. However, the fact that a California sea lion is also capable of hydrodynamic trail following raised the question how this is possible with a whisker that is much more cylindrical and lacks the undulated surface structure. Miersch et al. (2011) therefore compared the capability of isolated vibrissae of the harbor seal and the California sea lion in a rotational flow channel to decode vortex trails shedding from a

Fig. 6.8 Numerical simulation of the vortex street behind a harbor seal vibrissa, Reynolds number Re = 500 (corresponding to a swim speed around 0.5 ms⁻¹). *Left side* the vibrissa and its surface flow, showing a wavy separation line. *Right side* the separated flow behind the vibrissa, vortex cores depicted as iso-surfaces using the Q-criterion. *Shades of gray* cross-stream vorticity Ω_z. Vortices develop at some distance from the vibrissa; clockwise and counterclockwise rotating vortices are staggered spanwise (adapted from Hanke et al. 2010)

Fig. 6.9 Time history of the lift coefficient C_L and the drag coefficient C_D. The *dashed line* shows the results for the vibrissa, the *solid line* for a cylinder of equal hydraulic diameter (adapted from Hanke et al. 2010)

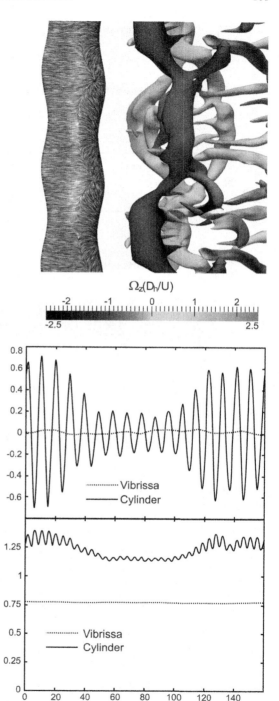

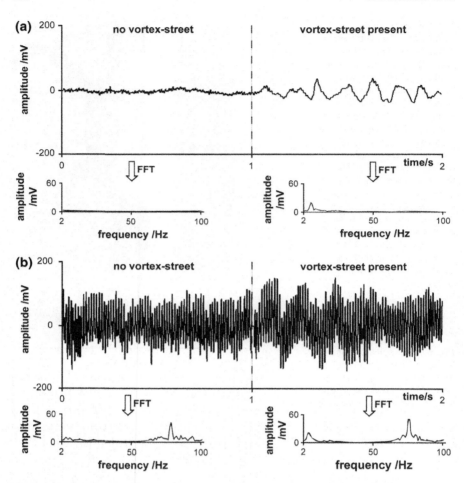

Fig. 6.10 The vibrations of a seal whisker (**a**) and a sea lion whisker (**b**) are plotted in the time domain and below the respective frequency spectrums. The *left* parts were recorded while the vibrissae were exposed to laminar flow. In the *right* parts a hydrodynamic disturbance was added upstream the whisker in form of a Kármán vortex street generated by an immersed cylinder. **a** Demonstrates the direct measurement of the vortex street (flow speed: 35 cms^{-1}; cylinder diameter: 8 mm) with a frequency of vortex shedding $f_{VS} = 7$ Hz (*right part*) and the absence of any vortex-induced vibrations (VIV) during laminar flow (*left part*). **b** Reveals the occurrence of VIVs during laminar flow (*left part*) with a frequency of 75 Hz and the additional modulation of that kind of carrier frequency by the vortex shedding of the Kármán vortex street $f_{VS} = 7$ Hz (flow speed: 34 cms^{-1}; cylinder diameter: 8 mm) (*right part*)

cylinder immersed in the flow tank. Above the water surface a single whisker was attached to a piezoceramic holder so that the hair extended into the water. The piezoceramic holder transformed mechanical vibrations into an electrical signal. Although both whisker types were able to detect the vortex shedding frequency generated by the cylinder, the signal detected by sea lion whiskers was considerably masked by noise (Fig. 6.10b). In contrast, harbor seal whiskers showed a much higher signal-to-noise ratio (SNR) due to largely reduced noise (Fig. 6.10a), which is consistent with the findings by Hanke et al. (2010) and Witte et al. (2012). However, a further analysis of the high noise signals of sea lion whiskers revealed that each noise signal contained a dominant frequency that correlated with flow velocity. Thus instead of being noise, this steady fundamental frequency may serve as a carrier signal that is modulated by each hydrodynamic event impinging on the hair. This modulation may be the crucial information used by sea lions and is suggested to represent the mechanism underlying hydrodynamic reception in the fast swimming sea lion.

Apparently, two types of hydrodynamic sensors developed in the course of pinniped evolution and it seems that the harbor seal whisker is the more effective one. In sea lions and perhaps other eared seals each hydrodynamic event should cause the deterioration of the carrier signal, which has to recover before a new modulation can occur so that the temporal resolution of the whisker should be reduced. Moderate temporal resolution should result in an impaired spatial resolution of, e.g., vortex patterns in a moving sea lion. This might explain why hydrodynamic trail following in the California sea lion is more affected by the aging of trails (Gläser et al. 2011). However, as the carrier frequency changes with the animal's swimming speed, it may provide information that enables the animal to determine its swimming velocity. Knowing its own swimming speed might be advantageous for underwater orientation. Thus, in the sea lion the mechanism underlying hydrodynamic reception might be a trade-off between these different functions. However, more results are needed to elucidate the mechanism of hydrodynamic reception in seals possessing the otarid-like vibrissal hair shaft.

The investigation of hydrodynamic perception as a sensory modality in seals is a new topical research field recently started with the dipole study of Dehnhardt et al. (1998b). Taking the extensive afferent innervation of the vibrissae into account, and considering that in the seal's environment hydrodynamic detection may often be crucial for finding food, we believe that we have only started to understand what kinds of hydrodynamic stimuli a seal can sense, analyze, and discriminate with its vibrissae.

References

Akamatsu T, Okumura T, Novarini N, Yan HY (2002) Empirical refinements applicable to the recording of fish sounds in small tanks. J Acoust Soc Am 112:3073–3082. doi:10.1121/1.1515799

Aksnes DL, Giske J (1993) A theoretical model of aquatic visual feeding. Ecol Model 67:233–250. http://dx.doi.org/10.1016/0304-3800(93)90007-F

Aksnes DL, Utne ACW (1997) A revised model of visual range in fish. Sarsia 82:137–147

Andersen SM, Lydersen C, Grahl-Nielsen O, Kovacs KM (2004) Autumn diet of harbor seals (*Phoca vitulina*) at Prins Karls Forland, Svalbard, assessed via scat and fatty-acid analyses. Canad J Zool 82:1230–1245

Au WWL (1993) The sonar of dolphins. Springer, New York

Baumann KI, Moll I, Halata Z (2003) The merkel cell: structure, development, function, and cancerogenesis. Springer, Berlin, Heidelberg, New York

Blake RW (1983) Functional design and burst-and-coast swimming in fishes. Can J Zool 61:2491–2494

Bleckmann H (1994) Reception of hydrodynamic stimuli in aquatic and semiaquatic animals. Progress in zoology 41. Gustav Fischer, Stuttgart, Jena, New York

Blickhan R, Krick CM, Zehren D, Nachtigall W (1992) Generation of a vortex chain in the wake of a subundulatory swimmer. Naturwissenschaften 79:220–221

Bodson A, Miersch L, Mauck B, Dehnhardt G (2006) Underwater auditory localization by a swimming harbor seal (*Phoca vitulina*). J Acoust Soc Am 120:1550–1557

Bodson A, Miersch L, Dehnhardt G (2007) Underwater localization of pure tones by harbor seals (*Phoca vitulina*). J Acoust Soc Am 122:2263–2269. doi:10.1121/1.2775424

Bolanowski SJ, Gescheider GA, Verrillo RT, Checkosky CM (1988) Four channels mediate the mechanical aspects of touch. J Acoust Soc Am 84:1680–1694. doi:10.1121/1.397184

Dehnhardt G (1990) Preliminary results from psychophysical studies on the tactile sensitivity in marine mammals. In: Thomas JA, Kastelein RA (eds) Sensory abilities of cetaceans. Plenum Press, New York, pp 435–446

Dehnhardt G (1994) Tactile size discrimination by a California sea lion (*Zalophus californianus*) using its mystacial vibrissae. J Comp Physiol A 175:791–800

Dehnhardt G, Dücker G (1996) Tactual discrimination of size and shape by a California sea lion (*Zalophus californianus*). Anim Learn Behav 24:366–374

Dehnhardt G, Kaminski A (1995) Sensitivity of the mystacial vibrissae of harbor seals (*Phoca vitulina*) for size differences of actively touched objects. J Exp Biol 198:2317–2323

Dehnhardt G, Mauk B (2008) Mechanoreception in secondarily aquatic vertebrates. In: Thewissen JGM, Nummela S (eds) Sensory evolution on the threshold: adaptations in secondarily aquatic vertebrates. University of California Press, Berkeley, Los Angeles, pp 295–314

Dehnhardt G, Sinder M, Sachser N (1997) Tactual discrimination of size by means of mystacial vibrissae in harbor seals: in air versus underwater. Zeitschrift für Säugetierkunde 62:40–43

Dehnhardt G, Mauck B, Hyvärinen H (1998a) Ambient temperature does not affect the tactile sensitivity of mystacial vibrissae of harbor seals. J Exp Biol 201:3023–3029

Dehnhardt G, Mauck B, Bleckmann H (1998b) Seal whiskers detect water movements. Nature 394:235–236

Dehnhardt G, Mauck B, Hanke W, Bleckmann H (2001) Hydrodynamic trail following in harbor seals (*Phoca vitulina*). Science 293:102–104

Dehnhardt G, Mauck B, Hyvärinen H (2003) The functional significance of the vibrissal system of marine mammals. In: Baumann KI, Moll I, Halata Z (eds) The merkel cell: structure, development, function, and cancerogenesis. Springer, Berlin, Heidelberg, New York, pp 127–135

Drucker EG, Lauder GV (2002) Experimental hydrodynamics of fish locomotion: functional insights from wake visualization. Integr Comp Biol 4:243–257

Dykes RW (1975) Afferent fibers from mystacial vibrissae of cats and seals. J Neurophysiol 38:650–662

Ebara S, Kumamoto K, Matsuura T, Mazurkiewicz JE, Rice FL (2002) Similarities and differences in the innervation of mystacial vibrissal follicle-sinus complexes in the rat and cat: a confocal microscopic study. J Comp Neurol 449:103–119. doi:10.1002/cne.10277

Fish FE, Innes S, Ronald K (1988) Kinematics and estimated thrust production of swimming harp and ringed seals. J Exp Biol 137:157–173

Folkow LP, Blix AS (1989) Thermoregulatory control of expired air temperature in diving harp seals. Am J Physiol 257:R306–R310

Gentry RL (1967) Underwater auditory localization in the California sea lion (*Zalophus californianus*). J Aud Res 7:187–193

Gescheider GA, Thorpe JM, Goodarz J, Bolanowski SJ (1997) The effects of skin temperature on the detection and discrimination of tactile stimulation. Somatosens Mot Res 14:181–188

Ginter CC, Marshall CD (2010) Morphological analysis of the bumpy profile of phocid vibrissae. Mar Mammal Sci 26:733–743. doi:10.1111/j.1748-7692.2009.00365.x

Ginter CC, DeWitt TJ, Fish FE, Marshall CD (2012) Fused traditional and geometric morphometrics demonstrate pinniped whisker diversity. PLoS One 7:1–10

Gläser N, Wieskotten S, Otter C, Dehnhardt G, Hanke W (2011) Hydrodynamic trail following in a California sea lion (*Zalophus californianus*). J Comp Physiol A 197:141–151

Grant R, Wieskotten S, Wengst N, Prescott T, Dehnhardt G (2013) Vibrissal touch sensing in the harbor seal (*Phoca vitulina*): how do seals judge size? J Comp Physiol A 199:521–533. doi:10.1007/s00359-013-0797-7

Hanke W, Bleckmann H (2004) The hydrodynamic trails of *Lepomis gibbosus* (Centrarchidae), *Colomesus psittacus* (Tetraodontidae) and *Thysochromis ansorgii* (Cichlidae) investigated with scanning particle image velocimetry. J Exp Biol 207:1585–1596. doi:10.1242/jeb.00922

Hanke W, Brücker C, Bleckmann H (2000) The ageing of the low-frequency water disturbances caused by swimming goldfish and its possible relevance to prey detection. J Exp Biol 203:1193–1200

Hanke FD, Dehnhardt G, Schaeffel F, Hanke W (2006a) Corneal topography, refractive state, and accommodation in harbor seals (*Phoca vitulina*). Vis Res 46:837–847. doi:10.1016/j.visres.2005.09.019

Hanke W, Römer R, Dehnhardt G (2006b) Visual fields and eye movements in a harbor seal (*Phoca vitulina*). Vis Res 46:2804–2814

Hanke FD, Hanke W, Hoffman KP, Dehnhardt G (2008a) Optokinetic nystagmus in harbor seals (*Phoca vitulina*). Vis Res 48:304–315

Hanke FD, Kröger RHH, Siebert U, Dehnhardt G (2008b) Multifocal lenses in a monochromat: the harbor seal. J Exp Biol 211:3315–3322. doi:10.1242/jeb.018747

Hanke FD, Hanke W, Scholtyssek C, Dehnhardt G (2009a) Basic mechanisms in pinniped vision. Exp Brain Res 199:299–311. doi:10.1007/s00221-009-1793-6

Hanke FD, Peichl L, Dehnhardt G (2009b) Retinal ganglion cell topography in juvenile harbor seals (*Phoca vitulina*). Brain Behav Evol 74:102–109. doi:10.1159/000235612

Hanke W, Witte M, Miersch L, Brede M, Oeffner J, Michael M, Hanke F, Leder A, Dehnhardt G (2010) Harbor seal vibrissa morphology suppresses vortex-induced vibrations. J Exp Biol 213:2665–2672. doi:10.1242/jeb.043216

Hanke FD, Scholtyssek C, Hanke W, Dehnhardt G (2011) Contrast sensitivity in a harbor seal (*Phoca vitulina*). J Comp Physiol A 197:203–210. doi:10.1007/s00359-010-0600-y

Hauksson E, Bogason V (1997) Comparative feeding of gray (*Halichoerus grypus*) and common seals (*Phoca vitulina*) in coastal waters of Iceland, with a note on the diet of hooded (*Cystophora cristata*) and harp seals (*Phoca groenlandica*). J Northwest Atl Fish Sci 22:125–135

Hepper PG, Wells DL (2005) How many footsteps do dogs need to determine the direction of an odour trail? Chem Senses 30:291–298. doi:10.1093/chemse/bji023

Hinch SG, Standen EM, Healey MC, Farrell AP (2002) Swimming patterns and behaviour of upriver-migrating adult pink (*Oncorhynchus gorbuscha*) and sockeye (*O. nerka*) salmon as assessed by EMG telemetry in the Fraser River, British Columbia. Canada Hydrobiol 483:147–160

Hobson ES (1966) Visual orientation and feeding in seals and sea lions. Nature 210:326–327. doi:10.1038/210326a0

Hokkanen JEI (1990) Temperature regulation of marine mammals. J Theor Biol 145:465–485. doi:10.1016/s0022-5193(05)80482-5

Hyvärinen H (1989) Diving in darkness: whiskers as sense organs of the ringed seal (*Phoca hispida saimensis*). J Zool 218:663–678

Hyvärinen H (1995) Structure and function of the vibrissae of the ringed seal (*Phoca hispida* L.). In: Kastelein RA, Thomas JA, Nachtigall PE (eds) Sensory systems of aquatic mammals. De Spil Publishers, Woerden, pp 429–445

Hyvärinen H, Katajisto H (1984) Functional structure of the vibrissae of the ringed seal (*Phoca hispida* Schr.). Acta Zoologica Fennica 171:27–30

Irving L (1969) Temperature regulation in marine mammals. In: Andersen HT (ed) The biology of marine mammals. Academic Press, New York, pp 147–174

Käkelä R, Hyvärinen H (1993) Fatty acid composition of fats around the mystacial and superciliary vibrissae differs from that of blubber in the Saimaa ringed seal (*Phoca hispida saimensis*). Comp Biochem Physiol B 105:547–552

Käkelä R, Hyvärinen H (1996) Site-specific fatty acid composition in adipose tissues of several northern aquatic and terrestrial mammals. Comp Biochem Physiol B 115:501–514. doi:10.1016/s0305-0491(96)00150-2

Kastak D, Schusterman RJ (1998) Low-frequency amphibious hearing in pinnipeds: methods, measurements, noise, and ecology. J Acoust Soc Am 103:2216–2228

Kastelein RA, van Gaalen MA (1988) The sensitivity of the vibrissae of a pacific walrus (*Odobenus rosmarus divergens*). Part 1. Aquat Mammals 14:123–133

Kvadsheim PH, Gotaas ARL, Folkow LP, Blix AS (1997) An experimental validation of heat loss models for marine mammals. J Theor Biol 184:15–23

Ladygina TF, Popov VV, Supin AY (1985) Somatotopic projections in the cerebral cortex of the fur seal (*Callorhinus ursinus*). Acad Sci Moskow 17:344–351

Lavigne DM, Bernholz CD, Ronald K (1977) Functional aspects of pinniped vision. In: Harrison RJ (ed) Functional anatomy of marine mammals, vol 3. Academic Press, London, pp 135–173

Levenson DH, Schusterman RJ (1999) Dark adaptation and visual sensitivity in shallow and deep-diving pinnipeds. Mar Mammal Sci 15:1303–1313

Ling JK (1966) The skin and hair of the southern elephant seal, *Mirounga leonina* (Linn.). Aust J Zool 14:855–866

Ling JK (1977) Vibrissae of marine mammals. In: Harrison RJ (ed) Functional anatomy of marine mammals. Academic Press, London, pp 387–415

Marshall CD, Amin H, Kovacs KM, Lydersen C (2006) Microstructure and innervation of the mystacial vibrissal follicle-sinus complex in bearded seals, *Erignathus barbatus* (Pinnipedia: Phocidae). Anat Rec A 288:13–25

Mauck B, Eysel U, Dehnhardt G (2000) Selective heating of vibrissal follicles in seals (*Phoca vitulina*) and dolphins (*Sotalia fluviatilis guianensis*). J Exp Biol 203:2125–2131

Miersch L, Hanke W, Wieskotten S, Hanke FD, Oeffner J, Leder A, Brede M, Witte M, Dehnhardt G (2011) Flow sensing by pinniped whiskers. Phil Trans R Soc B 366:3077–3084. doi:10.1098/rstb.2011.0155

Mills FHJ, Renouf D (1986) Determination of the vibration sensitivity of harbor seal *Phoca vitulina* (L.) vibrissae. J Exp Mar Biol Ecol 100:3–9

Møhl B (1964) Preliminary studies on hearing in seals. Vidensk Medd Dansk Naturh Foren 127:283–294

Møhl B (1968) Auditory sensitivity of the common seal in air and water. J Aud Res 8:27–38

Moore PWB, Au WWL (1975) Underwater localization of pulsed pure tones by the California sea lion (*Zalophus californianus*). J Acoust Soc Am 58:721–727

Nauen JC, Lauder GV (2002) Hydrodynamics of caudal fin locomotion by chub mackerel, *Scomber japonicus* (Scombridae). J Exp Biol 205:1709–1724

Popper AN, Fay RR (1993) Sound detection and processing by fish: critical review and major research questions (part 2 of 2). Brain Behav Evol 41:26–38

Renouf D (1979) Preliminary measurements of the sensitivity of the vibrissae of harbor seals (*Phoca vitulina*) to low frequency vibrations. J Zool 188:443–450

Rice FL, Mance A, Munger BL (1986) A comparative light microscopic analysis of the sensory innervation of the mystacial pad. I. Innervation of vibrissal follicle-sinus complexes. J Comp Neurol 252:154–174

Ryg M, Lydersen C, Knutsen LO, Bjorge A, Smith TG, Oritsland NA (1993) Scaling of insulation in seals and whales. J Zool 230:193–206

Scholtyssek C, Kelber A, Dehnhardt G (2008) Brightness discrimination in the harbor seal (*Phoca vitulina*). Vis Res 48:96–103. http://dx.doi.org/10.1016/j.visres.2007.10.012

Schulte-Pelkum N, Wieskotten S, Hanke W, Dehnhardt G, Mauck B (2007) Tracking of biogenic hydrodynamic trails in harbor seals (*Phoca vitulina*). J Exp Biol 210:781–787
Standen EM, Lauder GV (2007) Hydrodynamic function of dorsal and anal fins in brook trout (*Salvelinus fontinalis*). J Exp Biol 210:325–339
Standen EM, Hinch SG, Rand PS (2004) Influence of river speed on path selection by migrating adult sockeye salmon (*Oncorhynchus nerka*). Can J Fish Aquat Sci 61:905–912. doi:10.1139/F04-035
Stephens RJ, Beebe IJ, Poulter TC (1973) Innervation of the vibrissae of the California sea lion, *Zalophus californianus*. Anat Rec 176:421–442. doi:10.1002/ar.1091760406
Terhune JM (1974) Directional hearing of a harbor seal in air and water. J Acoust Soc Am 56:1862–1865
Tytell ED, Standen EM, Lauder GV (2008) Escaping Flatland: three-dimensional kinematics and hydrodynamics of median fins in fishes. J Exp Biol 211:187–195
Videler JJ, Weihs D (1982) Energetic advantages of burst-and-coast swimming of fish at high speeds. J Exp Biol 97:169–178
Wahlberg M, Westerberg H (2003) Sounds produced by herring (*Clupea harengus*) bubble release. Aquat Living Resour 16:271–275. http://dx.doi.org/10.1016/S0990-7440(03)00017-2
Walls GL (1942) The vertebrate eye and its adaptive radiation. The Cranbrook Institute of Science, Bloomfield Hills. doi:10.5962/bhl.title.7369
Watkins WA, Wartzok D (1985) Sensory biophysics of marine mammals. Mar Mammal Sci 1:219–260. doi:10.1111/j.1748-7692.1985.tb00011.x
Watts P, Hansen S, Lavigne DM (1993) Models of heat loss by marine mammals: thermoregulation below the zone of irrelevance. J Theor Biol 163:505–525. doi:10.1006/jtbi.1993.1135
Weiffen M, Möller B, Mauck B, Dehnhardt G (2006) Effect of water turbidity on the visual acuity of harbor seals (*Phoca vitulina*). Vis Res 46:1777–1783
Wieskotten S, Dehnhardt G, Mauck B, Miersch L, Hanke W (2010a) The impact of glide phases on the trackability of hydrodynamic trails in harbor seals (*Phoca vitulina*). J Exp Biol 213:3734–3740
Wieskotten S, Dehnhardt G, Mauck B, Miersch L, Hanke W (2010b) Hydrodynamic determination of the moving direction of an artificial fin by a harbor seal (*Phoca vitulina*). J Exp Biol 213:2194–2200
Wieskotten S, Mauck B, Miersch L, Dehnhardt G, Hanke W (2011) Hydrodynamic discrimination of wakes caused by objects of different size or shape in a harbor seal (*Phoca vitulina*). J Exp Biol 214:1922–1930
Williams TM, Kooyman GL (1985) Swimming performance and hydrodynamic characteristics of harbor seals *Phoca vitulina*. Physiol Zool 58:576–589
Wilson B, Batty RS, Dill LM (2004) Pacific and Atlantic herring produce burst pulse sounds. Proc R Soc London Ser B 7:95–97
Witte M, Hanke W, Wieskotten S, Miersch L, Brede M, Dehnhardt G, Leder A (2012) On the wake flow dynamics behind harbor seal vibrissae—a fluid mechanical explanation for an extraordinary capability. In: Tropea C, Bleckmann H (eds) Nature-inspired fluid mechanics, vol 119. Springer, Berlin, Heidelberg, pp 271–289. doi:10.1007/978-3-642-28302-4_17
Woolsey TA, Van Der Loos H (1970) The structural organization of layer IV in the somatosensory region (SI) of mouse cerebral cortex. The description of a cortical field composed of discrete cytoarchitectonic units. Brain Res 17:205–242. doi:10.1016/0006-8993(70)90079-x
Worthy GAJ (1991) Insulation and thermal balance of fasting harp and gray seal pups. Comp Biochem Physiol A 100:845–851. doi:10.1016/0300-9629(91)90302-s

Chapter 7
The Slightest Whiff of Air: Airflow Sensing in Arthropods

Friedrich G. Barth

Abstract The perception of medium flows has received ever increasing attention during the last two decades and has increasingly been recognized as a sensory capacity of its own. A combination of experimental work and physical–mathematical modeling has deepened our understanding of the workings of airflow sensors, mainly represented by insect filiform hairs and arachnid trichobothria, both as individual sensors and sensor arrays. This chapter points to the diversity of arthropod airflow sensors and stresses the importance of comparative studies. These should include animal groups so far largely neglected by sensory biology and neuroethology. Another need is to analyze biologically relevant flow patterns and to relate these to the functional properties of the various patterns of sensor arrangement found in different animal taxa. Finally, the capture of a freely flying fly by a wandering spider is taken to illustrate the challenges and promises of studies that aim to reveal the relation between a particular airflow pattern and a specific behavior.

7.1 Introduction

The sensory biology of medium flow perception learned a lot from both behavioral studies and mathematical/physical modeling. In addition, bio-inspired technical approaches to the many challenging problems involved in this particular sensory capacity helped to deepen our understanding. It seems that the reception of airflow

---

Dedicated to the memory of Joseph AC Humphrey.

F. G. Barth (✉)
Department of Neurobiology, Faculty of Life Sciences, University of Vienna,
Althanstr. 14, 1090 Vienna, Austria
e-mail: friedrich.g.barth@univie.ac.at

H. Bleckmann et al. (eds.), *Flow Sensing in Air and Water*,
DOI: 10.1007/978-3-642-41446-6_7, © Springer-Verlag Berlin Heidelberg 2014

169

and water flow has by now been established as its own sensory modality, although, unfortunately, some authors still consider it just a special kind of "hearing" (sound/acoustic reception) or a "sense of touch at a distance." Airflow reception is neither "touch" nor "hearing" in the sense of human experience and the reception of the pressure aspect of airborne sound. Instead the adequate input stimulus is fluid medium (air, water) flow, with its velocity and acceleration and the frictional forces set up by them being the crucial parameters.

Living organisms at all levels of complexity and specialization, from bacteria to mammals, are immersed in medium flow, be it air or water, which are hardly ever motionless. Being exposed to abiotic and biotic sources of medium flow almost permanently, nearly all animal groups, both aquatic and terrestrial, do have flow sensing organs (Budelmann 1989; Bleckmann 1994). Among these organs the highly developed fish lateral line systems, the seal whiskers and the filiform hairs of the arthropods are particularly prominent and well known. However, one also finds medium flow sensitivity in protozoa, coelenterates, cephalopods, and echinoderms, to give a few examples only (Coombs et al. 1989; Budelmann 1989; Bleckmann 1994). In all these cases, it is the medium flow the animals use as an important source of information about their environment. The oscillating movements of air particles (or water for aquatic animals) heard by us humans due to sound pressure oscillations are an exceptional case of a much broader range of natural stimuli which follow physical laws different from those applied to sound. The underlying physics (fluid mechanics) should therefore not be camouflaged by anthropocentric thinking. What this chapter then is referring to as a particular sensory capacity is the sensation of absolute or relative motion between air and the animal under consideration. Although much can be learned by the application of oscillating flows to stimulate airflow sensors in the laboratory (Shimozawa and Kanou 1984b; Barth et al. 1993), a typical natural situation is a flow that does not reverse its direction but accelerates and decelerates in the direction of motion and as a function of time. The sensors (hairs) are immersed in a pulsating boundary layer formed by the substrate supporting the hairs (Humphrey and Barth 2008). The last section of this chapter will deal with a spider catching a freely flying fly from the air and illustrate an effort to characterize a biologically relevant flow and its relation to a specific behavior (Barth et al. 1995; Klopsch et al. 2012, 2013). From this and other examples, it will become clear that we are neither dealing with "hearing" nor with "touch" but with a sense whose range of up to some 70 cm (as far as present knowledge goes) is somewhere between that of a tactile sensor (like an arthropod hair deflected by the direct exposure to force) and an arthropod hearing organ like a tympanal organ (stimulated by far field sound pressure).

From the models developed for airflow sensors, a lot can be derived for sensors working in water as well (Devarakonda et al. 1996; Humphrey and Barth 2008). This chapter, however, will concentrate on terrestrial arthropods and not try to be encyclopedic. One of its goals is to point the physiologist and neuroethologist to the widespread occurrence of airflow sensors. There is a wide open and promising field of research which has much to offer, in particular for comparative studies extending the number of taxonomic samples and building on the few case studies already available. One of the main gaps still to be filled is our ignorance about the details of most of the biologically relevant flow patterns as they occur in the respective animal's natural environment. As will be seen from the report on the spider–fly interaction, the measurement of these spatiotemporal flow patterns is not an easy task. It is urgently needed, however, because it is these patterns to which the respective sensory systems are adapted. Fortunately, the technologies now available in fluid mechanics, in particular the optical ones, which leave the flows undisturbed, will be of great help to master the task.

The reader interested in the comprehensive physical–mathematical modeling of the biomechanics of arthropod filiform hairs, initiated by pioneering studies such as those by Fletcher (1978), Tautz (1977, 1979) and in particular Shimozawa and Kanou (1984a, b), is referred to an extensive review by Humphrey and Barth (2008), which also discusses some bio-inspired artificial medium motion sensors. The inspiration of engineers by arthropod hair-like medium flow sensors to develop synthetic sensor arrays is also illustrated by several other recent publications like those of Große and Schröder (2012), Izadi and Krijnen (2012), and McConney and Tsukruk (2012). The reader interested in biomimetic aspects of natural flows will find a book on "Nature-inspired fluid mechanics" stimulating. It highlights the outcome of a large interdisciplinary research program (Tropaea and Bleckmann 2012). Finally, a recent review article by Casas and Dangles (2010) focuses on ecological aspects of medium flow sensing.

7.2 The Diversity of Sensor Occurrence and Arrangement

The airflow sensors of terrestrial arthropods are cuticular hair-like structures. Whereas in insects these are usually referred to as filiform hairs they are called trichobothria in arachnids. Their occurrence is widespread and they are known for insects (among them in particular the orthopterans like crickets and cockroaches), pseudoscorpions, scorpions, mites, and spiders (Reißland and Görner 1985). A common characteristic of all these hair-like airflow sensors is their small size and the tiny mass (according to Tautz 1977 the mass of a filiform hair of the *Barathra* caterpillar is 3.6×10^{-9} g) of their hair shaft, which represents a lever arm and whose suspension in the general cuticle of the exoskeleton is exquisitely flexible. As a consequence, the hair shaft is deflected and the sensor stimulated by the viscous forces exerted even by the slightest whiff of air.

In some arthropod taxa, one finds a rather dispersed arrangement of the sensors instead of a single occurrence and concentration on a small area of the body surface. Figure 7.1 shows the wandering spider *Cupiennius salei* as an example (for more details, see Barth et al. 1993). Its nearly 1,000 trichobothria are found on the tarsi, metatarsi, and tibiae of all of its eight walking legs and the pedipalps. They form a circular array of sensors in a widely spread-out arrangement. As a consequence, airflow signals from all horizontal directions are received equally well, providing the spider with a behaviorally relevant all-around "view"

Fig. 7.1 *Patterns of sensor arrangement. Left* Adult female of the spider *Cupiennius salei* with an average of 104 trichobothria on the tarsus, metatarsus and tibia of each of its walking legs and an average of 53 trichobothria on the tibia of each pedipalp. The *circular arrangement* of the sensors provides an "all-around view" regarding airflows. *Right* The scorpion, *Smeringurus mesaensis*, with a total of 48 trichobothria on each of its pedipalps. In the cockroach, *Periplaneta americana*, ca. 208 filiform hairs on each of the two abdominal cerci found in adults are even more localized than in the scorpion. Photographs by FG Barth (*left*) and R Müllan (*right*)

(Brittinger 1998; Barth 2002). In this regard, one is reminded of the fish lateral line system, the electrosensors of weakly electric fish, the touch receptors in the human skin, and other sensory systems for which spatial resolution and the analysis of spatiotemporal stimulus patterns are important. The spider's ability to catch flying prey from the air with a well-oriented and timed jump (see below) is a behavior hardly feasible without such a capacity. Different from the spider, in the scorpion and the cockroach shown in Fig. 7.1 the sensory hairs only occur on the front and rear side of the body, respectively.

So far we know only little about the details of biologically relevant flow fields. One may nevertheless speculate that there is a difference in the behavioral significance between the widely spread-out arrays of airflow sensors and the highly clustered ones or those consisting of a few strongly localized hairs only (Fig. 7.2). The filiform hairs on the cerci of cockroaches and crickets are examples of the highly clustered type. They mainly mediate quick oriented escape reactions likely to be less demanding from a sensory point of view (not saying that they are NOT demanding) than prey capture of a fly passing by in flight (Dumpert and Gnatzy 1977; Camhi 1984; Landolfa and Jacobs 1995; Gnatzy 1996; Paydar et al. 1999; Dangles et al. 2006; Casas et al. 2008). A similar argument may apply to the four airflow sensors on the prothorax of the cabbage white caterpillar (Tautz 1977). These are considered threshold detectors providing the animal with a warning system against a predatory wasp (Tautz and Markl 1978). Of course, a widely dispersed arrangement of trichobothria does not exclude an additional role in escape behavior (Hergenröder and Barth 1983; Suter 2003). Thus *Cupiennius* may well also use its trichobothria as a warning system against predatory wasps (Pompilidae) and parasitic neuropterans (Mantispidae). In the field they respond to the approach of these enemies by raising their front legs in a defensive way.

A careful comparative neuroethological approach, based on the animals' behavior in their natural environment, would be highly rewarding, in particular by the inclusion of the many animal groups so far neglected in regard to their sensory biology in general and airflow sensing in particular.

At a closer look, there is a bewildering diversity of numbers and distributions of airflow sensors in different animal groups. According to present knowledge, the large (leg span of adults >10 cm) wandering spider *C. salei* (Ctenidae) is the champion among spiders regarding number, with a total of ca. 950 trichobothria (Barth et al. 1993). True spiders all have trichobothria, but there are great differences correlated with size and way of life. Thus, orb weavers like *Araneus cornutus*, *Meta reticulata*, and *Linyphia triangularis* have only 7–11 trichobothria on each of their legs. Even the golden web spider, *Nephila clavipes*, of similar size as *C. salei*, carries only 40 trichobothria on each of its legs (Fig. 7.3) (Lehtinen 1980; Peters and Pfreundt 1986; Barth 2002). These findings seem to indicate that airflow sensing has a less prominent role in the behavior of web spiders than in that of wandering spiders. It was indeed impossible to elicit prey capture behavior by airflow stimulation in *Zygiella x-notata* and *N. clavipes*, which both are orb weavers, whereas web-borne vibrations were highly effective. The behavior elicited by airflow stimuli instead was a defensive lifting of the front legs (Klärner and Barth 1982). Among the arachnids, the wind scorpions (Solifugae), harvestmen (Opiliones), and Ricinulei do not have any trichobothria at all (Reissland and Görner 1985).

Animals with a minimum number of trichobothria may lend themselves particularly well to neuroethological studies. *Typopeltis crucifer*, a whip scorpion (Uropygi), has only four trichobothria, two of them on the distal end of the tibia of each of its elongated first legs, which it uses like feelers. Remarkably, the two trichobothria are arranged at an angle of almost exactly 90° to each other on the dorsal and lateral aspects of the tibia, respectively (Fig. 7.2g; Haupt 1996). What do the adequate flow fields look like? A close spatial relationship between the sensory hairs and the stimulus source seems likely. What exactly is the information the whip scorpion needs to have when probing its environment with its feeler legs and catching prey? Whip spiders (Amblypygi) use their antenniform forelegs like feelers as well, constantly moving them when alerted and probing their environment. They have seven trichobothria on the tibiae of their antenniform legs, not however on the tarsus, which is made up of 74 segments and has a total length of ca. 10 cm in *Heterophrynus elaphus* (Igelmund 1987). According to Weygoldt (1995, 2000) for the capture of moving prey the much more numerous trichobothria on the walking legs are the most important sensilla. An interesting case of a specialized function of trichobothria was found by Santer and Hebets (2008). Following their study, the two trichobothria on the patella of the whip spider *Phrynus marginemaculatus* are specifically stimulated during ritualized agonistic interactions. During these interactions, air movements caused by the vibrations of the antenniform legs (20–30 Hz, amplitude ca. 2 mm) are aimed at them by the predominantly male opponent.

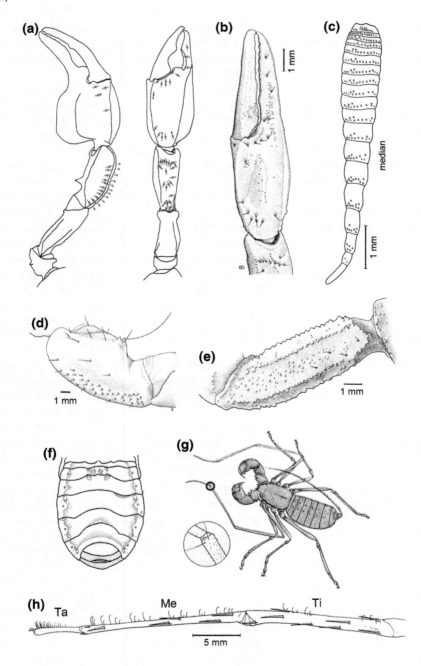

◄ **Fig. 7.2** *Patterns of sensor arrangement* in more detail. **a** Scorpion (*Euscorpius carpathicus*), ventral (*left*) and lateral (*right*) aspect of left pedipalp; *short lines* indicate preferred planes of oscillation (from Hoffmann 1967, modified). **b** Scorpion (*Euscorpius sp.*), external surface of chela (tibia); note trichobothria on fixed finger and patella. **c** Cockroach (*Periplaneta americana*) cercus with filiform hairs shown as *filled circles* on its segments (from Nicklaus 1965, modified). **d** Scorpion (*Pandinus imperator*), ventral view of pedipalpal patella. **e** Scorpion (*Buthus occitanus*) with just four trichobothria dorsally on the pedipalpal femur. **f** Firebug (*Pyrrhocoris apterus*), with 28 filiform hairs on the ventral side of its abdomen (length ca. 5.7 mm) (from Draslar 1973). **g** Whip scorpion (*Typopeltis crucifer*) with only two trichobothria on each of its tactile first legs (from Haupt 1996). **h** Spider (*Cupiennius salei*, adult) walking leg with the majority of its trichobothria on the dorsal side (from Barth et al. 1993, modified). **b, d, e** drawings by H Grillitsch, based on data by R. Müllan

Insect filiform hairs are particularly numerous on the orthopteran cerci. To give an example, the cricket *Acheta domesticus* has about 500 filiform hairs on each of its cerci, which are about 5 mm long (Gnatzy 1996). The corresponding value found for the cockroach *Periplaneta americana* is 208 (Nicklaus 1965).

7.3 Convergence Versus Homology or the Power of Physical Selection Pressures

Considering the diversity of the occurrence and topography of cuticular hairs responding to air particle movement, one might suspect that these are not homologous structures but instead represent a case of convergent evolution and have been shaped by the same physical selection pressures. Several structural differences indeed indicate such a convergent evolution for insect filiform hairs and arachnid trichobothria (Barth 2002). (1) Whereas insect filiform hairs are supplied by one sensory cell only, we have three or four of them in spider trichobothria. There are as many as 11 in the whip scorpion *Typopeltis crucifer*. For unknown reasons only six of these send their dendrites up to the dendritic sheath, which is coupled to the inner hair shaft (Haupt 1996). Even 16 sensory cells were found in some myriapods (Haupt 1970). (2) The coupling of the dendrites to the hair shaft, the way the hair shaft is suspended in the general cuticle (Fig. 7.3a, d), and the hair's mechanical directionality clearly differ in the airflow sensors of different phylogenetic origin (spiders: Görner 1965; Anton 1991 quoted in Barth 2002; scorpions: Hoffmann 1967; Messlinger 1987; mites: Haupt and Coineau 1975; Alberti et al. 1994). (3) In addition the number of sheath cells contributes to the differences in the structure and developmental processes between the filiform airflow sensors in different groups of arthropods. The minimum number of sheath cells is three, forming the socket, the dendritic sheath, and the hair shaft, respectively (cricket: Schmidt and Gnatzy 1971; scorpion: Altner 1977; Messlinger 1987).

Even within the Arachnida, like in spiders and scorpions, a homology of the trichobothria is at least doubtful (Schuh 1975; Weygoldt and Paulus 1979; Haupt

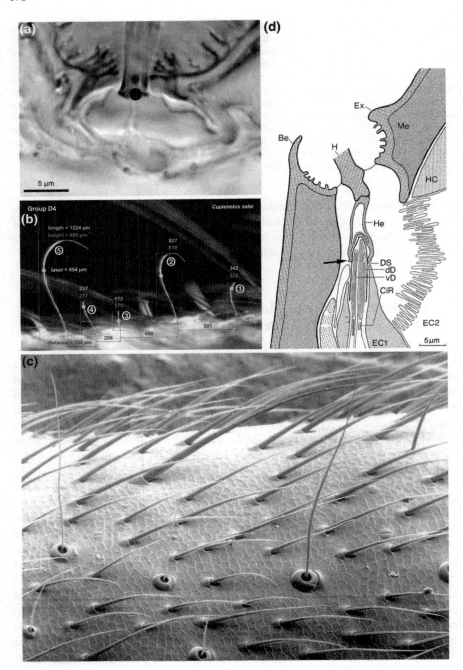

◄ **Fig. 7.3** *Structure of trichobothrium.* **a** Slice preparation of living trichobothrium on scorpion (*Euscorpius sp.*) pedipalp showing its suspension; note location of axis of rotation (*black dot*) and shortness of the hair shaft's inner lever arm as well as the sensory cells' dendrites attached to it (*courtesy* CF Schaber). **b** Example showing length, height, and s/d ratio (s, distance; d, diameter) for intact array of metatarsal trichobothria of a wandering spider (*Cupiennius salei*) (Barth, Klopsch and Schaber, unpublished). **c** Trichobothria on the walking leg of the orb-weaver *Nephila clavipes*. Note upright orientation of hair shafts (diameter at base ca. 3.5 μm) and cuticular cup (outer diameter ca. 30 μm). **d** Longitudinal section of basal part of spider trichobothrium (*C. salei*) with details of the dendrites' coupling to hair shaft (from Anton 1991 and Humphrey and Barth 2008). *H* hair shaft, suspended by joint membrane; *Be* cuticular cup; *He* helmet-like structure with the dendrites of three to four sensory cells reaching its inner end; *CiR* ciliary region; *DS* dendrite sheath; *dD, vD* dorsal and ventral dendrite; *EC1, 2* sheath cells; *Ex* exocuticle; *HC* hypodermal cell; *Me* mesocuticle

1980; Reissland and Görner 1985). A particularly bizarre fine structure to be mentioned was reported for the trichobothria of oribatid mites (*Acrogalumna longipluma*) (Alberti et al. 1994). Their socket is curiously twisted several times and its lumen divided into six chambers by thin cuticular lamellae. The basal section of the hair shaft follows the twists of the socket. The socket's presumed protective function and the relation of its unusual structure to the hair shaft's mobility still remain to be studied.

Notwithstanding all structural differences all arthropod airflow sensors are hair-like lightweight lever arms (the hair shaft) flexibly suspended in the cuticle (Fig. 7.3). In all cases the length ratio between the outer lever arm, which pro-trudes into the air, and the inner lever arm (the part of the hair shaft extending from the point of rotation inwards) coupled to the sensory cells is large, with values up to more than 1,000:1. This implies that the tip displacement of the hair shaft is scaled down considerably on the way to the dendrites and that the force is scaled up correspondingly. One is reminded of the role played by the middle ear in impedance matching in mammalian hearing. However, the effect is much larger in case of the airflow sensors (see also Shimozawa et al. 2003 and Sect. 7.4 below).

One or several sensory cells at the base of the hair shaft pick up the mechanical stimulus passed on to them by the inner end of the hair shaft (short lever arm). As can be seen from the physics of the air–hair interaction and the predictions derived from models reviewed in Humphrey and Barth (2008) the similarity of the mechanical properties of the hairs in different animal taxa are striking. Impor-tantly, the lengths of the airflow sensor hairs found in roaches, crickets, spiders, scorpions, etc., are all within the same range of about 100–2,000 μm. This remarkable finding is interpreted to reflect the role played by the boundary layer in all these cases (Fig. 7.4). The data available suggest that hair length scales with the thickness of the boundary layer flow that the hair is best matched to detect (Humphrey and Barth 2008). Longer hairs would not only be a waste of energy and material but also detract from the match between boundary layer thickness and frequency tuning by a change of the total moment of inertia (Barth et al. 1993; Devarakonda et al. 1996; Steinmann et al. 2006; Humphrey and Barth 2008).

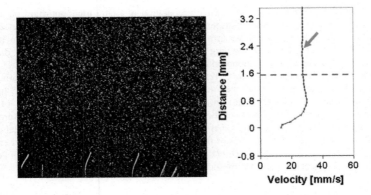

Fig. 7.4 *Boundary layer* above spider leg. *Left* Seeding particles above trichobothria exposed to a sinusoidally oscillating flow parallel to leg long axis; far field velocity 30 mm/s, frequency 50 Hz. *Right* Boundary layer thickness (*dashed line*) above leg derived from the movement of the seeding particles using digital particle image velocimetry (DPIV). *Arrow* free field velocity (FG Barth, C Klopsch, and CF Schaber, unpublished)

7.4 A Remarkable Mechanical Sensitivity

The mechanical sensitivity of spider and cricket airflow sensory hairs is outstandingly high. Trichobothria and filiform hairs are easily identified among all other arthropod cuticular hairs. They are the only ones deflected by the slightest movement of air, the main reason being the small elastic restoring force at their suspension. In the spider *C. salei,* the elastic restoring force of the hair suspension was found to differ by up to four powers of ten between trichobothria and tactile hairs (Barth and Dechant 2003). Whereas in trichobothria the spring stiffness or torsional restoring constant S of the hair suspension was on the order of 10^{-12} Nm/rad, it measured about 10^{-9} to 10^{-8} Nm/rad in spider tactile hairs (Albert et al. 2001; Barth and Dechant 2003). This value is close to that found for fly macrochaetae (Theiß 1979). The small force resisting the deflection of the hair shaft in spiders and orthopteran insects together with the hair shaft's stiffness (bending stiffness of spider trichobothria: ca. 0.18 N/m; McConney et al. 2009) explain why filiform hairs/trichobothria do not bend while being deflected by the torque resulting from the viscous forces of the airflow and before touching the rim of the socket (which is at deflection angles between 25 and 35° in *C. salei*). Significantly, this is not the case for spider tactile hairs because of their suspension's much larger torsional restoring force (Dechant et al. 2001; Barth and Dechant 2003). The smallness of the forces needed to deflect the hair also explains why insect filiform hairs and spider trichobothria are among the most sensitive of all sensors known, a circumstance which has attracted the attention of engineers and physicists. To express this sensitivity in numbers, the energy (work) needed to deflect a trichobothrium of *Cupiennius* at its resonance frequency through one movement cycle at a deflection threshold angle of 0.01° (eliciting a nervous response from the sensory cells) is between 2.5×10^{-20} and 1.5×10^{-19} J. This is close to kBT (4.1×10^{-21} J), where kB is the Boltzmann constant and T the absolute temperature. A very similar sensitivity applies to the filiform hairs on the cricket cerci (Thurm 1982; Humphrey et al. 2003). According to the same calculations, cricket hairs require just slightly larger far field velocities and work to attain the same threshold displacement (Humphrey et al. 2003). At threshold cricket hairs work close to the thermal noise of Brownian motion, too. The signal was shown to be enhanced by stochastic resonance, both in the sensor itself and the first-order interneurons (Levin and Miller 1996; Shimozawa et al. 2003; for stochastic resonance in crayfish mechanoreceptors see Douglass et al. 1993). This means an improvement of signal encoding by external noise close to the threshold. The frictional forces acting on the hair shaft at its threshold deflection are in the order of $0.4\text{–}4 \times 10^{-6}$ N. There are responses to airflow as slow as 0.15 mm/s and, in an extreme case, to deflections as small as 0.001° (Barth and Höller 1999). Note that the sensitivity values we are dealing with correspond to a fraction of the energy contained in one single quantum of visible light. In the frequency range of their operation, arthropod airflow sensors rival with the sensitivity of the receptor cells in our human eyes and ears (Gitter and Klinke 1989).

Another way to point to the outstanding sensitivity of arthropod airflow sensors brings us back to their characteristics as lever systems (Fig. 7.3). Remember that the length relation between the outer and the inner lever arms measures up to 1,000:1 [according to live slice preparations of spider, cricket, scorpion, and cockroach filiform hairs, Schaber and Barth in prep., and electron microscopic data by Christian (1971, 1972), Gnatzy and Tautz (1980), and Anton (1991)]. This implies that the axis of rotation is located only a few micrometers below the cuticular surface (Fig. 7.3a). This in turn leads to a massive scaling down of the hair's deflection at its inner end and a corresponding scaling up of force. The deflection of the inner end of the hair shaft's inner lever arm in spider trichobothria (*C. salei*) was calculated to be just 0.07 nm at a deflection threshold of 0.01° (Barth and Höller 1999; for the cricket see Gnatzy and Tautz 1980).

As a note of caution, it should be stressed that not all sensilla referred to as trichobothria or filiform hairs in the literature (mainly defined morphologically as a filiform hair shaft inserted in a cup-like socket; see Fig. 7.3) necessarily exhibit the sensitivity established for spiders and crickets. Thus according to Alberti et al. (1994), the trichobothria of moss mites did not visibly vibrate when exposed to the air current generated with a Pasteur pipette. Possibly then, some trichobothria are comparatively insensitive airflow sensors with an additional or even primary function as sensors responding to touch or substrate vibration. Candidates for the latter function are the trichobothria of oribatid mites that live in the soil. There are also trichobothria with unusual shapes of the hair shaft. In mites one finds "comb-like, fan-like, globose, capitate, or fusiform" trichobothria (Alberti et al. 1994). All these trichobothria await experimental examination from a functional and neuroethological point of view.

7.5 Sensor Arrays and the Complexity Added by Viscosity-Mediated Coupling

As already seen (Fig. 7.2), arthropod airflow sensors typically do not occur singly but in groups, forming a diversity of patterns that await functional interpretation. In some animal groups like scorpions and pseudoscorpions, these patterns are used as taxonomic characters (Vachon 1973; Mahnert 1976).

When the shafts of hairs forming a group differ in length (Fig. 7.3b), the frequency range covered by the group is larger than that of a single hair. The mechanical frequency response was actually measured for groups of up to five hairs on the tibia of *C. salei*. The length of the shortest of these hairs was 400 μm whereas that of the longest one was 1,150 μm (Barth et al. 1993). The mechanical tuning of all hairs was broad but a difference in the range of best frequencies nevertheless was clearly seen. Whereas the shortest hair responded maximally to frequencies around 600 Hz, the longest one had its maximum response at around 50 Hz. A length difference of only 750 μm correlated with a shift of the "best frequency" by 550 Hz. Interestingly, physical–mathematical modeling predicts that changing hair length (as compared to diameter, mass, or the elastic restoring constant) is the most effective way to change the hair's mechanical tuning (Humphrey et al. 1993; Devarakonda et al. 1996).

Apart from their tuning characteristics, hairs of different lengths differ in regard to absolute mechanical sensitivity, the long ones being more sensitive than the short ones (Görner and Andrews 1969). As an example, the deflection of tibial trichobothria of *C. salei* by an airflow of 50 mm/s at the respective best frequencies was about 1° in hairs 300 μm long and roughly 6° in hairs 1,200 μm long (Barth et al. 1993). However, deflection does not depend on hair length alone, but also on S and R, as it is evident that the trichobothria of corresponding length on the tarsus and metatarsus are more mechanically sensitive than those on the tibia (Barth et al. 1993). Since boundary layer thickness increases with decreasing frequency, very short hairs are insensitive to low frequencies, when they are immersed in a zone of reduced flow velocity. This effect is closely associated with the lower limit of a hair's frequency range, whereas the hair's inertia limits the frequency response range at its upper end. Since the moment of inertia increases with the third power of ten of hair length, the hair becomes increasingly immobile with increasing stimulus frequency (Shimozawa et al. 1998, 2003).

When exposing a group of trichobothria showing length gradation to the wake generated by a flying fly, the differences in the mechanical responses of the hairs could be clearly seen (Fig. 7.5). According to high-speed video analysis, shorter hairs responded more quickly (shorter latency) to higher frequency components than long hairs. This was to be expected considering their smaller mass and total inertia, torsional restoring constant, and total damping constant. At a distance of 28 cm from the fly, the fly's wake deflected the long hairs only, because the airflow did not contain its higher frequency components anymore (Fig. 7.5b). For the complex theoretical treatment of the effect of short (ca. 7–70 ms) airflow

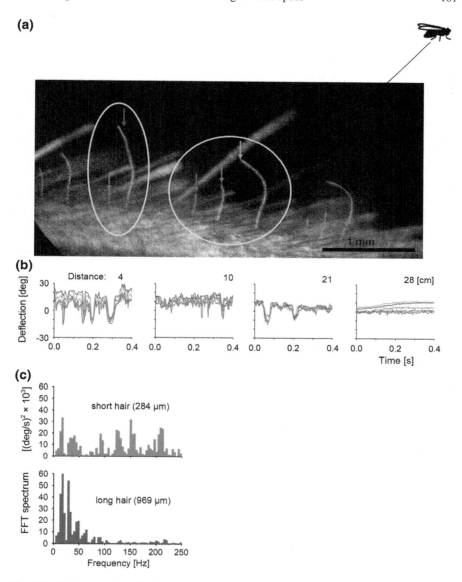

Fig. 7.5 *Deflection of spider trichobothria* (*Cupiennius salei*) *in an array* responding to the airflow generated by the wake of a flying blowfly. **a** Five trichobothria differing in length and marked by arrows of different colors. **b** Deflection of hairs measured with a laser vibrometer at increasing distance from the fly. See text for details. **c** Spectra of deflection frequencies; note mechanical response of short hair to higher frequency components in airflow (FG Barth, C Klopsch, CF Schaber unpublished)

pulsations on the deflection of spider trichobothria and cricket cercal filiform hairs, the reader is referred to Kant and Humphrey (2009).

The question whether neighboring airflow sensors in an array respond to airflow as truly independent units or instead are subject to viscosity-mediated interactions between the hairs was recently studied in several laboratories. Considering both the small size of filiform hairs and trichobothria and the low Reynolds number flow we are dealing with in the typical case, such a coupling between hairs is a realistic possibility. In a pioneering study, both experimental and theoretical, Bathellier et al. (2005, 2010) examined whether the flow field around a hair changes the dynamics of neighboring hairs, taking trichobothria of intact spiders (*C. salei*) as an example.

Viscosity-mediated coupling may be expected if the boundary layers of the two neighboring hairs overlap significantly ($s \leq 2\delta$; s, distance between hairs; δ, boundary layer thickness). This might happen at low velocities for a steady flow and at low frequencies of an oscillatory flow. According to Bathellier et al. (2005), viscous coupling between a pair of trichobothria is at most very small at the biologically relevant airflow frequencies tested (30–300 Hz). The main reason is the wide spacing of 20–50 hair diameters between the hairs and the tuning of the hairs to the frequencies mentioned. Freely moving long hairs are not or hardly affected by the flow perturbation caused by freely moving shorter hairs, because their distal hair shafts protrude well above the zone of perturbation and the torque is the largest at the hair tip.

The predictions of the theoretical model agree well with the experimental results. They also agree with earlier theories (Humphrey et al. 1993) developed for single hairs when theoretically turning grouped hairs into single hairs by a corresponding increase of the distance between them (Bathellier et al. 2005, 2010). The degree of coupling decreases with increasing s/δ ratio. This relation, however, depends on the frequency (stronger coupling at low than at high frequencies due to larger values for δ). Nevertheless, for $s/\delta > 50$ the coupling is very weak for all tested frequencies (see also Cheer and Koehl 1987). Interactions between freely moving hairs of similar length are particularly small at their best frequencies (maximum deflection speed).

Considering their spacing and the specific distribution of hair lengths in a group, the perturbation of a hair's deflection by a neighboring hair was found to be less than 5 % for the trichobothria of *C. salei* within the behaviorally most relevant range of frequencies (up to 200 Hz) (Brittinger 1998; Barth and Höller 1999). Not being coupled preserves the differences in the response characteristics of the individual hairs in a group and may therefore be advantageous.

Following a recent study, the s/δ values for the 48 trichobothria on the pedipalp of a scorpion (*Smeringurus mesaensis*, Vaejovidae) are too large to induce coupling of any significance (Müllan 2012) (see also examples of scorpions in Fig. 7.2). Thus the spider (*Cupiennius*), the scorpion (*Smeringurus*), and the cabbage white caterpillar (*Barathra*) (Markl and Tautz 1975) all show at most very little or negligible coupling. The situation is different on the cockroach and cricket cercus (Magal et al. 2006; Dangles et al. 2006). According to Dangles et al. (2006), the density of filiform hairs on the cerci of the wood cricket *Nemobius silvestris* is >400 hairs/mm^2, implying close neighborhood. *Acheta domesticus*, another cricket, has about 500 filiform hairs on each cercus that in adults is only about 5 mm long (Gnatzy 1996). Not surprisingly, coupling was calculated to be significant for the closely spaced cricket hairs (Cummins et al. 2007).

The reader interested in the most recent developments of theory is also referred to Heys et al. (2008) and Lewin and Hallam (2010) who applied computational fluid dynamics (CFD) techniques. So far there are no studies on the effects of close neighborhood in intact hairs except that of Bathellier et al. (2005, 2010). This also applies to the recent experimental measurement of the airflow around and between biomimetic MEMS hairs, which in addition to being synthetic were fixed (Casas et al. 2010). Following their study, Casas et al. (2010) predicted the potential of strong hydrodynamic coupling within entire natural hair canopies like those found on the insect cerci. The most recent review on the response of groups of filiform hairs to simple and more complex stimuli is that by Cummins and Gedeon (2012). Despite the progress made in modeling during the last decade, there is a lot still to be done. The emphasis of future studies should be on intact arrays of flow sensors and stimuli at least approaching the natural conditions.

Among the cases that offer themselves for further study are the duplex trichobothria near the tip of the fixed finger of the pseudoscorpion chela (Heterophyronida; Feaelloidea). The distance between the basal sections of these trichobothria measures <10 μm. A substantial coupling between the hairs is very likely. So far there is neither a functional analysis available nor is the fine structure of the hairs known (Harvey 1992; Judson 2007).

7.6 Biologically Relevant Airflows

Considering the many animal species that use medium flow signals to acquire information about their environment, our knowledge of the physical properties of airflow stimuli is very limited. To learn more about the biologically relevant flows is indispensable because it is these flows the sensory systems have to cope with. Synthetic sinusoidal oscillatory flows are well suited to study the basic properties of the sensors. Usually, however, they are not the natural flows. These are characterized by accelerating and decelerating dc-flows and often will be complicated by turbulence, vortices, changing boundary layers, etc.

The task to be accomplished is demanding. Ideally, the measurements should be done in the field with the actors in their natural habitat and during their natural activity period. Quasi-natural situations created in the laboratory are helpful once it is known what they should be like. Valuable new technologies providing high spatial and temporal resolution and not disturbing the flow will be increasingly applied. High-speed video and digital particle image velocimetry (DPIV), mapping the velocity distribution in a relatively large area, will be of particular value as exemplified in the following section. The reader is also referred to the pioneering studies of older times when these technologies were not or hardly available yet. Examples are the study by Tautz and Markl (1978) on the interaction between cabbage white (*Barathra*) caterpillars and their flying wasp predators, the study of Gnatzy (review 1996) and his associates on the predator–prey interaction between digger wasps (*Liris*) and crickets, the seminal experiments by Camhi (review 1984) and his associates on the detection of natural predators by the cockroach (*Periplaneta americana*) and the analysis of the role played by the trichobothria when the semi-aquatic spider *Dolomedes* escapes from a predator frog (Suter 2003). An elegant recent study by Casas et al. (2008) on the escape reaction of a cricket, which can be triggered by the airflow generated by an approaching wolf spider, used DPIV to analyze the airflow upstream a running spider. In all these cases, filiform airflow sensors were essential in eliciting escape responses. This most likely is also true for the cercal system of the praying mantis, which (apart from the detection of bat echolocation by its auditory system) detects wind generated by flying bats at least 50 cm away (Triblehorn and Yager 2006).

Interestingly, the spatial ranges covered by the sensors as seen from behavioral reactions show similarity in all these cases. The distances from the signal source at which behavior (motion) is elicited varied between a few centimeters and about 70 cm. The distance of a fly whose wake still deflects a trichobothrium of *C. salei* above threshold was found to be 50–70 cm (Barth and Höller 1999). In comparison, a behavioral reaction could be elicited with the fly only up to 17–27 cm away, depending on its altitude above the substrate and the wake's frequency composition (review: Barth 2002).

7.7 The Spider and the Fly

This final section shortly summarizes recent work (Klopsch 2010; Klopsch et al. 2012, 2013) on the wandering spider *C. salei*. It illustrates both the difficulties and opportunities of the analysis of a behavior guided by airflow stimuli.

Cupiennius can frequently be observed to catch a flying insect prey like a fly by a well oriented and timed jump into the air (Fig. 7.6). This is a remarkable behavior considering the tasks to be mastered. The spider has to detect the prey-generated airflow signal from within the noise typical of its environment and to recognize it as prey-like. It also has to ensure proper timing to reach the fast moving signal source and to activate the proper motor program ensuring the oriented jump.

Given the multimodal nature of the natural stimulus, one first has to make sure that airflow sensing is indeed involved. The roles played by visual stimulation, substrate vibration, and sound pressure have to be specified or excluded.

Fig. 7.6 *The spider's (Cupiennius salei) jump* toward a flying insect prey. A humming fly *(Calliphora erythrocephala)* attached to a holder is moved from *left* to *right*. **a** The spider in its resting position with its front side pointing away from the fly. **b** Even before the approaching fly is above the spider, the spider has already turned toward the fly. **c** The prey capture jump, elicited when the fly is above the spider leg closest to it. Leg span of spider 7.5 cm; see text for details (FG Barth, C Klopsch, CF Schaber, unpublished)

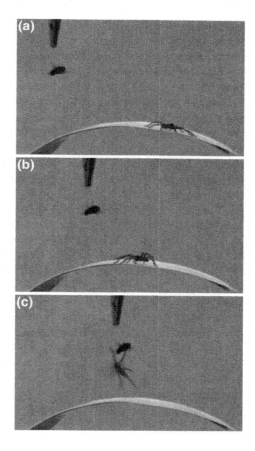

7.7.1   The Problem of Multimodal Stimuli

The information contained in the airflow generated by the flying fly is indeed necessary and sufficient to guide the spider's remarkable behavior. (1) *Vision*: It had been known for a while (Barth et al. 1995; Brittinger 1998) that spiders (*C. salei*) deprived of their trichobothria never jump into the air to catch flying prey but do it with their eyes blinded (as in the experiments reported below) and their trichobothria intact.[1] (2) *Substrate vibration*: The flying fly might also induce substrate vibrations in the spider's dwelling plant, potentially stimulating its highly sensitive vibration receptors (Barth and Geethabali 1982; Schaber et al. 2012).

---

[1] It has recently been shown that *C. salei* also jumps at dark disks slowly moving across a computer screen. Attack behavior can therefore be elicited independently by substrate vibrations, airflow stimuli, and visual stimuli (Fenk et al. 2010). The significance of the latter under field conditions, taking into account the spider's nocturnal activity, still remains to be shown.

Using laser-Doppler-vibrometry, both the vertical and horizontal vibrations of a bromeliad leaf (bromeliads being typical dwelling plants of *C. salei*) were found to be below the threshold value for the spider vibration receptor organ, even below its lowest threshold known at 1 kHz (Barth and Geethabali 1982). Substrate vibrations, therefore, do not play a role here either. This conclusion is supported by the finding that the spider also jumps from a heavy and strongly damped 20 kg metal plate, which the fly could not induce to vibrate to any measurable (laser vibrometry) extent (Klopsch et al. 2013). (3) *Airborne sound*: Lastly, it remains to be shown that the trichobothria do not detect the air particle velocity due to the pressure fluctuations associated with fly-generated airborne sound. The sound pressure owing to the first (strongest) harmonic (corresponding to wing beat frequencies between 130 and 200 Hz) ranged from about 64 dB SPL (1 cm below fly) to about 25 dB SPL (10 cm ahead and 7 cm below fly). It radiated from the fly in a roughly dipole-like pattern. The adequate stimulus for trichobothria is air movement (velocity). However, sound pressure in the acoustic near-field cannot be simply transformed into velocity values. Therefore, the angular deflection and velocity of the trichobothria due to the drag forces, which result from the airborne sound emitted by the fly, were measured directly using laser-Doppler vibrometry (Klopsch et al. 2013). The values found for the tarsal trichobothrium most sensitive to the frequencies contained in the first harmonic (peak ranges between 130 and 200 Hz) were far below the threshold values (previously determined electrophysiologically) at most positions relative to the fly. With the fly 5 cm above the spider, where it was caught successfully in behavioral experiments, both angular deflection and velocity reached just 8 and 6 % of the physiological thresholds, respectively. It was concluded that the spiders were unable to sense the fly by making use of the sound it emitted (Klopsch et al. 2013).

7.7.2 The Relevant Airflow Pattern

Which properties of the "fly-flow" provided the cues supporting the spider's successful jump? To answer this question, the airflow generated by a freely flying fly had to be examined. Both the flows generated by tethered and stationary humming flies and by flies tethered, humming, and experimentally moved differed from the flow generated by a freely flying fly (Klopsch et al. 2012). Using freely flying flies for DPIV experiments was not an easy task. The most relevant results were the following.

(1) Air circulated around the front half of a freely flying fly (*Calliphora erythrocephala*) (Fig. 7.7a). The wake behind a horizontally flying fly extended obliquely downward, moving with the fly. The basic structure of the flow remained the same within a wide range of flight velocities (measured between 13 and 81 cm/s; maximum flight speed of blowfly ca. 1.2 m/s according to Schilstra and van Hateren 1999) but changed significantly with speed in case of the artificially pulled humming fly (Klopsch et al. 2012). (2) From the spider's point of view, the relevant question is

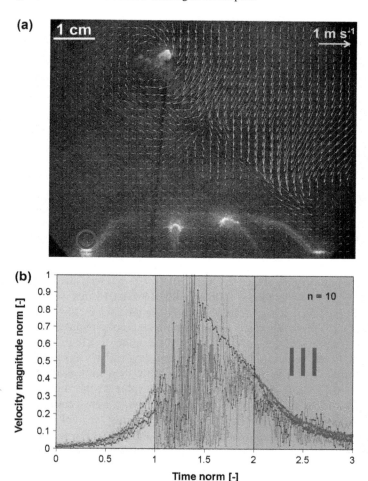

Fig. 7.7 *Airflow signal* generated by a blowfly (*Calliphora erythrocephala*) flying freely from right to left and measured using digital particle image velocimetry (DPIV). **a** Airflow velocity vector map within vertical laser light sheet around the fly and also cutting the four right legs of the spider (*blue circle* marks leg 4). **b** The three phases of the fly-generated airflow (velocity magnitude) above the spider's tarsal trichobothria. The velocity signals (5 animals, 10 measurements) were normalized (maximum representing 1) and superimposed. See text for details. The time was normalized as well, with Phase I ranging from 0 to 1, Phase II from 1 to 2, and Phase III from 2 to 3 (from Klopsch 2010, modified)

what happens close to its airflow sensors when the fly is approaching and passing by. Importantly, the flow close to the sensors was subdivided into three successive phases, which greatly differed from each other (Fig. 7.7b). *Phase I* was characteristic of the fly's approach. Its strength (velocity) increased nearly exponentially with time and decreasing distance to the fly. The largest flow velocity was 0.16 ± 0.05 m/s ± S.D. and fluctuation was small (0.013 ± 0.006 m/s). Judging

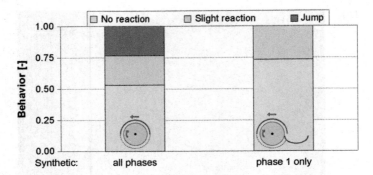

Fig. 7.8 *The spider's behavioral reactions* toward synthetic airflows imitating the entire (all phases, *left*) fly signal or only its Phase I (*right*). *Note* schematic cross sections of rotating cylinder device used to generate the synthetic signals. *Arrow* above cylinder indicates direction in which the device was moved. See text for details (from Klopsch 2010, modified)

from its behavioral reaction the spider detected this predominantly horizontal flow ahead of the approaching fly when it was still 38.4 ± 5.6 mm away. The fluctuation values of Phase I flow increased from about 0.010 to 0.018 m/s when the altitude of the fly increased from about 20 to 60 mm (above the spider). Phase II started when the fly was still a few millimeters ahead of the spider tarsus or directly above it. Then the fly's wake exposed the airflow sensors of the spider to a highly fluctuating flow with an increased vertical velocity component. Phase II always abruptly followed Phase I (Fig. 7.7b). The rms values of the flow then increased by an order of magnitude (0.113 ± 0.050 m/s), and the mean value of the relative fluctuations (70 ± 23 %) more than doubled. The peak of the power spectrum shifted from ca. 8 Hz (Phase I) to ca. 18 Hz and, more importantly (see below), the spectrum contained frequencies of up to at least 250 Hz. Phase III only started when the fly was already ca. 3 cm beyond the tarsus. It was characterized by a gradually decaying flow (Fig. 7.7b) and need not be further discussed here because at that point of time the spider consistently had already jumped.

7.7.3 What does this Mean to the Spider?

During all its phases, the strength of the fly-generated airflow close to the sensors was above 1 mm/s and thus above their physiological threshold (Barth and Höller 1999).

Behavioral studies with blinded spiders exposed to natural or controlled synthetic flows helped to recognize the significance of particular flow features (Klopsch et al. 2013). The spiders responded to Phase I flow with orientation behavior, never, however, with a jump. Jumps could be elicited by the onset of Phase II flow, both when exposing the spider to natural airflows or to a synthetic Phase II flow (Fig. 7.8). The jump always occurred before Phase III flow hits the sensors.

7.7.4 Again: The Tasks to be Mastered

One should now ask again how the spider solves the various tasks associated with a successful prey capture jump and specified at the outset.

7.7.4.1 Sensitivity

Given the outstanding sensitivity of its trichobothria, the spider will have sensory access to many flows, including those of a prey fly and a predator bat, wasp, or bird. Airflow may also be relevant at a later stage of courtship communication in C. salei (Barth 1997) and other spiders. Since the trichobothrial system is a highly dynamic/phasic one which responds to fluctuating instead of steady flow (Friedel and Barth 1997; Barth and Höller 1999; Humphrey and Barth 2008; McConney et al. 2009), it appears to be well adapted to the reception of highly fluctuating prey-generated signals like that of the blowfly.

7.7.4.2 Detection of the Signal and the Problem of Background Noise

Like in all sensory systems, the signal-to-noise ratio is a significant issue in regard to signal detection. According to measurements in the natural habitat of C. salei in Central America at the time of its nocturnal activity (Barth et al. 1995), this spider has to deal with low frequency background air motion and high frequency components superimposed. Typically, the background airflow is composed of very low frequencies only (below 10 Hz, often even below 1 Hz) and characterized by both a narrow frequency spectrum and a low degree of fluctuation (2–3 %). Shortly after sunset, the wind speeds in the field and the likelihood of gusts usually decrease significantly. This is one of the advantages that Cupiennius and other animals responding to airflow gain by being night active, beginning shortly after sundown. Trichobothria are much less sensitive to the low background frequencies than to the higher frequency components contained in the prey-generated airflows (Barth and Höller 1999).

7.7.4.3 Recognition of the Signal

All available data point to the particular biological importance of transient as opposed to static flow effects and to that of the higher frequency components usually not found in the background flow. The jump of Cupiennius is triggered right at the beginning of the highly fluctuating flow that contains frequencies up to 250 Hz and more. As pointed out already, the biological relevance of these fluctuations is reflected by a corresponding adaptedness of the trichobothrial system to transient events. This can be seen in the sensors' structure, and both their biomechanical and physiological properties. The strictly phasic response type was also found in central nervous interneurons integrating the input received from the sensory periphery (Friedel and Barth 1997; review Barth 2002; Humphrey and Barth 2008).

7.7.4.4 Localization and Orientation

The spider's jump not only needs to be well oriented but also has to take the fly's translational velocity into account. We are still far from understanding the underlying neuronal operations in any detail but there are some fragments already worth being reported.

1. The spider grabs its prey from a wide space with its front legs, which span 10 cm and more in adults. This implies that the necessary precision of the jump is much lower than it would have to be in case of prey capture with the chelicerae.
2. The horizontal distance to the fly and its approach is likely to be indicated by the succession of the three phases of the airflow. The onset of Phase I occurs when the fly is 20–60 mm away, independent of its altitude. Phase I only elicits a turning reaction toward the source of the flow. The jump is triggered at the beginning of Phase II. Therefore, the transition between the phases obviously informs the spider that the distance to the prey is short enough to warrant a jump. The high frequency components contained in the wake of the fly signal quickly disappear with distance. The spider does not jump anymore when the wake looks like background noise (very narrow frequency spectrum, very low frequencies). This occurs at a distance of about 25 cm from the fly (Barth and Höller 1999). In this case then the lack of high frequency components means "too far away," equivalent to "just background noise."
3. The spider always turns into the direction of the leg stimulated first (Brittinger 1998; Barth 2002). Any azimuth of the stimulus source can be identified equally well; thanks to the ring-shaped arrangement of the airflow sensors all around the spider (Fig. 7.1). When stimulating two legs with air puffs with time delays ≥ 50 ms, the spider turned toward the leg stimulated first with high significance. Even at a Δt of only 10 ms, the spider turned more often toward the leg stimulated first (Brittinger 1998; Barth 2002). The airflows generated by freely flying flies arrived at different legs (leg span of the spiders used was 7.5 cm) at times differing by about 90 ms (Klopsch et al. 2012).
4. In addition to time differences *Cupiennius* may also use stimulus intensity differences at different legs for its orientation toward the stimulus source. The airflow is strongest where it appears first, that is above the leg pointing toward the stimulus source. Klopsch et al. (2013) found velocity ratios of up to 6.5:1. A detailed analysis has still to be carried out.
5. How does the spider know about the fly's altitude and does it need to know it at all? It is tempting to assume that the spider only needs to know whether and

when the fly is within reach of its jump. This occurs at the transition between Phase I and Phase II of the airflow. Information on the prey's altitude is contained in the degree of fluctuation of Phase I but not in the transition from Phase I to Phase II. It has not been shown yet, however, that the spider makes use of this information.

6. The timing task of the spider is complicated by the movement of its prey. Independent from the fly's altitude and direction of approach, information on the horizontal velocity of the fly is contained in the differences of the airflow's time of arrival at the different legs as well as in the steepness of the velocity increase in Phase I. It is not known whether the spider indeed uses these measures. The problem of timing may be solved in a simpler way using the trigger effect of the beginning of Phase II. At that point in time, the prey is near enough which may be all the spider has to know. Remarkably, the spider is able to adjust its leg and body position even when already in the air, and to thereby compensate for imprecision.

7.8 Outlook

Based on earlier pioneering studies, the perception of airflow by arthropods has recently received increased attention by biologists, engineers, and physicists. The functional morphology of the hair-like sensors and their responses to the viscous forces exerted by airflow were the subject of both experimental studies and physical–mathematical modeling. The outstanding sensitivity of at least some of the airflow sensors, which rivals that of the human ear and eye, has been fully confirmed. At least in the cases so far studied in some detail flow fluctuation and the "high" frequency tail in the spectrum of the stimuli and usually absent from background noise were shown to be a particular marker of biological relevance. The question of viscosity-mediated coupling between neighboring sensors in an array has attracted the interest of several authors. The gaps still to be filled in our understanding of arthropod airflow sensing cover a wide field of research, which promises a rich harvest. Future studies should include some of the diversity of airflow sensing found in many arthropod groups, not just crickets, cockroaches, and spiders, the favorite experimental animals so far. In order to understand the principles at work, it will help to extend our knowledge of the relevant airflow patterns to which a particular arthropod is exposed in its habitat. Likewise, more electrophysiological studies are needed in order to uncover the physiological diversity of filiform hairs and trichobothria. Neuroethological case studies in animals so far not having enjoyed the interest of physiologists will greatly widen our view of what airflow sensing means in arthropod biology and how it is adapted to specific behavioral and ecological demands.

Acknowledgments The author's research and that of his associates was generously supported by the DARPA BIOsenSE program grant no. FA9550-05-1-0459 and by several earlier grants of the Austrian Science Fund, FWF. The kind help of Clemens Schaber with the preparation of the figures is gratefully acknowledged.

References

Albert JT, Friedrich OC, Dechant H-E, Barth FG (2001) Arthropod touch reception: spider hair sensilla as rapid touch detectors. J Comp Physiol A 187:303–312

Alberti G, Moreno AI, Kratzmann M (1994) The fine structure of trichobothria in moss mites with special emphasis on *Acrogalumna longipluma* (Berlese, 1904) (Oribatida, Acari, Arachnida). Acta Zool (Stockholm) 75(1):57–74

Altner H (1977) Insect sensillum specificity and structure: an approach to a new typology. In: LeMagnen J, MacLeod P (eds) Olfaction and taste, vol VI., Paris Information retrieval, London, Washington DC, pp 295–303

Anton S (1991) Zentrale Projektionen von Mechano- und Chemoreceptoren bei der Jagdspinne *Cupiennius salei* Keys. Doctoral thesis, University of Vienna

Barth FG (1997) Vibratory communication in spiders: adaptation and compromise at many levels. In: Lehrer M (ed) Orientation and communication in arthropods. Birkhäuser, pp 247–272

Barth FG (2002) A spider's world: senses and behavior. Trichobothria: the measurement of air movement, Chap. IX. Springer, Berlin Heidelberg, pp 85–109

Barth FG, Dechant H-E (2003) Arthropod cuticular hairs: tactile sensors and the refinement of stimulus transformation. In: Barth FG, Humphrey JAC, Secomb TW (eds) Sensors and sensing in biology and engineering. Springer, New York, pp 159–171

Barth FG, Geethabali (1982) Spider vibration receptors: threshold curves of individual slits in the metatarsal lyriform organ. J Comp Physiol A 148:175–185

Barth FG, Höller A (1999) Dynamics of arthropod filiform hairs. V. The response of spider trichobothria to natural stimuli. Phil Trans R Soc Lond B 354:183–192

Barth FG, Wastl U, Humphrey JAC, Devarakonda R (1993) Dynamics of arthropod filiform hairs. II. Mechanical properties of spider trichobothria (*Cupiennius salei* Keys.). Phil Trans R Soc Lond B 340:445–461

Barth FG, Humphrey JAC, Wastl U, Halbritter J, Brittinger W (1995) Dynamics of arthropod filiform hairs. III. Flow patterns related to air movement detection in a spider (*Cupiennius salei* Keys.). Phil Trans R Soc Lond B 347:397–412

Bathellier B, Barth FG, Albert JT, Humphrey JAC (2005) Viscosity-mediated motion coupling between pairs of trichobothria on the leg of the spider *Cupiennius salei*. J Comp Physiol A 191:733–746. See also Erratum (2010) J Comp Physiol A 196:89

Bleckmann H (1994) Reception of hydrodynamic stimuli in aquatic and semiaquatic animals. In: Rathmayer W (ed) Progress in Zoology, vol 41. G Fischer, Stuttgart, p 115

Brittinger W (1998) Trichobothrien, Medienströmung und das Verhalten der Jagdspinnen (*Cupiennius salei* Keys.). Doctoral thesis, University of Vienna

Budelmann BU (1989) Hydrodynamic receptor systems in invertebrates. In: Coombs S, Görner P, Münz H (eds) The mechanosensory lateral line: neurobiology and evolution. Springer, New York, pp 607–632

Camhi J (1984) Neuroethology: nerve cells and the natural behaviour of animals. Sinauer, Sunderland MA

Casas J, Dangles O (2010) Physical ecology of fluid flow sensing in arthropods. Annu Rev Entomol 55:505–520

Casas J, Steinmann T, Dangles O (2008) The aerodynamic signature of running spiders. PLoS ONE 3(5):e2116 (p 6)

Casas J, Steinmann T, Krijnen G (2010) Why do insects have such a high density of flow-sensing hairs? Insights from the hydromechanics of biomimetic MEMS sensors. J R Soc Interface. doi:10.1098/rsif.2010.0093

Cheer AYL, Koehl MAR (1987) Paddles and rakes: fluid flow through bristled appendages of small organisms. J Theor Biol 129:17–39

Christian UH (1971) Zur Feinstruktur der Trichobothrien der Winkelspinne Tegenaria derhami (Scopoli), (Agelenidae, Araneae). Cytobiologie 4:172–185

Christian UH (1972) Trichobothrien, ein Mechanoreceptor bei Spinnen. Elektronenmikroskopische Befunde bei der Winkelspinne Tegenaria derhami (Scopoli), (Agelenidae, Araneae). Verh Dtsch Zool Ges 66:31–36

Coombs S, Görner P, Münz H (1989) The mechanosensory lateral line: neurobiology and evolution. Springer, New York

Cummins B, Gedeon T (2012) Assessing the mechanical response of groups of arthropod filiform flow sensors. In: Barth FG, Humphrey JAC, Srinivasan MV (eds) Frontiers in sensing from biology to engineering. Springer, New York, pp 239–250

Cummins B, Gedeon T, Klapper I, Cortez R (2007) Interaction between arthropod filiform hairs in a fluid environment. J Theor Biol 247:266–280

Dangles O, Pierre D, Vannier F, Casas J (2006) Ontogeny of air-motion sensing in cricket. J Exp Biol 209:4363–4370

Dechant H-E, Rammerstorfer FG, Barth FG (2001) Arthropod touch reception: stimulus transformation and finite element model of spider tactile hairs. J Comp Physiol A 187:313–322

Devarakonda R, Barth FG, Humphrey JAC (1996) Dynamics of arthropod filiform hairs. IV. Hair motion in air and water. Phil Trans R Soc Lond B 351:933–946

Douglass JK, Wilkens L, Pantazelou E, Moss F (1993) Noise enhancement of information transfer in crayfish mechanoreceptors by stochastic resonance. Nature 365:337–340

Draslar K (1973) Functional properties of trichobothria in the bug Pyrrhocoris apterus (L.). J Comp Physiol 84:175–184

Dumpert K, Gnatzy W (1977) Cricket combined mechanoreceptors and kicking response. J Comp Physiol 122:9–25

Fenk LM, Hoinkes T, Schmid A (2010) Vision as a third sensory modality to elicit attack behavior in a nocturnal spider. J Comp Physiol A 196:957–961

Fletcher NH (1978) Acoustical response of hair receptors in insects. J Comp Physiol 127:185–189

Friedel T, Barth FG (1997) Wind-sensitive interneurons in the spider CNS (Cupiennius salei): directional information processing of sensory inputs from trichobothria on the walking legs. J Comp Physiol A 180:223–233

Gitter AH, Klinke R (1989) Die Energieschwellen von Auge und Ohr in heutiger Sicht. Naturwissenschaften 76:160–164

Gnatzy W (1996) Digger wasp vs. cricket: Neuroethology of a predator-prey interaction. In: Lindauer M (ed) Information processing in animals, vol 10. G Fischer, Stuttgart, 92 pp

Gnatzy W, Tautz J (1980) Ultrastructure and mechanical properties of an insect mechanoreceptor: stimulus-transmitting structures and sensory apparatus of the cercal filiform hairs of Gryllus. Cell Tissue Res 213:441–463

Görner P (1965) A proposed transducing mechanism for a multiply innervated mechanoreceptor (trichobothrium) in spiders. Cold Spring Harbor Symp Quant Biol 30:69–73

Görner P, Andrews P (1969) Trichobothrien, ein Ferntastsinnesorgan bei Webspinnen (Araneen). Z vergl Physiol 64:301–317

Große S, Schröder W (2012) Deflection-based flow field sensors: examples and requirements. In: Barth FG, Humphrey JAC, Srinivasan MV (eds) Frontiers in sensing—from biology to engineering. Springer, New York, pp 393–403

Harvey MS (1992) The phylogeny and classification of the Pseudoscorpionida (Chelicerata: Arachnida). Invertebr Taxon 6:1373–1435

Haupt J (1970) Beitrag zur Kenntnis der Sinnesorgane von Symphylen (Myriapoda). I. Elektronenmikroskopische Untersuchung des Trichobothriums von *Scutigerella immaculata* Newport. Z Zellforsch mikr Anat 110:588–599

Haupt J (1980) Phylogenetic aspects of recent studies on myriapod sense organs. In: Camatini M (ed) Myriapod biology. Academic Press, London, pp 391–406

Haupt J (1996) Fine structure of the trichobothria and their regeneration during moulting in the whip scorpion *Typopeltis crucifer* Pocock, 1894. Acta Zool (Stockholm) 77(2):123–136

Haupt J, Coineau Y (1975) Trichobothrien und Tastborsten der Milbe *Microcaeculus* (Acari, Prostigmata; Caeculidae). Z Morph Tiere 81:305–322

Hergenröder R, Barth FG (1983) The release of attack and escape behavior by vibratory stimuli in a wandering spider (*Cupiennius salei* Keys). J Comp Physiol 152:347–359

Heys J, Gedeon T, Knott BC, Kim Y (2008) Modeling arthropod hair motion using the penalty immersed boundary method. J Biomech Eng 41:977–984

Hoffmann C (1967) Bau und Funktion der Trichobothrien von *Euscorpius carpathicus* L. Z vergl Physiol 54:290–352

Humphrey JAC, Barth FG (2008) Medium flow-sensing hairs: biomechanics and models. In: Casas J, Simpson SJ (eds) Insect mechanics and control. Adv Insect Physiol 34:1–81

Humphrey JAC, Barth FG, Reed M, Spak A (2003) The physics of arthropod medium-flow sensitive hairs: biological models for artificial sensors. In: Barth FG, Humphrey JAC, Secomb TW (eds) Sensors and sensing in biology and engineering. Springer, Wien, New York, pp 129–144

Humphrey JAC, Devarakonda R, Iglesias J, Barth FG (1993) Dynamics of arthropod filiform hairs. I. Mathematical modeling of the hair and air motions. Phil Trans R Soc Lond B 340:423–444

Igelmund P (1987) Morphology, sense organs and regeneration of the forelegs (whips) of the whip spider *Heterophrynus elaphus* (Arachnida, Amblypygi). J Morphol 193:75–89

Izadi N, Krijnen GJM (2012) Design and fabrication process for artificial lateral line sensors. In: Barth FG, Humphrey JAC, Srinivasan MV (eds) Frontiers in sensing—from biology to engineering. Springer, Wien, New York, pp 405–421

Judson MLI (2007) A new and endangered pseudoscorpion of the genus *Lagynochthonius* (Arachnida, Chelonethi, Chthoniidae) from a cave in Vietnam, with notes on chelal morphology and the composition of the Tyrannochthoniini. Zootaxa 1627:53–68

Kant R, Humphrey JAC (2009) Response of cricket and spider motion-sensing hairs to airflow pulsations. J R Soc Interface 6:1047–1064

Klärner D, Barth FG (1982) Vibratory signals and prey capture in orb-weaving spiders (*Zygiella x-notata, Nephila clavipes*; Araneidae). J Comp Physiol 148:445–455

Klopsch Ch (2010) The flow field around a flying blowfly: characteristics and guidance of spider prey capture behaviour. Vienna University of Technology, Doctoral thesis

Klopsch Ch, Kuhlmann HC, Barth FG (2012) Airflow elicits a spider's jump towards airborne prey. I. Airflow around a flying blowfly. J R Soc Interface 9:2591–2602. doi:10.1098/rsif.2012.0186

Klopsch Ch, Kuhlmann HC, Barth FG (2013) Airflow elicits a spider's jump towards airborne prey. II. Flow characteristics guiding behavior. J R Soc Interface 10:82. doi:10.20120820

Landolfa MA, Jacobs GA (1995) Direction sensitivity of the filiform hair population of the cricket cercal system. J Comp Physiol A 177:759–766

Lehtinen PT (1980) Trichobothrial patterns in high level taxonomy of spiders. In: Proceedings of 8th international conference on Arachnol. Egermann, Wien, pp 493–498

Levin JE, Miller JP (1996) Broadband neural encoding in the cricket cercal sensory system enhanced by stochastic resonance. Nature 380:165–168

Lewin GC, Hallam J (2010) A computational fluid dynamics model of viscous coupling of hairs. J Comp Physiol A 196:385–395

Magal C, Dangles O, Caporroy P, Casas J (2006) Hair canopy of cricket sensory system tuned to predator signals. J Theor Biol 241:459–466

Mahnert V (1976) Etude comparative des trichobothries de pseudoscorpions au microscope électronique à balayage. CR Séanc Soc Phys Hist Nat 11:96–99

Markl H, Tautz J (1975) The sensitivity of hair receptors in caterpillars of *Barathra brassicae* L (Lepidoptera, Noctuidae) to particle movement in a sound field. J Comp Physiol 99:79–87

McConney ME, Tsukruk VV (2012) Synthetic materials for bio-inspired flow-responsive structures. In: Barth FG, Humphrey JAC, Srinivasan MV (eds) Frontiers in sensing—from biology to engineering. Springer, Wien, New York, pp 341–349

McConney ME, Schaber CF, Julian MD, Eberhardt WC, Humphrey JAC, Barth FG, Tsukruk VV (2009) Surface force spectroscopic point load measurements and viscoelastic modelling of the micromechanical properties of air flow sensitive hairs of a spider (*Cupiennius salei*). J R Soc Interface 6(37):681–694

Messlinger K (1987) Fine structure of scorpion trichobothria (Arachnida, Scorpiones) Zoomorph 107:49–57

Müllan R (2012) Air-flow sensing in *Smeringurus mesaensis* (Scorpiones: Vaejovidae). Sensor arrangement, behavioral significance and oscillation characteristics of scorpion trichobothria. Doctoral thesis, University of Vienna

Nicklaus R (1965) Die Erregung einzelner Fadenhaare von *Periplaneta americana* in Abhängigkeit von der Größe und Richtung der Auslenkung. Z vergl Physiol 50:331–362

Peters W, Pfreundt C (1986) Die Verteilung von Trichobothrien und lyraförmigen Organen an den Laufbeinen von Spinnen mit unterschiedlicher Lebensweise. Zool Beitr N F 29:209–225

Paydar S, Doan CA, Jacobs CA (1999) Neural mapping of direction and frequency in the cricket cercal sensory system. J Neurosci 19:1771–1781

Reissland A, Görner P (1985) Trichobothria. In: Barth FG (ed) Neurobiology of arachnids. Springer, Heidelberg, pp 138–161

Santer RD, Hebets EA (2008) Agonistic signals received by arthropod filiform hair allude to the prevalence of near-field sound communication. Proc R Soc B 275:363–368

Schaber CF, Gorb SN, Barth FG (2012) Force transformation in spider strain sensors: white light interferometry. J R Soc Interface 9(71):1254–1264

Schilstra C, van Hateren JH (1999) Blowfly flight and optic. I. Thorax kinematics and flight dynamics. J Exp Biol 202:1481–1490

Schmidt K, Gnatzy W (1971) Die Feinstruktur der Sinneshaare auf den Cerci von *Gryllus* Deg (Saltatoria, Gryllidae). II. Die Häutung der Faden- und Keulenhaare. Z Zellforsch 122:210–226

Schuh RT (1975) The structure, distribution, and taxonomic importance of trichobothria in the Miridae (Hemiptera). Am Mus Novit 2585:1–26

Shimozawa T, Kanou M (1984a) Varieties of filiform hairs: range fractionation by sensory afferents and cercal interneurons of a cricket. J Comp Physiol A 155:485–493

Shimozawa T, Kanou M (1984b) The aerodynamics and sensory physiology of range fractionation in the cercal filiform sensilla of the cricket *Gryllus bimaculatus*. J Comp Physiol A 155:495–505

Shimozawa T, Murakami J, Kumagai T (1998) Cricket wind receptor cell detects mechanical energy of the level of kT of thermal fluctuation. Abstract 112, International Society of Neuroethology Conference, San Diego

Shimozawa T, Murakami J, Kumagai T (2003) Cricket wind receptors: thermal noise for the highest sensitivity known. In: Barth FG, Humphrey JAC, Secomb TW (eds) Sensors and sensing in biology and engineering, Chap 10. Springer, Wien, New York, pp 145–157

Steinmann T, Casas J, Krijnen G, Dangles O (2006) Air-flow sensitive hairs: boundary layers in oscillatory flows around arthropod appendages. J Exp Biol 209:4398–4408

Suter RB (2003) Trichobothrial mediation of an aquatic escape response: directional jumps by the fishing spider, *Dolomedes triton*, foil frog attacks. J Insect Sci 3:19–25

Tautz J (1977) Reception of medium vibration by thoracal hairs of caterpillars of *Barathra brassicae* L. (Lepidoptera, Noctuidae). I. Mechanical properties of the receptor hairs. J Comp Physiol 118:13–31

Tautz J (1979) Reception of particle oscillation in a medium: an unorthodox sensory capacity. Naturwissenschaften 66:452–461

Tautz J, Markl H (1978) Caterpillars detect flying wasps by hairs sensitive to medium vibration. Behav Ecol Sociobiol 4:101–110

Theiß J (1979) Mechanoreceptive bristles on the head of the blowfly: mechanics and electrophysiology of the macrochaetae. J Comp Physiol 132:55–68

Thurm U (1982) Biophysik der Mechanorezeption. In: Hoppe W, Lohmann W, Markl H, Ziegler H (eds) Biophysik, 2nd edn. Springer, Berlin Heidelberg, pp 691–696

Triblehorn JD, Yager DD (2006) Wind generated by an attacking bat: anemometric measurements and detection by the praying mantis cercal system. J Exp Biol 209:1430–1440

Tropaea C, Bleckmann H (2012) Nature-inspired fluid mechanics. Springer, Heidelberg, 372 pp

Vachon M (1973) Etude des caractères utilisées pour classer les familles et les genres de scorpions (Arachnides). 1. La trichobothriotaxie en arachnologie. Sigles trichobothriaux et types de trichobothriotaxie chez les scorpions. Bull Mus Hist Nat 3 Ser 140:857–958

Weygoldt P (1995) A whip spider that ate rolled oats, with observations on prey-capture behaviour in whip spiders. Newsl Br Arachnol Soc 74:6–8

Weygoldt P (2000) Whip spiders (Chelicerata: Amblypygi): their biology, morphology and systematics. Apollo Books, Stenstrup

Weygoldt P, Paulus H (1979) Untersuchungen zur Morphologie, Taxonomie und Phylogenie der Chelicerata. I. Morphologische Untersuchungen. Z zool Systematik Evolutionsforschung 17:85–116

Chapter 8
Air Flow Sensing in Bats

Susanne J. Sterbing-D'Angelo and Cynthia F. Moss

Abstract Bats are the only mammals capable of powered flight, and impress with complicated aerial maneuvers like tight turns, hovering, or perching upside-down. The bat wing membrane is covered with microscopically small hairs that are associated with a variety of tactile receptors at the follicle. The directionality profile of neuronal responses to air flow—as measured in the somatosensory cortex of the bats—indicates that the hairs respond strongest to reverse airflow, and might therefore act as stall detectors. We found that depilation of different functional regions of the wing membrane alters flight behavior in obstacle avoidance tasks by reducing aerial maneuverability, as indicated by wider turning angles and increased flight speed. We provide here for the first time electrophysiological and behavioral data showing that bat wing hairs are involved in sensorimotor flight control by providing aerodynamic feedback.

Abbreviations

CX   Cerebral cortex
D   Digit
IC   Inferior colliculus
IFM Interfemoral membrane
K20 Monoclonal keratin antibody
OB   Olfactory bulb
S1   Primary somatosensory cortex
SC   Superior colliculus

S. J. Sterbing-D'Angelo (✉)
Institute for Systems Research, University of Maryland, Building 144,
College Park, MD 20742, USA
e-mail: ssterbin@umd.edu

C. F. Moss
Department of Psychology, Institute for Systems Research, University of Maryland,
Building 144, College Park, MD 20742, USA
e-mail: cmoss@psyc.umd.edu

H. Bleckmann et al. (eds.), *Flow Sensing in Air and Water*,
DOI: 10.1007/978-3-642-41446-6_8, © Springer-Verlag Berlin Heidelberg 2014

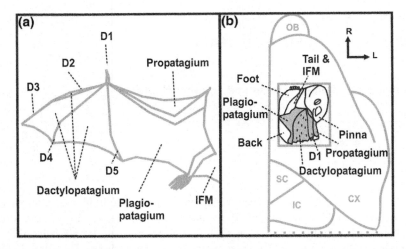

Fig. 8.1 **a** Nomenclature for parts of the bat wing. *D1–D5* digits (1: thumb), *IFM* interfemoral membrane, **b** Schematic brain surface view of the right hemisphere of the Big Brown Bat, *Eptesicus fuscus CX* cerebral cortex, *IC* inferior colliculus, *OB* olfactory bulb, *SC* superior colliculus, *R* rostral, *L* lateral. The *rectangle* indicates the approximate location of the primary somatosensory cortex, S1. The inserted sketch depicts the body representation mapped onto the brain surface. The *wing* representation is marked in *grey*. Note that the *wing* representation covers approximately 1/3 of the entire S1 area surface (see also Sterbing-D'Angelo et al. 2011; Chadha et al. 2010)

8.1 Introduction

Bat flight—the only true, powered flight found in mammals—is characterized by remarkable aerial maneuvers like steep banking, hovering, and landing upside-down. Skeletal specializations (see Fig. 8.1a), muscular control of wing shape, e.g., camber, and the highly compliant characteristics of the wing membrane are the basis of maneuverability and energy efficiency (Swartz et al. 1996; Winter et al. 1998; Voigt and Winter 1999; Stockwell 2001). Moreover, bat flight is very robust in turbulent, gusty, and low Reynolds number air flow conditions, for example during low-speed flight and hovering. While earlier studies relied on high-speed video tracking and modeling to characterize flight (Rayner 1979a, b), recent particle image velocimetry (PIV) experiments showed that these animals produce complex aerodynamic wake patterns (Hedenström et al. 2007; Muijres et al. 2008). However, the sensory-motor mechanisms that underlie the robustness of bats' flight have not been studied in detail, and despite the fact that the wing is well represented in the primary somatosensory cortex (Fig. 8.1b) of echolocating bats (Big Brown Bat—*Eptesicus fuscus (E.f.)*: Chadha et al. 2010, Pallid Bat—*Anthrozous pallidus (A.p.)*: Zook and Fowler 1986; Ghost Bat—*Macroderma gigas (M.g.)*: Wise et al. 1986), we know only little about the nature and function of the cutaneous tactile receptors located in the wing membrane. To the naked eye, the bat's wing membrane appears hairless, in contrast to the head and body of the animals, which are densely covered with fur. At first thought this appears odd, because fur surfaces are known to stabilize (microlaminarize) the boundary layer airflow by breaking up large vortices into microturbulences (Nachtigall 1979). However, a sparse grid of microscopically small hairs, many of which are protruding from domed structures, is found on both dorsal and ventral surfaces of the bat wing. These hairs were described first in the early twentieth century (Maxim 1912), but their role for bat flight has never been studied until recently (Sterbing-D'Angelo et al. 2011; Zook and Fowler 1986).

Bats typically have a layer of soft vellus hair (undercoat) overlaid with a layer of guard hair that is straighter and coarser. Except for sensory whiskers, bat pelage hair structure is quite uniform over the entire head and body. Big Brown Bat hair, like most microchiropteran bat species pelage (Debelica and Thies 2009), has a spiny coronal scale pattern, i.e., one scale forms a "ring-like" structure. This is only possible because bat pelage hair is amongst the finest of all mammalian hair with a diameter of only 10–20 µm. The main functions of pelage hair is heat insulation, and possibly improving aerodynamics by forming ripplets that reduce parasitic drag associated with skin friction and airflow separation (Bullen and McKenzie 2008). Parasitic drag is maximized or minimized depending upon the extent of turbulence in the boundary layer of air in immediate proximity to the body. In contrast to the pelage found on head and body of bats, the hairs on the wing membrane are too sparsely distributed and too short (100–600 µm) to be involved in heat insulation or in physically influencing the airflow over the wing surface. For the same reason a hypothetical function to avoid wetting of the wing membrane that has been described for insect wing hair (Wagner et al. 1996) can be excluded. The possible functional role for the domed wing hairs of bats has been speculated about for quite a while (Welwitsch's bat, *Myotis welwitschii*, Maxim 1912). But only recently researchers have begun to systematically study the properties of the hairs, and the properties of the tactile receptors that surround them (Zook and Fowler 1986; Zook 2006; Sterbing-D'Angelo et al. 2011). The domes from which the bat wing hairs protrude resemble touch domes (Merkel cell neurite complex) in the skin of non-flying mammals, which are rarely associated with hairs (Pinkus 1902; Smith 1977). Preliminary histological studies (Zook and Fowler 1986; Zook 2006) revealed a considerably large population of presumptive Merkel cells concentrated at the basement membrane along the dome surface and surrounding the hair follicle of the Pallid Bat. In congruence with the classic description of Merkel cells (Pinkus 1905; Smith 1977; Halata et al. 1993; Haeberle et al. 2004), these cells were typically large, clear cells with lobulated nuclei restricted to the epidermal basal lamina. Staining of these cells with both Merkel-specific quinacrine fluorescence (Nurse et al. 1983) and a cell-specific antibody to the cytokeratin protein, CK20 (Moll et al. 1995) further confirmed the identification as Merkel cells. Nerve fibers within the dome complex can be traced to individual Merkel cells in *A.p.* and *E.f.* (Zook 2005). Other tactile receptors have been identified in early anatomical studies (e.g., Ackert 1914). Recently, a fluorescent marker study confirmed that free nerve endings and lanceolate receptors are present at the wing hair base of the Big Brown Bat (Chadha et al. 2012). In this species, it turns out that only a subset of wing hairs stained positive for Merkel cells, while the majority of hair follicles were surrounded by lanceolate endings. Since Merkel cells are known as "slowly-adapting" tactile receptors, and lanceolate endings as "rapidly adapting" receptors (Adrian 1941), it is likely that both receptor populations code for different air flow parameters. The sensory hairs on the bat wing are very stiff, with an average taper of 10 and a transversal elastic modulus of 500 MPa, typical for α-keratin. During deflection, the mechanical impact on the closely surrounding receptor cells, lanceolate endings, as well as Merkel cell neurite complexes, has been modeled to be so substantial at biological flight speeds, providing evidence that the classification of the wing hairs as airflow sensors is warranted. Hence, our hypothesis is that the domed wing hairs provide aerodynamic feedback to the somatosensory brain, and are therefore involved in flight stabilization. Sensilla on the wing and other body parts of insects have been shown to play a role in flight control (e.g., Haskell 1958; Pflueger and Tautz 1982; Dickinson 1990; Ai et al. 2010), as have vibrotactile receptors at the feather base of birds (Necker 1985; Hoerster 1990). Hence, we studied how neurons in the primary somatosensory cortex respond to experimental air flow stimuli, and how the removal of wing hairs influences flight behavior.

8.2 Methods

Detailed methods of the experimental procedures, data acquisition, and analysis have been published elsewhere (Sterbing-D'Angelo et al. 2011; Chadha et al. 2012). Therefore, we only provide a short summary here.

8.2.1 Animals

Eptesicus fuscus (*E.f.*) were wild-caught in Maryland. *Carollia perspicillata* (*C.p.*) and *Glossophaga soricina* (*G.s.*) were donated (Montréal Biodôme, Canada), and bred in a captive colony. All bats were housed under reversed 12 h light/dark conditions with appropriate temperature and humidity levels for each species. *C.p.* and *G.s.* were maintained on a diet of various fruits, nectar, and water. *E.f.* were maintained on a diet of mealworms, *Tenebrio molitor*, and water. All procedures were approved by the University of Maryland Institutional Animal Care and Use Committee.

8.2.2 Scanning Electron Microscopy

Circular wing membrane samples (13 mm dia.) from 24 different locations of 2 *E.f.* (12 each), 8 samples from the wing of each one *C.p.* and *G.s.* at corresponding locations, except tail membrane, were fixed in 2.5 % glutaraldehyde solution, washed in phosphate buffer, and then fixed in 1 % buffered osmium tetroxide. After standard washing and dehydration procedure, the samples were dried in a critical point dryer (Denton DCP-1), and mounted onto metal pedestals with silver paste, hardened at 50 °C, and then coated with gold palladium alloy (Denton DV-502/502 Vacuum Evaporator). The samples were viewed in a scanning electron microscope (Amray AMR-1610). For immunohistochemistry, bat wing was fixed in 4 % paraformaldehyde, cryoprotected with sucrose, frozen, and sectioned at 20 μm.

8.2.3 Experimental Procedures

For surgery and electrophysiological recordings, *E.f.* were anesthetized with 1–3 % isoflurane mixed with 700 ml/min O_2. Breathing rate was monitored visually, and body temperature was maintained at about 37 °C by placing the animal on a heating pad. Standard sterile surgical procedures were followed throughout the experiment. After exposing the skull, a custom-made stainless steel head-post was glued close to Bregma using cyanoacrylate glue. Bats were allowed to recover for 2–3 days before electrophysiological recordings were initiated. A craniotomy (intact Dura mater) measuring approximately 2×2 mm was made to expose the somatosensory cortical region, and sterile saline/silicone oil (Fluka Analytical, DC 200) was used to prevent the exposed brain surface from desiccation. Either a high impedance recording electrode (15–20 MΩ tungsten, FHC Inc.) was used to record extracellularly from multi-neuron clusters or a silicon probe was used (Neuronexus). The electrode/probe entered the cerebral cortex perpendicularly to the surface and was positioned using 3 digital microdrives (Mitutoyo). Recordings were made from multiple electrode penetrations from depths of 50–250 μm, ensuring that the recordings were made mostly within the supragranular layers of the cortex according to a standard brain atlas of *E.f.* (Covey, unpublished). The contralateral wing was spread and secured to the recording table. Tactile receptive fields were measured by using a set of calibrated monofilaments (von Frey hairs, North Coast), applying pressure on a logarithmic scale from 0.008 to 300 g (equals 0.08–2,943 mN), with a 5 % standard deviation. Both wing surfaces were tested. For stimulation with air puffs a syringe was directed at the center of the receptive field from different angles in 45° or 90°. steps. Air puff stimuli were generated by a dispensing workstation (EFD Ultra®2400), and electronically varied in duration and amplitude. Air flow was calibrated (Datametrics 100VT-A). At each recording site, the magnitude of the air puff was adjusted to be just above the neuronal threshold, verified by microscopic inspection that no indentation of the membrane occurred. The workstation trigger also started the data acquisition board that recorded the waveform of the neural responses after amplification (Bak Electronics, Plexon Omniplex). Each stimulus was presented 20 times with an interstimulus interval of 10 s. Single neuron activity could be extracted by standard offline spike sorting (Neuroexplorer v.3,

Offline Sorter v.3, Plexon). At the end of the recording sessions, bats were given a lethal dose (0.05 ml) of sodium pentobarbital (390 mg/ml, i.p.). Data collected under research protocol, "Somatosensory signaling for flight control," approved by the University of Maryland Institutional Animal Care and Use Committee.

8.2.4 Flight Testing

Flight path recordings were conducted in a carpeted flight room (7 × 6 × 2.5 m), with acoustic foam on walls and ceiling (Sonex) under low-intensity, long-wavelength light conditions high-speed (>650 nm, incandescent bulbs filtered through Plexiglas G2711; Atofina Chemicals). Two high speed (250 frames per second) infrared video cameras (FASTCAM-PCI-R2) were used to record 3D position of bat and obstacles. *E.f.* and *C.p.* were trained to perform rewarded obstacle flight tasks (see Fig. 8.7a, b) that required flight maneuvers (details in Sterbing-D'Angelo et al. 2011, and in results section). The flight data were collected before and after depilation of the wing hairs (Veet®). Flight speed and turn angles were calculated from the high-speed videos.

8.3 Results

8.3.1 Hair Morphology and Distribution

The morphology and distribution of wing hairs were examined for three echo locating species, the Big Brown Bat (*Eptesicus fuscus, E.f.*), the Short-Tailed Fruit Bat (*Carollia perspicillata, C.p.*), and Pallas's Long-Tongued Bat (*Glossophaga soricina, G.s.*) using scanning electron microscopy. The ecological niches and diets of these three bat species differ and consequently impact requirements for flight control. In particular, the insectivorous *E.f.* must make sharp turns in flight to pursue and capture evasive insect prey, the frugi-/nectarivorous *C.p.* must maneuver through dense vegetation to find fruit, and the nectarivorous *G.s.* must hover over flowers to take nectar. In all three species, the short hairs are sparsely distributed along dorsal and ventral surfaces of the wing, and are morphologically distinct from the long pelage hairs. The pelage hairs were only found on SEM samples that were cut close to the limbs. They were up to several milimeters long, relatively thick at the base (6–18 μm) found close to the ventral forearm, around the leg, and on the tail membrane, sometimes referred to as inter-femoral membrane (IFM) or uropatagium (Fig. 8.1).

In all three species, on the membraneous parts of the wing, a second type of hair was found, which is invisible to the naked eye. These hairs are so thin that only one follicle cell builds each segment of the hair, resulting in a coronal scale

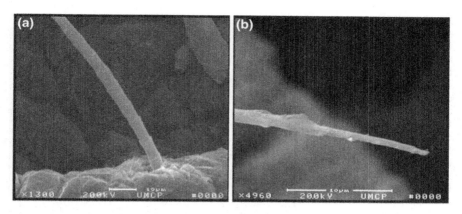

Fig. 8.2 Scanning electron microscope images of hairs from the wing membrane of the Big Brown Bat, *Eptesicus fuscus*. **a** Base of a hair (diameter ∼5 μm) protruding from a dome-like structure. Note the coronal scale pattern. **b** Tip of a hair (diameter ∼400 nm). The calibration bars located on the bottom of both images indicate 10 μm

pattern. The tip diameter of these hairs is only 200–900 nm, if they are intact (Fig. 8.2). These small hairs are typically found in rows, generating a sparse grid of about one hair per mm². This finding confirms early studies, e.g., (Ackert 1914), who described that "the proximal parts of the membranes are covered with fine hairs similar of those of the pelage, while over the distal areas extremely fine, more or less *modified* hairs occur sparsely."

In the two phyllostomid species, *C.p.* and *G.s.*, the distribution of the hairs, as well as their length and thickness, are similar to *E.f.* except that in some areas of their wing membrane, particularly on the dorsal plagiopatagium at the trailing edge and on the propatagium (part of the leading edge of the wing), several hairs—typically three to five-protrude from one dome in clusters (Fig. 8.3a, b), a finding that has been previously described for *A.p.* (Zook 2006). In *E.f.*, only very rarely two hairs can be found right next to each other (Fig. 1.4). Theoretical considerations and modeling of boundary layer detection reveals that the measured hair lengths shown here are in very good agreement with the theoretical ideal length of hair for maximum shear-force sensitivity to boundary layer shape and avoidance of viscous coupling, may be with the exception of hairs within a multi-hair cluster (Dickinson 2010).

8.3.2 Tactile Receptors in the Wing Membrane

A variety of receptors have been found in the bat wing membrane. Figure 8.4 shows rings of fluorescent-marked Merkel cells around wing hairs in *E.f.*, stained with topically applied Rhodamine. Furthermore, preliminary data of our collaborators at Columbia University (E. A. Lumpkin and K. L. Marshall; see Chadha et al. 2012) revealed that many hairs are co-localized with lanceolate nerve

Fig. 8.3 a Scanning electron microscope image of hairs from the wing membrane of the Short-Tailed Fruit Bat, *Carollia perspicillata*. This sample was taken from the dorsal plagiopatagium close to the trailing edge. Three to five hairs protrude from each domed structure (calibration *bar*: 100 μm); **b** Image of hairs from the same region on the wing membrane of the Pallas's Long-Tongued Bat, *Glossophaga soricina*. Note that the membrane of this extracted sample is folded up due to embedded elastin bands. During flight, the membrane would be stretched out flat

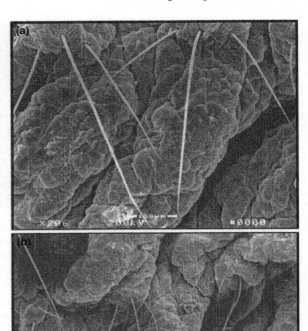

endings, a finding that confirms an earlier report for a different bat species (Ackert 1914). For a subset of wing hairs, the K20 antibody additionally marked Merkel cells (Chadha et al. 2012) confirming preliminary data by Zook (2005). Free nerve endings and other endings marked by Peripherin were found throughout the epidermis (Chadha et al. 2012).

8.3.3 Cortical Neuronal Responses to Air Flow: Directionality

Single neuron as well as multi-neuron cluster responses in the primary somato-sensory cortex of the big brown bat varied with spatially restricted (diameter of stimulated wing area <8 mm) air flow from different direction (Sterbing-D'Angelo et al. 2011; Chadha et al. 2012). Air puffs were delivered to the dorsal wing surface from eight directions in 45° steps for the multi-units and from the four major

Fig. 8.4 Microphotograph of Rhodamine-stained fluorescent Merkel cells in a whole wing mount of an isoflurane-anesthetized Big Brown Bat, *Eptesicus fuscus* Merkel cells form a well-organized partial or full circle around the hair follicle. *Arrows* point to some of the hairs that emerge from the ring-like structures

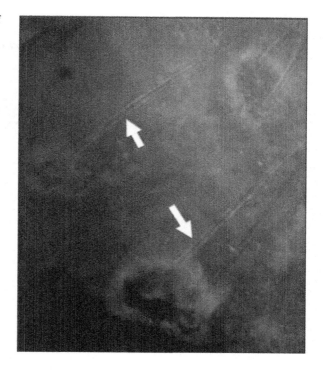

directions for the single units. The multi-unit responses were half-wave rectified and averaged across the 20 trials. The responses for each stimulation direction were normalized to maximum, and plotted as polar diagrams. Minimum–maximum ratio between the best (preferred) and worst direction was calculated to quantify the strength of directionality of the air puff response of 20 multi-units from 4 *E.f.* Most units, 9/9 single units and 15/20 multi-units—although their receptive fields were located on different areas of the dorsal wing membrane—favored air flow from the rear (135–225°). The only exceptions were multi-units with receptive fields located on the extreme leading edge of the wing, and those with receptive fields close to the body. The responses of the latter might have been influenced by turbulences caused by the proximity of elevated body features. Figure 8.5a shows polar plots of normalized multi-neuron responses to air flow from different directions.

8.3.4 Cortical Neuronal Responses to Air Flow: Temporal and Spatial Characteristics

Recordings with silicon probes (Neuronexus) collected responses from various sites on the wing membrane simultaneously. The multi-neuron responses at each probe were sorted to extract single neuron activity. Figure 8.5b shows responses of single neurons to 40 ms air puffs (stimulated wing area ∼8 mm in diameter)

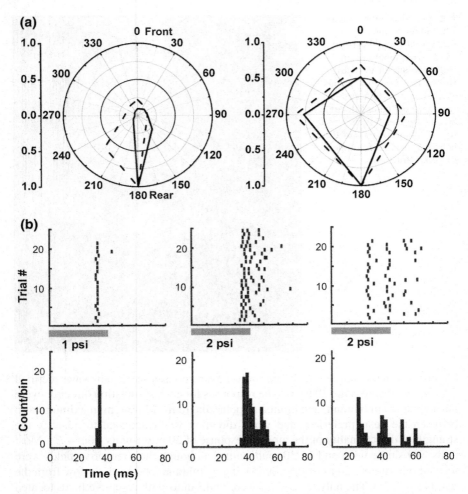

Fig. 8.5 a Polar plots depicting the directionality of *wing* hair responses in primary somatorensory cortex of *Eptesicus fuscus* Units could show sharp (*left*) or wider tuning (*right*) to pseudo-randomly chosen directional air flow. The *hatched lines* indicates results from a repeated set of stimulations minutes after the initial data set (*solid lines*) was recorded, showing that the directionality of the units was well preserved over time. **b** Spike raster plots of a phasic, a sustained, and a complex firing single neuron, recorded from primary somatosensory cortex of *Eptesicus fuscus* (Chadha et al. 2012)

presented from the neuron's preferred direction at different airflow velocities. The single neuron responses were phasic, sustained, or "complex" with periodic spike patterns that could indicate a preference for deflection close to the resonance frequency of the hairs, an effect that has been described for a whisker model (Williams and Kramer 2010).

When a receptive field was identified on the wing membrane, two blunt syringe needles were positioned so that their openings pointed to the same location on receptive field center, one from the dorsal and one from the ventral side, at a fixed measured distance (for most units 3 mm) from the wing surface. An air flow magnitude that just elicited a weak response, and therefore was close to threshold was chosen and delivered first to either side of the membrane alone, and then to both sides simultaneously. For most units (7/9 recorded from two animals) we found evidence for facilitation when both surfaces were stimulated simultaneously, which means that the unit's response was stronger than the sum of the responses to stimulation of either side alone. Figure 8.6 shows examples of facilitation in S1 recordings collected from two bats. The facilitation effect appears to be strongest close to threshold of the respective neuronal cluster. At high supra-threshold air stimulation levels, it cannot be ruled out that cutaneous receptors other than the receptors at the hair base are contributing to the response, because the entire wing membrane is increasingly indented with stimulus intensity, in addition to the deflection of the hairs.

8.3.5 Flight Behavior

The schematics in Fig. 8.7a, b illustrate the setups used to study the role of wing hairs in flight control for obstacle avoidance in two bat species, *E.f.* and *C.p.*. Several *E.f.* were trained to fly through an artificial forest (Fig. 8.7a) and *C.p.* were trained to fly though openings in a series of nets which created a maze (Fig. 8.7b). The artificial trees were nearly cylindrical, constructed out of visually transparent nets, to allow us to continuously monitor the bat's flight path with video cameras. Both species gained access to a food reward for successfully maneuvering around obstacles. Flight behavior was monitored with two high-speed video cameras mounted in corners of the room. With the stereo video recordings, we are able to reconstruct the 3-D flight paths of the bats as they performed the tasks. In all of the behavioral experiments, bats were run under baseline, and experimental conditions. Baseline recordings were conducted over a minimum of twenty trials, to establish the norms of the bat's flight behavior. Then, the experimental trials were conducted, in which hairs are removed from selected regions of the wings, using a depilatory cream (Veet®). Absence of hairs was confirmed by microscopic inspection.

Removing the tactile wing hairs of the trailing edge resulted in higher average flight speed and reduced angular turn angle (i.e., wider turns) as the treated bat approached an obstacle, compared to baseline. Although comparing bat flight to fixed-wing aircraft flight is problematic, increasing air speed is also recommended to pilots to recover from stalls. We interpret the depilated bats' increase average flight speed as a result of lacking input from the domed hair receptors to the somatosensory system. Our findings that neural responses to airflow are directional suggest that wing sensors may play a role in stall detection. In the behavioral task,

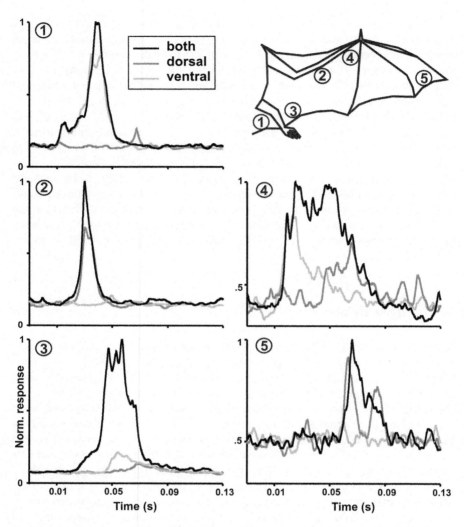

Fig. 8.6 Somatosensory (S1) responses of five multi-neuron clusters with receptive fields located on different *wing* areas at near-threshold stimulation with air puffs of 40 ms duration. The lines indicate averages of 20 trials for each condition (*both, dorsal* only, *ventral* only). In case of stimulation of both surfaces, the air flow was split with valves to ensure that the total air mass reaching the *wing* membrane was kept constant. Two syringes were directed at the center of the receptive field at the same distance and angle from both the *dorsal* and *ventral* side of the membrane. Air puffs stimulated either the *dorsal* surface of the wing alone (*medium gray*), the *ventral* surface of the *wing* alone (*light gray*), or both the *dorsal* and *ventral* surfaces of the wing simultaneously (*black*). For most units, responses were facilitated by simultaneous stimulation of *both* surfaces, although the air mass reaching each surface was split in half to keep the overall air mass that reaches the membrane constant

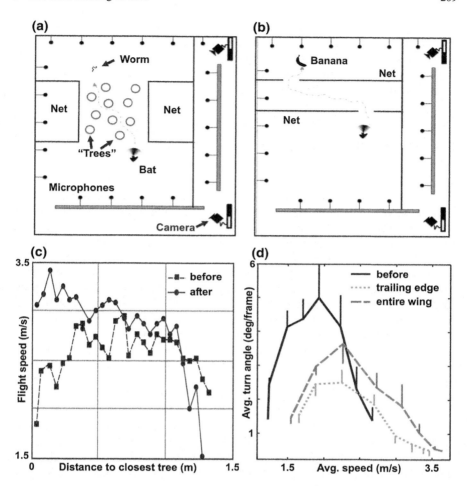

Fig. 8.7 a, b Schematic of setups to record flight behavior of *Eptesicus fuscus* through an artificial forest (**a**), and of *Carollia perspicillata* through a net maze (**b**). Both tasks required turning in flight to avoid obstacles. In both setups, two high-speed video cameras recorded flight behavior, and 3-D flight paths were reconstructed. **c** Flight speed versus distance to the closest tree is plotted for one *Eptesicus fuscus*, before and after depilation (mean of 10 trials from one animal). Note that the overall flight speed increased. **d** Flight speed versus average turn angle for *Carollia perspicillata,* before depilation, and after depilation of the hairs along the trailing edge only, and later the entire wing (2 animals, 117 trials, mean ± SE). Also in this species, and for this task, flight speed and average turn angle changed after depilation. However, the depilation of the trailing edge only seems to have a greater effect than the depilation of the entire wing surface

a depilated animal may fly faster and make wider turns compared to baseline in an attempt to avoid a stall by speeding up, because reverse airflow signals have been disrupted, and the treated animal may have experienced that it is more vulnerable to stall. Alternatively, the hairs may simply function as flight velocity sensors, and the depilated bat would interpret a lack of input from the hairs as low-speed, and consequently change its flight speed. Of course, also kinesthetic and proprioceptive inputs are still available to the bat, and it remains an open question the extent to which a bat can adapt to the absence of wing hairs over time. We tested the flight performance within two days of depilation. It is unclear if, and in which time frame, the domed hairs grow back.

8.4 Discussion

The sparse distribution of sensory hairs on the bat flight membrane, as well as their small size, has eluded formal demonstration of their function for decades; however, these very anatomical features provide compelling evidence for their role in air flow sensing for flight control. At the microscopic scale viscous force is dominant, and viscous coupling occurs especially at the low Reynolds numbers (Re) (down to only $Re = 5,000$ for hovering flight), typically observed in bat flight. Viscous coupling can be observed over a distance of 50 hair diameters or even more (Casas et al. 2010; Heys et al. 2008; Cummins et al. 2007). Coupling leads to sensory "dead zones" between hairs, and a more dense distribution of hairs on the membrane would not only add weight, but also increase highly non-linear ensemble reactions to air flow that might "blur" the information carried by the tactile input. Hence, a sparse distribution of air flow sensing hairs seems to be advantageous in many respects. Interestingly, the diameter, length and average distance of hairs in the bats of our study are quite similar to those found for filiform hairs on the cerci of crickets (diameter: 1–9 μm, length: 30–1,500 μm, Shimozawa and Kanou 1984), but less variable. The cercal filiform hairs sense velocity, acceleration, and direction of air flow. Measurement of viscous coupling between these filiform hairs revealed that hairs might influence each other up to a distance of about 400 μm. Arthropod filiform hairs, however, have much lower thresholds for air flow velocity (0.03 mm/s) than the bat wing hairs (20–30 mm/s, under isoflurane anesthesia), obviously due to the fundamentally different sensory receptor properties between mammals and arthropods, but also as adaptation to physiological flight speeds (upto 8–10 m/s). At a given distance, viscous coupling is stronger for hairs of similar length than different length. We found that the length of the tactile hairs systematically varied along the rostro-caudal axis with the longest ones located close to the proximal forelimb (300–600 μm) and the shortest close to the trailing edge of the wing (100–300 μm). In some bat species, apart from *E.f.*, groups of sensory hairs protruding from a single dome were reported (*A.p.*: Zook 2006, *C.p.* and *G.s.*: our study). The hairs within the "tufts" tend to have very different lengths, possibly to minimize viscous coupling effects. The sharp taper of the hairs on the wing membrane, in comparison to pelage hairs, is very similar to the taper of whisker (sinus) hairs of rodents (~ 10), although the wing hairs are morphologically not sinus hairs, and of much smaller scale than whiskers. The taper reduces the maximum deflection angle of the entire hair, which leads to higher spatial acuity than estimates for more bluntly tapered hairs

(Williams and Kramer 2010). Since periodic spiking was frequently observed in our single neuron responses recorded in S1, a sharp taper would also improve the robustness for changes in resonance frequency, which would keep periodicity coding in the somatosensory cortex in a stable range. Preliminary laser scanning vibrometer tests performed on hairs from the Big Brown Bat revealed natural frequencies between 60 and 80 Hz. Hence, the role of the sensors at the base of the hairs might serve different purposes. Firstly, they could provide sensory feedback for lift control by sensing the size and location of the leading edge vortex. The importance of the leading edge vortex for generating additional lift has been pointed out by (Muijres et al. 2008). Secondly, by detecting reverse air flow that compromises flight stability, signals from the hairs could be used to prevent stall. Finally, the rapidly adapting receptors, in particular, would provide information of sudden changes in air flow direction. Thus, we propose that the rapid-adapting and the slow-adapting receptors at the hair base have different functional roles: the fast-adapting lanceolate endings are well suited for detecting sudden changes of airflow conditions, e.g., during wind gusts or turbulence. In contrast, the slow-adapting receptors like Merkel cells are better suited to detect overall air flow patterns across the wing, and monitor the dimension of the leading edge vortex. Of course, Merkel cells and lanceolate cells are not the only receptor types that potentially could influence flight performance: stretch receptors, especially those in the plagiopatal muscles most likely influence the shape of the wing membrane, and possibly also signal local membrane oscillations caused by turbulences (Swartz et al. 2005). In our depilation experiments, the resulting changes in flight maneuverability were measurable but not dramatic, at least for the geometric setups that were used. One reason is that the bats still have the information from all types of tactile receptors that are embedded in the skin, and which are unaffected by the depilatory cream. Besides nerve endings on hairs, Ackert (1914) and others described free nerve endings, special sensory end organs like "end bulbs," "terminal corpuscles," motor nerve endings on striated muscles, and nerve endings on modified sweat glands in the wing membrane of bats. The removal of the hairs which act as levers merely causes a reduction of the most effective way to stimulate the receptors at the hair base. Also, general kinesthetic information is still available to the bat after depilation. It is known that bats become familiar with the environment very quickly, and that they may also orient using kinesthetic memory. Further investigation will be aimed at the different contributions to flight behavior of the individual receptor types.

Acknowledgments This study was sponsored by air force office of scientific research (AFOSR), MURI grant "Bio-inspired flight for micro-air vehicles." Data collected under research protocol, "Somatosensory signaling for flight control," approved by the University of Maryland Institutional Animal Care and Use Committee. We thank Mohit Chadha, Wei Xian, Ben Falk, and Aaron Reynolds for contributions.

References

Ackert JE (1914) The innervations of the integument of chiroptera. J Morphol 25:301–334

Adrian ED (1941) Afferent discharges to the cerebral cortex from peripheral sense organs. J Physiol 100:159–191

Ai H, Yoshida A, Yokohari F (2010) Vibration receptive sensilla on the wing margins of the silkworm moth *Bombyx mori*. J Insect Physiol 56:236–246

Bullen RD, McKenzie NL (2008) Aerodynamic cleanliness in bats. Austral J Zool 56:281–296

Casas J, Steinmann T, Krijnen G (2010) Why do insects have such a high density of flow-sensing hairs? Insights from the hydromechanics of biomimetic MEMS sensors. J R Soc Interface 7:1487–1495

Chadha M, Moss CF, Sterbing-D'Angelo SJ (2010) Organization of the primary somatosensory cortex and wing representation in the Big Brown Bat, *Eptesicus fuscus*. J Comp Physiol A 197:89–96

Chadha M, Marshall KL, Sterbing-D'Angelo SJ, Lumpkin EA, Moss CF (2012) Tactile sensing along the wing of the echolocating bat. Eptesicus fuscus. Soc Neurosci Abstr 523:03

Cummins B, Gedeon T, Klapper I, Cortez R (2007) Interaction between arthropod filiform hairs in a fluid environment. J Theor Biol 247:266–280

Debelica A, Thies ML (2009) Atlas and key to the hair of terrestrial texas mammals. In: Robert J Baker (ed) Special publications of the Museum of Texas Tech University, vol 55. Museum of Texas Tech University, Lubbock, USA

Dickinson MH (1990) Comparison of encoding properties of campaniform sensilla on the fly wing. J Exp Biol 151:245–261

Dickinson BT (2010) Hair receptor sensitivity to changes in laminar boundary layer shape. Bioinspir Biomim 5:1–11

Haeberle H, Fujiwara M, Chuang J et al (2004) Molecular profiling reveals synaptic release machinery in merkel cells. Proc Natl Acad Sci 101:14503–14508

Halata Z (1993) Sensory innervation of the hairy skin (light-and electronmicroscopic study). J Invest Dermatol 101:75S–81S

Haskell PT (1958) Physiology of some wind-sensitive receptors of the desert locust (*Schistocerca gregaria*). XVth Int Zool Congr, London

Hedenström A, Johansson LC, Wolf M, von Busse R, Winter Y, Spedding GR (2007) Bat flight generates complex aerodynamic tracks. Science 316:894–897

Heys J, Gedeon T, Knott B, Kim Y (2008) Modeling arthropod filiform hair motion using the penalty immersed boundary method. J Biomech 41:977–984

Hoerster W (1990) Histological and electrophysiological investigations on the vibration-sensitive receptors (Herbst corpuscles) in the wing of the pigeon (*Columba livia*). J Comp Physiol A 166:663–673

Maxim H (1912) The sixth sense of the bat. Sir Hiram's contention. The possible prevention of sea collisions. Sci Am 27:80–81

Moll I, Kuhn C, Moll R (1995) Cytokeratin-20 is a general marker of cutaneous merkel cells while certain neuronal proteins are absent. J Invest Dermat 104:910–915

Muijres FT, Johansson LC, Barfield R, Wolf M, Spedding GR, Hedenström A (2008) Leading-edge vortex improves lift in slow-flying bats. Science 319:1250–1253

Nachtigall W (1979) Gliding flight in petaurus-breviceps-papuanus. Model measurements of the influence of fur cover on flow and generation of aerodynamic force components. J Comp Physiol 133:339–349

Necker R (1985) Receptors in the wing of the pigeon and their possible role in bird flight. In: Nachtigall W (ed) Biona Rep 3: Vogelflug. Fischer, New York

Nurse CA, Mearow KM, Holmes M, Visheau B, Diamond J (1983) Merkel cell distribution in the epidermis as determined by quinacrine fluorescence. Cell Tissue Res 228:511–524

Pflueger HJ, Tautz J (1982) Air movement sensitive hairs and interneurons in locusta migratoria. J Comp Physiol A 145:369–380

Pinkus F (1902) Ueber einen bisher unbekannten nebenapparat am haarsystem des menschen: haarscheiben. Derm Z 9:465–499

Pinkus F (1905) Ueber Hautsinnesorgane neben dem menschlichen Haar (Haarscheiben) and ihre vergleichend-anatomische bedeutung. Arch Mikrosk Anat 65:121–179

Rayner JMV (1979a) Vortex theory of animal flight. 1. vortex wake of a hovering animal. J Fluid Mech 91:697–730

Rayner JMV (1979b) Vortex theory of animal flight. 2. Forward flight of birds. J Fluid Mech 91:731–763

Shimozawa T, Kanou M (1984) Variety of filiform hairs: range fractionation by sensory afferents and cercal interneurons of a cricket. J Comp Physiol A 155:485–493

Smith KR (1977) The haarscheibe. J Invest Dermat 69:68–74

Sterbing-D'Angelo S, Chadha M, Chiu C, Falk B, Xian W, Barcelo J, Zook JM, Moss CF (2011) Bat wing sensors support flight control. Proc Natl Acad Soc 108:11291–11296

Stockwell EF (2001) Morphology and flight maneuverability in new world leaf-nosed bats (chiroptera: phyllostomidae). J Zool 254:505–514

Swartz SM, Bishop K, Ismael-Aguirre MF (2005) Dynamic complexity of wing form in bats: implications for flight performance. In: Zubaid A, McCracken G, Kunz T (eds) Functional and evolutionary ecology of bats. Oxford Press, Oxford

Swartz SM, Groves MS, Kim HD, Walsh WR (1996) Mechanical properties of bat wing membrane skin. J Zool 239:357–378

Voigt CC, Winter Y (1999) Energetic cost of hovering flight in nectar-feeding bats (phyllostomidae: glossophaginae) and its scaling in moths, birds and bats. J Comp Physiol B 169:38–48

Wagner P, Neinhuis C, Barthlott W (1996) Wettability and contaminability of insect wings as a function of their surface sculptures. Acta Zool 77:213–225

Williams CM, Kramer EM (2010) The advantages of a tapered whisker. PLoS ONE 5: Article Number: e8806. doi:10.1371/journal.pone.0008806

Winter Y, Voigt C, Von Helversen O (1998) Gas exchange during hovering flight in a nectar-feeding bat, *Glossophaga soricina*. J Exp Biol 201:237–244

Wise LZ, Pettigrew JD, Calford MB (1986) Somatosensory cortical representation in the Australian ghost bat, *Macroderma gigas*. J Comp Neurol 248:257–262

Zook JM, Fowler BC (1986) A specialized mechanosensory array of the bat wing. Myotis 23–24:1–36

Zook JM (2005) The neuroethology of touch in bats: cutaneous receptors of the wing. Soc Neurosci Abstr 78:21

Zook JM (2006) Somatosensory adaptations of flying mammals, In: Kaas JH (ed) Evolution of nervous systems vol 3. Academic Press, Oxford

Chapter 9
Flies, Optic Flow and Multisensory Stabilization Reflexes

Holger G. Krapp

Abstract Besides sensing flow in air and water by means of mechanosensory mechanisms, visually oriented animals, including humans, exploit *optic flow* to control their behavior. The rather intuitive term optic flow has been coined more than six decades ago by J. J. Gibson in his accounts on human visual perception and describes the direction and magnitude of image motion due to movements of the visual system relative to a static environment. Ever since Gibson proposed the potential significance of optic flow for the control of behavioral action, a large number of studies on animal models across phyla were dedicated to identify the neuronal mechanisms underlying the processing of optic flow. A necessary prerequisite for using optic flow to estimate and ultimately control locomotion and balance in space is the capability of the nervous system to analyse the direction of visual motion. In this chapter I will describe fundamental properties of optic flow, review the functional structure of elementary movement detectors which analyse visual motion, and outline how the visual system overcomes notorious problems of local sensor measurements, i.e. their noisiness and ambiguity. I will focus on a specific species, the dipteran blowfly, which has been studied for many years, in particular in the context of multimodal stabilization reflexes, both at the behavioral and neuronal levels.

Keywords Optic flow · Motion detection · Matched filters · Self-motion estimation · Multisensory integration · Sensorimotor transformation · Sensorimotor control · Stabilization reflexes · State-dependence · Neuronal mechanisms · Behavioral systems analysis · Insect vision

H. G. Krapp (✉)
Department of Bioengineering, Imperial College London, London, UK
e-mail: h.g.krapp@imperial.ac.uk

H. Bleckmann et al. (eds.), *Flow Sensing in Air and Water*,
DOI: 10.1007/978-3-642-41446-6_9, © Springer-Verlag Berlin Heidelberg 2014

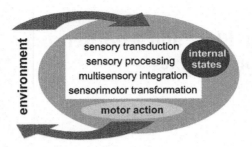

Fig. 9.1 Action–perception cycle for biological systems. The acquisition, processing and integration of sensory signals, as well as their transformation into commands controlling motor activity, take place under closed-loop conditions. Any behavioral action will result in a set of new sensory inputs which are again processed along the different stages. Local feedback and other adaptation mechanisms combined with internal states related to physiological parameters and the current locomotor state of the animal may have an impact on any of the processing stages. Further explanations in text. (Redrawn from Krapp and Wicklein 2008)

9.1 Introduction

The key function of the nervous system is to enable adaptive behavior. To understand the neural mechanisms underlying behavioral control we may want to consider the constraints under which the nervous system operates. Otherwise, the phenomena observed at the neuronal and behavioral level are difficult to interpret and may result in somewhat arbitrary conclusions regarding a system's functional organization. One of the constraints is that behavioral control—in most cases— takes place under closed-loop conditions. While this appears to be a trivial point, it is surprising how many studies on sensory processing and motor control are performed in separation, even though an animal's motor action largely determines the inputs received by its various sensory systems.

The 'action-perception cycle' illustrates the closed-loop nature of animal behavior, including the processing steps completed by the nervous system (Fig. 9.1). Information about the external environment is provided by different modalities, which convert physical energy of environmental stimuli into electrical signals—a process called *sensory transduction*. This is followed by *sensory processing* within the different modalities, including filtering the signals and combining them to extract behaviorally relevant features, e.g. the direction, distance and size of a landmark. Next, *multisensory integration* takes place where signals from different modalities encoding related information are combined to increase the probability of a particular stimulus actually being detected. Finally, the integrated signals are subject to a *sensorimotor transformation* where they are converted into motor commands which control the behavioral output. The latter is normally associated with movements of the animal which result in changes of the sensory inputs and the whole cycle starts all over again.

While the action–perception cycle makes a strong point regarding the closed-loop nature of behavioral control, it still is a considerable simplification of the processes involved. The sequence of processes involved does not suggest any local feedback mechanisms, which we know exist basically at all processing levels, causing state-dependent modulations in signal processing. In flies, for instance, such *internal states* may reflect current locomotor activity (Maimon et al. 2010; Longden and Krapp 2009), sensory adaptation (e.g. Madess and Laughlin 1985; Harris et al. 2000) or the physiological state of the animal, for instance, in terms of nutrition (Longden et al. submitted). An intuitive example would be a fly facing a visual expansion pattern during flight when approaching an object in the environment. If the glucose level in its hemolymph is low, an expansion pattern may trigger the fly's landing behavior followed by searching for food. If the glucose level is high, the same expansion pattern may just initiate a collision avoidance manoeuvre. In a sense the systems properties are not time-invariant, but rather depend on context as well as on physiological and locomotor state variables. In addition, higher level processes—associated with learning, memory and decision making as discussed for primates (Gilbert and Sigman 2007)—will certainly impact on lower level information processing, as well. Altogether, state-dependent information processing combined with prior experience enables more flexible and adaptive behavior, but in turn, may be modified by another fundamental constraint: the limited energy supply in the nervous system (Attwell and Laughlin 2001).

9.2 Estimating Self-Motion from Optic Flow

9.2.1 Optic Flow

In its generality, the action–perception cycle and the qualifications introduced in the previous section apply to all species and sensory systems. This obviously includes sensory systems providing information about flow in air and water based on mechanosensory transduction mechanisms. The significance of these mechanisms for the control of animal behavior across phyla and the functional properties of the specific systems are comprehensively covered by all other chapters of this volume.

In this chapter, I will outline a very different sensory mechanism that—despite its semiotic relationship to 'flow sensing'—exploits specific aspects of visual information which proved to be highly valuable when it comes to behavioral control. *Optic flow* refers to a formal concept rather than a physiological mechanism. In the first instance, optic flow is a way of describing image shifts in terms of direction and magnitude as a consequence of relative motion between a given visual system and its surroundings (reviewed in Gibbson 1950; Nakayama and Loomis 1974; Barron et al. 1994). It is often presented as a vector field where—in case of an animal—each vector indicates the geometrical projection of relative motion as a function of position within the visual field and the animal's self-motion components, its translation and rotation. In their seminal work describing 'Facts on Optic Flow' Koenderink and van Door (1987) formalized the dependence of local flow or parallax vectors, p_i, as follows:

$$p_i = -(T - (T \cdot d_i)d_i)/D_i - (R \times d_i), \tag{9.1}$$

where the unit vectors d_i denotes the direction at which a given p_i is observed, D_i is a scalar that gives the distance to the closest object, and T and R indicate the translation—and rotation components of self-motion in vector notation. The operators '·'and '×' indicate the scalar—and vector products, respectively.

Referring back to the action–perception cycle, the ability to maintain balance or to stabilize its gaze requires an animal to retrieve information about its current self-motion components, T and R. From Eq. (9.1) it is obvious that the visual system is challenged with the inverse task: it needs to extract the current self-motion components from a given optic flow field. Koenderink and van Doorn (1987) worked out that given a sufficient number of local flow vector observations, an iterative least square approach allows us to retrieve the rotation component, R, and the direction of translation, $T/|T| = t$. The translation component cannot be fully retrieved because it depends on both, distance and speed. Motion parallax vectors induced during fast translation relative to an objects in the far distance result in the same magnitude as those induced by slow translation in the face of objects close by. I will return to the ambiguity between speed and proximity later when describing another well-established function of processing optic flow, the estimation of relative distance (see below). In the context of stabilization reflexes, the rotation component is of primary interest, which—as Eq. (9.1) indicates—is independent of distance.

9.2.2 Flies and Optic Flow

The algorithm proposed by Koenderink and van Door (1987) can be applied to both, engineered and biological vision systems. When it comes to identifying the neural mechanisms underlying the estimation of self-motion, however, dipteran flies have proven to be extremely useful (reviewed in Hausen 1993; Krapp 2000; Borst and Haag 2002). In these animals, more than six decades of anatomical, physiological and quantitative behavioral studies on visually guided behaviors have gathered a massive body of data enabling us to finally close the loop between sensory processing and motor outputs (reviewed in Krapp and Wicklein 2008). Flies arguably belong to the few animal species where the different processing stages outlined in the action–perception cycle (Fig. 9.1) may be associated with identified pathway, circuits and even cells (revs.: Strausfeld 1984; Egelhaaf and Borst 1993; Hausen 1993). For this reason, I will focus on a particular species, the blowfly *Calliphora*, to present a case study of how optic flow processing and self-motion estimation is

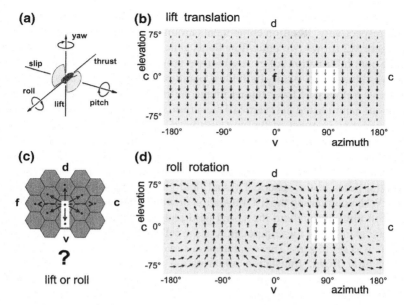

Fig. 9.2 Self-motion and optic flow. **a** Fly self-motions may be described in terms of translations and rotations along and around the animal's major body axes, referred to as thrust, side-slip, lift and roll, pitch and yaw, respectively. Lift translation (**b**) and roll rotation (**d**) result in optic flow fields easily distinguishable at the global scale. Each vector indicates the direction and velocity of relative motion between the visual systems and its environment as a function of azimuth and elevation. Small letters f, d, v and c denote frontal, dorsal, ventral and caudal. **c** A set of local detectors analysing motion along the ommatidial rows within the hexagonal eye lattice of a fly. Each detector is responding best to a different direction of visual motion (*black arrows*). Horizontal motion is either analysed by next-but neighbouring facets (*dashed arrows*) or by a combination of two obliquely oriented detectors. The detector marked in *light grey*, if facing motion in equatorial region of the right visual field (*shaded grey* regions in **b** and **d**), is activated during lift translation and roll rotation alike. Its activity is ambiguous with respect to the self-motion component that caused it. Further explanations in text (Redrawn from Krapp 2000)

implemented at the neuronal level. Like in many other flying insects, stabilization reflexes in *Calliphora* are mainly concerned with compensating for external disturbances cause by turbulent air, to maintain a default flight attitude and a level gaze. Most flies are endowed with nearly 4π vision (rev.: Nilsson and Land 2012). Except for a tiny fraction that is obstructed by the animal's own body, the whole world is mapped onto the spherical visual field of their compound eyes. This is a perfect prerequisite for exploiting the global properties of optic flow fields which are tightly linked to the translational and rotational components of self-motion (Koenderink and van Doorn 1987; Dahmen et al. 2001).

To illustrate the global properties of optic flow fields we may apply Eq. (9.1) and compute the distribution of local parallax vectors for two special cases. A fly's self-motion can be described in terms of translation—and rotation components along and around its cardinal body axes (Fig. 9.2a). Translation components along the longitudinal, transverse and vertical body axes are referred to as thrust, side-slip and lift, respectively. Rotation components around the same axes are roll, pitch and yaw. Figure 9.2 shows two optic flow fields generated during a pure lift translation (b) and a pure roll rotation (d) in a cylindrical projection of the spherical visual field into a two-dimensional plane. Individual parallax vector, p_i, indicates the direction and relative velocity of local retinal image shifts as a function of two angles, the (horizontal) azimuth and the (vertical) elevation, which may as well be expressed as the x, y and z component of the unit vectors d_i in Eq. (9.1). The point $0° =$ azimuth $=$ elevation is right in front of the animal (f = frontal, Fig. 9.2b, d), while d, v and c stand for dorsal, ventral and caudal, respectively. It should be noted that due to the reduction from three to two dimensions the spatial representation of the dorsal and ventral poles in the spherical visual field are expanded by a factor 1/sin(elevation). When computing the lift translation flow field the distances, D_i, in Eq. (9.1) were set to unity throughout the visual field and the rotation vector, R, was set to zero. Equivalently, for the computation of the roll rotation flow field the translation vector, T, was set to zero. These assumptions may appear to be unrealistic because during natural flight, translation and rotation components are likely to occur simultaneously. For the purpose of illustrating the global differences between translational—and rotational optic flow, however, these simplifications are justified.

What are the global differences between the two flow fields then? In case of a pure translation of the animal, all parallax vectors are aligned along great circles connecting the so-called pole of expansion in the flow field, i.e. the direction the animal is moving to, with the pole of contraction, which is the direction the animal is coming from (Fig. 9.2b). Both poles are directly connected by a line that is parallel to the translation vector, T. A flow field generated during pure rotation has a markedly different global appearance (Fig. 9.2d). Here, all parallax vectors are aligned along parallel circles centred at the axis of rotation. In both cases, there is no relative motion at the location of the self-motion axes—which defines the two singularities within a flow field. Maximum parallax vectors are observed half way between the singularities. For the lift flow field, the singularities correspond to the positions labelled d and v, while for the roll flow field they are found at f and c in Fig. 9.2d.

At the global level, the coherent structure of translational and rotational optic flow fields can be easily distinguished based on the overall distribution of the local parallax vectors. By analysing the pattern of retinal image shifts it should be possible to retrieve the unknown parameters t and R, required to control visually guided behavior and stabilization reflexes. The logical question to ask next is how does the visual system compute visual motion information?

9.2.3 The Elementary Movement Detector

From quantitative behavioral experiments on beetles 60 years ago, Hassenstein and Reichardt (1953) derived the functional structure of a mechanism that explains how the nervous system distinguishes between visual motions in opposite directions.

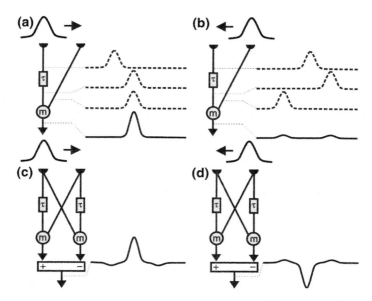

Fig. 9.3 Simplified functional structure of an elementary movement detector (*EMD*). **a** The signal at the output of a 'half-detector' depends on the direction of motion. Here, the detector's preferred direction is motion from *left to right* which results in a prominent output signal. **b** Motion in the detectors null direction (*right to left*) induces only tiny outputs. Fully opponent directional responses—i.e. positive responses for preferred direction motion (**c**) and negative responses for motion in the null direction (**d**)—are achieved by combining two half-detectors in a mirror-symmetrical arrangement and subtracting the outputs from one another. τ = delay stage, **m** = multiplication stage. Further explanations in text (Redrawn from Borst and Egelhaaf 1989)

Their phenomenological model known as the elementary movement detector (EMD), or Reichardt detector, is based on a spatio-temporal correlation of light intensities measured at neighbouring positions within the retinal photoreceptor array (reviewed in Reichardt 1961). Figure 9.3 presents a simplified version of an EMD (Borst and Egelhaaf 1989) to illustrate its operation. The upper left panel (a) shows a Gaussian light intensity distribution moving from left to right inducing a signal in the left input channel of the detector. At an intermediate stage, 'τ', the signal is delayed before it further propagates to a multiplication stage, **m**. If the time it takes the light intensity distribution to reach the right input channel is perfectly compensated for by the delay τ, the full amplitudes of the signals will be multiplied at stage **m**, resulting in a maximum output. If the light intensity distribution moves from right to left, the EMD will hardly produce any output at all because the signals within the two input channels will effectively be de-correlated in time (Fig. 9.3b). So the preferred motion direction of the EMD shown in Fig. 9.3a is from left to right, while motion from right to left is in the detector's anti-preferred or null direction (Fig. 9.3b). Such a so-called half-detector only produces a positive response upon motion in its preferred direction. To create a fully opponent movement detector the outputs of two mirror-symmetrical half-detectors with motion preferences in opposite directions are subtracted from one another at a common integration stage. The outputs of the full-detector upon motion in its preferred and anti-preferred direction are shown in Fig. 9.3c, d. To a certain degree the subtraction stage reduces signal components due to stationary luminance changes and enables average responses that are positive and negative for motion in the preferred and anti-preferred direction, respectively.

The EMD structure fulfils the three necessary and sufficient conditions to analyse directional motion: (1) two inputs, for instance from photoreceptors with neighbouring receptive fields, (2) an asymmetric processing of the two input signals—here, only one of the two input signals is delayed in time and (3) the signals have to be combined in a non-linear operation—which in most cases is modelled as a multiplication operation. These conditions are already met by a half-detector. As a consequence of the spatio-temporal correlation underlying its function, the EMD does not provide an output signal that is proportional to image velocity. This becomes obvious if we consider the delay, τ, of one of the input signals in relation to the velocity of the moving light intensity distribution: If τ is a fixed time delay, then there will only be one particular velocity of the light distribution that results in maximum signal correlation at the multiplication stage—trivially, for zero and infinite image velocities the detector will output a zero signal. In addition, its output is strongly dependent on image contrast and also on the ratio between the distance of neighbouring input channels and the spatial frequencies contained in the moving image (rev.: Reichardt 1987). I will later on discuss these factors in the context of processing optic flow.

There has been a tremendous effort over the last 30 years or so, to identify within the fly visual system the individual stages of an EMD. It was obvious that the input each individual channel would be provided by signals from photoreceptors sharing the same optical axis, and thus measuring light intensity changes at one and the same location in the visual field. But where exactly the time delay would be imposed on one of the signals and in which part of the motion vision pathway the multiplication of the two signals takes place were mostly a matter of speculation. Only when several labs recently started to take advantage of genetically modified *Drosophila* lines, was some of the first experimental evidence provided in support of earlier suggestions that the second visual neuropile, the medulla, could host the core stages of the EMD. Another somewhat older prediction based on electrophysiological studies by Riehle and Franceschini (1984), was that EMDs integrate signals processed from separate on and off channels: this prediction has now been supported by work in fruitflies (Eichner et al. 2011). A recent publication compares the computational structure of EMDs found in insects with those assumed to be implemented in vertebrate visual systems to show general principles and specific adaptations in the detection of visual motion (Borst and Euler 2011).

After this mechanistic explanation of directional motion detection we can now return to the fundamental problem of how the fly visual system retrieves self-motion parameters from analysing optic flow.

9.2.4 Extracting Self-Motion Components: Matched Filters for Optic Flow

Recall that the distinction between translation- and rotation-induced optic flow fields relies on differences in global features (Fig. 9.2). Global features, however, are not immediately accessible to the visual system. Instead, EMDs analyse the direction of retinal image shifts on a local basis, which may result in ambiguous signals. The problem is nicely illustrated by the grey region in Fig. 9.2b, d. In this equatorial region of the lateral visual field, centred at azimuth = 90° and elevation = 0°, the local parallax vectors are pointing downwards during both lift translation and roll rotation. Imagine a set of EMDs within the hexagonal eye lattice of a fly's compound eye analysing directional motion at exactly that position (Fig. 9.2c). The preferred directions of the EMDs are indicated by vectors connecting neighbouring facets along the ommatidial rows. Clearly, the EMD analysing vertical downward motion would be stimulated in both case, during lift translation and roll rotation. Thus, based on the activity of the response generated by only one, or even a few local EMDs analysing downward motion, the animal would not be able to infer whether it was lift or roll causing the EMD activity (Fig. 9.2b, d).

The ambiguity of individual EMD output signals is made even worse by another well-known issue regarding local motion detection, the so-called aperture problem (e.g. Nakayama and Silverman 1988). Independent of the self-motion component causing it, the activity of an EMD induced by a moving contour is always ambiguous with respect to direction and speed. This can be understood from the functional structure of an EMD. An elongated contour moving at a given velocity along a direction that is 45° off the detector's preferred direction may induce the same output as the contour moving exactly along the detector's preferred direction, but at slower velocity. In addition, EMD outputs may have a substantial degree of variability due to system noise at various processing stages (e.g. White et al. 2000). And finally, EMD signals depend on potentially noisy image parameters such as random fluctuations of local contrast and spatial frequency contents—which may even result in sign inversions if time-dependent terms of the responses are considered (e.g. Egelhaaf et al. 1989; Borst et al. 1995).

Like in many other sensory systems where the time-varying signal of a single receptor is insufficient for guiding behavior, many of the problems mentioned above may be resolved by spatially integrating the outputs of multiple local signals. How this may be achieved in the context of retrieving self-motion parameters from optic flow is shown in Fig. 9.4—for the case of an optic flow field induced during body roll. From the local sets of EMDs, those with preferred directions that match the the local parallax vectors within the roll optic flow field are integrated by a hypothetical filter-neuron. As a result of this selective integration the filter-neuron should produce the strongest response when the fly performs a roll rotation (Fig. 9.4). Even though a lift translation may as well induce a certain level of excitation, the receptive field of the filter-neuron best matches a roll optic flow field (Krapp et al. 1998).

Setting up matched filters in the spatial or temporal domain by integrating local ambiguous sensory signals to extract specific global stimulus features or parameters, is a fairly general principle in sensory neuroscience. Classic examples include the detection of polarized light patterns for navigation (e.g. Wehner 1987), identification of species-specific calling-songs in crickets (e.g. Schildberger 1984) and even face recognition in humans by cells in the inferior temporal cortex (Quiroga et al. 2005).

9.2.5 Fly Lobula Plate Tangential Cells

Based on the qualitative matched filter model shown in Fig. 9.4 and from intuition, there are a few predictions that can be made regarding the response properties of matched filter-neurons for optic flow (Franz and Krapp 2000). First, the output of the cells should be motion sensitive and directional selective. Second, the preferred direction of the cell, when probed with local visual motion, should depend on the stimulus position within the visual field. And finally, the receptive field size of such matched filter neuron should be very large, a prediction that will be qualified later on.

From the mid- to late 1970s researchers at Caltech in Pasadena and at the Max-Planck Institute of biological Kybernetik in Tübingen had begun recording from interneurons in the posterior part of the third optic lobe, the lobula plate (Strausfeld 1976). The so-called lobula plate tangential cells (LPTCs), showed at least two out of three predicted response properties for matched-filters for optic flow. They were directionally selective and had large receptive fields (reviewed in Hausen 1984, 1993). They also presented an ideal opportunity to study the properties of EMDs at one of the key stages along the motion vision pathway (reviewed in e.g. Egelhaaf and Borst 1993). Anatomical and electrophysiological studies suggested that LPTCs corresponded to the subtraction stage in the EMD model (e.g. rev.: Borst and Egelhaaf 1993). LPTCs had extensive dendritic arborizations that were thought to integrate the signals of hundreds (*Drosophila*) or even thousands (*Calliphora*) of retinotopically arranged half-detectors scanning retinal image shifts throughout the visual field (reviewed in Hausen 1984). Amongst the 50–60 LPTCs within each lobula plate, two subpopulations of tangential cells were distinguished based on their morphological appearance and their overall directional selectivity: Three cells of the Horizontal System (HS; e.g. Hausen 1982) and the 10 cells of the Vertical System (VS; e.g. Pierantoni 1976; Hengstenberg 1982). The main dendrites of the HS-cells were found to be horizontally oriented and the cells responded predominantly to horizontal motion. VS-cell had vertically oriented main dendrites and mainly responded to vertical downward motion.

A combination from anatomical and lesion studies in blowflies (Geiger and Naessle 1981; Hausen and Wehrhahn 1983) and work on the *Drosophila* mutant, 'optomotor blind'—omb^{H31} (Heisenberg et al. 1978), suggested that LPTCs provided signals to various motor systems in the fly's thorax and to the neck motor

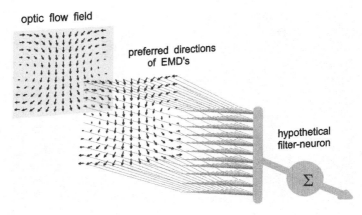

optic flow field

preferred directions
of EMD's

hypothetical
filter-neuron

Σ

Fig. 9.4 Neuronal matched filter for optic flow. At any position within the visual field, from the sets of EMDs analysing motion only those are chosen the preferred directions of which coincides with the direction of the local parallax vectors of a specific optic flow field. Here the optic flow field during a roll rotation is matched by an array of EMDs the output of which is integrated by a hypothetical filter-neuron. This neuron would be strongly activated when the fly performs a roll rotation. Further explanations in text (Redrawn from Krapp 2000)

system. Work in the lab of Axel Borst, for instance, using cell-specific Gal4 *Drosophila* lines in combination with optogenetics—which allows for the expression of light-controllable ion channels (e.g. Berndt et al. 2009)—currently establishes the significance of LPTCs for stabilization reflexes in fruitflies.

Up until nearly 20 years ago it was still believed that measuring horizontal and vertical components of motion—information that the HS-and VS-cells could in principle provide—was sufficient for the control of optomotor behavior. A fast local motion stimulus to characterize in detail the local motion preferences across the whole LPTC's receptive fields, however, produced a slightly different picture. Figure 9.5a shows a schematic of an experimental setup that delivered such a local motion stimulus. The fly in the centre of the setup was challenged with a black disc moving along a small circular trajectory while either extra- or intracellular recordings from LPTCs were performed. When the tangent motion direction of the black disc coincided with the local preferred direction of the LPTC, the cell's activity was strongly increased, while disc motion in the opposite direction caused the cell's activity to be inhibited. Altogether, the response of the neuron to an entire motion cycle was very similar to a cosine-shaped function (Fig. 9.5b). The average across consecutive response cycles (Fig. 9.5c) was used to compute two parameters: the local preferred direction (LPD) and local motion sensitivity (LMS), which could then be plotted as vectors at the position within the cell's receptive field where the stimulus was presented (Krapp and Hengstenberg 1997). Such an analysis was initially carried out on some spiking LPTCs (loc. cit.), all individually identified VS-cell (Fig. 9.5d), and HS-cells (Krapp and Hengstenberg 1996; Krapp et al. 1998; Krapp et al. 2001; rev: Taylor and Krapp 2007).

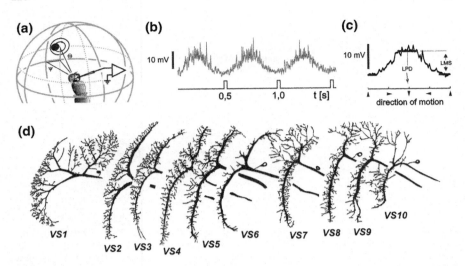

Fig. 9.5 Determining local preferred directions and motion sensitivities in identified VS-cells. **a** Cartoon of experimental setup to measure directional preferences of visual interneurons in the blowfly. The stimulus (back disc, 7.8° diameter) is moving at constant speed (2 cycles per second) on a circular path (10.5° diameter), while the activity of an interneuron is recorded. **b** The membrane potential of the neuron is sinusoidally modulated by the movement of the disc. **c** From the averaged response cycles the local preferred direction (*LPD*) and the local motion sensitivity (*LMS*) are can be determined. **d** Morphological reconstruction of the 10 cells of the Vertical System. Further explanations in text (Redrawn from Krapp et al. 2001; Krapp et al. 1998)

The results looked rather intriguing in the context of analysing optic flow fields. The VS-cells showed preferences for all possible directions, with smooth spatial transitions between vertical and horizontal motion (Krapp and Hengstgenberg 1996; Krapp et al. 1998). The distributions of motion preferences of three VS-cells are shown in Fig. 9.6. What all VS-cells have in common is that they show strong sensitivity to vertical downward motion, but at different azimuthal positions within their ipsilateral receptive field. At other positions, the local preferred directions are well aligned with the orientation of parallax vectors within optic flow fields induced by different horizontal body rotations (Krapp et al. 1998; rev: Krapp 2000). The receptive field of the VS6 cell, for instance, shows a striking similarity to an optic flow field caused by a roll rotation (cf. Figs. 9.2d and 9.6b). The VS1-cell is most sensitive to downward motion in the frontal, and to upward motion in the dorso-caudal receptive field, suggesting this cell to respond particularly well to nose-up pitch rotations (Fig. 9.6a). The VS8-cell, finally, shows strong downward motion preferences at an azimuth of 135° and slightly weaker upward preferences in the frontal visual field with a clear-cut singularity found at an azimuth of 45°. This indicates that the VS8-cell would strongly respond during a combination of pitch and roll (Fig. 9.6c).

Additional studies on the receptive field organization of many more LPTCs in combination with theoretical studies further supported the hypothesis that LPTCs

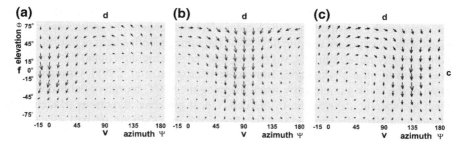

Fig. 9.6 Average ipsilateral receptive field organization of three VS-cells ($N \geq 5$). Local preferred directions and motion sensitivities are plotted as vectors as a function of azimuth and elevation within the visual field. The distribution of all three cells is reminiscent of the global structure of rotatory optic flow fields. The VS1-cell (**a**), VS6-cell (**b**), and VS8-cell (**c**) receptive fields are matched to flow fields induced by nose-up pitch, roll to the right, and a rotation in between nose-down pitch and roll, respectively. Further explanation in text (Redrawn from Krapp and Hengstenberg 1996)

function as *matched filters for optic flow* to estimate self-motion components (Franz and Krapp 2000; Karmeier et al. 2003, 2005; revs.: Krapp 2000; Taylor and Krapp 2007; Krapp and Wicklein 2008). Applying the matched filter approach to robotic platforms demonstrated its usefulness to estimate self-motion components under open-loop conditions (Franz et al. 2004). In addition, the receptive field properties of the LPTCs have been interpreted within a control engineering framework and successfully implemented on an autonomously flying quadrupter to stabilize lateral distance to corridor walls, yaw orientation and forward speed (Hyslop et al. 2010).

Consequently, LPTCs such as the HS-cells have been suggested to serve two different functions exploiting specific properties of optic flow fields. As originally assumed, these cells are involved in supporting stabilization reflexes such as optomotor yaw responses (e.g. Wehrhahn and Hausen 1993) and compensatory head rotations around the yaw axis (e.g. Huston and Krapp 2008). Depending on the current flight phase, however, the activity of the HS-cells may indicate relative distance. In free flight blowflies perform sharp banked turns, followed by a drift-like horizontal translation due to the animal's inertia (e.g. Lindemann et al. 2005). Only during the translation phase does the current optic flow field contain information about relative distance to the objects in the surroundings. Assuming that the rotation vector, R, reduces to zero during the translation phase and that the HS-cells integrate a large number of p_i at different locations d_i, the overall response of the HS-cells should be roughly proportional to the average distance $\langle D_i \rangle$, as can be derived from Eq. (9.1). Comparing the responses between the so-called equatorial HSE-cells in the left and the right part of the animal's visual system would enable the fly to turn away from the side where the HSE-cell response is stronger, as it indicates closer proximity to potential obstacles (e.g. Lindemann et al. 2005; Karmeier et al. 2006). Other LPTCs are involved in neural circuits enabling figure-ground discrimination (e.g. Egelhaaf 1985, Warzecha et al. 1993).

One of the most important properties of the LPTCs in the context of optic flow processing is the distribution of the local preferred directions. Experiments on flies raised with no exposure to optic flow show the same receptive field organization as their normally raised siblings, suggesting that the specific directional distributions have evolved on a phylogenetic time scale (Karmeier et al. 2001). Changing the speed of the black stimulus disc when mapping the LPTC receptive fields does not affect the specific distribution of the cells' local preferred directions (Baden and Krapp, unpublished). The fact that the directional distributions are basically set in stone is a necessary condition for the matched filter hypothesis to work (Franz and Krapp 2000). This is very different for the local motion sensitivities which are anything but constant over time. They depend on a huge number of factors which include non-linear processing steps like dendritic gain control (Borst et al. 1995), sub-linear spatial integration (Haag et al. 1992), contrast gain control (Harris et al. 2000), motion adaptation (e.g. Maddess and Laughlin 1985) and the locomotor state of the animal (Maimon et al. 2010; Longden and Krapp 2010). From the classical control engineering point of view it sounds rather bizarre to use signals that are not representing an absolute velocity measure for the control of angular rotation rates. The whole point of those non-linear mechanisms is, however, to keep the operating range of the visual system in a regime where the output depends in a monotonic way on the changes of the input signal. This makes sense in that it relaxes a problem I mentioned earlier: the EMD output does not depend linearly on the speed of visual motion, but reaches its maximum at a particular temporal frequency of the stimulus pattern—i.e. the speed of the pattern divided by its spatial wavelength. Lower and higher temporal frequencies result in smaller output signals of the EMD (see also Sect. 3.6). Under closed-loop feedback control this does not necessarily result in a problem, as long as the dynamic range of the inputs is mapped onto the output range by a monotonic function that avoids signal saturation. In such case absolute values are not necessarily needed for control (Taylor and Krapp 2007).

9.2.6 Transforming Sensory Signals into Motor Commands

Based on the experimental evidence accumulated so far in combination with model simulations, a convincing case can be made for the LPTCs to play a major role in analysing optic flow and supporting stabilization reflexes. The logical next step in the functional analysis of the motion vision pathway was to study the response properties of motor neurons and descending neurons receiving input from LPTCs. A convenient system for study was the neck motor mediating compensatory head movements whenever the body attitude is affected by external perturbations due to turbulent air. The comparison of LPTC and motor neuron receptive fields provided interesting insights into the nature of the sensorimotor transformation along the motion vision pathway (Huston and Krapp 2008). At first glance, the distribution of local preferred directions and motion sensitivities showed little differences between LPTCs and neck motor neurons. A thorough data analysis, however, showed that the neck motor neuron receptive fields had stronger binocular inputs than the LPTCs, resulting in a higher specificity for rotational rather than translational optic flow fields (Huston and Krapp 2008). This result made perfect sense as compensatory head movements are first of all dealing with rotations. Higher rotation specificity in the neck motor neurons was established by increasing contralateral response contributions. In the context of estimating self-motion components from optic flow, it is worth formally explaining the consequences of an increased binocular input.

In Sect. 9.3.1, the Koenderink and von Doorn (1987) algorithm was proposed for retrieving self-motion parameters from optic flow fields. Below a slight modification of the equations Koenderink and van Doorn derived for an initial estimate of the translation and rotation parameters, t, and R., before the iteration procedure kicks in. Assuming that all distances, D_i, in Eq. (9.1) are set to unity, the unit vector t may be replaced with the actual translation vector T. The ability to determine translation- and rotation-induced contributions to local parallax vectors, and thus to reliably estimate R and T, is increased with expanding the area in which the parallax vectors are measured. This can be substantiated by inspecting and interpreting the equations describing the initial estimates of R and T:

$$R = (I - M)^{-1}[(T \times \langle d_i \rangle) + \langle p_i \times d_i \rangle] \tag{9.2a}$$

$$T = (I - M)^{-1}[(R \times \langle d_i \rangle) + \langle p_i \rangle] \tag{9.2b}$$

where I is the 3×3 identity matrix and M is a 3×3 weighting matrix that gives the average contributions along the x, y and z directions. Variables p_i and d_i, as before, denote the parallax vectors observed in the directions of the unit vectors d_i, while '$\langle \ldots \rangle$' and '$\times$' indicate the average and the cross product, respectively.

It may be surprising to find the estimate of R to depend on T and vice versa. T and R on the right hand side of the equations indicate the 'apparent' translation and rotation contributions due to rotation and translation, respectively. This is easily understood when re-visiting Fig. 9.2, where the grey regions illustrate that local directional motion does not allow us to infer whether translation or rotation caused the direction of the observed local parallax vectors. The equations, however, also provide a clear-cut mechanism of how to avoid this ambiguity. The apparent translation and rotation vectors are applied to the average of the observation directions, $\langle d_i \rangle$, by means of a cross product. If the d_i are chosen so their average becomes zero, the cross products become zero and the apparent rotation and translation terms vanish, in which case the wanted self-motion components R and T (left part of equations) can be correctly estimate directly from the averages $\langle p_i \times d_i \rangle$ and $\langle p_i \rangle$.

This is exactly what is happening at the level of the neck motor neurons. The body rotations of the fly, R, the populations of LPTCs and neck motor neurons are tuned to detect are almost the same. Neck motor neurons, however, receive stronger binocular input which basically allows them to increase their rotation specificity by minimizing translation-induced response contributions (Huston and Krapp 2008).

Intriguingly, the response properties did not indicate any sophisticated *sensorimotor transformation* along the gaze stabilization pathway, which—according to the action–perception cycle illustrated in Fig. 9.1—should be one of the processing steps required for behavioral control. The sensorimotor transformation is supposed to resolve a general issue that needs to be dealt with, no matter which sensory modality is considered. This is, sensory information is initially sampled by individual receptors measuring physical quantities within their modality-specific, local coordinates (e.g. Masino and Knudsen 1990; Graf et al. 1993). For motion vision, the measuring axes would be described in local retinal coordinates aligned with the orientation of ommatidial rows within the hexagonal eye lattice (cf. Sect. 3.3). The outputs of the LPTCs, however, are directly passed on to the motor systems without major modifications, except for an increased specificity to rotation as mentioned above. Where did the sensorimotor transformation take place then? The most consistent interpretation of the experimental findings is that the transformation from retinal coordinates into motor coordinates is established by the selective integration of local motion signals on the dendrites of the LPTCs. This step replaces any high-level computation or signal transformation, and results immediately in LPTC outputs that are perfectly well suited to control motor action (Huston and Krapp 2008; Krapp 2010). The specific choice of preferred rotation axes each LPTC covers may be related to the species-specific dynamic properties of the motor systems to be controlled. I will return to this point later in the conclusions.

9.3 Multisensory Integration

Most animals do not trust vision alone when estimating their self-motion components. This is particularly true for animals with high degrees of manoeuvrability such as most dipteran flies but also for vertebrates. The reasons for the need of other modalities are related to two commonly known factors: first, the visual system is notoriously slow. The photo transduction cascade takes way more time to convert photons into changes of photoreceptor potentials than it would take physical pressure changes to affect the potential in a mechanoreceptor. And secondly, the analysis of visual motion takes extra time due to the delay τ in one of the channels in the EMD, which is necessary for the spatio-temporal correlation of its two input signals. These factors, together with the computational structure of the EMD limit the bandwidth of the motion vision pathway.

Across phyla, visual mechanisms supporting stabilization reflexes, which rely on immediate compensatory motor action, are combined with fast mechanosensory mechanisms. Primates combine visual information with information from the vestibular system providing linear and angular acceleration information to control their posture, head orientation and gaze (rev.: Angelaki and Cullen 2008). Flying insects do not have a vestibular system. But two-winged flies, or dipterans, have evolved a gyroscope-like mechanosensory system that provides information about angular rotations (e.g. Nalbach and Hengstenberg 1994). The former pair of hind wings positioned at the posterior part of the thorax has transformed into tiny club-like structures, called halteres, which move up and down at the same frequency as the front wings, but in anti-phase (Fig. 9.7a). During locomotion the beating halteres form a stable oscillating system, which, during rapid body turns, experiences inertial forces picked up by fields of campaniform sensilla (Fig. 9.7a, bottom). The integrated activation patterns of the sensilla provide the fly with angular rotation rates (Nalbach 1994). Like the vestibular system in vertebrates the halteres are way faster than the visual system. They initiate compensatory head movements within <10 ms, where the motion visual pathway requires more than twice as long (rev.: Hengstenberg 1993). A third sensory system to introduce in the context of stabilization reflexes are the call-called ocelli, the second visual system in many insects. In blowfies it is formed from three little lens eyes located on top of the head (Fig. 9.7a) which measure light intensity changes in the dorsal visual hemisphere. Sudden attitude changes of the animal induce differences in the illumination pattern of the three ocelli which enable the fly to assess horizontal body rotations, though at comparatively crude spatial resolution (revs.: Hengstenberg 1993; Krapp 2009; see below).

There are several other mechanosensory mechanisms involved in insect flight and gaze control the function of which has been comprehensively reviewed recently (rev.: Taylor and Krapp 2007). Here, I will focus on the role of the compound eyes, estimating self-motion based on optic flow, the ocelli indicating fast attitude changes, and the halteres measuring body rotation rates.

The question is: how is information from the different systems combined at the neuronal level to support stable gaze and flight? It is important to realize that the stabilization against externally caused body rotations and maintaining a level gaze are tightly linked. Clearly, when a gust of wind is inducing a sudden change of body attitude, the role of the neck motor system is to immediately compensate for it so the visual input remains in its default orientation. A large number of visual tasks, including the analysis of optic flow, require an animal to keep a level gaze which also ensures all head-based sensor systems stay aligned with the inertial vector in a common frame of reference (Krapp et al. 2012).

Recent electrophysiological studies on LPTCs, the ocellar pathway, neck motor neurons, as well as descending neurons, which connect the fly brain with the thoracic motor centres, have aimed to understand the neural basis of multisensory stabilization reflexes. Neuroanatomical investigations suggest the lateral protocerebrum, an area where the synaptic terminals of the LPTCs are found, is massively involved in multisensory integration (Strausfeld and Seyan 1985). It receives inputs from several sensory nerves, more central processing areas and ascending signals from the thorax (Haag et al. 2010). We also know from neuroanatomy that many of the inputs to the integrating neurons are mediated by mixed chemical/electrical synapses (Strausfeld and Bassemir 1983). The electrical connections probably explain why VS-cell and the V1-cell—a spiking LPTC postsynaptic to the VS-cells—change their activity upon stimulation of the ocelli (Parsons et al. 2006). Subsequent intracellular recordings from VS-cells have indeed demonstrated that

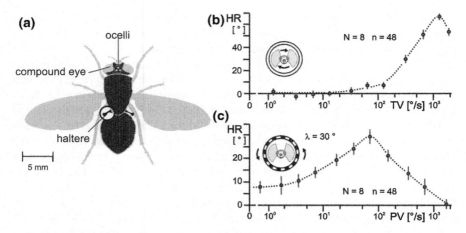

Fig. 9.7 Visual systems and halteres in the blowfly contribute to compensatory head roll. **a** The fly carries three small under-focussed lens eyes on top of its head, called ocelli, that sense fast luminance changes in the dorsal visual hemisphere. The compound eye analyse optic flow and the halteres measure angular rotation rates by processing the outputs of campaniform sensilla arranged close to their hinge-joints. **b** Head roll response (*HR*) plotted against angular velocity of the thorax (*TV*) in a homogeneous visual surrounding, mainly stimulating the halteres (*upper plot*). **c** Head roll response (*HR*) plotted against angular velocity of a visual pattern (*PV*) stimulating the motion vision pathway through the compound eyes in a stationary fly. λ = spatial wavelength of the visual grating. Further explanations in text (Redrawn from Hengstenberg 1993)

ocellar input induces fast activity changes as a function of horizontal body rotations mimicked by differential illumination of the ocelli (Parsons et al. 2010; Fig. 9.8b). Unlike stimulation of the motion vision pathway via the compound eyes, the resolution of different horizontal rotations is rather poor. For optic flow inputs from the compound eyes, each VS-cell prefers a slightly different horizontal rotation, i.e. the VS-cells set up a non-orthogonal, multi-dimensional system to estimate self-motion around multiple axes (Fig. 9.8c). Upon stimulation of the ocelli, however, the VS-cells covered only three different horizontal axes: the roll axis and a rotation axis in between roll and pitch (Fig. 9.8d). The interpretation of this somewhat surprising result is that ocellar input, although not very accurate at resolving the exact horizontal self-rotation, provides a fast response after a short delay to speed up the more fine-grained but slower signals conveyed by the motion vision pathway (Parsons et al. 2010). Strategies for reducing response delays have been identified as one of the most important features of sensory pathways providing feedback for stabilization reflexes. This is because long response delays in combination with fast modulations of the input signals and a high feedback gain are seriously prone to control insta-bilities (Elzinga et al. 2012), which biological control systems need to avoid.

The interaction of visual and mechanosensory signals from the halters was studied in electrophysiological experiments on blowfly neck motor neurons (Huston and Krapp 2009). For some motor neurons, visual motion stimuli alone turned out to be insufficient to induced action potentials—a necessary condition for

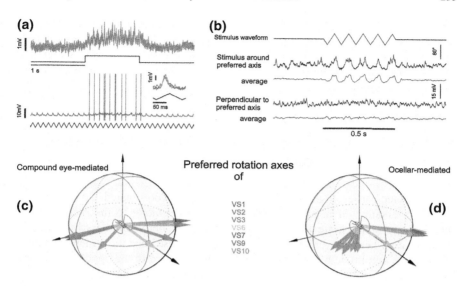

Fig. 9.8 Neuronal integration of visual and mechanosensory information. **a** Intracellular recording from a neck motor neuron in the blowfly. Visual motion in the neuron's preferred direction on its own induces only a sub-threshold membrane depolarization (*upper recording trace*). Periodic movement of the halteres results in brief EPSPs which are tightly phase-locked to the haltere movements (inset). Only if both stimuli are applied simultaneously, the motor neuron generates action potentials. Stimulus time courses are plotted below the recording traces. **b** Ocellar stimulation mimicking back and force rotations of the fly around different horizontal rotation axes induce membrane potential changes in VS-cells. Rotations around the preferred axis result in maximum responses, while rotations around an axis perpendicular to the preferred axis have no effect. **c** Preferred rotation axes of VS-cells upon compound eye and (**d**) ocellar stimulation. Further explanations in text (Redrawn from Huston and Krapp 2009; Parsons et al. 2010)

the contraction of neck muscles. Instead, visual motion depolarized the motor neurons' membrane only by a few millivolts, not exceeding the threshold potential (Fig. 9.8a). Similarly, if the halteres were moved up and down, as a result the motor neurons generated a brief excitatory postsynaptic potential that was tightly phase-coupled to the periodic haltere movement. But again the membrane potential changes did not exceed threshold. Only when both the visual motion and the haltere stimuli were applied simultaneously, were action potentials generated in the motor neuron (Fig. 9.8a, bottom recording trace). In these motor neurons, visual and haltere signal integration provides a non-linear gating mechanism that only enables head movements if both modalities are active at the same time (Huston and Krapp 2009). Similar non-linear integration effects were demonstrated in experiments on a neck motor neuron that received input from the halteres, the wind-sensitive Johnston organ, optic flow processing tangential cells and from a central source signalling flight activity (Haag et al. 2010).

Based on the electrophysiological evidence so far no clear conclusion can be reached, as to whether or not the signals from different sensory systems are linearly combined when supporting fly stabilization reflexes. Haag et al. (2007) reported some evidence of linear combinations at the level of an identified descending neuron, but in later studies on another descending neuron, Wertz et al. (2008) observed a non-linear integration already when motion information from both eyes is combined.

9.3.1 Gaze Control as a Paradigm for Studying Multisensory Stabilization Reflexes

As mentioned earlier, flight and gaze control must be coordinated. To maintain the default orientation of head-based sensory systems relative to the inertial vector flies have to compensate for any externally caused body rotations by means of compensatory head movements. The amplitude of these head movements would be equal to the amplitude of the body rotation, but executed in the opposite direction. The kinematic relationship between head and body movements suggests that studying compensatory head movements can indirectly tell us something about how flight control may be organized upon stimulation of several modalities. This argument is further supported by anatomical and physiological evidence that sensorimotor pathways supplying head movements are also involved in the control of flight (e.g. Gilbert et al. 1995; Gronenberg et al. 1990).

From the mid-1980s to early 1990s, Hengstenberg and colleagues studied gaze stabilization in blowflies (Hengstenberg et al. 1986; reviewed in Hengstenberg 1991, 1993). They were aiming to understand how different sensory mechanisms contribute to compensatory head roll when the fly is rotated around its longitudinal body axis. Systematic work was done on several systems including the halteres (Nalbach and Hengstenberg 1994), ocelli (Schuppe and Hengstenberg 1993), and the prosternal organ, which measures the angular position of the head relative to the body (Preuss and Hengstenberg 1992).

The results of those studies convincingly demonstrated one of the general design principles of biological sensorimotor control: individual biological sensor systems are not specified to cover the entire dynamic range of possible inputs. Unpredictable external disturbances may induce very subtle attitude changes in the low angular velocity range but could potentially cause body rotations up to a few thousands of degrees per second. Sensory mechanisms supporting stabilization reflexes, however, normally have low-pass or band-pass characteristics with optimal performances and different response delays depending on the individual modality (reviewed in Hengstenberg 1991). Halteres, for instance do not sense low velocity drifts in gaze stabilization experiments but show peak sensitivity from about 1,000°/s (Fig. 9.7b). The motion vision pathway, on the other hand, covers the low dynamic range, with the highest responses measured for angular velocities of about 100°/s, followed by a steep roll-off and very low sensitivity for angular velocities >1,000°/s (Fig. 9.7c). In a way different sensory systems—or mechanisms—complement each other enabling the fly to deal with external disturbances over an extended dynamic range.

I mentioned earlier that the integration of neuronal signals may either be linear or non-linear at different processing stages along the sensorimotor pathways. Based on his behavioral work on gaze stabilization, however, Hengstenberg came to the conclusion that signals from different sensory systems may be internally scaled and then linearly combined to control the movements of the head (reviewed in Hengstenberg 1993). Although the experimental data support this conclusion, there are still a few key issues remaining with respect to the nature of the underlying neuronal mechanisms. It is quite possible that several non-linear processes compensate each other to produce an overall linear behavioral output. Another point is that gaze stabilization had not yet been evaluated within a control engineering framework. Judging the design of a system based on response characteristics observed during the stimulation of individual sensor system and then adding or subtracting them may miss out on the interactions between the systems under closed-loop condition.

In an attempt to address this issue, recent studies applying a linear systems analysis approach in combination with closed-loop simulations have provided new insight into how the integration of visual and mechanosensory signals may be interpreted. The control engineering approach has the advantage that it enables the identification and validation of the system and allows for firm quantitative predictions of the systems behavior under closed-loop conditions (Schwyn et al. 2011).

To assess the systems properties in a straight forward way, flies were attached to a step motor (Fig. 9.9a) that oscillated the animals around their roll axis at different amplitudes and frequencies. The rotations of the body (TR), corresponding to the input, and the compensating head rotations (HR), corresponding to the output, were retrieved from high speed videographs. The performance of the system was quantified as the difference between the two parameters (Fig. 9.9b), where perfect compensation would be achieved if $TR-HR = 0$. In general, the behavioral output, i.e. the head roll angel, was highly correlated with the sinusoidal waveform of the stimulus (Fig. 9.9c). Two experimental groups were tested: intact animals and animals where the halteres were removed. The Bode plot summarizes the performance of the experimental groups stimulated with oscillation frequencies in the range from 0.05 to 15 Hz (Fig. 9.9e). On average, intact animals performed at comparatively high gain (-3 dB) up to frequencies of around 3 Hz. Beyond that frequency the gain dropped slightly and the phase lag of the response increased by about 20°. The compensatory head movements of flies without halteres were very similar up to frequencies of about 1 Hz. The roll-off at higher frequencies, however, is clearly steeper and the phase lag at the highest stimulus frequency reaches around 150°, indicating that the halteres system extends the frequency range of the response.

These not particularly surprising results were in qualitative agreement with the studies by Hengstenberg and colleagues (reviewed in e.g. Hengstenberg 1993), which had already demonstrated the range fractioning of the motion vision and haltere systems. However, the ability to fit transfer functions to the experimental

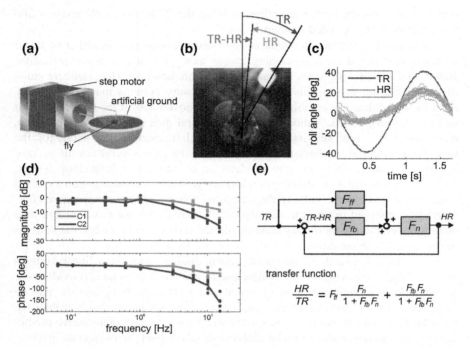

Fig. 9.9 Behavioral systems analysis of the blowfly gaze stabilization system. **a** Schematized setup showing a step motor that oscillates a fly over a dark artificial ground in an otherwise illuminated area. Head movements of the fly are recorded by a high speed video camera. **b** In a frame-by-frame analysis the body rotation (*TR*, stimulus) and the head angle (*HR*) are retrieved. **c** *TR* and *HR* are plotted against time. Perfect compensation for the imposed body rotation would result in a flat line of *HR*, i.e. the head is kept level at all times, though the body is rotated. **d** Bode plot of the frequency response gain (*upper plot*) and phase (*lower plot*) in flies with compound eyes and halteres intact (C1, *light grey*) and with the halters removed (C2, *dark grey*). **e** Control architecture and transfer functions of the fly gaze stabilization system considering the contribution of the compound eyes (F_{fb}) and the halteres (F_{ff}) in combination with the dynamics of the neck motor system (F_n). Further explanation in text (Redrawn from Schwyn et al. 2011 and unpublished results)

data and thereby derive a linear control architecture (Fig. 9.9e), revealed two interesting findings. The first, quite obvious point was that no matter what the experimental stimulus conditions, the system's behavior has always to be described as a combined transfer function that takes into account the properties of the sensors (F_{ff} and F_{fb}), but also the dynamic properties of the neck motor system (F_n). When formalizing the closed-loop gain (*HR/TR*) based on the transfer functions it also becomes clear that the mutual interactions between the sensory systems have to be taken into account (Fig. 9.9e, bottom). The second point concerns the control architecture (Fig. 9.9e). The two blocks labelled F_{ff} and F_{fb} give the transfer functions of the haltere and motion vision system, where the subscripts *ff* and *fb* indicate feed-forward and feedback component, respectively.

Such architecture is well established in control engineering know as a *two degree of freedom controller* (Doyle et al. 1992). Here, the degrees of freedom refer to feed-forward and feedback. The general operation of the gaze stabilization system can be interpreted in the following way. Fast body rotations are immediately picked up by the haltere system (F_{ff}) which induces a fast, low latency response in the neck motor system (F_n), resulting in an initial compensatory head roll. This halter-induced response reduces the motion of the head relative to the environment, and thus shifts the effective input to the motion vision pathway towards the lower velocity range. This clever control design, in a way, adjusts the dynamic input range to better fits the slower response properties of the motion vision pathway which has longer response delays and a peak performance in the lower velocity range (Schwyn et al. in preparation).

Further experiments where two frequencies were superimposed to analyse harmonic distortion, supported the assumption that the neck motor system operates linearly to an amazingly high degree (97–98 %, Schwyn et al. 2011). This value is even more impressive if compared to an assessment of the optomotor system in *Drosophila* which, in a white-noise systems analysis study, reached linearity levels of about 70 % (Theobald et al. 2010). Given these results, the more intriguing question becomes how the reported nonlinearities in the sensory pathways of flies compensate each other to produce a nearly linear output.

One of the drawbacks when analysing fly gaze stabilization behavior is that there is no independent information on the dynamics of the neck motor system itself. Current projects apply μ-CT techniques enabling three-dimensional renderings of the neck motor system based on which biomechanical models may be derived. The idea is to use finite element models to predict head movements induced by the activation of the 21 muscles pairs in the blowfly neck motor system and validate the models by individual electrical stimulation of the different muscles. Both, earlier anatomical studies (Strausfeld et al. 1987; Milde et al. 1987) and pioneering electro-stimulation experiments (Gilbert et al. 1995) will be used as guidance when fully reconstructing the functional organization of the system.

9.4 Conclusions

Optic flow sensing certainly is one of the key functions of the visual system in support of locomotor and gaze control. It demonstrates nicely how *task-specific integration* transforms noisy, and ambiguous local sensor signals into outputs that are directly suited for motor control. The resulting preferred rotation axes of the optic flow processing LPTCs form a non-orthogonal, multi-dimensional coordinate system within which changes in specific states may be sensed and ultimately controlled. It is tempting to assume that the exact choice of these axes depends on the species-specific *modes of motion* a given animal is likely to encounter (reviewed in Taylor and Krapp 2007; Krapp et al. 2011). Comparative studies across species of flying insect with different flight dynamics and sensor equipment will be required to support such *mode sensing hypothesis*. It may potentially provide a more general framework for understanding the functional relationship between sensory systems and the motor output they control. Sensory and motor systems certainly evolved in parallel where changes at the sensor side would most likely result in corresponding modifications of the motor side or vice versa.

If we want to understand general design principles of sensorimotor control it may be worth considering the constraints under which both sensory and motor systems have evolved and operate. Animals generally control their behavior under closed-loop conditions. Any sensory feedback, therefore, is directly affected and constrained by the animal's dynamic interaction with its physical environment. As the dynamics of motor systems do normally not change—at least not for some stages during their life cycle—animals may reduce the overall complexity of sensory inputs, by adjusting their sensing capabilities and, in a way, predict the stimulus distributions they are likely to encounter (Laughlin 1987). Establishing matched filters to extract relevant sensory information under these conditions reflects such a strategy (Wehner 1987; Franz and Krapp 2000). Limited bandwidth of individual sensor systems is another common constraint in sensory processing. Across phyla sensory information is typically fractioned into slow visual and fast mechanosensory channels, which are then integrated to enhance behavioral performance. In fly stabilization reflexes, a combination of fast feed-forward and slow feedback pathways is likely to increase sensory bandwidth without the risk of control instabilities. The exact neural strategies implemented when fusing information from different modalities—in terms of linear or non-linear combinations—are still being studied. Finally, nervous systems across phyla face another common constraint: limited energy resources (Attwell and Laughlin 2001; Niven and Laughlin 2008). More recent work on processing visual motion information indicates that the neuronal activity in the motion vision pathway not only depends on the locomotor state of an animal (Longden and Krapp 2009; Maimon et al. 2010; Jung et al. 2011), but also on its nutritional state (Longden et al. submitted). These findings suggest that the retrieval of sensory information can hardly be treated as time-invariant processes, but that the amount of energy invested in sensory processing may be subtly balanced depending on the current metabolic state of the animal.

In the end, a complete understanding of sensorimotor control mechanisms requires investigation of the entire action–perception cycle, keeping in mind the constraints that shape the different stages. Interdisciplinary collaborations where behavioral systems analysis, functional anatomy, electrophysiology and modelling are combined, followed by implementation and testing of our experimentally derived hypotheses on robotic systems will move the field forward. After all, understanding complex systems—and that is what animals really are—requires complex approaches.

Acknowledgments   Some work presented in this chapter was supported by grants from the US AFRL and AFOSR through EOARD (grants: FA8655-05-1-3066, FA8655-09-1-3022, FA8655-09-1-3067, FA8655-09-1-3083). I would like to thank Karin Bierig for performing expert behavioral experiments, help with preparing the figures, and her patience while I was writing this chapter. I would also like to thank Simon Laughlin, Graham Taylor and Sean Humbert for many inspiring discussions and my current and previous lab members for their excellent work. Special thanks to Sheryl Coombs who provided many insightful comments that helped to improve an earlier version of this chapter.

References

Angelaki DE, Cullen KE (2008) Vestibular system: the many facets of a multimodal sense. Annu Rev Neurosci 31:125–150

Attwell D, Laughlin SB (2001) An energy budget for signaling in the grey matter of the brain. J Cereb Blood Flow Metab 21:1133–1145

Barron JL, Fleet DJ, Beauchemin SS (1994) Performance of optical flow techniques. Int J Comp Vis 12:43–77

Berndt A, Yizhar O, Gunaydin LA, Hegemann P, Deisseroth K (2009) Bi-stable neural state switches. Nat Neurosci 12:229–234

Borst A, Egelhaaf M (1993) Detecting visual motion: theory and models. In: Miles FA, Walman J (eds) Visual motion and its role in the stabilization of gaze. Rev Oculomotor Res, vol 5. Elsevier, Amsterdam, pp 3–27

Borst A, Egelhaaf M, Haag J (1995) Mechanisms of dendritic integration underlying gain control in fly motion-sensitive interneurons. J Comput Neurosci 2:5–18

Borst A, Haag J (2002) Neural networks in the cockpit of the fly. J Comp Physiol A 188:419–437

Borst A, Egelhaaf M (1989) Principles of visual motion detection. TINS 12:297–306

Borst A, Euler T (2011) Seeing things in motion: models, circuits, and mechanisms. Neuron 71:974–994

Dahmen H, Franz MO, Krapp HG (2001) Extracting egomotion from optic flow: limits of accuracy and neural matched filters. In: Zanker MJ, Zeil J (eds) Motion vision. Computational, neural, and ecological constraints. Springer, Berlin, pp 143–168

Doyle JC, Francis BA, Tannenbaum A (1992) Feedback control theory. Dover Publications, New York

Egelhaaf M, Borst A, Reichardt W (1989) Computational structure of a biological motion-detection system as revealed by local detector analysis in the fly's nervous system. J Opt Soc Am A: 6:1070–1087

Egelhaaf M, Borst A (1993) Movement detection in arthropods. In: Miles FA, Walman J (eds) Visual motion and its role in the stabilization of gaze. Rev Oculomotor Res, vol 5. Elsevier, Amsterdam, pp 53–77

Egelhaaf M, Kern R, Krapp HG, Kretzberg J, Kurtz R, Warzecha AK (2002) Neural encoding of behaviourally relevant visual-motion information in the fly. TINS 25:96–102

Egelhaaf M (1985) On the neuronal basis of figure-ground discrimination by relative motion in the visual-system of the fly.2. Figure-detection cells, a new class of visual interneurones. Biol Cybern 52:195–209

Eichner H, Joesch M, Schnell B, Reiff DF, Borst A (2011) Internal structure of the fly elementary motion detector. Neuron 70:1155–1164

Elzinga MJ, Dickson WB, Dickinson MH (2012) The influence of sensory feedback delays on the yaw dynamics of insect flight. Int Comp Biol 52:E53

Franz MO, Chahl JS, Krapp HG (2004) Insect-inspired estimation of egomotion. Neural Comput 16:2245–2260

Franz MO, Krapp HG (2000) Wide-field, motion-sensitive neurons and matched filters for optic flow fields. Biol Cybern 83:185–197

Geiger G, Nassel DR (1981) Visual orientation behaviour of flies after selective laser beam ablation of interneurones. Nature 293:398–399

Gibson JJ (1950) The perception of the visual world. Houghton Mifflin, Boston

Gilbert C, Gronenberg W, Strausfeld NJ (1995) Oculomotor control in calliphorid flies: head movements during activation and inhibition of neck motor neurons corroborate neuroanatomical predictions. J Comp Neurol 361:285–297

Gilbert CD, Sigman M (2007) Brain states: top-down influences in sensory processing. Neuron 54:677–696

Graf W, Baker J, Peterson BW (1993) Sensorimotor transformation in the cats vestibuloocular reflex system.1. Neuronal signals coding spatial coordination of compensatory eye-movements. J Neurophysiol 70:2425–2441

Gronenberg W, Strausfeld NJ (1990) Descending neurons supplying the neck and flight motor of Diptera: physiological and anatomical characteristics. J Comp Neurol 302:973–991

Gronenberg W, Milde JJ, Strausfeld NJ (1995) Oculomotor control in calliphorid flies: organization of descending neurons to neck motor neurons responding to visual stimuli. J Comp Neurol 361:267–284

Haag J, Wertz A, Borst A (2010) Central gating of fly optomotor response. PNAS 107:20104–20109

Haag J, Egelhaaf M, Borst A (1992) Dendritic integration of motion information in visual interneurons of the blowfly. Neurosci Lett 140:173–176

Haag J, Wertz A, Borst A (2007) Integration of lobula plate output signals by DNOVS1, an identified premotor descending neuron. J Neurosci 27:1992–2000

Harris RA, O'Carroll DC, Laughlin SB (2000) Contrast gain reduction in fly motion adaptation. Neuron 28:595–606

Hassenstein B, Reichardt W (1953) Der Schluss von Reiz-Reaktions-Funktionen auf System-Strukturen. Z Naturforsch B 8:518–524

Hausen K (1993) Decoding of retinal image flow in insects. In: Miles FA, Walman J (eds) Visual motion and its role in the stabilization of gaze. Rev Oculomotor Res, vol 5. Elsevier, Amsterdam, pp 203–235

Hausen K, Wehrhahn C (1983) Microsurgical lesion of horizontal cells changes optomotor yaw responses in the blowfly Calliphora-erythrocephala. Proc Royal Soc B 219:211–216

Hausen K (1982) Motion sensitive interneurons in the optomotor system of the fly.1. The horizontal cells—structure and signals. Biol Cybern 45:143–156

Hausen K (1984) The lobula-complex of the fly: structure, function and significance in visual behaviour. In: Ali MA (ed) Photoreception and vision in invertebrates. Plenum Press, New York, pp 523–559

Heisenberg M, Wonneberger R, Wolf R (1978) Optomotor-Blind[H31]—Drosophila mutant of lobula plate giant neurons. J Comp Physiol A 124:287–296

Hengstenberg R (1982) Common visual response properties of giant vertical cells in the lobula plate of the blowfly Calliphora. J Comp Physiol A 149:179–193

Hengstenberg R, Sandeman DC, Hengstenberg B (1986) Compensatory head roll in the blowfly Calliphora during flight. Proc Royal Soc B 227:455–482

Hengstenberg R (1991) Gaze control in the blowfly Calliphora: a multisensory two-stage integration process. Neurosci 3:19–29

Hengstenberg R (1993) Multisensory control in insect oculomotor systems. In: Miles FA, Walman J (eds) Visual motion and its role in the stabilization of gaze. Rev Oculomotor Res, vol 5. Elsevier, Amsterdam, pp 285–298

Huston SJ, Krapp HG (2009) Nonlinear integration of visual and haltere inputs in fly neck motor neurons. J Neurosci 29:13097–13105

Huston SJ, Krapp HG (2008) Visuomotor transformation in the fly gaze stabilization system. PLoS Biol 6:1468–1478

Hyslop A, Krapp HG, Humbert JS (2010) Control theoretic interpretation of directional motion preferences in optic flow processing interneurons. Biol Cybern 103:353–364

Jung SN, Borst A, Haag J (2011) Flight activity alters velocity tuning of fly motion-sensitive neurons. J Neurosci 31:9231–9237

Karmeier K, Tabor R, Egelhaaf M, Krapp HG (2001) Early visual experience and the receptive-field organization of optic flow processing interneurones in the fly motion pathway. Vis Neurosci 18:1–8

Karmeier K, van Hateren JH, Kern R, Egelhaaf M (2006) Encoding of naturalistic optic flow by a population of blowfly motion-sensitive neurons. J Neurophysiol 96:1602–1614

Karmeier K, Krapp HG, Egelhaaf M (2005) Population coding of self-motion: applying bayesian analysis to a population of visual interneurons in the fly. J Neurophysiol 94:2182–2194

Karmeier K, Krapp HG, Egelhaaf M (2003) Robustness of the tuning of fly visual interneurons to rotatory optic flow. J Neurophysiol 90:1626–1634

Koenderink JJ, van Doorn AJ (1987) Facts on optic flow. Biol Cybern 56:247–254

Krapp HG, Hengstenberg R (1997) A fast stimulus procedure to determine local receptive field properties of motion-sensitive visual interneurons. Vis Res 37:225–234

Krapp HG, Hengstenberg R, Egelhaaf M (2001) Binocular contributions to optic flow processing in the fly visual system. J Neurophysiol 85:724–734

Krapp HG, Wicklein M (2008) Central processing of visual information in insects. In: Basbaum AI, Kenako A, Shepherd GM, Westheimer G (eds) The senses: a comprehensive reference, vol 1. Academic Press, San Diego, Vision I, pp 131–204. Masland IR, Albright TD (eds)

Krapp HG, Hengstenberg B, Hengstenberg R (1998) Dendritic structure and receptive-field organization of optic flow processing interneurons in the fly. J Neurophysiol 79:1902–1917

Krapp HG, Hengstenberg R (1996) Estimation of self-motion by optic flow processing in single visual interneurons. Nature 384:463–466

Krapp HG (2000) Neuronal matched filters for optic flow processing in flying insects. Int Rev Neurobiol 44:93–120

Krapp HG (2009) Ocelli. Curr Biol 19:R435–R437

Krapp HG (2010) Sensorimotor transformation: from visual responses to motor commands. Curr Biol 20:R236–R239

Krapp HG, Taylor GK, Humbert JS (2012) The mode-sensing hypothesis: matching sensors, actuators and flight dynamics. In: Barth FG, Humphrey JAC, Srinivasan MV (eds) Fontiers in sensing: from biology to engineering. Springer, Wien, pp 101–114

Land MF, Nilsson DE (2012) Animal eyes. Oxford University Press, Oxford

Laughlin SB (1987) Form and function of retinal processing. TINS 10:478–483

Lindemann JP, Kern R, van Hateren JH, Ritter H, Egelhaaf M (2005) On the computations analyzing natural optic flow: quantitative model analysis of the blowfly motion vision pathway. J Neurosci 25:6435–6448

Longden KD, Krapp HG (2011) Sensory neurophysiology: motion vision during motor action. Curr Biol 21:R650–R652

Longden KD, Krapp HG (2009) State-dependent performance of optic-flow processing interneurons. J Neurophysiol 102:3606–3618

Longden KD, Muzzu T, Cook DJ, Schultz SR, Krapp HG (submitted) Nutritional state modulates the neural processing of vision motion

Maddess T, Laughlin SB (1985) Adaptation of the motion-sensitive neuron H-1 is generated locally and governed by contrast frequency. Proc Royal Soc B 225:251–275

Maimon G, Straw AD, Dickinson MH (2010) Active flight increases the gain of visual motion processing in *Drosophila*. Nat Neurosci 13:393–399

Masino T, Knudsen EI (1990) Horizontal and vertical components of head movement are controlled by distinct neural circuits in the barn owl. Nature 345:434–437

Milde JJ, Seyan HS, Strausfeld NJ (1987) The neck motor system of the fly *Calliphora-erythrocephala*. 2. Sensory organization. J Comp Physiol A 160:225–238

Nakayama K, Loomis JM (1974) Optical velocity patterns, velocity-sensitive neurons, and space perception: a hypothesis. Perception 3:63–80

Nakayama K, Silverman GH (1988) The aperture problem–I. Perception of nonrigidity and motion direction in translating sinusoidal lines. Vision Res 28:739–746

Nalbach G (1994) Extremely non-orthogonal axes in a sense organ for rotation: behavioural analysis of the dipteran haltere system. Neuroscience 61:149–163

Nalbach G, Hengstenberg R (1994) The halteres of the blowfly Calliphora. 2. 3-dimensional organization of compensatory reactions to real and simulated rotations. J Comp Physiol A 175:695–708

Niven JE, Laughlin SB (2008) Energy limitation as a selective pressure on the evolution of sensory systems. J Exp Biol 211:1792–1804

Parsons MM, Krapp HG, Laughlin SB (2006) A motion-sensitive neurone responds to signals from the two visual systems of the blowfly, the compound eyes and ocelli. J Exp Biol 209:4464–4474

Parsons MM, Krapp HG, Laughlin SB (2010) Sensor fusion in identified visual interneurons. Curr Biol 20:624–628

Pierantoni R (1976) O look into the cock-pit of a fly. Cell Tissue Res 171:101–122

Preuss T, Hengstenberg R (1992) Structure and kinematics of the prosternal organs and their influence on head position in the blowfly Calliphora-Erythrocephala Meig. J Comp Physiol A 171:483–493

Quiroga RQ, Reddy L, Kreiman G, Koch C, Fried I (2005) Invariant visual representation by single neurons in the human brain. Nature 435:1102–1107

Reichardt W (1961) Autocorrelation, a principle for the evaluation of sensory information by the central nervous system. In: Rosenblith WA (ed) Sensory communication. MIT Press, Cambridge, pp 303–317

Reichardt W (1987) Evaluation of optical motion information by movement detectors. J Comp Physiol A 161:533–547

Riehle A, Franceschini N (1984) Motion detection in flies: parametric control over ON–OFF pathways. Exp Brain Res 54:390–394

Schildberger K (1984) Temporal selectivity of identified auditory neurons in the cricket brain. J Comp Physiol A 155:171–185

Schuppe H, Hengstenberg R (1993) Optical-properties of the ocelli of Calliphora-erythrocephala and their role in the dorsal light response. J Comp Physiol A 173:143–149

Schwyn DA, Hernadez Heras FJ, Bolliger G, Parsons MM, Krapp HG, Tanaka RI (2011). Interplay between feedback and feed forward control in fly gaze stabilization. In: 18th world congress of international federation of automated control (IFAC), Milan, pp 9674–9679

Schwyn DA, Hernadez Heras FJ, Bolliger G, Parsons MM, Laughlin SB, Tanaka RI, Krapp HG (in preparation) Blowfly gaze stabilization in response to external disturbances: a two-degree-of-freedom control system.

Strausfeld NJ (1976) Atlas of an insect brain. Springer, Berlin

Strausfeld NJ, Bassemir UK (1983) Cobalt-coupled neurons of a giant fibre system in diptera. J Neurocytol 12:971–991

Strausfeld NJ, Seyan HS (1985) Convergence of visual, haltere, and prosternal inputs at neck motor neurons of Calliphora-Erythrocephala. Cell Tissue Res 240:601–615

Strausfeld NJ (1984) Functional neuroanatomy of the bowfly's visual system. In: Ali MA (ed) Photoreception and vision in invertebrates. Plenum Press, New York, pp 483–522

Strausfeld NJ, Seyan HS, Milde JJ (1987) The neck motor system of the fly Calliphora-Erythrocephala. 1. muscles and motor neurons. J Comp Physiol A 160:205–224

Taylor GK, Krapp HG (2007) Sensory systems and flight stability: what do insects measure and why? Adv Insect Physiol: Insect Mechanics and Control 34:231–316

Theobald JC, Ringach DL, Frye MA (2010) Dynamics of optomotor responses in Drosophila to perturbations in optic flow. J Exp Biol 213:1366–1375

Warzecha AK, Egelhaaf M, Borst A (1993) Neural circuit tuning fly visual interneurons to motion of small objects. I. Dissection of the circuit by pharmacological and photoinactivation techniques. J Neurophysiol 69:329–339

Wehner R (1987) Matched-filters—neural models of the external world. J Comp Physiol A 161:511–531

Wertz A, Borst A, Haag J (2008) Nonlinear integration of binocular optic flow by DNOVS2, a descending neuron of the fly. J Neurosci 28:3131–3140

White JA, Rubinstein JT, Kay AR (2000) Channel noise in neurons. TINS 23:131–137

Part III
Evolution and Development of Flow Sensors

Chapter 10
Lateral Line Morphology and Development and Implications for the Ontogeny of Flow Sensing in Fishes

Jacqueline F. Webb

Abstract This chapter considers the morphological diversity of lateral line canals and neuromast receptor organs among fishes and how the pattern and timing of lateral line development can inform an understanding of the ontogeny of flow sensing. The morphology (and presumably the function) of the lateral line system changes considerably as a fish develops. Morphogenesis of the lateral line system starts before hatch and continues through the larval and juvenile stages, and thus, may take up to several months to complete during which fish size increases considerably. The appropriate course and timing of the development are critical for the development of feeding, swimming, and predator avoidance behaviors, and the ability to orient to environmental flows, which all ensure survival of young fishes. It is predicted that lateral line function, and thus flow sensing, is affected by a combination of both ontogenetic changes in the morphology of the lateral line system and the nature of the changing hydrodynamic regime in which a developing fish lives. Several aspects of lateral line development are predicted to have important effects on the functional ontogeny of the system: (1) increase in neuromast number, (2) changes in the relative number of superficial and canal neuromasts (CN), (3) changes in neuromast morphology (size, shape, hair cell number), and (4) variation in the pattern and timing of canal development.

10.1 Introduction

The mechanosensory lateral line system mediates prey detection, predator avoidance, navigation (including rheotaxis), and communication in freshwater and marine fishes that experience a wide range of hydrodynamic regimes in all types of

J. F. Webb (✉)
Department of Biological Sciences, Center for Biotechnology and Life Sciences,
University of Rhode Island, 120 Flagg Road, Kingston, RI 02881, USA
e-mail: Jacqueline_webb@mail.uri.edu

H. Bleckmann et al. (eds.), *Flow Sensing in Air and Water*,
DOI: 10.1007/978-3-642-41446-6_10, © Springer-Verlag Berlin Heidelberg 2014

aquatic habitats. Furthermore, a fish's survival depends on the ability to receive and respond to biologically relevant stimuli from both biotic and abiotic sources throughout its life history. The lateral line system has been the subject of study in several different research communities—ichthyology, sensory biology, and developmental biology. Each community has approached the system from a different perspective, seeking explanations for how the system has evolved among taxa (Webb 1989a, 2014), its role in critical behaviors in larval, juvenile, and adult fishes (Mogdans and Bleckmann 2012; Coombs and Montgomery, this volume; Windsor, this volume; Gardiner and Atema, this volume), and how patterns and mechanisms of lateral line development may inform our understanding of vertebrate development (Dambly–Chaudière et al. 2003; Ghysen et al., this volume). Taken together, these data can contribute to a robust understanding of the ontogeny of structure–function relationships in the lateral line system.

The neuromast receptor organs that compose the lateral line system are located in either pored, fluid-filled lateral line canals (canal neuromasts, CN), or on the skin (superficial neuromasts, SN). The neuromasts on the head, trunk, and tail are innervated by branches of the anterior, middle, and posterior lateral line nerves (Webb 2014). The morphological diversity of the lateral line system is well-defined (Coombs et al. 1988; Webb 2014) and a good deal of this variation can be explained as the result of variation in the pattern and timing of its development (Webb 1989a; Webb et al. in press). The appropriate course and timing of lateral line development are critical for the development of feeding, swimming, and predator avoidance behaviors, and the ability to orient to environmental flows, which all ensure survival of young fishes in their natural (and exotic) habitats, and in the context of aquaculture efforts (Blaxter and Fuiman 1989; Blaxter 1991; Fuiman et al. 2004; Pankhurst 2008). The morphology of the lateral line system changes considerably as fish develop through the embryonic, larval, and juvenile life history stages. This chapter will consider the diversity of lateral line morphology among fishes and how the pattern and timing of lateral line development in the context of the life history of teleost fishes can inform an understanding of the ontogeny of flow sensing.

10.2 Structural Organization and Diversity of the Lateral Line Canals

Neuromasts are found in stereotyped locations on the dorsal, lateral, and ventral sides of the head in bony fishes, which are established during the embryonic and larval stages of development (Fig. 10.1; see Sect. 10.3). In juveniles and adults, a subset of these neuromasts (the canal neuromasts, CNs) is found within the lateral line canals, which are integrated within an evolutionarily conserved subset of dermal (dermatocranial) bones (Fig. 10.2). These canals are positioned above the eye (supraorbital canal), below the eye (infraorbital canal), on the cheek (preopercular canal), and on the ventral surface of the lower jaw (mandibular canal). The canals are

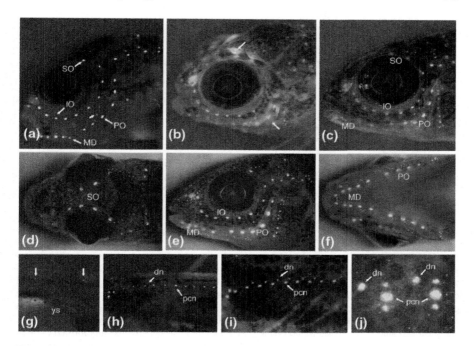

Fig. 10.1 Distribution of neuromasts on the head (**a–f**) and anterior portion of the trunk (**g–j**) in larvae and juveniles of a cichlid, *Aulonocara stuartgranti* (Percomorpha: Cichlidae) using fluorescent staining of hair cells (DASPEI and 4-Di-2-ASP). Note that presumptive CNs and enclosed CNs are significantly larger than the smaller SNs found in lines or clusters. **a** Larva (8 mm SL); CNs of the supraorbital (*SO*), infraorbital (*IO*), preopercular (*PO*), and mandibular (*MD*) neuromast series sit on skin (as presumptive CNs) and a *vertical line* of small SN's on the cheek is already present. **b** Larva (9.5 mm SL) in which SO and PO neuromasts sit in partially formed canal segments (*arrows*), IO neuromasts (with the exception of the three anterior-most IO neuromasts) are still superficial. **c** Small juvenile (17 mm SL), all CNs have been enclosed with the exception of the IO CNs. **d–f** Larger juvenile (31 mm SL) with SO, IO, and PO, and MD CNs, clearly visualized in dorsal (*left*), lateral (*middle*), and ventral (*right*) views, respectively. **g** Yolk-sac larva (7 mm SL) with two sparsely placed presumptive CNs (*arrows*) on the skin over the yolk sac (*ys*). **h** Yolk-sac larva (8 mm SL) with a full line of presumptive CNs (pcn, in register with each trunk muscle segment), and some additional SNs dorsal to them (*dn*). **i** Larva (9.5 mm SL) with full line of presumptive CNs (*pcn*) and a full line of dorsal SNs (*dn*) forming a series of neuromast pairs. **j** Juvenile (15 mm) showing enlargement of two groups of neuromasts (each in the epithelium over one lateral line scale) composed of a presumptive CN (*pcn*) flanked by two smaller neuromasts, and one neuromast (*dn*) rostral and dorsal to the presumptive CN. Rostral is to the *left* in all images

bilaterally symmetrical, and the right and left supraorbital canals may come together in a common pore (Fig. 10.2b), or more caudally, a supratemporal commissure may join the right and left canals on the dorsal surface of the head. The supraorbital, infraorbital, and preopercular canals typically meet just caudal to the eye continuing caudally as the otic canal and the postotic canal. The postotic canal continues through the post-temporal and supracleithral bones and into the trunk canal. CNs are found in stereotyped locations within the cranial canals, with one neuromast in the

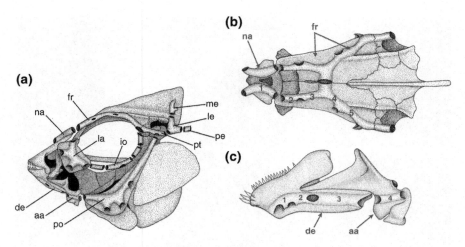

Fig. 10.2 Skull of the convict cichlid (*Amatitlania nigrofasciata* = *Archocentrus nigrofascia-tus*) illustrating the pored *lateral line* canals that are integrated into dermal bones of the skull. **a** *Lateral view* showing supraorbital canal in the nasal (*na*) and frontal (*fr*) bones, infraorbital canals in the lacrimal (*la*) and series of tubular infraorbital bones (*io*), preopercular canal in the preoperculum (*po*), mandibular canal in the dentary (*de*) and anguloarticular (*aa*), otic canal in the pterotic (*pt*), supratemporal commissure in the lateral and medial extracapsular bones (*le, me*), and postotic canal in the post-temporal bone (*pe*). **b** *Dorsal view* showing the supraorbital canal in the nasal (*na*) and frontal (*fr*) bones. **c** *Ventro-lateral view* showing the mandibular canal in the dentary (*de*) and anguloarticular (*aa*) bones. *Numbers* indicate the location of canal neuromasts within the supraorbital (**b**) and mandibular (**c**) canals. **a** From Webb (2000) reprinted with permission of Academic Press/Elsevier, Inc. **b, c** from Tarby and Webb (2003), reprinted with permission of Wiley and Sons, Inc

epithelium that lines the canal between adjacent canal pores (Fig. 10.2; see Sect. 10.3.2; Webb and Northcutt 1997). The trunk canal is contained within a horizontal (rostrocaudal) series of overlapping lateral line scales (Fig. 10.3b) that extends to the caudal peduncle (base of the caudal fin). Each lateral line scale contains a hollow, cylindrical lateral line canal segment (with anterior and posterior pores, Fig. 10.3b, e), lined by a thin epidermal epithelium in which one neuromast is generally found (Webb 1989c; Wonsettler and Webb 1997).

Structural diversity of the lateral line canal system is defined by the morphology and degree of development of the cranial canals, and the number, placement, and degree of development of the trunk canals (Webb 2014). Five cranial lateral line canal patterns are found among teleost fishes—three variations on narrow canals (narrow-simple, narrow-branched, narrow with widened tubules), reduced canals, and widened canals (Fig. 10.4a–e). Eight trunk canal patterns are defined by the placement and degree of development of a single canal, the presence of multiple trunk canals, or the absence of all canals (Fig. 10.4f–m; Webb 1989b, 2014). In addition, one or more canals or, more typically, one or more lines of SN also extend along the membranes between the fin rays of the caudal fin (e.g., Peters 1973; Webb 1989b; Wada et al. 2008; Asaoka et al. 2011a, b).

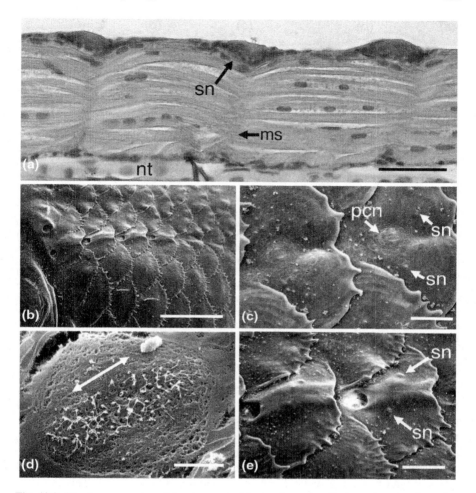

Fig. 10.3 Trunk neuromasts and development of the trunk canal. **a** *Horizontal* section (rostral to *left*) of the trunk in a young larva of zebrafish (*Danio rerio*) showing three superficial neuromasts (*sn*) each positioned at a myoseptum (*ms*, the boundary between adjacent muscle segments [myomeres]) at the level of the notochord (*nt*). **b–e** Scanning electron micrographs of the trunk canal in the convict cichlid, *Amatitlania nigrofasciata* (=*Archocentrus nigrofasciatus*). **b** Anterior portion of the trunk canal (just behind the operculum) showing the series of tubed *lateral line* scales (as in **e**) followed by *lateral line* scales in which the canal has not yet formed (as in **c**). **c** Close-up of two *lateral line* scales prior to canal formation, showing presumptive canal neuromasts (*pcn*) and dorsal and ventral superficial neuromasts (*sn*) on each scale. **d** Close-up of a presumptive canal neuromast in (**c**) with *arrow* indicating axis of best physiological sensitivity defined by orientation of the population of hair cells in *center* of neuromast. **e** Two *lateral line* scales after canal is formed (note *pores* at either end of each canal segment) with superficial neuromasts (*sn*) remaining on the skin dorsal and ventral to the canal. *Scale bars* **a** ∼10 μm, **b** =1 mm, **c** 100 μm, **d** 10 μm, **e** 1 mm. **b–e** from Webb (1989c) reprinted with permission of John Wiley and Sons

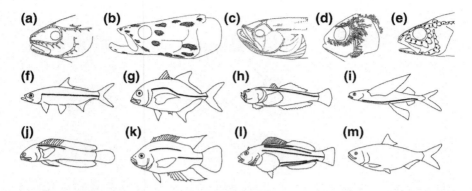

Fig. 10.4 Diversity of cranial (**a–e**) and trunk canal (**f–m**) morphology among teleost fishes. **a** Narrow-simple canals (saithe, *Pollachius virens*; modified from Marshall 1965, reprinted with permission from Elsevier, Inc.). **b** Narrow canals with widened tubules (*Arapaima*, from Nelson 1969, courtesy of The American Museum of Natural History). **c** Reduced canal system with lines of superficial neuromasts (*dots*) in the plainfin midshipman (*Porichthys notatus*, Greene 1899). **d** Narrow with branched canal system (Atlantic menhaden, *Brevoortia* tyrannus, from Hoss and Blaxter 1982), reprinted with permission by Wiley & Sons, Inc. **e** Widened canal system in common percarina (*Percarina demidoff*, from Jakubowski 1967), reprinted with permission of the author. **f** Complete straight canal. **g** Complete arched canal. **h** Dorsally placed canal. **i** Ventrally placed canal. **j** Incomplete canal. **k** Disjunct canal. **l** Multiple canal. **m** Absent (lack of canal). **f–m** Modified from Webb (1989b), reprinted with permission of S. Karger AG, Basel

10.2.1 Structure and Diversity of Neuromasts

Neuromasts are small epithelial organs (~ 10–500 μm diameter) composed of mechanosensory hair cells and non-sensory cells (supporting and mantle cells; Figs. 10.3d, 10.5). Each hair cell has a ciliary bundle on its apical surface that is composed of a single kinocilium (a true cilium) that may be quite long relative to the bundle of multiple stereocilia (microvilli; especially in SN), which are graded in length and located to one side of the kinocilium (Flock 1965a). This morphological polarization defines the physiological polarization (the axis of best sensitivity to flows) of each hair cell. Within a neuromast the hair cells are orientated 180° to one another, thus defining the neuromast's single axis of best physiological sensitivity. The morphology of the hair cell bundles (absolute and relative length of kinocilium and stereocilia, number of stereocilia) may vary within a neuromast (among hair cells at the periphery versus the center of a neuromast), between CNs and SNs in an individual (e.g., longer kinocilia in SNs), among species (e.g., fish with narrow versus widened canals), and among developmental stages of a given species, but the functional significance of such variation needs more study.

In neuromasts of juvenile and adult fishes, hair cells (especially those in CNs) are typically restricted to a smaller, round, or oval region (the "sensory zone," Jakubowski 1967, or "sensory strip," Coombs et al. 1988) that is surrounded by a population of non-sensory mantle cells that define neuromast shape (Figs. 10.3d, 10.5, 10.6). The ciliary bundles of all of the hair cells extend into an elongate,

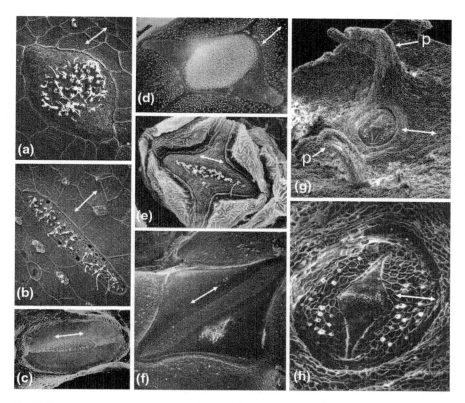

Fig. 10.5 Morphological diversity of canal neuromasts (*CN*) and superficial neuromasts (*SN*) among teleost fishes. **a** SN on trunk of windowpane flounder, *Scophthalmus aquosus*. **b** CN in narrow cranial canal of zebrafish, *Danio rerio*. **c** CN in narrow canal of mottled sculpin, *Cottus bairdi*. **d** CN in widened cranial canal in clown knifefish, *Notopterus chitala*. **e** SN on blind side of the head of California tongue sole, *Symphurus atricauda*. **f** CN in widened canal in rex sole (a flounder), *Glyptocephalus* sp. **g** SN from the dorsal most of five *horizontal lines* of SNs on the trunk of an adult plainfin midshipman, *Porichthys notatus*, showing the pair of papillae (*p*) that accompany each of the neuromasts. **h** Close-up of the SN in **g**, showing the elongate sensory strip containing hair cells. Hair cell orientation (*arrows*) is parallel to the axis of the canal in CNs. **a–f** from Webb (2011) reprinted with permission of Elsevier, Inc.

gelatinous cupula (composed of an inner and an outer layer, Münz 1979) whose base has the same overall shape as the neuromast. The cupula grows continuously (e.g., Vischer 1989; Mukai and Kobayashi 1992) and may be many times longer than the diameter of the neuromast (Teyke 1990; van Trump and McHenry 2008), especially in SNs where cupular length is not constrained by the diameter of the canal lumen.

CNs are typically larger than SNs in juvenile and adult fishes (Fig. 10.7f vs. c–e; Blaxter 1987; Münz 1989; Webb 1989c; Song and Northcutt 1991; Webb and Shirey 2003). One CN is found between adjacent pore positions as the result of a stereotyped process of canal development (Sect. 10.3.2). As a result of this process,

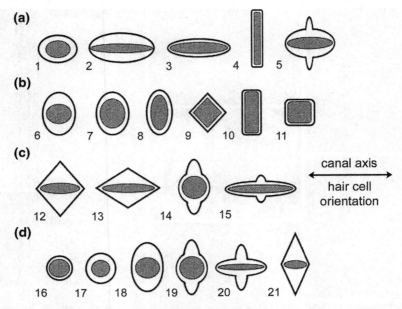

Fig. 10.6 Diversity in neuromast morphology among teleost fishes. **a** Canal neuromasts in narrow canals. **b–c** Canal neuromasts in widened canals. **d** Superficial neuromasts. *Shaded area* indicates the sensory strip containing sensory hair cells. Drawings reflect variation in overall shape of neuromasts, and in the shape of the sensory strip within the neuromast, but are not drawn to scale. *Double-headed arrow* represents the canal axis (for **a–c**) and hair cell orientation for all neuromasts (**a–d**). 1 *Amatitlania (= Cichlasoma) nigrofasciata* (Webb 1989c). 2 *Cottus bairdi* (Fig. 10.5c; Webb 2011). 3 *Perca fluviatilis, Sardina pilchardus, Cottus gobio, Mullus barbatus, Spicara smaris, Fundulus heteroclitus* (Jakubowski 1966). 4 *Danio rerio* (Fig. 10.5b; Webb and Shirey 2003). 5 *Dicentrarchus labrax* (Faucher et al. 2003). 6 *Gymnocephalus (= Acerina) cernuua* (Jakubowski 1966, 1967), *Apogon cyanosoma* (Rouse and Pickles 1991). 7/8 *Esox lucius, Cobitis taenia, Noemacheilus barbatulus, Abramis brama* (Jakubowski 1966). 9 *Lota lota* (Jakubowski 1966). 10 *Aspro zingel* (Jakubowski 1967), *Notropis buccatus* (unpublished observation). 11 *Acerina streber* (Jakubowski 1967). 12 *Aulonocara stuartgranti, Tramitichromis* sp. (Becker et al., in preparation). 13 *Glyptocephalus* spp. (unpublished data) 14 *Percarina demidoff* (Jakubowski 1967). 15 Sander (= *Lucioperca) lucioperca* (Jakubowski 1967). 16/17 *Danio rerio* (Webb and Shirey 2003). 18 *Carassius auratus* (Schmitz et al. 2008). 19 *Scophthalmus aquosus* (Fig. 10.5a; Webb 2011). 20 *Symphurus atricauda* (Fig. 10.5e; Webb 2011). 21 *Porichthys notatus* (Fig. 10.5h; Webb 2011)

the location of CNs within a canal can be predicted by noting the location of canal pores. CN shape is generally correlated with canal morphology such that CNs in narrow cranial canals (Fig. 10.6a) tend to be round or oval with a major axis parallel to the canal axis (with some exceptions, e.g., Webb and Shirey 2003; Schmitz et al. 2008), while CNs in widened cranial canals tend to be larger with a secondary morphological axis perpendicular to the canal axis resulting in a diamond shape (Fig. 10.6b; Garman 1899; Jakubowski 1963, 1967, 1974). Hair cell orientation is always parallel to the length of the canal regardless of neuromast shape (Figs. 10.5, 10.6, 10.7).

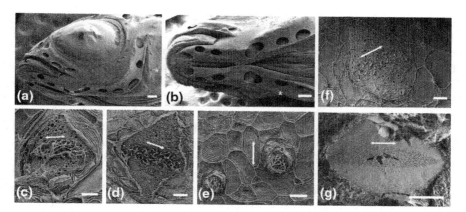

Fig. 10.7 Scanning electron micrographs of canals, and canal (*CN*) and superficial neuromast (*SN*) morphology in the cichlid, *Aulonocara stuartgranti* (**a–f** small juveniles, ~11–12 mm SL; (**g**) from adult, ~70 mm SL). **a, b** *Lateral* and *ventral views* of the head of a small juvenile. **c** SN just rostral to the eye. **d** SN from vertical series on operculum (*asterisk* in **a**). **e** Two SNs from a cluster on the skin overlying the ventral surface of the mandible (*asterisk* in **b**). **f** Mandibular presumptive CN from small juvenile. **g** Mandibular CN in adult (dissected canal). *Double-headed arrows* = hair cell orientation, which is parallel to the long axis of the sensory strip in all neuromasts. Rostral is to *left* in all images. *Scale bars* **a–b** 200 μm, **c–f** 10 μm, **g** 200 μm

SNs tend to be much more diverse in their distribution and topography, but less variable in their morphology than CNs. They tend to be small, either round or diamond-shaped (Figs. 10.5a, e, h; 10.6c) and occur singly, in lines (e.g., pit lines, stitches), or in clusters on the head, trunk, and tail (Münz 1979; Webb 1989c; Song et al. 1995). They typically sit flush with the skin but may sit in pits, depressions, or grooves, or on top of stalks, filaments, or papillae, especially in deep-sea and cave-dwelling fishes (e.g., Moore and Burris 1956; Marshall 1996; Pietsch 2009). The placement of SNs on structures that extend above the surface of the skin can be interpreted to be an adaptation to enhance the sensitivity to flows in taxa that do not swim actively and/or live in low flow environments. In some bottom-associated fishes, SNs may be highly proliferated (e.g., California tongue sole, *Symphurus atricauda*, see Fig. 10.5e) and/or found between paired non-sensory structures (e.g., the flaps or papillae in oyster toadfish, *Opsanus tau* and plainfin midshipman, *Porichthys notatus*, Fig. 10.5g, h; Webb 2014). These structures appear to protect the SNs (Applebaum and Schemmel 1983; Nakae and Sasaki 2010), and may alter (or filter) the hydrodynamic stimuli available to them (Schwartz et al. 2011). In those species in which canals are not completely developed (due to the evolutionary process of heterochrony), neuromasts may sit in canal grooves or flat on the skin in the location where canals are found in related species; they tend to retain the hair cell orientation of their CN homologues (Webb 2014).

SNs that are found adjacent to, or in the epithelium overlying, lateral line canals are termed "accessory neuromasts" (Coombs et al. 1988; but see an alternate use of this term by Ghysen et al., this volume). These tend to have hair cell orientations either parallel to or perpendicular to the axis of the canals they accompany (Münz 1979; Marshall 1986; Webb 1989c). On the trunk, accessory neuromasts with orientations perpendicular to one another ("orthogonal pairs," Marshall 1986; Coombs et al. 1988; Webb 1989c) may be found adjacent to the trunk canal. Together, these neuromasts are able to respond to stimuli from multiple directions. On the head, the orientation of accessory SNs tends to be either parallel or perpendicular to the canal they accompany. However, orientation tends to be more variable with reference to body axis especially if they are associated with a canal that does not follow a straight course, such as the infraorbital canal, which follows the circumference of the orbit (Webb 1989c; Schmitz et al. 2008).

SN proliferation has reached an extreme on the head and/or trunk in several taxa—in those with well-developed cranial canals (e.g., goldfish, *Carrasius auratus*, Schmitz et al. 2008; Mexican tetra [blind cavefish], *Astyanax mexicanus*; van Trump et al. 2010). Other taxa with significantly reduced head canals (e.g., gobioid fishes) have lines of proliferated SNs that may number in the thousands characterized by complex innervation patterns (e.g., Nakae et al. 2006, 2012; Asaoka et al. 2011b).

10.3 Early Life History and the Development of the Lateral Line System

The development of the lateral line system of teleost fishes is a prolonged process that starts before hatch and continues through the larval stage and into the juvenile stage. As such, this process may take up to several months to complete and it occurs in three phases (Webb 1989a): (1) differentiation of neuromasts from placode-derived primordia (described in detail in zebrafish; see key references below), (2) neuromast maturation, growth, and proliferation, and (3) canal formation, and enclosure of presumptive CNs in lateral line canals.

Neuromasts and the sensory neurons that innervate them arise from a series of cranial ectodermal placodes located just rostral and caudal to the developing ear on the head of embryos (reviewed by Schlosser 2010). Elongation of placode-derived sensory ridges establishes neuromast distributions on the head (Northcutt 2003; Gillis et al. 2012), and migration of one or more placode-derived neuromast primordia establishes initial neuromast distributions on the trunk (López-Shier et al. 2004; Chitnis et al. 2011; Ghysen et al., this volume). As the neuromast primordia start elongating and migrating, some placodal cells stay behind and differentiate into the bipolar sensory neurons whose cell bodies form the lateral line ganglia, and whose neurites project to the lateral line centers in the hindbrain (e.g., Liao and Haehnel 2012; Fritzsch and López-Schier). Peripherally, axons of these sensory neurons comigrate with neuromast primordia and form the anterior, middle, and posterior lateral line nerves that innervate the hair cells in neuromasts on the head, trunk, and tail (Metcalfe 1989; Gilmour et al. 2004).

At hatch, teleost larvae typically have a small number of functional SN that gradually increase in number and become more broadly distributed in particular patterns on the skin on the head, trunk, and tail (Poling and Fuiman 1997; Diaz et al. 2003; Nunez et al. 2009; Ghysen et al., this volume). The duration of the larval stage varies tremendously among taxa, from a few days to several weeks or even a year or more, especially in marine species (Disler 1971; Fuiman et al. 2004). During this period, fish length typically increases two to tenfold (e.g., $\sim$3–6 mm SL at hatch to $\sim$12–20 mm SL at metamorphosis). The development of the cranial canals and trunk canals occurs in late-stage larvae and in newly transformed juvenile fishes (Tarby and Webb 2003; Webb and Shirey 2003), and the proliferation of additional SNs may continue through the juvenile stage. The larval to juvenile transformation (metamorphosis) is typically followed by a shift to the juvenile habitat "larval settlement," an accompanying suite of behavioral changes, and ecological "recruitment" into the juvenile/adult population.

10.3.1 Neuromasts in Larval Fishes

The SNs of newly hatched larvae are quite small ($\sim$5 μm diameter) and round, and contain just a few hair cells with well-formed ciliary bundles (Figs. 10.3a, 10.8a; Raible and Kruse 2000; López-Schier et al. 2004) embedded in an elongate, cylindrical cupula (Mukai et al. 1994; Mukai and Kobayashi 1995; van Trump and McHenry 2008). Physiological polarization of the hair cells is already evident. The single axis of best physiological sensitivity, which is defined by the polarization of the hair cells, appears to be maintained as the neuromast increases in size with the addition of new hair cells (Münz 1989; Webb 1989c; Vischer 1989; Webb and Shirey 2003). Interestingly, in some species, the first neuromast to differentiate on the head is located in the postotic region (behind the eye), and is described as having radial polarity (Shardo 1996; Otsuka and Nagai 1997) or multiple polarities (Kawamura et al. 2003). The presence of this unusual neuromast appears to be short lived (Okamura et al. 2002; Diaz et al. 2003). It is not known if this neuromast maintains its radial hair cell orientation (or if it transitions to bi-directional polarity with hair cell turnover, for instance), if it is incorporated into a lateral line canal, or if it degenerates. Whether the radial orientation of this one neuromast is an adaptation for survival just after hatch, or a remnant of the deep evolutionary history of the lateral line system, remains to be seen.

In young larvae, the SNs that will remain on the skin and the presumptive CNs that will subsequently become enclosed in canals tend to be similar in size and shape and are not immediately distinguishable (e.g., Webb and Shirey 2003). However, towards the end of the larval period, presumptive CNs tend to increase in size and change shape as canals start developing (Webb 1999; Tarby and Webb 2003; Webb and Shirey 2003) and become quite distinct from the SNs that will remain on the skin (Figs. 10.1, 10.7; Janssen et al. 1987; Diaz et al. 2003; Webb and Shirey 2003). The question remains as to whether these two subsets of

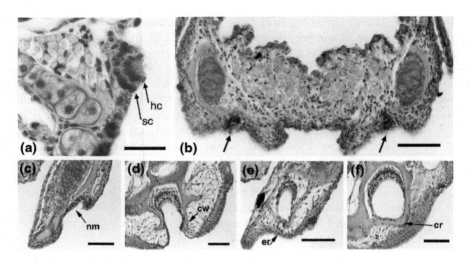

Fig. 10.8 Neuromast morphology and morphogenesis of the mandibular lateral line canal (**b–f** Stages I–IV) in larval zebrafish (*Danio rerio*). **a** Superficial neuromast on day of hatch, illustrating hair cells (*hc*) and support cells (*sc*). **b** Stage I—transverse section through mandible showing Meckel's cartilage (*blue*) and presumptive CNs (*arrows*). **c** Stage IIa—presumptive CN (*nm*) at *bottom* of depression (11 mm SL). **d** Stage IIb—presumptive CN in canal groove with ossified canal walls (*pink, cw*; 20 mm SL). **e** Stage III—CN enclosed by ossified canal walls (*pink*) and epithelial roof (*er*, 12 mm SL). **f** Stage IV—CN enclosed with ossified canal roof (*pink, cr,*; 22 mm SL). *Scale bars* **a** = 10 μm, **b–f** = 50 μm. **a–b**. From Webb (2011) reprinted with permission of Elsevier, Inc. **c–f** From Webb and Shirey (2003) reprinted with permission of John Wiley & Sons Inc

neuromasts have distinctive functional attributes in early larvae, or if their divergent functional attributes arise later with ontogenetic changes in morphology and/or in the hydrodynamic environment in which they subsequently function (e.g., on the skin, in open grooves, in ossified, pored canals).

10.3.2 Morphogenesis of Lateral Line Canals

The development of the lateral line canals is characterized by formation and ossification of cylindrical canal segments and the end-to-end fusion of adjacent segments to form pored canals (Fig. 10.1a–c). The development of canal segments is initiated in the vicinity of individual presumptive CNs and occurs as a series of four sequential stages (Tarby and Webb 2003; Webb and Shirey 2003) in which individual presumptive CNs are each gradually enclosed in a tubular canal segment (Figs. 10.1b, 10.8). As the canal segments enclose and begin to ossify, they have also begun to increase in diameter (Tarby and Webb 2003). At the same time, adjacent canal segments grow toward one another and fuse, leaving a common

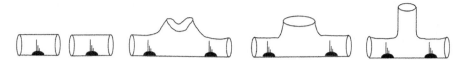

Fig. 10.9 Formation of pored lateral line canals from canal segments. After adjacent tubular canal segments enclose individual neuromasts (see Stage I–IV in Fig. 10.8), they grow toward one another and the terminal pores of adjacent segments form a common pore. This pattern accounts for the alternating positions of pores and neuromasts in lateral line canals (after Allis 1889)

pore between them (Fig. 10.9), thus explaining the alternating positions of CN and canal pores typical of teleost (and most bony) fishes (Webb and Northcutt 1997).

The development of the lateral line system on the trunk starts with the differentiation of SNs (presumptive CNs and other SNs) from several migrating lateral line primordia in embryos and larvae (Ghysen et al., this volume). Most teleosts have a full complement of scales on the body and a single trunk canal. Formation of the scales typically starts along the horizontal myoseptum at the caudal peduncle and proceeds rostrally in late-stage larvae (reviewed by Sire and Akimenko 2004). There is a correlated (but unexplored) patterning of presumptive CNs and scales such that one scale "lateral line scale" forms under each presumptive CN after which trunk canal segments enclose the presumptive CNs in a rostral to caudal wave (but see Webb 1990; Fig. 10.1g–j; Wonsettler and Webb 1997).

10.4 Functional Implications of the Pattern and Timing of Lateral Line Development

Studies of lateral line function in developing fishes (and of sensory system function more generally) have traditionally taken two approaches. Experimental work has examined correlations between changes in behavior and developmental events such as the appearance of the lateral line canals in various teleosts, or development of the *recessus lateralis* in clupeoid fishes, in particular (Poling and Fuiman 1997; Diaz et al. 2003; Fuiman et al. 2004). Experimental analysis and modeling of neuromast function and biomechanics in young zebrafish (*Danio rerio*) larvae (4–5 days post-fertilization) have predicted and/or have experimentally demonstrated the role of the lateral line system in prey detection, prey avoidance, and rheotactic behavior (Pirih et al., this volume; Liao, this volume; Suli et al. 2012). These studies have contributed fundamental knowledge to our understanding of neuromast and lateral line function in larval fishes. The next step requires the synthesis of our understanding of the pattern and timing of lateral line development and an interpretation of the functional ontogeny of neuromasts from hatch through metamorphosis.

Several aspects of lateral line development are predicted to have important effects on the functional ontogeny of the system: (1) increase in neuromast number, (2) changes in the relative number of SNs and CNs, (3) changes in neuromast morphology (size, shape, hair cell number), and (4) variation in the pattern and timing of canal development. Such a consideration of individual variables, as presented below, is a necessary starting point, but complexities introduced by the interaction of these variables should also be considered. Several important factors for the interpretation of the functional ontogeny of the lateral line system will not be considered here: (1) The role of the changing hydrodynamic regime (e.g., Reynolds Number, Re) experienced with increases in body size (Blaxter and Fuiman 1989), and with both swimming abilities (e.g., transition from planktonic to nektonic, Leis 2006) and swimming speed (Blaxter 1991; Leis et al. 2006, 2007). (2) The ways in which increases in neuromast size and canal diameter define the changing hydrodynamic environment in which neuromasts function (Windsor and McHenry 2009). (3) The way in which patterns of peripheral innervation and central projections may change during ontogeny (Liao, this volume). (4) Integraton of multimodal inputs and the development of specialized linkages of the swim bladder and lateral line system and/or ear—the *recessus lateralis* of clupeoid fishes (herrings and relatives, Higgs and Fuiman 1996; Fuiman et al. 2004) and the laterophysic connection in a genus of chaetodontids (coral reef butterflyfishes, Webb et al. 2012).

10.4.1 Changes in Neuromast Number

After initial differentiation of embryonic neuromasts, the number of neuromasts on the head and trunk increases through the larval stage (Kawamura et al. 2003; Fuiman et al. 2004; Faucher et al. 2005; Nunez et al. 2009; Ghysen et al., this volume). An increase in SNs is predicted to increase overall sensitivity of the system to water flows, and differential addition of SNs to the head or trunk are thought to enhance regional sensitivity to flows generated by prey (e.g., Faucher et al. 2005). The addition of SNs may also result in an increase in neuromast density (if the rate of the neuromast addition is greater than the rate of increase in body surface area). This would presumably increase the spatial resolution capabilities of the system, but decreases in neuromast density have been noted (Poling and Fuiman 1997).

The relative number of SNs (which act as velocimeters) and CNs (which act as accelerometers) also changes through ontogenetic time (e.g., Higgs and Fuiman 1996; Fig. 10.10). Total neuromast number initially increases as SNs differentiate in embryos and larvae, but once all of the presumptive CNs have differentiated (in larvae) their number stabilizes and they are enclosed in canals toward the end of the larval period (Webb and Shirey 2003; Fuiman et al. 2004; Becker and Webb unpublished). Additional SNs typically differentiate on the head and/or trunk (Blaxter and Fuiman 1989; Higgs and Fuiman 1996; Fuiman et al. 2004; Faucher et al. 2005)

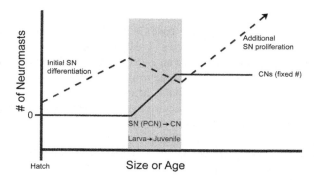

Fig. 10.10 Schematic representation of ontogenetic change in the relative numbers of superficial neuromasts (SN = SNs + presumptive CNs [PCNs]) and canal neuromasts (*CN*) in larvae and juvenile fishes. *Gray bar* represents the size/age interval during which metamorphosis (larval to juvenile transformation) occurs and presumptive CNs become enclosed in canals, thus decreasing the number of superficial neuromasts (*SN*). See text for additional details

further increasing the number of SNs relative to the number of CNs. The addition of SNs, especially those with orientations complementary to (e.g., perpendicular to) that of the CNs, would further enhance the functional breadth of the system.

10.4.2 Ontogeny of Neuromast Size and Shape

Neuromasts increase in size (Fuiman et al. 2004) and change shape during the larval stage. Presumptive CNs, in particular, exhibit dramatic changes in size and shape (and hair cell number) as canal formation commences and they become enclosed in canals (Webb 1989a; Tarby and Webb 2003; Webb and Shirey 2003; Diaz et al. 2003; Bird and Webb unpublished). The morphological diversity of CNs found among adult fishes (Figs. 10.5b, c, d, f, 10.6a, b) can be explained as the result of a simple increase in hair cell number with a disproportionately larger increase in the population of non-sensory mantle cells surrounding the hair cells, which restricts hair cells to a sensory strip in the center of the neuromast (Figs. 10.5, 10.6, 10.7). One interesting exception are the CNs of late larval and juvenile zebrafish (*Danio rerio*; Figs. 10.5b, 10.6a) in which the sensory hair cells are distributed throughout their long and narrow apical surface and an obvious surrounding population of non-sensory cells is not present (Webb and Shirey 2003).

Biomechanical models of CN function have assumed that neuromast are round with hemispherical cupulae, that hair cells are distributed throughout the apical surface of the neuromast, and that hair cell density is constant (hair cell number proportional to neuromast size; MJ McHenry pers. comm.). Such models predict that an increase in hair cell number will increase the stiffness of the cupula in which the kinocilia and stereocilia are embedded (van Netten and McHenry 2014) thus decreasing neuromast sensitivity to water flows. However, a change in cupula size (from 100 to 1,000 μm radius at base) is predicted to result in a proportional decrease in sensitivity at low frequencies (1–10 Hz) due to the contribution of each hair cell bundle to cupular stiffness (van Netten 2006). This prediction assumes a constant density of hair cells over the neuromast, and thus a constant density of hair cell bundles throughout the cupula. More hair cells would translate into a stiffer cupula, which would reduce sensitivity. However, the hair cells of CNs in juvenile and adult fishes are typically restricted to a sensory strip (Figs. 10.3d, 10.5c, d, f, 10.6a). Perhaps an ontogenetic increase in neuromast size without a proportional increase in hair cell number, where hair cells are restricted to a sensory strip, may provide a functional tradeoff. Further, a predicted decrease in sensitivity with an increase in hair cell number would be offset by the presence of a larger cupula, which can present more surface area as a target for hydrodynamic forces thus increase the probability of stimulation by water flows further enhancing sensitivity. This logic would also provide a functional explanation for the evolution of the large diamond-shaped neuromasts, especially those in species with widened lateral line canals (Figs. 10.5d, f, 10.6b, 10.7f, g), which are more sensitive than those in the narrow canals of comparably sized fish (discussed in Webb 2014). In contrast to CNs, SNs tend to remain small and round with a small number of hair cells (Figs. 10.6c, 10.7c, e) preserving neuromast sensitivity, especially at low frequencies (van Trump and McHenry 2008; van Netten and McHenry 2014). If SNs remain relatively small, with a sensory strip that limits hair cell number (Figs. 10.5a, 10.6c), but may change from a round to a diamond shape (Figs. 10.5g, h, 10.6c), such that an increase in cupular surface area with a change in shape may provide additional enhancement of neuromast sensitivity.

Correlated increases in hair cell number and canal neuromast size may be accompanied by changes in hair cell density (as per Blaxter and Fuiman 1989), if the rate of hair cell addition exceeds the rate at which the neuromast (or the sensory strip) increases in size. However, the effect of changes in neuromast density in models of neuromast function and biomechanics has not yet been explored.

10.4.3 Variation in Timing of Lateral Line Canal Development

Both theoretical and experimental studies have demonstrated that CNs and SNs have distinctive functional attributes and behavioral roles (Mogdans and Bleckmann 2012), so the ontogenetic transition during which presumptive CNs become enclosed in canals is of particular functional significance. Thus it follows that, the timing of the onset of canal development and the relative duration of the intermediate stages of canal development are predicted to affect neuromast function.

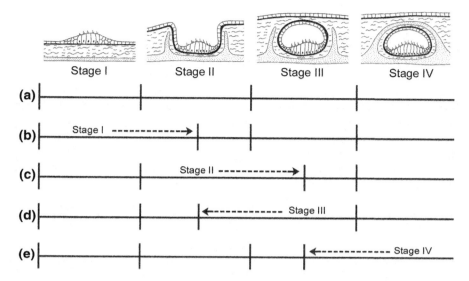

Fig. 10.11 Hypothetical variation in the relative duration of stages of development of lateral line canal segments (Stages I–IV, as in Fig. 10.8) in late-stage larvae and early juveniles. Note that fish size increases as canal segments proceed through Stages I–IV. **a** Equal duration of Stages I, II, and III. **b** Prolonged duration of Stage I and shortening of duration of Stage II. **c** Prolonged duration of Stage II and shortening of duration of Stage III. **d** Accelerated onset of Stage III, with shortening of duration of Stage II. **e** Accelerated onset of Stage IV with shortening of duration of Stage III. Additional combinations of changes in duration of more than one stage are hypothetically possible. Drawings of developmental stages from Tarby and Webb (2003), reprinted with permission of John Wiley and Sons

10.4.3.1 Onset of Canal Development

The developmental period (and the associated size growth interval) during which only SNs are present in fish larvae is defined by the timing of the onset of canal development (transition from Stage I to II), which typically occurs just prior to metamorphosis. Among species, the duration of the larval stage may range from several days to many months, and thus the onset of canal formation occurs in larvae of very different sizes and swimming capabilities (e.g., Leis 2006; Leis et al. 2006, 2007).

10.4.3.2 Relative Duration of the Intermediate Stages of Canal Development

The relative timing and duration of the transitions through the four stages of canal development are predicted to affect neuromast function (Fig. 10.11). For instance, variation in the relative duration of Stage II (Fig. 10.11a, b, c), during which CNs sit in depressions/grooves, will determine the growth interval during which water is channeled through small grooves in order to stimulate a neuromast (e.g., Schwartz et al. 2011). Similarly, the duration of Stage III (soft tissue canal roof; Fig. 10.11a, c, d, e), during which the canal roof is present but not ossified, is likely to also affect CN function because of stimulation by movements of the pliable canal roof. At Stage IV, water flows can only access the CNs in the canal lumen through small bony pores. Canal development occurs as a fish grows, at relatively low Reynolds numbers as defined by the dimensions of the cupula, diameter of the canal groove or canal lumen (on the order of a fraction of a millimeter), and the velocity of water flows (biotic and/or abiotic in origin) encountered by the fish. Evoutionary changes in the relative duration of the four stages of canal development relative to fish size is predicted to affect lateral line function.

Comparative data on three freshwater species with different reproductive strategies and life histories—zebrafish (*Danio rerio*, Webb and Shirey 2003), and two cichlids (convict cichlid, *Amatitlania nigrofasciatus* [= *Archocentrus nigrofasciatus*], Tarby and Webb 2003, and peacock cichlids, *Aulonocara* spp., unpublished observation)—provide interesting insights into the importance of the timing of canal development on functional ontogeny of the lateral line system. In zebrafish, the yolk is absorbed quickly and exogenous feeding starts 5 days postfertilization (dpf), but the onset of canal development does not start until 3 weeks later (at ~28 dpf) after the fish have doubled in size (to ~11–12 mm SL). CNs then sit in canal grooves (Stage IIb) for a prolonged period as individuals again double in size. Finally, with the exception of the supraorbital canals, which remain at Stage III (soft canal roof), the cranial canals quickly transition from Stage III to Stage IV (Webb and Shirey 2003). In contrast, in the South American convict cichlid (which lays eggs in nests), yolk is absorbed a bit later (at ~7 days dpf) as active swimming commences. Canal formation starts in small fish (~6 mm SL), and the canals quickly transition through stages III and IV (over a growth interval of just 2–3 mm) so that juveniles of 9–11 mm SL already have completely ossified canals (Tarby and Webb 2003). In African peacock cichlids (*Aulonocara* spp.), in which the mother broods the eggs and larvae in her mouth, canal development starts at a smaller larval body size (~6 mm SL). After about 3 weeks in the mother's mouth, and a prolonged yolk-sac stage, the yolk is absorbed, cranial canals quickly transition through Stages II–IV over a short growth interval, and juveniles emerge with a well-developed set of lateral line canals at about 11–12 mm SL ready to feed (Webb et al. unpublished). Thus, upon the initiation of feeding, late-stage zebrafish larvae still have a cranial lateral line canal system composed of presumptive CNs in grooves (Stage IIb), while juvenile cichlids of comparable size already have CNs in well-ossified canals (Stage IV). In both cases, it is likely that the developmental condition of the lateral line canals in zebrafish and convict cichlids as they commence feeding is well suited (and perhaps adaptive) for detecting their particular prey types.

It is also important to note that in an individual, development among canal segments within a canal, and among canals, occurs asynchronously. For instance, the supraorbital, preopercular, and mandibular canals are typically the first to start forming while the infraorbital canal is typically the last to form (ostariophysans: Lekander 1949; zebrafish: Webb and Shirey 2003; convict cichlid: Tarby and Webb 2003; Figs. 10.1b, c, e, 10.7a, b). Furthermore, the development of canal segments within a cranial canal series does not necessarily occur sequentially (e.g., in either a rostrocaudal or caudorostral sequence) along the length of a canal. Formation of the trunk canal in the lateral line scales starts after development of the head canals starts, and its development occurs in a rostral to caudal wave. In addition, SNs on the head and trunk continue to differentiate in larvae and juveniles. Thus, SNs on the skin (Stage I), presumptive CNs in open grooves (Stage II), and CNs in canals (Stage III, IV) will be present in an individual at any one point in time between the onset of canal development and completion of canal development over several weeks and a considerable growth interval. It is thus suggested that the asynchronicity of development among canal segments on both the head and trunk will add to the breadth of lateral line functionality (the ways in which water flows may stimulate different types of neuromasts as the canals develop) at any one point in ontogenetic time. This may buffer transforming larvae from the challenges they face (e.g., the need to continue to effectively evade predators and to detect prey), which make metamorphosis a particularly vulnerable time in a fish's life history.

10.5 Summary and Questions for Future Research

Lateral line function is predicted to be affected by a combination of ontogenetic changes in the morphology of the lateral line system and in the nature of the changing hydrodynamic regime in which a developing fish functions. Future studies of the ontogeny of flow sensing in fishes should consider the following questions and conceptual issues:

1. The pattern and timing of lateral line development need to be considered in individual species and in a comparative context informed by a knowledge of phylogenetic relationships in order to fully understand the evolution of lateral line development (and its functional consequences, as determined experimentally) in larval and juvenile fishes.
2. Patterns and mechanisms of ontogenetic change in neuromast morphology (patterns of hair cell and mantle cell proliferation), which result in the diversity in neuromast size and shape found among juvenile and adult fishes, need to be considered in future analyses of the functional ontogeny of the lateral line system.
3. Physical constraints may serve as selection pressures for evolutionary changes in the timing of canal development. For instance, is there a minimum canal diameter below which the Reynolds Number (Re) is so low as to render CNs ineffective in responding to water flows? Could the relative timing of the different phases of canal development be constrained by the hydrodynamics (e.g., Re) under which the lateral line system functions?

4. Neuromast patterning in the posterior lateral line system has been studied extensively in zebrafish and a comparison of this process in a few other species is shedding light on the relationship between initial neuromast patterning and adult morphology in fishes (Ghysen et al., this volume). On the head and trunk, the relationship of the process of neuromast patterning to subsequent processes of both neuromast maturation and canal development deserves additional study. Furthermore, the structural and functional diversity found in the lateral line canals among teleosts has been attributed to differences in the pattern and timing of canal development (Webb 1989a, 1990), but comparative studies of development (from neuromast patterning to canal development) in closely related species with divergent adult morphologies are still needed.

5. The correlation between lateral line morphology and the onset of behavior has been examined in detail in a limited range of species (at one or a just a few points in developmental time) but few studies have considered the relationship between lateral line morphology and behavior throughout the entire life history of a single species. The development of the lateral line system is a prolonged process that occurs over weeks (or even months) during which the lateral line system develops from an array of SNs to a combination of SNs and CNs in lateral line canals (with changes in the functional attributes of the system). Thus, the dynamics between structural and functional changes in the system and the ontogeny of behavior will be quite interesting, especially in those species in which lateral line-mediated behaviors are suspected or known to be important.

6. The study of neuroethology, sensory ecomorphology, sensory ecology, or neuroecology (Zimmer and Derby 2011) starts with the analysis of individual sensory systems. The analysis of multimodal inputs, with selective elimination of different sensory modalities in adult fishes, is gradually gaining more attention. Studies are needed that consider the contribution of all developing sensory systems to multimodal sensory integration in order to understand the ontogeny of flow sensing are still needed.

Acknowledgments I thank Dr. Matthew J. McHenry for his insights on the functional implications of ontogenetic changes in neuromast morphology derived from his modeling studies. Members of the Webb lab contributed to figures and provided helpful comments on an earlier version of the manuscript. The writing of this chapter was partially carried out while the author was a Whitman Summer Investigator at the MBL (Woods Hole) and was supported by funds from the College of the Environment and Life Sciences, University of Rhode Island and NSF grant # IOS-0843307.

References

Appelbaum S, Schemmel C (1983) Dermal sense organs and their significance in the feeding behaviour of the common sole *Solea vulgaris*. Mar Ecol Prog Ser 13:36–39

Asaoka R, Nakae M, Sasaki K (2011a) Description and innervation of the lateral line system in two gobioids, *Odontobutis obscura* and *Pterobobius elapoides* (Teleostei: Perciformes). Ichthyol Res 58:51–61

Asaoka R, Nakae M, Sasaki K (2011b) The innervation and adaptive significance of extensively distributed neuromasts in *Glossogobius olivaceus* (Perciformes: Gobiidae). Ichthyol Res 59:143–150

Blaxter JHS (1987) Structure and development of the lateral line. Biol Rev 62:471–514

Blaxter JHS (1991) Sensory systems and behavior of larval fish. In: Mauchline J, Nemoto T (eds) Marine biology—its accomplishments and future prospect. Hokusensha, Tokyo, pp 15–38

Blaxter JHS, Fuiman LA (1989) Function of the free neuromasts of marine teleost larvae. In: Coombs S, Görner P, Münz H (eds) The mechanosensory lateral line: neurobiology and evolution. Springer, New York, pp 481–499

Chitnis AJ, Nogare DD, Matsuda M (2011) Building the posterior lateral line system in zebrafish. Dev Neurobiol 72:234–255

Coombs S, Janssen J, Webb JF (1988) Diversity of lateral line systems: phylogenetic, and functional considerations. In: Atema J, Fay RR, Popper AN, Tavolga WN (eds) Sensory biology of aquatic animals. Springer, New York, pp 553–593

Dambly-Chaudière C, Sapede D, Soubiran F, Decorde K, Gompel N, Ghysen A (2003) The lateral line of zebrafish: a model system for the analysis of morphogenesis and neural development in vertebrates. Biol Cell 95:579–587

Diaz JP, Prié-Granié M, Kentouri M, Varsamos S, Connes R (2003) Development of the lateral line system in the sea bass. J Fish Biol 62:24–40

Disler NN (1971) Lateral line sense organs and their importance in fish behavior. Israel Program for Scientific Translations, Jerulalem, p 328

Faucher K, Aubert A, Lagardére J-P (2003) Spatial distribution and morphological characteristics of the trunk lateral line neuromasts of the sea bass (*Dicentrarchus labrax*, L; Teleostei, Serranidae). Brain Behav Evol 62:223–232

Faucher K, Lagardére J-P, Aubert A (2005) Quantitative aspects of the spatial distribution and morphological characteristics of the sea bass (*Dicentrarchus labrax* L.; Teleostei, Serranidae) trunk lateral line neuromasts. Brain Behav Evol 65:231–243

Flock A (1965a) Electron microscopic and electrophysiological studies on the lateral line canal organ. Acta Otolaryngologica Supplementum S199:7–90

Fuiman LA, Higgs DM, Poling KR (2004) Changing structure and function of the ear and lateral line system of fishes during development. Amer Fish Soc Symp 40:117–144

Garman S (1899) Reports on an exploration off the west coasts of Mexico, Central and South America, and off the Galapogos Islands, in charge of Alexander Agassiz, by the US fish commission steamer "Albatross" during 1891, Lieut. Commander Z.L. Tanner, USN, Commanding. XXVI—the fishes. Mem Mus Comp Zool 24:1–431 + 97 plates

Gillis JA, Modrell MS, Northcutt RG, Catania KC, Luer C, Baker CVH (2012) Electrosensory ampullary organs are derived from lateral line placodes in cartilaginous fishes. Devel 139:3142–3146

Gilmour D, Knaut H, Maischein HM, Nusslein-Volhard C (2004) Towing of sensory axons by their migrating target cells in vivo. Nat Neurosci 7:491–492

Higgs DM, Fuiman LA (1996) Ontogeny of visual and mechanosensory structure and function in Atlantic menhaden *Brevoortia tyrannus*. J Exper Biol 199:2619–2629

Hoss DE, Blaxter JHS (1982) Development and function of the swimbladder-inner ear-lateral line system in the Atlantic menhaden, *Brevoortia tyrannus* (Latrobe). J Fish Biol 20:131–142

Jakubowski M (1963) Cutaneous sense organs of fishes. I. The lateral-line organs in the stone-perch (*Acerina cernua* L.). Acta Biologica Cracoviensia Ser Zoologia 6:59–78

Jakubowski M (1966) Cutaneous sense organs of fishes. V. Canal system of lateral-line organs in *Mullus barbatus ponticus* Essipov and *Spicara smaris* L. (topography, innervation, structure). Acta Biologica Cracoviensia Ser Zoologia 9:225–237

Jakubowski M (1967) Cutaneous sense organs of fishes. Part VII. The structure of the system of lateral-line canal organs in the Percidae. Acta Biologica Cracoviensia Ser Zoologia 10:69–81

Jakubowski M (1974) Structure of the lateral-line canal system and related bones in the berycoid fish *Hoplostethus mediteranneus* Cuv. et Val. (Trachichthyidae, Pisces). Acta Anat 87:261–274

Janssen J, Coombs S, Hoekstra D, Platt C (1987) Anatomy and differential growth of the lateral line system of the mottled sculpin, *Cottus bairdi* (Scorpaeniformes: Cottidae). Brain Behav Evol 30:210–229

Kawamura G, Masuma S, Tezuka N, Koiso M, Jinbo T, Namba K (2003) Morphogenesis of sense organs in the bluefin tuna *Thunnus orientalis*. In: Browman HI, Skiftesvik AB (eds) The fish big bang, 26th annual larval fish conference. Institute of Marine Research, Bergen, Norway, pp 123–135

Leis JM (2006) Are larvae of demersal fishes plankton or nekton? Adv Mar Biol 51:57–141

Leis JM, Hay AC, Trnski T (2006) In situ ontogeny of behavior in pelagic larvae of three temperate, marine, demersal fishes. Mar Biol 148:655–669

Leis JM, Hay AC, Lockett MM, Chen J-P, Fang L-S (2007) Ontogeny of swimming speed in larvae of pelagic-spawning, tropical, marine fishes. Mar Ecol Prog Ser 349:257–269

Lekander B (1949) The sensory line system and the canal bones in the head of some Ostariophysi. Acta Zool 30:1–131

Liao JC, Haehnel M (2012) Physiology of afferent neurons in larval zebrafish provides a functional framework for lateral line somatotopy. J Neurophys 107:2615–2623

López-Shier H, Starr CJ, Kappler FA, Kollmar R, Hudspeth AJ (2004) Directional cell migration establishes the axes of planar polarity in the posterior lateral-line organ of the zebrafish. Devel Cell 7:401–412

Marshall NJ (1965) Systematic and biological studies of the Macrourid fishes (Anacanthini-Teleostii). Deep Sea Res 12:299–322

Marshall NJ (1986) Structure and general distribution of free neuromasts in the black goby, *Gobius niger*. J Mar Biol Assoc UK 66:323–333

Marshall NJ (1996) The lateral line systems of three deep-sea fish. J Fish Biol 49:239–258 (Suppl)

Metcalfe WK (1989) Organization and development of the zebrafish posterior lateral line. In: Coombs S, Görner P, Münz H (eds) The mechanosensory lateral line: neurobiology and evolution. Springer, New York, pp 147–159

Mogdans J, Bleckmann H (2012) Coping with flow: behavior, neurophysiology and modeling of the fish lateral line system. Biol Cyber 106:627–642

Moore GA, Burris WE (1956) Description of the lateral-line system of the pirate perch, *Aphredoderus sayanus*. Copeia 1956:18–20

Mukai Y, Kobayashki H (1992) Cupular growth rate of free neuromasts in three species of cyprinid fish. Nippon Suisan Gakkaishi 58:1849–1853

Mukai Y, Kobayashki H (1995) Development of free neuromasts with special reference to sensory polarity in larvae of the willow shiner, *Gnathopogon elongates caerulescens* (Teleostei, Cyprinidae). Zool Sci 12:125–131

Mukai Y, Yoshikawa H, Kobayashi H (1994) The relationship between the length of the cupulae of free neuromasts and feeding ability in larvae of the willow shiner *Gnathopogon elongates caerulescens* (Teleostei, Cyprinidae). J Exp Biol 197:399–403

Münz H (1979) Morphology and innervation of the lateral line system in *Sarotherodon niloticus* (L.) (Cichlidae, Teleostei). Zoomorph 93:73–86

Münz H (1989) Functional organization of the lateral line periphery. In: Coombs S, Görner P, Münz H (eds) The mechanosensory lateral line: neurobiology and evolution. Springer, New York, pp 285–297

Nakae M, Sasaki K (2010) Lateral line system and its innervation in Tetraodontiformes with outgroup comparisons: descriptions and phylogenetic implications. J Morphol 271:559–579

Nakae M, Asai S, Sasaki K (2006) The lateral line system and its innervation in *Champsodon snyderi* (Champsodontidae): distribution of approximately 1000 neuromasts. Ichthyol Res 53:209–215

Nakae M, Asaoka R, Wada H, Sasaki K (2012) Fluorescent dye staining of neuromasts in live fishes: An aid to systematic studies. Ichthyol Res, 59:286–290

Northcutt RG (2003) Development of the lateral line system in the channel catfish. In: Browman HI, Skiftesvik AB (eds) The fish big bang, 26th annual larval fish conference. Institute of Marine Research, Bergen, Norway, pp 137–159

Nuñez VA, Sarrazin AF, Cubedo N, Allende ML, Dambly-Chaudière C, Ghysen A (2009) Postembryonic development of the posterior lateral line in the zebrafish. Evol Devel 11:391–404

Okamura A, Oka HP, Yamada Y, Utoh T, Mikawa N, Horie N, Tanaka S (2002) Development of lateral line organs in leptocephali of the freshwater eel Anguilla japonica (Teleostei, Anguilliformes). J Morphol 254:81–91

Otsuka M, Nagai S (1997) Neuromast formation in the prehatching embryos of the cod-fish, Gadus macrocephalus Tilesius. Zool Sci 14:475–481

Pankhurst PM (2008) Mechanoreception. In: Finn RN, Kapoor BG (eds) Fish larval physiology. Science Publishers, Enfield, pp 305–329

Peters HM (1973) Anatomie und Entwicklungsgeschichte des Laterallissystems von Tilapia (Pisces, Cichlidae). Zeitschrift fur Morphologie der Tiere 74:89–161

Pietsch TW (2009) Oceanic anglerfishes—extraordinary diversity in the deep sea. University of California Press, Berkeley

Poling KR, Fuiman LA (1997) Sensory development and concurrent behavioural changes in Atlantic croaker larvae. J Fish Biol 51:402–421

Raible DW, Kruse GJ (2000) Organization of the lateral line system in embryonic zebrafish. J Comp Neurol 421:189–198

Rouse GW, Pickles JO (1991) Ultrastructure of free neuromasts of Bathygobius fuscus (Gobiidae) and canal neuromasts of Apogon cyanosoma (Apogonidae). J Morphol 209:111–120

Schlosser G (2010) Making senses: development of vertebrate cranial placodes. Int Rev Cell Mol Biol 283:129–234

Schmitz A, Bleckmann H, Mogdans J (2008) Organization of the superficial neuromast system in goldfish, Carrasius auratus. J Morphol 269:751–761

Schwartz JS, Reichenbach T, Hudspeth AJ (2011) A hydrodynamic sensory antenna used by killifish for nocturnal hunting. J Exp Biol 214:1857–1866

Shardo JD (1996) Radial polarity of the first neuromast in embryonic American shad, Alosa sapidissima (Teleostei: Clupeomorpha). Copeia 1996:226–228

Sire JY, Akimenko MA (2004) Scale development in fish: a review, with description of sonic hedgehog (shh) expression in the zebrafish (Danio rerio). Int J Devel Biol 48:233–247

Song J, Northcutt RG (1991) Morphology, distribution and innervation of the lateral-line receptors of the Florida gar, Lepisosteus platyrhincus. Brain Behav Evol 37:10–37

Song J, Yan HY, Popper AN (1995) Damage and recovery of hair cells in fish canal (but not superficial) neuromasts after gentamicin exposure. Hear Res 91:63–71

Suli A, Watson GM, Rubel EW, Raible DW (2012) Rheotaxis in larval zebrafish is mediated by lateral line mechanosensory hair cells. PLoS ONE 7:1–6

Tarby ML, Webb JF (2003) Development of the supraorbital and mandibular lateral line canals in the cichlid, Archocentrus nigrofasciatus. J Morphol 254:44–57

Tekye T (1990) Morphological differences in neuromasts of the blind cave fish Astyanax hubbsi and the sighted river fish Astyanax mexicanus. Brain Behav Evol 35:23–30

Van Netten SM (2006) Hydrodynamic detection by cupulae in a lateral line canal: functional relations between physics an physiology. Biol Cybern 94:67–85

Van Netten SM, McHenry MJ (2014) The biophysics of the fish lateral line. In: Coombs S, Bleckmann H (eds) The lateral line system. Springer, New York, pp. 99–120

Van Trump WJ, McHenry MJ (2008) The morphology and mechanical sensitivity of lateral line receptors in zebrafish larvae (Danio rerio). J Exp Biol 211:2105–2115

Van Trump WJ, Coombs S, Duncan K, McHenry MJ (2010) Gentamicin is ototoxic to all hair cells in the fish lateral line system. Hear Res 261:42–50

Vischer HA (1989) The development of lateral-line receptors in Eigenmannia (Teleostei, Gymnotiformes). I. The mechanoreceptive lateral-line system. Brain Behav Evol 33:205–222

Wada H, Hamaguchi S, Sakaizumi M (2008) Development of diverse lateral line patterns on the teleost caudal fin. Devel Dyn 237:2889–2902

Webb JF (1989a) Developmental constraints and evolution of the lateral line system in teleost fishes. In: Coombs S, Görner P, Münz H (eds) The mechanosensory lateral line: neurobiology and evolution. Springer, New York, pp 79–97

Webb JF (1989b) Gross morphology and evolution of the mechanosensory lateral line system in teleost fishes. Brain Behav Evol 33:34–53

Webb JF (1989c) Neuromast morphology and lateral line trunk ontogeny in two species of cichlids: an SEM study. J Morphol 202:53–68

Webb JF (1990) Ontogeny and phylogeny of the trunk lateral line system in cichlid fishes. J Zool Lond 221:405–418

Webb JF (1999) Diversity of fish larvae in development and evolution. In: Hall BK, Wake MH (eds) Origin and evolution of larval forms. Academic Press, San Diego, pp 109–158

Webb JF (2000) Mechanosensory lateral line: Functional morphology and neuroanatomy. In Ostrander G (ed) Handbook of experimental animals-The laboratory fish London: Academic Press, pp 236–244

Webb JF (2011) Lateral line structure. In Farrell AP (ed) Encyclopedia of fish physiology: From genome to environment, Vol 1. Academic Press, San Diego, pp 336–346

Webb JF (2014) Morphological diversity, development, and evolution of the mechanosensory lateral line system. In: Coombs S, Bleckmann H (eds) The lateral line system. Springer, New York, pp 17–72

Webb JF, Northcutt RG (1997) Morphology and distribution of pit organs and canal neuromasts in non-teleost bony fishes. Brain Behav Evol 50:139–151

Webb JF, Shirey JE (2003) Post-embryonic development of the lateral line canals and neuromasts in the zebrafish. Dev Dyn 228:370–385

Webb JF, Walsh RM, Casper B, Mann DA, Kelly N, Cicchino N (2012) Ontogeny of the ear, hearing capabilities, and laterophysic connection in the spotfin butterflyfish (*Chaetodon ocellatus*). Env Biol Fishes 95:275–290

Windsor SP, McHenry MJ (2009) The influence of viscous hydrodynamics on the fish lateral-line system. Integr Comp Biol 49:691–701

Wonsettler AL, Webb JF (1997) Morphology and development of the multiple lateral line canals on the trunk in two species of *Hexagrammos* (Scorpaeniformes: Hexagrammidae). J Morphol 233:195–214

Zimmer RK, Derby CD (2011) Neuroecology and the need for a broader synthesis. Integr Comp Biol 51:751–755

Chapter 11
Evolution of Polarized Hair Cells in Aquatic Vertebrates and Their Connection to Directionally Sensitive Neurons

Bernd Fritzsch and Hernán López-Schier

Abstract The mechanosensory hair cells enable aquatic vertebrates to maintain body position with respect to gravity, as well as to detect a wide variety of hydrodynamic (including hydroacoustic) stimuli. The evolution of hair-cell bearing sensory organs has been driven by the physical properties of the medium, which directs the adaptation of their molecular developmental program to extract information from different stimuli. Mutation and selection have shaped hair cell bearing organs to extract information about the distance, size, and movement direction to elicit motor behaviors, including avoidance, approach, or schooling in fish swarms. Here we will review some molecular, cellular, and developmental steps that outline a plausible evolution of hair cells and their use as hydrodynamic sensors in aquatic vertebrates. We suggest an evolutionary progression (from simple to complex) through multiplication of genes of the mechanosensory hair cell followed by cellular and organ diversification. We posit that this cell evolved, through morphological intermediates, which transformed the kinocilium surrounded by microvilli of the unicellular ancestor of metazoans into the polarized stereocilia of vertebrate hair cells. Anaxonic sensory cells of vertebrates have evolved, after an ancestral gene duplication, into both neurons and hair cells. Evolution of novel genes allowed the formation of discrete sensory organs such as the inner ear and the lateral line. An interesting but incompletely understood aspect of this evolution is the generation of hair-cell polarization and their distribution within sensory epithelia and their innervation.

Keywords Neuromast · Hair cell · Evolution · Polarity · Directional sensitivity · Molecular evolution · Organ formation · Planar cell polarity

B. Fritzsch
Department of Biology, University of Iowa, 143 BB, Iowa City, IA 52242, USA
e-mail: bernd-fritzsch@uiowa.edu

H. López-Schier (✉)
Research Unit Sensory Biology and Organogenesis, Helmholtz Zentrum München,
85764 Munich, Neuherberg, Germany
e-mail: hernan.lopez-schier@helmholtz-muenchen.de

H. Bleckmann et al. (eds.), *Flow Sensing in Air and Water*,
DOI: 10.1007/978-3-642-41446-6_11, © Springer-Verlag Berlin Heidelberg 2014

11.1 Introduction

The inner ear and the lateral line are the main receptor organs that non-mammalian aquatic vertebrates use to detect, respectively, whole body movement and local hydrodynamic events at the body surface. The sensory organs of the lateral line and inner ear consist of specialized cells found only in vertebrates, the mechanosensory hair cells and their surrounding supporting cells. Hair cells are mechanosensitive transducers, which convert shearing forces generated by sound, water movements, gravity, and angular acceleration (mediated by different structures in different organs) into electric activity. These mechanically induced changes in their resting potential are conducted to the brain by associated afferent neurons. Hair cells are directionally sensitive, responding maximally by membrane depolarization when their apical hair bundles are displaced in the excitatory direction, and with a hyperpolarizing response when bundles are displaced in the opposite direction. The directional sensitivity of the hair cell is dictated by the asymmetry of the hair bundle, which consists of a kinocilium adjacent to a staircase arrangement of specialized, actin-rich microvilli. These actin-rich microvilli are usually referred to as "stereocilia" and we will retain this name here for easy comparison with existing literature. The stereocilia are interconnected by tip links that allow mechanically gated channels to open or close upon stimulation toward (depolarizing response) or away (hyperpolarizing response) from the kinocilium. This functional polarization of a hair cell is repeated across the plane of sensory epithelia. All mechanosensory organs have hair cells of either identical, opposing or, rarely, random planar orientations. How hair-cell polarities arise during development within an organ varies greatly and seems to correlate with specific developmental programs for each organ. For example, the utricle and saccule of the inner ear have opposing populations of hair cells, but they may be either polarized toward or away from the area of polarity reversal. Each hair-cell population is furthermore innervated by separate nerve fibers, which preserves information about the excitatory direction of hair-bundle displacement. Thus, a sensory organ with two opposing populations of hair cells will have a minimum of two afferent fibers. Despite this fundamental principle of hair-cell polarity and appropriate innervation, comparatively little is known about the central representation and processing of the signals transduced by the differently polarized hair cells. In addition, hair-cell anlagen can develop into the non-polarized electroreceptor hair cells (Northcutt et al. 1995) and other non-polarized mechanoreceptive cells such as Merkel cells in whiskers seem to share the molecular basis of their development (Bermingham et al. 2001).

This chapter will summarize the molecular basis of hair-cell polarity, the development of polarity-specific innervation, the afferent projections to the brain, and some aspects of polarity-specific central processing. Since molecular evidence indicates that the ancestry of the mechanosensory hair cell predates the evolution of either lateral line or inner ear (Pierce et al. 2008; Short et al. 2012; Candiani et al. 2011; Joyce Tang et al. 2013), it is likely that this cell type was simply recruited when these organs evolved for the detection of hydrodynamic events.

11.2 Evolving Flow Detection Abilities: An Assessment of Mechanosensory Hair-Cell Distribution and its Implication for the Evolution of Hydrodynamic Detection

All aquatic metazoans, with the exception of sponges have cells or organs that show structural evidence for possible mechanical stimulation and thus for flow sensing. They can be broadly characterized into gravistatic (acoustic) sensors and lateralis-like organs. Whereas the former function to detect whole body orientation (gravistatic reception), sound, or angular acceleration of inner ear fluid during rotations (semicircular canals), the latter detects hydrodynamic events at the body surface. Nearly every free swimming aquatic multicellular organism has a gravistatic organ or statocyst (Markl 1974). In contrast, multicellular organs known or hypothesized to detect hydrodynamic events are rare in free swimming organisms. These organs have been identified in derived mollusks (Budelmann and Bleckmann 1988), vertebrates (Northcutt 1988), and possibly in some chordate ancestors of vertebrates such as the lancelet (Fritzsch 1996) and perhaps ascidians (Burighel et al. 2011). In general, non-vertebrate deuterostomes have single cells or small group of cells that can be identified by signature genes known for vertebrate mechanosensory organs (Joyce Tang et al. 2013; Candiani et al. 2011; Pierce et al. 2008; Pan et al. 2012b) but lack clearly identifiable lateral line or inner ear equivalents (Short et al. 2012; Bouchard et al. 2010). An interesting case are coelenterates. Sessile coelenterates like the corals have lateralis-like organs that seem to function much like those in vertebrates (Repass and Watson 2001). In contrast, free swimming jelly fish have complex statocysts to provide, together with associated eyes, information about the direction of gravity and light as opposing vectors to orient in space (Arkett et al. 1988; Kozmik et al. 2003).

How can this puzzling diversity of shape and position of organs and structural diversity of cells be aligned with the adaptive radiation of multicellular organisms? Obviously, there are two complementary approaches: one approach is using structural (Burighel et al. 2011) and molecular evidence (Fritzsch et al. 2007) to test deep homology of all mechanosensory cells and organs. Alternatively, one can combine the structural diversity of organs with the common function of movement detection and argue for parallel evolution (Markl 1974).

For example, molecular evidence has clarified for various eyes that a single transcription factor, Pax6, is present in all eyes regardless of their receptor cell type (rhabdomeric or ciliogenic) or developmental origin (ectodermal or neuroectodermal) (Arendt 2003; Vopalensky et al. 2012). This suggests to some that all complex eyes are derived from a common ancestor (Gehring 2011), consisting of a pair of clonally related sensory/pigment cells. Below we explore such a scenario for the evolution of hydrodynamic surface sensors of the lateralis type and the inner ear/statocyst.

All mechanosensation transforms shearing forces generated at the surface of a single cell or organ into electric signals. At the level of the receptor cell, all lateralis-like surface detectors as well as inner ear gravistatic, flow sensing angular accelerometers or sound pressure receptors function alike. That is, the hair cell responds to displacement of the hair bundle. However, the nature of the appropriate stimuli that generate the shearing force depend on the position of the sensor (in the body or on the surface) and the structures surrounding the sense organ. For example, dense otoconia or otoliths can detect linear acceleration due to gravity or sound-induced vibration of the body (fish ears), hydrogel cupulae and fluid-filled semicircular canals detect angular acceleration, and cupulae and fluid-filled canals serve in the lateral line as flow sensors. For the vestibular part of the vertebrate ear, the quality of the stimulus is clear: gravity or any linear acceleration displace otoconia, whereas angular acceleration through lagging endolymph stimulates the semicircular canal cristae. Hydrodynamic sensing on the surface requires a reference for determining self-movement, so that flows generated by self-generated movement can be distinguished from body motions induced from exogenous sources. For sessile organisms tied to a substrate, such a reference is unnecessary because flow past the skin surface is never self-generated. Such a system might best be represented by the lateralis-like organs of corals (Repass and Watson 2001), the nematocytes of hydrozoans (Loosli et al. 1996), or even the lateral line system of some benthic fish that remain mostly stationary and coupled to the substrate for long periods of time (e.g., flatfish). In contrast, if an animal is suspended in the water column, any hydrodynamic event generated by monopoles, dipoles, or multipoles (Lewis and Fay 2004) will generate both a total body movement as well as flow along the body surface. In these cases, the stimulus to the flow sensors on the body surface (e.g., lateral line system of fishes) is the difference between the motion of the animal's body and the local surrounding water motions (Denton and Gray 1983), whereas the stimulus to the inertial sensors of the ear is whole body acceleration (Kalmijn 1988). Thus, in order to extract a faithful representation of the local surrounding water motions from lateral line responses, the central nervous system must subtract out the motions of the body, as encoded by the inner ear inertial sensors, from surface hydrodynamic events. If, in addition, the animal is self-propelled, the animal's own self-generated movements will also stimulate both the ear and the lateral line, a phenomenon known as sensory re-afference (Lambert et al. 2012). In this case, independent knowledge of self-generated body motion is required so that it can be distinguished from body motion induced by exogenous sources. All of this, in turn, requires some computational analyses performed by well-developed central nervous systems, such as those found in crustaceans (Budelmann 1992), cephalopods (Bleckmann et al. 1991), and vertebrates (Punzo et al. 2004; Fritzsch and Glover 2007). In conclusion, lateralis-like organs in sessile organisms can faithfully encode surrounding water motions without additional information from inertial detecting ears and do not face the problems created by sensory re-afference due to self-generated motion. In contrast, lateralis organs in motile organisms require additional information from inertial senses about body motion in order to faithfully reconstruct the pattern of water motions created by external sources and information about sensory information generated by own motions.

Based on these considerations, lateralis organs could have evolved multiple times in both sessile and motile organisms. We propose that among motile organisms, evolution of a lateral line system would have followed the evolution of a gravistatic sensor or ear and would precede it during evolution only if motility is secondarily acquired. This should be so because a reference system to provide information about hydrodynamics-induced whole body movements is needed before a lateral line could provide useful information for water movement on the surface for a free swimming organism without a sophisticated brain. Viewed from this perspective, the poorly organized lateralis-like organs found in hagfish might present a primitive organization of the lateral line and not a secondary reduction (Braun and Northcutt 1997). Consistent with our preposition, among mollusks, only the most derived cephalopods have both an statocyst and lateral line organs while other mollusks have only a statocyst (Budelmann and Bleckmann 1988). While this correlation among mollusks could be a chance event, it could also reflect the causality argued for above. More work is needed to establish which of the two possibilities reflects the adaptive history among mollusks. Moreover, how much of this example of mollusks can be transferred onto vertebrate evolution remains to be seen. Clear is that ears and lateral line organs are only found in vertebrates among deuterostome animals. In contrast to vertebrates, neither the chordate nor non-chordate outgroup has either of these organs or the molecular machinery to make anything but single sensory cells with molecular and anatomical similarities to hair cells (Joyce Tang et al. 2013; Short et al. 2012; Candiani et al. 2011; Pierce et al. 2008). Irrespective of this uncertainty, evolution of all hair cell bearing mechanosensory organs was logically only possible after mechanosensory hair cells had evolved. We therefore will next analyze what is known about the evolution of these hair cells, the molecular mechanisms of their polarity development, and the evolution and development of neurons that connect them to the brain.

11.2.1 Evolution of Mechanosensory Hair Cells

Multicellular organisms derive from single cell or colonial choanoflagellates, where each cell has a central motile kinocilium that is driving water and thus particles through a surrounding collar of interconnected microvilli (Fairclough et al. 2013). Particles, such as bacteria, caught with this "net" are digested by the cell. A similar organization is found in sponges with the choanoflagellate-like choanocytes driving fluid through chambers (Fig. 11.1). In cnidarians and deuterostomes, there is a tendency for this basic metazoan feature to be transformed into a different arrangement of an eccentric kinocilium and asymmetrically arranged microvilli of variable thickness (Fritzsch et al. 2007). This is particularly diverse in ascidians where multiple configurations of what appears to be eccentric assemblies of 'stereovilli' have been described (Burighel et al. 2011; Jorgensen 1989). Unfortunately, it has not been verified by electrophysiological means

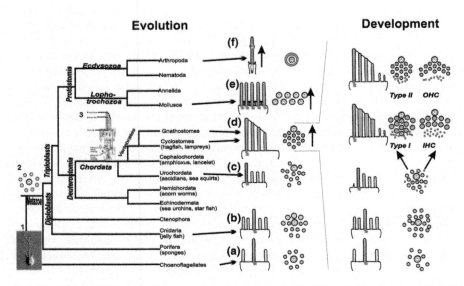

Fig. 11.1 Evolution and development of mechanosensory cells. Kinocilia (*red*) and microvilli/ stereocilia (*green*) of known or suspected mechanosensory cells in various eukaryotic unicellular (*1*) and multicellular (*3*) organisms are shown. Orthologs of structural genes relevant for mechanosensation or for development of polarity and several transcription factors are known in protists, Diploblasts (*2*), and various triploblasts. The existence of homologous genes across phyla indicates that they are ancestral to vertebrates. Note that the single-celled ancestor of all multicellular animals, the choanoflagellates (*1*), has a single, kinocilium surrounded by microvilli utilizing an actin core (**a**). In some diploblasts, the central kinocilium is surrounded by an asymmetric assembly of microvilli, potentially providing directional sensitivity (**b**). Among deuterostomes, urochordates have various presumed mechanosensory cells that have a kinocilium with asymmetrically arranged microvilli (**c**). Vertebrates are unique in that a highly polarized, organ-pipe assembly of actin-rich stereocilia are attached via tip links to each other (**d**). **e**, **f** Display the very different organization of protostome mechanosensors, either a single kinocilium that is pulled along its length (**f**) or multiple kinocilia, connected by links (**e**). *Right Hand Panel* Mammalian hair cells develop their stereocilia in a process that starts with a central kinocilium surrounded by few microvilli. As the number of microvilli increases the kinocilium moves into an off-center position and eventually toward one end of the developing hair cell. Microvilli in front of the moving kinocilium become reduced and eventually all disappear. In contrast microvilli trailing the kinocilium grow in length, thickness, and actin content to turn into stereocilia. As the kinocilium reaches its eccentric position, the distinction between kinocilium and microvilli has established the polarity with the longest stereocilia being next to the kinocilium. Development diverges through unknown molecular means to generate the four different hair cells found in the mammalian sensory epithelia. *Type I* and inner hair cells develop thick stereocilia and the characteristic bundles, which are *C-shaped* for inner hair cells. Other vestibular hair cells develop as *Type II* and outer hair cells with thinner stereocilia. In addition, the organization of the stereocilia in the outer hair cells forms a characteristic *M shape* with the kinocilium in the inflection of the *M*. Later in development small microvilli all but disappear and the kinocilium is resorbed in inner and outer hair cells of the organ of Corti. It appears that ampullary electroreceptive cells could be viewed as developmentally truncated mechanosensory hair cells, which adopt a different phenotype without stereocilia development, possibly also true for the "lateral line" organs of hagfish. Taken from (Duncan and Fritzsch 2012a)

whether these cells are mechanosensitive (Fig. 11.1). In an apparent form of ontogenetic recapitulation, the hair cells of vertebrates develop with a central kinocilium surrounded by microvilli (Fig. 11.1) which eventually moves into an eccentric position and microvilli transform into the stereovilli/stereocilia (Duncan and Fritzsch 2012a; Schwander et al. 2010). Given that microvilli already are linked in choanoflagellates it appears that the variable links found among stereocilia may be derived from those ancestral links serving a different function. Mammalian stereocilia would thus be distinguishable from ancestral microvilli by their graded height, there dense core and tapering near the base, and the molecularly unknown mechanoelectric transducer channel.

While the development and evolution of the hair-cell stereocilia is obvious in a series of transformational similarities across phyla and during hair-cell development (Fig. 11.1), it remains unclear how this morphological change is regulated during development at a cellular level. Below we will address this issue in the light of newly generated data in regenerating lateral line organs that indicate a molecular process. Before we focus on this issue we need to establish first the molecular basis of hair-cell development as revealed by mutational analysis mostly in mice and zebrafish.

An analysis of loss of hair-cell function in mice and fish has revealed the bHLH gene Atoh1 to be essential for hair-cell differentiation (Bermingham et al. 1999). The homologous gene of insects, atonal, is likewise needed for mechanosensory cell development of proprioreceptors such as chordotonal organs (Hassan and Bellen 2000). Replacing the fly *atonal* gene by the mammalian *Atoh1* gene rescues fly organ development and the *atonal* gene can rescue mouse hair-cell development (Wang et al. 2002). Moreover, atonal-like genes exist in coelenterates (Seipel et al. 2004) but whether or not they drive mechanosensory cell development in the statocysts or lateralis organs of these animals has not yet been studied. Beyond *atonal/Atoh1*, other homologous genes have been identified that are apparently conserved in their developmental function for mechanosensory cells across phyla. Foremost among these are the POU domain factor *Pou4f3* and the zinc-finger protein *Gfi1* (Wallis et al. 2003; Hertzano et al. 2004) and many other genes that relate to hearing loss in flies and mammals alike (Senthilan et al. 2012).

For example, a set of extremely conserved microRNAs has been identified that is expressed in known or suspected mechanosensory cells across phyla. The miR-183 of flies, and that of vertebrates differ only in a single nucleotide and thus can be used to identify putative homologous mechanosensory cells across phyla (Pierce et al. 2008; Candiani et al. 2011). While not as conserved as miR-124, which is identical in all neurons of all triploblasts, the nearly 100 % conservation of miR-183 supports the notion that cells expressing this microRNA are channeled toward mechanosensory cell development, possibly enhancing the function of other genes expressed in developing mechanosensory cells such as *Atoh1*, *Pou4f3*, and *Gfi1*. Most importantly, miR-183 is also found in single cells in collar region of an outgroup of vertebrates, the acorn worms (Pierce et al. 2008) and has recently been identified in chordates such as the lancelet (Candiani et al. 2011) and ascidians (Joyce Tang et al. 2013). Nevertheless, the distribution of all these genes and their expression in single putative mechanosensory cells of protostomes and deuterostome alike suggests that the molecular basis for mechanosensory cell development evolved with bilateral triploblasts and possibly already with diploblastic animals. What needs to be worked out now is how these conserved molecules interact during development of the various cell types and how this interaction evolved to generate the vertebrate hair cell.

11.3 Evolving the Molecular Basis for Stereocilia Formation and Apical Polarity Development

During both the evolution and development of hair cells, a mechanism has to operate to drive the kinocilium of the vertebrate hair cell into an eccentric position (Fig. 11.1). In addition, there must be a mechanism for organizing the development of a given cell's polarity in a coordinated fashion across an entire epithelium. Various epithelia of the lateral line and the inner ear either have a uniform polarity (canal cristae, organ of Corti), or two opposing polarities (saccule, utricle, lateral line organs), or have multiple polarities (saccule, auditory organs, and lagena in certain species). Genetic analyses have shown that the generation of polarity across otic epithelia is a complex process that requires a molecular signal, which will orient the intracellular machinery between cells to deliver uniform polarity across epithelia. The molecular basis for this phenomenon is the planar cell polarity pathway, PCP (Kelly and Chen 2007; Dabdoub and Kelley 2005). If this pathway is defective, individual hair cells are properly polarized internally but assume a random orientation within the epithelium, indicating that different mechanisms operate to organize the eccentric development of the stereocilia and kinocilium within a cell and to organize the pattern of hair-cell orientation across cells within a sensory epithelium.

In the zebrafish lateral line, for example, the global orientation of hair cells in neuromasts with respect to the animal's main body axes appears to be a consequence of the direction of migration of the lateral line primordium (Lopez-Schier et al. 2004). The homeostasis of planar polarity in neuromasts depends on the activity of the core planar-polarity pathway. In *trilobite/Vangl2* mutant zebrafish, the planar polarization of the hair bundles is profoundly affected. The misorientation of hair bundles is not a consequence of the defective cell migration, however, and reflects a specific requirement of Vangl2 in neuromast polarity signaling, comparable to the mammalian inner ear (Kelley et al. 2009). Furthermore, the discovery and continuous monitoring of *bona fide* hair-cell progenitors in the lateral line showed that the randomized polarization of hair cells in *trilobite* mutants is not a consequence of defects in the oriented division of the progenitors (Lopez-Schier and Hudspeth 2006). A dominant genetic interaction between Vangl2 and components of the Bardet–Biedl Syndrome (BBS) complex (Tayeh et al. 2008; Loktev et al. 2008) produces rotated hair bundles in mice, and

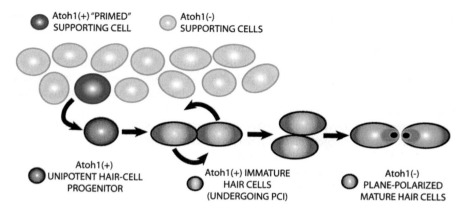

Fig. 11.2 Development of epithelial planar polarization in the zebrafish neuromast. Scheme of cellular identities in the neuromast (color-coded for the different cellular status or identity). *Black arrows* indicate the temporal progression along the differentiation path. This model of cell-fate acquisition in the neuromast proposed here combines the temporal hierarchy of gene expression of Atoh1 with the transition of hair cells from the division of their unipotent progenitor. We suggest that the planar cell inversions (PCI), indicated by two curved arrows, leads to the eventual realignment and the planar polarization of the resulting hair cells. Hair-cell polarity is shown by the *black dot* that represents the kinocilium and the *red "horseshoe"* that represents the stereocilia. Primed supporting cells (*dark blue*) arise within polar territories devoid of Notch signaling. These cells immediately stabilize Atho1a expression and become the immediate progenitors of hair cells. Progenitors divide to produce two hair cells (*red + yellow oblong cells*). Immature sibling hair cells undergo PCI, and stabilize planar polarity afterward. Hair cells eventually mature to become mechanoreceptive (*green*)

enhances the gastrulation defects in *trilobite* mutant zebrafish. Because Vangl2 is a constituent of basal bodies and ciliary axonemes, it may directly govern the behavior of basal bodies, instructing the localization of kinocilia and consequently the orientation of the hair bundles. It was originally suggested that the combined activity of the oriented division of hair-cell progenitors, and a Vangl2-dependent localization of basal bodies or kinocilia underlies the fidelity of the mechanism that governs hair-cell planar polarization during development and regeneration in the zebrafish lateral line. However, later studies using live visualization of the regenerative process, combined with sensors of fate commitment, showed that the majority of hair-cell siblings rotate around their contact point immediately after the cytokinesis of their progenitor cell (Fig. 11.2). This process, called "planar cell inversion," demonstrated that the orientation of hair-cell progenitor division is not essential for planar-polarity development of homeostasis in the lateral line (Wibowo et al. 2012).

Other studies suggest that microRNAs (miRs), 21–22 nucleotide-long RNAs play a role in the establishment of planar polarity in hair cells. MicroRNA interferes with the translation of targeted mRNA, preventing or blocking protein formation. If the major enzyme for the maturation of microRNA and other small RNA is deleted, some hair cells may form but they develop long stereocilia around a central kinocilium (Soukup et al. 2009). What specific mRNAs are disabled in their translation and how this relates to the possibly enhanced expression of other mRNAs remains unclear. It also remains unclear how a process that leads to the formation of apical specialization and its asymmetric development is tied into overall hair-cell development. Delayed deletion of *Atoh1* leads to incomplete formation of stereocilia (Pan et al. 2012a), indicating that a certain amount and/or time of Atoh1 protein presence is needed to activate all relevant pathways for normal stereocilia development with the characteristic staircase arrangement. Consistent with the suggestion that polarity differentiation at a cellular level is part of the bHLH signaling pathway is the defective polarity of in hair cells when the *Atoh1* gene is replaced by *Neurog1* (Jahan et al. 2012). Overall, there is progress in understanding molecular mechanisms in both polarity development of stereocilia within a hair cell as well as across cells within an organ. However, there are still large gaps to mechanistically link the cellular with the intercellular pathways.

11.4 Development and Evolution of Neurons and Their Central Projections in Vertebrates

11.4.1 Doubling Genes and Implementing a New Program

Without any doubt, the original chordate mechanosensory cell had its own cellular expansion (e.g., an axon-like process) that connected directly to the nervous system, much like in most invertebrates, including coelenterates (Jorgensen 1989). In contrast, vertebrates and certain mollusks have evolved a new set of cells, sensory neurons, that serve to connect mechanosensory cells to the central nervous system (Budelmann 1992), possibly splitting the ancestral single cell into cells with a different function (Pan et al. 2012b). Some mollusks have both types of cells situated next to each other within the same sensory epithelium, indicating that both developmental mechanisms can coexist within the same developing organ (Budelmann 1992). Details of the evolution of this second type of cell with the unique function of conducting information collected by the mechanosensory cells to the CNS is unclear, but it appears that at least in vertebrates a second set of bHLH transcription factors (*Neurog1* and *Neurod1*) have been recruited to specify and execute neuronal development (Pan et al. 2012b). Based on the distribution of the bHLH genes and the known distribution of neurons connecting mechanosensory cells to the CNS, it appears plausible that gene duplication and diversification evolved prior to novel functional assignment in the developmental program of mechanosensors to form complex organs (Czerny et al. 1999). How was this aggregation of cells accomplished molecularly during development and evolution?

Mice that lack *Neurog1* or *Neurod1* show a reduction in hair cell formation in certain epithelia (Matei et al. 2005). In addition to the loss of neurons and hair cells, *Neurod1* null mice show a transformation of neurons into vestibular like hair cells, which come about through loss of *Neurod1* mediated suppression of *Atoh1* expression in neurons (Jahan et al. 2010). These hair cells form within the ganglia and suggest that neurons, after their delamination from the forming ear, are still plastic to the extent that altering their program through loss of a single gene can convert them to hair cells. In addition to this neuronal specific cross-regulation of Neurod1 there is also cross-regulation between Neurog1 and Atoh1 with both positive and negative feedback/feed-forward loops (Pan et al. 2012b). The detailed regulation of neuronal and hair cell formation is a much more complicated process than anticipated based on the initial simple findings that all neurons disappear in *Neurog1* null mice and that all hair cells die through apoptosis in *Atoh1* null mice (Chen et al. 2002; Pan et al. 2011, 2012a). Indeed, it has been proposed that neurosensory precursors exist that sequentially give rise to neurons followed by hair cells (Fritzsch et al. 2006a). Recent evidence in mice (Jahan et al. 2010) and zebrafish supports this notion (Sapede et al. 2012). More recent data suggest a possible clonal relationship between sensory neurons of the ear and the lateral line, providing clear molecular evidence for the close association of these two functionally distinct sensory systems in zebrafish (Hans et al. 2013).

11.4.2 Timing of Afferent Neuron Formation and Projections to the CNS Reflect Different Developmental Adaptations

From the above considerations, it follows that a hair cell with an axon must have evolved before the axonless hair cell/neuron combination of vertebrates. However, various animals show temporal shifts with respect to hair cell and neuron formation, either developing both simultaneously or in some cases, even developing sensory neurons prior to hair cells. For example, in the migratory neuromast, the afferents form first and their processes accompany the migratory lateral line promordium (Lopez-Schier and Hudspeth 2005). In mice, initial formation and growth of afferents to sensory epithelia and into the CNS is independent of the formation of hair cells and thus proceeds nearly normal in the absence of hair cells (Fritzsch et al. 2005b; Pan et al. 2011), but afferents soon die due to the loss of hair cell expressed neurotrophins (Fritzsch et al. 2004; Yang et al. 2011).

The central projection of inner ear, lateral line, and ampullary electroreceptive afferents seem to follow a simple scheme (Fig. 11.3). The vestibular part of the ear is the first to develop central projections, adjacent to the trigeminal system. Consistent with the delay in formation of lateral line organs and their differentiation (Northcutt et al. 1995; Schlosser 2010), there is a delayed formation of central projections of lateralis afferents. In addition, these afferents project dorsal to the vestibular, inner ear, and other placode-derived organs such as the paratympanic organ of birds project within the vestibular fibers (O'Neill et al. 2012). The latest developing organ system of vertebrates, the ampullary electroreceptors,

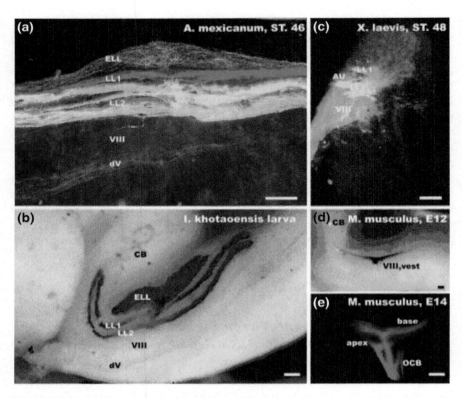

Fig. 11.3 The organization of the lateral line and inner ear afferent projections target specific nuclei. Shown are afferent lateral line projections for the axolotl (**a**), a gymnophionan (**b**), a frog (**c**), and the vestibular and cochlear projections of a mouse embryo (**d, e**). In **a**, note that older salamander and gymnophionan larvae show a clear segregation of mechanosensory afferents of a given peripheral lateral line nerve into two distinct fascicles (*LL1, LL2*). Labeling two peripheral nerve branches (the anterodorsal and anteroventral lateral line nerve) with differently colored dextran amines results in labeling of two discrete fascicles for each nerve. In contrast, no such fasciculation is apparent in the electrosensory afferent projections (*ELL*). In **b**, gymnophionans can be seen to have a similar organization, including a segregation of lateral line fibers into two fascicles and no such segregation of electrosensory fibers. In **c**, frog lateral line projections are shown as two entering fascicles (*green*), but widespread ramification throughout the entire lateral line nucleus, whereas the frog inner ear projection shows a single t vestibular component (*VIII*) (*orange*) ventral to the lateral line projection, but also the formation of the auditory projection (*AU*) lateral to the lateral line projection. In mice, the vestibular (**d**) and cochlear (**e**) fibers are forming distinct projections from the earliest time they can be labeled and do so even for subdivisions derived from different parts of the cochlea (**e**). Labeling of all cochlear afferent fibers would result in a continuous projection with no distinct fascicles being apparent comparable to the electrosensory afferent projections in salamanders (**a**) and gymnophionans (**b**). *CB* cerebellum; *dV* descending tract of the trigeminal nerve; *LL1, LL2* lateral line afferent fascicles; *ELL* electroreceptive ampullary organ projection; *VIII* vest, vestibular component; *OCB* olivocochlear bundle. Bars indicate 100 μm in all images. Modified after (Fritzsch et al. 2005a)

project even further dorsal, ending adjacent to the choroid plexus (Fritzsch et al. 2005a). Development of central projections in salamanders may also recapitulate the sequence of evolution of the ear, the lateral line, and the ampullary electroreceptive system, assuming that hagfish with their underdeveloped lateral line and absence of ampullary electroreceptive system represent the primitive condition of vertebrates with only a functional ear. Similarly, in the ear, the late evolving sound pressure receiving auditory system of mammals (Fritzsch 1992; Duncan and Fritzsch 2012b) projects dorsal to the ancestral vestibular system (Fig. 11.3). Importantly, an auditory projection develops in amphibians whether lateral line afferents persist or are lost during metamorphosis (Fritzsch 1990; Fritzsch et al. 2006b). The transformation of the hindbrain to form a dorsal expansion of the alar plate that produces the neurons that receive the various afferents seems to correlate nicely with the formation of new sets of afferents from evolutionary new organs. How these two processes, each of which must be regulated independently in the placode and the hindbrain, are developmentally linked to ensure that correct connections are made remains unknown.

11.5 Polarity of Hair Cells is Linked to Afferent Organization and Their Central Projection

Beyond the general regulation of neuron and hair-cell differentiation, there is also a more specific regulation related in an as yet unexplored way to the formation of populations of hair cells with specific polarities. After the discovery of hair cells with opposing polarities in the lateral line and in many inner ear sensory epithelia, it was shown that all afferents respond only to hair cells of one polarity (Flock and Wersall 1962). Follow-up work showed that in some vertebrates such as lampreys, salamanders, gymnophionans, and basic frogs, all afferents of a given lateral line contribute to two fascicles (Fritzsch 1981a) which keep the same organization as they project centrally (Fritzsch et al. 2005a). The apparent correlation of two fascicles with the two opposing polarities of all hair cells in a given lateral line implied that each afferent of a given polarity project centrally in one of the two fascicles (Fig. 11.4). Follow-up work on the central projection of single neuromast showed indeed that each neuromast receives only two afferents and both project into two distinct fascicles (Fritzsch 1981b). All neuromasts within a given lateral line are oriented in the same direction relative to the line and the afferents running within each nerve contribute centrally to just two bundles (Fig. 11.4). Single afferent recordings with intracellular labeling of the recorded afferents are needed to verify that afferents in a given bundle receive input from hair cells of the same polarity only across all neuromasts of a given lateral line. Such an organization could provide simple means of detecting movement directions of a given stimulus (Fig. 11.4).

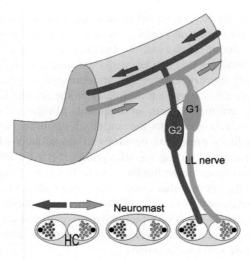

Fig. 11.4 Polarity-specific organization in the lateral line. Three lateral line neuromasts of an ideal lateral line are shown, each consisting of two hair cells (*HC*) with opposing polarities (*purple* and *green* stereocilia adjacent to the *black* kinocilium). Each hair cell is innervated by its own afferent fiber that projects into a discrete fasciclein the hindbrain. How different polarity-specific afferents run in the lateral line nerve, how the ganglion neurons that innervate differently polarized hair cells are distributed and whether all afferents contacting hair cells of a similar polarity course in the same fascicle remains unknown

   Studies on the developing lateral line of zebrafish have shown similar results indicating a segregation of information from differently oriented hair cells. During development and regeneration, a specific afferent fiber will sort out only hair cells of a given polarity in a certain neuromast (Faucherre et al. 2009). While the variations in polarity per neuromast seem to permit just two opposing polarities, the overall organization of neuromast in adjacent lines can be oriented 90° to one another. In addition, many neuromasts of different vertebrates receive more than two afferents making the pattern of innervation in those vertebrates more complex and less easily relatable to the two opposing polarities of hair cells. If the neuroanatomical predictions of opposite response polarities in salamanders, caecilians, lampreys (Figs. 11.3, 11.4), and other vertebrates with relatively simple afferent organizations are proven correct, it would raise an argument that somehow development of polarity of hair cells and the segregation of afferent projections is linked. How the two or more afferents that sort to hair cells of distinct polarities to deliver a given signal centrally to distinct polarity-specific second-order neurons remains to be worked out.

   An interesting parallelism to the lateral line system is found in the ear of vertebrates. All gravistatic sensors have two opposing polarities whereas semi-circular canals have one polarity (Fig. 11.5). In contrast to the various distributions of the two populations of hair cells in the lateral line system, the distribution of hair cells with opposing polarities in the utricle and saccule of most basic

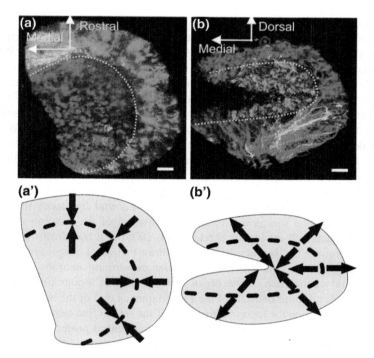

Fig. 11.5 Polarity-specific segregation of hair cells and their innervation in the ear. These images show the distribution of innervating fibers in the utricle (**a**) and saccule (**b**) of a mouse as well as the pattern of hair-cell orientations (*arrows*) and the line of polarity reversal, the striola (*dashed line* in **a'**, **b'**). Injection of the red lipophilic dye was in the cerebellum, injection of the green lipophilic dye was in the brainstem. After labeling, the maculae of the utricle and saccule of the same animal were dissected, mounted flat, and imaged. Note that most afferents that connect to the cerebellum project to hair cells of the same polarity in either epithelium. However, these hair cells are medial to the striola in one macula and lateral to the striola in the other. Similar distinct innervation according to hair-cell polarity exists in the lateral neuromasts. However, the distribution of hair cells with different polarities shows different patterns in different neuromast across vertebrates. Modified after (Maklad et al. 2010)

vertebrates is highly conserved (Lewis et al. 1985) with some more variation in highly derived bony fish. Interestingly, hair cells with different polarities in gravistatic sensors are segregated from each other into distinct halves of the epithelia with a sharp boundary separating them. Surprisingly, afferents to each epithelial subdivision projects somewhat distinct centrally: apparently one polarity only is connected to the cerebellum whereas the other is connected primarily to the hindbrain in mice (Maklad et al. 2010). As with developing lateral line afferents in salamanders (Fritzsch et al. 2005a), it appears that this polarity-specific segregation happens when afferents are growing into the brain, suggesting that a molecular mechanism specifies both hair-cell polarity and the segregated innervation pattern (Fig. 11.5).

In contrast to the lateral line neuromast, the ear of most vertebrates has two sets of organs with a single polarity, the hair cells of the cristae ampullaris in the semicircular canals of all vertebrates and the organ of Corti in mammals. Work by numerous researchers has demonstrated that canal cristae afferents from the two ears converge their signals onto vestibular nucleus neurons via connections between the two vestibular nuclei. These connections form a functionally yoked system that uses the opposing polarity of the hair cells of the matching canal of the other ear to fine tune angular acceleration reception (Szentagothai 1950; Beraneck and Straka 2011; Straka et al. 2009). In amphibians there is interconnection between matching positions of the lateral line nuclei that could mediate such interactions.

The other sensory epithelium with a single polarity of hair cells, the mammalian auditory organ of Corti, has a topographic map of sound frequency in the hindbrain. It appears that this is a simple transformation of the position of neurons according to their cell cycle exit (Matei et al. 2005) that transforms topology of hair cells along the basilar membrane and within the spiral ganglion into a topographical map in the cochlear nuclei. Earlier developing neurons in the high-frequency end of the base of the organ of Corti project topographically distinct from the late developing neurons of the low-frequency end of the apex (Fig. 11.3). There is also evidence for topographic maps of the lateral line (somatotopic maps) in the hindbrain of zebrafish. Despite this different central projection of afferents, second-order auditory nucleus neurons converge on a specific set of brainstem neurons, the superior olive, to generate a computational map of the sound field. How much interaction along these lines exist in the lateral line is unclear. What is known is that the clawed frog *Xenopus* has a topographical representation of sources of surface waves in the torus semicircularis (Zittlau et al. 1985) that is projected to the superior colliculus in line with other space maps.

Yet another organization is found in electroreceptive projections. In tuberous organs of weakly electric fish, afferents form multiple topographic surface map (Bullock and Heiligenberg 1986). No such map is known for non-teleost ampullary organs and topographic information may not be encoded in the disorganized afferents but may be a property of the signal extraction mediated by the complex assembly of electroreceptive second-order neurons of the medulla (Fig. 11.3). What can be stated is that the central organization of electroreceptive afferents is different from mechanosensory afferents and shows adaptation to the unique properties of this stimulus (Fig. 11.3).

In summary, afferent organization reflects both the polarity of afferents and the unique needs for the extraction of specific matching information inherent to each system. How much bilateral interaction of neuromasts of given polarity orientation on the two sides of the body form a yoked functional unit matching the organization of canal cristæ and the tonotopic organization of auditory system remains to be functionally verified.

11.6 Evolution of Mechanosensory Organs: A Repeat of Eye Evolution with a Different, Yet Closely Related Gene?

The seemingly convergent evolution of structurally very different eyes in metazoans has given way to the consideration that a single ancestral gene, the transcription factor Pax6, may orchestrate eye development across phyla no matter how diverse (Gehring 2011; Short et al. 2012). In other words, structurally different eye development nevertheless has a Pax6 gene as a major organizer that may be tied into different outcomes through the evolutionary recruitment of different sets of downstream genes. It has been hypothesized that this apparent conservation relates to the downstream activation of a conserved molecular cascade needed for the transduction process (Arendt 2003, 2005), a case of "deep molecular homology." Initial findings on another Pax gene, Pax2, seemingly indicated a somewhat different story in the ear with a loss of the cochlea in the absence of Pax2 (Torres and Giraldez 1998). Later work showed that Pax2 belongs to a family of three genes, Pax2/5/8, that typically are co-expressed and can act redundantly (Pfeffer et al. 2000; Short et al. 2012). Basic metazoans seem to have a fusion protein of Pax6 and Pax2/5/8 that is expressed in both the eyes and statocyst of box jellyfish (Kozmik et al. 2003) and apparently split in triploblasts into two distinct families. In triploblasts, Pax6 is associated with eye development (as well as the ganglia and part of the CNS) whereas Pax2/5/8 is associated with ear development (Bouchard et al. 2010), but also with kidney, thyroid and brain development (Short et al. 2012). Co-expression of Pax2 and Pax8 is needed for ear development and only a rudimentary vesicle forms without these two Pax genes (Bouchard et al. 2010). Chicken, which have lost the Pax8 genes, require Pax2 for otocyst invagination (Christophorou et al. 2010), supporting the notion that evolution of ears and eyes is linked to the evolution of specific Pax genes. Unfortunately, the multiple use of Pax2/5/8 in the development of other organs such as kidneys make it at best tentative to establish homology of Pax2/5/8 expressing organs across phyla without additional information how Pax2/5/8 is tied into mechanosensory cell and possibly mechanosensory neuron development. With these limitations in mind, it appears safe to conclude that the vertebrate ear depends on a set of Pax genes to orchestrate organ development much like the eye depend on closely related Pax transcription factor for its development. The added complexity of Pax function in ear development may relate to the apparent early multiplication that allowed the use of various Pax2/5/8 family members in the development of different organs (Short et al. 2012).

11.7 Evolution of Placodes: Evolving a Developmental Program to Specify Localized Epidermal Transformation into Various Sensory Systems

A ubiquitous transformation of ectoderm to form any sensory system is the specification of an ectodermal region that is undergoing enhanced proliferation and development of sensory systems (olfactory, lateral line, ear) or various sensory system associated structures (lens, trigeminal neurons). In vertebrates, this is accomplished through the formation of a pan-placodal region around the anterior end of the developing central nervous system (Streit 2007). This region is characterized by the ubiquitous expression of several transcription factors that ensure a neurosensory commitment such as *Eya1, Six1, Sox2*. Following and partially overlapping with the expression of these pan-placodal markers are genes that characterize specific areas (*Pax6* for the lens, *Pax2/8* and *Foxi3* for the ear). This more restricted expression of specific genes is an essential transition to turn the pan-placodal region into a region that activates various localized programs specific for the respective organ (Schlosser 2010). It is likely that the evolution of the pan-placodal region is tied to the evolution of the vertebrate CNS through a similar process of concentrating the ability to form sensory cells and neurons into a restricted region of the developing embryo (Fritzsch and Glover 2007) and the two processes are tied through inductive interactions (Schlosser 2010).

In this general scenario, the formation of the lateral line placode is thus far unexplained: no specific gene is currently known that is uniquely associated with the lateral line placode (Schlosser 2010) except for the recently discovered molecular dependency of lateral lline and ear sensory neurons (Hans et al. 2013). This is in stark contrast to the apparently highly conserved role of Pax genes in eye, ear, and trigeminal placode development. In addition, the lateral line placode appears comparatively late in development as a discrete entity and all vertebrates investigated thus far show a clear delay in lateral line development compared to inner ear development. This is particularly well studied in the zebrafish, salamanders, and frogs (Schlosser 2010; Fritzsch et al. 2005a), but seem to hold for other aquatic vertebrates as well. Overall, this developmental delay is even more pronounced in the development of the ampullary electoreceptive system, which is the last to form in salamanders (Fritzsch et al. 2005a). An ampullary electroreceptive system apparently evolved exclusively in vertebrates (Bullock and Heiligenberg 1986) and developmentally is tied into the lateral line placode, as demonstrated by grafting experiments (Northcutt et al. 1995). Unfortunately, the molecular basis of ampullary or tuberous electroreceptor development in bony fish is still unclear but it is likely that unique bHLH transcription factors will be identified as associated with the development of that system.

In summary, the molecular basis of otic and lateral line placode is associated with the general neurosensory forming area of the pan-placodal region. However, while the molecular mechanism that uniquely identifies the otic region is emerging, no unique marker of the later developing mechanosensory lateral line is known and even less is known molecularly for the development of electroceptive ampullary organs as none of the vertebrates thus far investigated molecularly possesses such a system. This lack of developmental molecular data limits our understanding how the mechanosensory hair cell could become diversified and associated with different organs and their functions.

11.8 Conclusion

The century-old idea of an octavolateral system that assumed that lateral line organs predate the evolution of the inner ear has been replaced in the last 30 years by novel insights. We propose a new "hair cell first" hypothesis (Duncan and Fritzsch 2012a; Pan et al. 2012b) which suggest that the evolutionary changes of ancient-ciliated mechanosensory cell to become a two-celled system of vertebrates (hair cell and neuron) is the precursor for the subsequent evolution of all mechano- and electrosensory organs. This ancestry of sensory cells is now strongly supported by molecular and morphological evidence. Once this two cell system had evolved, the developmental modules that can generate the vertebrate hair cells and associated neurons via a unique embryonic tissue (the placodes) could be assembled to form an inner ear. Based on the developmental delay relative to the ear and the incomplete development of a lateral line in haigfish, we propose that the mechanosensory lateral line evolved after the ear followed by the developmental and evolutionarily even further delayed ampullary electroreceptive system. Each organ system, as it evolved through developmental additions, adopted projections into discrete areas of the hindbrain for processing of their respective information that follows the developmental progress of organ formation. Where hair cells are polarized for directional stimuli detection, afferents contact only one hair cell type of one polarity and seem to project that information discretely into the brain, sometimes accompanied by a morphological segregation of fascicles possible derived from hair cells of a single polarity. The available data support the notion that the different subsystems of the so-called octavolateral system evolved more likely sequentially as the molecular developmental modules for hair cell/sensory neuron development multiplied and diversified during vertebrate evolution. The suggested molecular developmental distinction of the mechano- and electrosensory lateral line and inner ear is nowhere more obvious than in the metamorphic and evolutionary loss of the lateral line organs and their central connections: this loss is not accompanied by any obvious changes in the inner ear and its central projection to the evolving auditory nuclei. This long-held idea of transformation of lateral line nuclei into auditory nuclei is in sharp contrast to data on frogs that retain both a lateral line nucleus and develop auditory nuclei (Fig. 11.3).

Acknowledgments This work was supported by NIH grants R01 DC 005590 (BF) and P30 DC 010362 (BF) to BF, and by the European Research Council Starting Grant "Sensorineural" to HLS. We express our thanks to the Roy. J. Carver foundation for the purchase of the Leica TCS SP5 confocal microscope and the Office of the Vice President for Research for support. We wish to express our gratitude to the organizers of the lateral line symposium, in particular H. Bleckmann and S. Coombs for helping us streamlining this presentation.

References

Arendt D (2003) Evolution of eyes and photoreceptor cell types. Int J Dev Biol 47(7–8):563–571

Arendt D (2005) Genes and homology in nervous system evolution: comparing gene functions, expression patterns, and cell type molecular fingerprints. Theory Biosci 124(2):185–197

Arkett SA, Mackie GO, Meech RW (1988) Hair cell mechanoreception in a jellyfish Aglantha digitale. J Exp Biol 135:329–342

Beraneck M, Straka H (2011) Vestibular signal processing by separate sets of neuronal filters. J Vestib Res 21(1):5–19

Bermingham NA, Hassan BA, Price SD, Vollrath MA, Ben-Arie N, Eatock RA, Bellen HJ, Lysakowski A, Zoghbi HY (1999) Math1: an essential gene for the generation of inner ear hair cells. Science 284(5421):1837–1841

Bermingham NA, Hassan BA, Wang VY, Fernandez M, Banfi S, Bellen HJ, Fritzsch B, Zoghbi HY (2001) Proprioceptor pathway development is dependent on Math1. Neuron 30(2):411–422

Bleckmann H, Budelmann BU, Bullock TH (1991) Peripheral and central nervous responses evoked by small water movements in a cephalopod. J Comp Physiol [A] 168(2):247–257

Bouchard M, de Caprona D, Busslinger M, Xu P, Fritzsch B (2010) Pax2 and Pax8 cooperate in mouse inner ear morphogenesis and innervation. BMC Dev Biol 10:89

Braun CB, Northcutt RG (1997) The lateral line system of hagfishes (Craniata: Myxinoidea). Acta Zoologica 3:247–268

Budelmann B (1992) Hearing in Nonarthropod Invertebrates. In: Webster DB, Fay RR, Popper AN (eds) The evolutionary biology of hearing. Springer, New York, pp 141–155

Budelmann BU, Bleckmann H (1988) A lateral line analogue in cephalopods: water waves generate microphonic potentials in the epidermal head lines of Sepia and Lolliguncula. J Comp Physiol [A] 164(1):1–5

Bullock TH, Heiligenberg W (1986) Electroreception. Wiley and Sons, New York

Burighel P, Caicci F, Manni L (2011) Hair cells in non-vertebrate models: lower chordates and molluscs. Hear Res 273(1–2):14–24

Candiani S, Moronti L, De Pietri Tonelli D, Garbarino G, Pestarino M (2011) A study of neural-related microRNAs in the developing amphioxus. Evodevo 2:15

Chen P, Johnson JE, Zoghbi HY, Segil N (2002) The role of Math1 in inner ear development: uncoupling the establishment of the sensory primordium from hair cell fate determination. Development 129(10):2495–2505

Christophorou NA, Mende M, Lleras-Forero L, Grocott T, Streit A (2010) Pax2 coordinates epithelial morphogenesis and cell fate in the inner ear. Dev Biol 345:180–190

Czerny T, Halder G, Kloter U, Souabni A, Gehring WJ, Busslinger M (1999) Twin of eyeless, a second Pax-6 gene of Drosophila, acts upstream of eyeless in the control of eye development. Mol Cell 3(3):297–307

Dabdoub A, Kelley MW (2005) Planar cell polarity and a potential role for a Wnt morphogen gradient in stereociliary bundle orientation in the mammalian inner ear. J Neurobiol 64(4):446–457

Denton EJ, Gray J (1983) Mechanical factors in the excitation of clupeid lateral lines. Proc R Soc Lond B Biol Sci 218(1210):1–26

Duncan JS, Fritzsch B (2012a) Evolution of Sound & Balance Perception: innovations that aggregate single hair cells into the ear and transform a gravistatic sensor into the organ of Corti. J Anatomy 295:1760–1774

Duncan JS, Fritzsch B (2012b) Transforming the vestibular system one molecule at a time: the molecular and developmental basis of vertebrate auditory evolution. Adv Exp Med Biol 739:173–186

Fairclough SR, Chen Z, Kramer E, Zeng Q, Young S, Robertson HM, Begovic E, Richter DJ, Russ C, Westbrook MJ, Manning G, Lang BF, Haas B, Nusbaum C, King N (2013) Premetazoan genome evolution and the regulation of cell differentiation in the choanoflagellate Salpingoeca rosetta. Genome Biol 14(2):R15

Faucherre A, Pujol-Marti J, Kawakami K, Lopez-Schier H (2009) Afferent neurons of the zebrafish lateral line are strict selectors of hair-cell orientation. PLoS ONE 4(2):e4477

Flock A, Wersall J (1962) A study of the orientation of the sensory hairs of the receptor cells in the lateral line organ of fish, with special reference to the function of the receptors. J Cell Biol 15:19–27

Fritzsch B (1981a) Electroreceptors and direction specific arrangement in the lateral-line system of salamanders. Z Naturforsch 36:493–495

Fritzsch B (1981b) The pattern of lateral-line afferents in urodeles. A horseradish- peroxidase study. Cell Tissue Res 218(3):581–594

Fritzsch B (1990) On the coincidence of loss of electroreception and reorganization of brain stem nuclei. In: Finlay B, Innocenti GM, Scheich H (eds) The neocortex: ontogeny and phylogeny. Plenum Press, London, pp 103–109

Fritzsch B (1992) The water-to-land transition: evolution of the tetrapod basilar papilla, middle ear and auditory nuclei. In: Webster DB, Fay RR, Popper AN (eds) The evolutionary biology of hearing. Springer, New York, pp 351–375

Fritzsch B (1996) Similarities and differences in lancelet and craniate nervous systems. Isr J Zool 42:147–160

Fritzsch B, Glover JC (2007) Evolution of the deuterostome central nervous system: an intercalation of developmental patterning processes with cellular specification processes. In: Kaas JH (ed) Evolution of nervous systems, vol 2. Academic Press, Oxford, pp 1–24

Fritzsch B, Tessarollo L, Coppola E, Reichardt LF (2004) Neurotrophins in the ear: their roles in sensory neuron survival and fiber guidance. Prog Brain Res 146:265–278

Fritzsch B, Gregory D, Rosa-Molinar E (2005a) The development of the hindbrain afferent projections in the axolotl: evidence for timing as a specific mechanism of afferent fiber sorting. Zoology (Jena) 108(4):297–306

Fritzsch B, Matei VA, Nichols DH, Bermingham N, Jones K, Beisel KW, Wang VY (2005b) Atoh1 null mice show directed afferent fiber growth to undifferentiated ear sensory epithelia followed by incomplete fiber retention. Dev Dyn 233(2):570–583

Fritzsch B, Beisel KW, Hansen LA (2006a) The molecular basis of neurosensory cell formation in ear development: a blueprint for hair cell and sensory neuron regeneration? Bioessays 28(12):1181–1193

Fritzsch B, Pauley S, Feng F, Matei V, Nichols DH (2006b) The evolution of the vertebrate auditory system: transformations of vestibular mechanosensory cells for sound processing is combined with newly generated central processing neurons. Int J Comp Psychol 19:1–24

Fritzsch B, Beisel KW, Pauley S, Soukup G (2007) Molecular evolution of the vertebrate mechanosensory cell and ear. Int J Dev Biol 51(6–7):663–678

Gehring WJ (2011) Chance and necessity in eye evolution. Genome Biol Evol 3:1053–1066

Hans S, Irmscher A, Brand M (2013) Zebrafish Foxi1 provides a neuronal ground state during inner ear induction preceding the Dlx3b/4b-regulated sensory lineage. Development 140(9):1936–1945

Hassan BA, Bellen HJ (2000) Doing the MATH: is the mouse a good model for fly development? Genes Dev 14(15):1852–1865

Hertzano R, Montcouquiol M, Rashi-Elkeles S, Elkon R, Yucel R, Frankel WN, Rechavi G, Moroy T, Friedman TB, Kelley MW, Avraham KB (2004) Transcription profiling of inner ears from Pou4f3(ddl/ddl) identifies Gfi1 as a target of the Pou4f3 deafness gene. Hum Mol Genet 13(18):2143–2153

Jahan I, Pan N, Kersigo J, Fritzsch B (2010) Neurod1 suppresses hair cell differentiation in ear ganglia and regulates hair cell subtype development in the cochlea. PLoS ONE 5(7):e11661

Jahan I, Pan N, Kersigo J, Calisto LE, Morris KA, Kopecky B, Duncan JS, Beisel KW, Fritzsch B (2012) Expression of Neurog1 instead of Atoh1 can partially rescue organ of Corti cell survival. PLoS ONE 7(1):e30853

Jorgensen JM (1989) Evolution of octavolateralis sensory cells. In: Coombs S, Goerner P, Muenz H (eds) The mechanosensory lateral line. Neurobiology and evolution. Springer, New York, pp 99–115

Joyce Tang W, Chen JS, Zeller RW (2013) Transcriptional regulation of the peripheral nervous system in Ciona intestinalis. Dev Biol 378:183–193

Kalmijn AJ (1988) Electromagnetic orientation: a relativistic approach. Prog Clin Biol Res 257:23–45

Kelley MW, Driver EC, Puligilla C (2009) Regulation of cell fate and patterning in the developing mammalian cochlea. Curr Opin Otolaryngol Head Neck Surg 17(5):381–387

Kelly M, Chen P (2007) Shaping the mammalian auditory sensory organ by the planar cell polarity pathway. Int J Dev Biol 51(6–7):535–547

Kozmik Z, Daube M, Frei E, Norman B, Kos L, Dishaw LJ, Noll M, Piatigorsky J (2003) Role of pax genes in eye evolution. A cnidarian PaxB gene uniting Pax2 and Pax6 functions. Dev Cell 5(5):773–785

Lambert FM, Combes D, Simmers J, Straka H (2012) Gaze stabilization by efference copy signaling without sensory feedback during vertebrate locomotion. Curr Biol 22(18):1649–1658

Lewis ER, Fay RR (2004) Environmental variables and the fundamental nature of hearing. In: Manley GA, Popper AN, Fay RR (eds) Evolution of the vertebrate auditory system. Springer, New York, pp 27–54

Lewis ER, Leverenz EL, Bialek WS (1985) The vertebrate inner ear. CRC Press, Boca Raton

Loktev AV, Zhang Q, Beck JS, Searby CC, Scheetz TE, Bazan JF, Slusarski DC, Sheffield VC, Jackson PK, Nachury MV (2008) A BBSome subunit links ciliogenesis, microtubule stability, and acetylation. Dev Cell 15(6):854–865

Loosli F, Kmita-Cunisse M, Gehring WJ (1996) Isolation of a Pax-6 homolog from the ribbonworm Lineus sanguineus. Proc Natl Acad Sci U S A 93(7):2658–2663

Lopez-Schier H, Hudspeth AJ (2005) Supernumerary neuromasts in the posterior lateral line of zebrafish lacking peripheral glia. Proc Natl Acad Sci U S A 102(5):1496–1501

Lopez-Schier H, Hudspeth AJ (2006) A two-step mechanism underlies the planar polarization of regenerating sensory hair cells. Proc Natl Acad Sci U S A 103(49):18615–18620

Lopez-Schier H, Starr CJ, Kappler JA, Kollmar R, Hudspeth AJ (2004) Directional cell migration establishes the axes of planar polarity in the posterior lateral-line organ of the zebrafish. Dev Cell 7(3):401–412

Maklad A, Kamel S, Wong E, Fritzsch B (2010) Development and organization of polarity-specific segregation of primary vestibular afferent fibers in mice. Cell Tissue Res 340(2):303–321

Markl H (1974) The perception of gravity and of angular acceleration in invertebrates. In: Kornhuber HH (ed) Handbook of sensory physiology, vol VI/1 vestibular system. Springer, Berlin, pp 17–74

Matei V, Pauley S, Kaing S, Rowitch D, Beisel KW, Morris K, Feng F, Jones K, Lee J, Fritzsch B (2005) Smaller inner ear sensory epithelia in Neurog 1 null mice are related to earlier hair cell cycle exit. Dev Dyn 234(3):633–650

Northcutt RG (1988) The phylogenetic distribution and innervation of craniate mechanoreceptive lateral lines. In: Coombs S, Görner P, Münz H (eds) The mechanosensory lateral line: neurobiology and evolution. Springer, Heidelberg, pp 17–78

Northcutt RG, Brandle K, Fritzsch B (1995) Electroreceptors and mechanosensory lateral line organs arise from single placodes in axolotls. Dev Biol 168(2):358–373

O'Neill P, Mak SS, Fritzsch B, Ladher RK, Baker CVH (2012) The amniote paratympanic organ develops from a previously undiscovered sensory placode. Nat commun 3:1041

Pan N, Jahan I, Kersigo J, Kopecky B, Santi P, Johnson S, Schmitz H, Fritzsch B (2011) Conditional deletion of Atoh1 using Pax2-Cre results in viable mice without differentiated cochlear hair cells that have lost most of the organ of Corti. Hear Res 275(1–2):66–80

Pan N, Jahan I, Kersigo J, Duncan J, Kopecky B, Fritzsch B (2012a) A novel Atoh1 "self-terminating" mouse model reveals the necessity of proper Atoh1 expression level and duration for inner ear hair cell differentiation and viability. PLoS ONE 7(1):e30358

Pan N, Kopecky B, Jahan I, Fritzsch B (2012b) Understanding the evolution and development of neurosensory transcription factors of the ear to enhance therapeutic translation. Cell Tissue Res 349:415–432

Pfeffer PL, Bouchard M, Busslinger M (2000) Pax2 and homeodomain proteins cooperatively regulate a 435 bp enhancer of the mouse Pax5 gene at the midbrain-hindbrain boundary. Development 127(5):1017–1028

Pierce ML, Weston MD, Fritzsch B, Gabel HW, Ruvkun G, Soukup GA (2008) MicroRNA-183 family conservation and ciliated neurosensory organ expression. Evol Dev 10(1):106–113

Punzo C, Plaza S, Seimiya M, Schnupf P, Kurata S, Jaeger J, Gehring WJ (2004) Functional divergence between eyeless and twin of eyeless in Drosophila melanogaster. Development 131(16):3943–3953

Repass JJ, Watson GM (2001) Anemone repair proteins as a potential therapeutic agent for vertebrate hair cells: facilitated recovery of the lateral line of blind cave fish. Hear Res 154(1–2):98–107

Sapede D, Dyballa S, Pujades C (2012) Cell lineage analysis reveals three different progenitor pools for neurosensory elements in the otic vesicle. J Neurosci 32(46):16424–16434

Schlosser G (2010) Making senses development of vertebrate cranial placodes. Int Rev Cell Mol Biol 283:129–234

Schwander M, Kachar B, Muller U (2010) Review series: the cell biology of hearing. J Cell Biol 190(1):9–20

Seipel K, Yanze N, Schmid V (2004) Developmental and evolutionary aspects of the basic helix-loop-helix transcription factors Atonal-like 1 and Achaete-scute homolog 2 in the jellyfish. Dev Biol 269(2):331–345

Senthilan PR, Piepenbrock D, Ovezmyradov G, Nadrowski B, Bechstedt S, Pauls S, Winkler M, Mobius W, Howard J, Gopfert MC (2012) Drosophila auditory organ genes and genetic hearing defects. Cell 150(5):1042–1054

Short S, Kozmik Z, Holland LZ (2012) The function and developmental expression of alternatively spliced isoforms of amphioxus and Xenopus laevis Pax2/5/8 genes: revealing divergence at the invertebrate to vertebrate transition. J Exp Zool B Mol Dev Evol 318(7):555–571

Soukup GA, Fritzsch B, Pierce ML, Weston MD, Jahan I, McManus MT, Harfe BD (2009) Residual microRNA expression dictates the extent of inner ear development in conditional Dicer knockout mice. Dev Biol 328(2):328–341

Straka H, Lambert FM, Pfanzelt S, Beraneck M (2009) Vestibulo-ocular signal transformation in frequency-tuned channels. Ann N Y Acad Sci 1164:37–44

Streit A (2007) The preplacodal region: an ectodermal domain with multipotential progenitors that contribute to sense organs and cranial sensory ganglia. Int J Dev Biol 51(6–7):447–461

Szentagothai J (1950) The elementary vestibulo-ocular reflex arc. J Neurophysiol 13(6):395–407

Tayeh MK, Yen HJ, Beck JS, Searby CC, Westfall TA, Griesbach H, Sheffield VC, Slusarski DC (2008) Genetic interaction between Bardet-Biedl syndrome genes and implications for limb patterning. Hum Mol Genet 17(13):1956–1967

Torres M, Giraldez F (1998) The development of the vertebrate inner ear. Mech Dev 71(1–2):5–21

Vopalensky P, Pergner J, Liegertova M, Benito-Gutierrez E, Arendt D, Kozmik Z (2012) Molecular analysis of the amphioxus frontal eye unravels the evolutionary origin of the retina and pigment cells of the vertebrate eye. Proc Natl Acad Sci U S A 109(38):15383–15388

Wallis D, Hamblen M, Zhou Y, Venken KJ, Schumacher A, Grimes HL, Zoghbi HY, Orkin SH, Bellen HJ (2003) The zinc finger transcription factor Gfi1, implicated in lymphomagenesis, is required for inner ear hair cell differentiation and survival. Development 130(1):221–232

Wang VY, Hassan BA, Bellen HJ, Zoghbi HY (2002) Drosophila atonal fully rescues the phenotype of Math1 null mice: new functions evolve in new cellular contexts. Curr Biol 12(18):1611–1616

Wibowo I, Pinto-Teixeira F, Satou C, Higashijima S, Lopez-Schier H (2012) Compartmentalized Notch signaling sustains epithelial mirror symmetry. Development 138(6):1143–1152

Yang T, Kersigo J, Jahan I, Pan N, Fritzsch B (2011) The molecular basis of making spiral ganglion neurons and connecting them to hair cells of the organ of Corti. Hear Res 278(1–2):21–33

Zittlau KE, Claas B, Munz H, Gorner P (1985) Multisensory interaction in the torus semicircularis of the clawed toad Xenopus laevis. Neurosci Lett 60(1):77–81

Chapter 12
Patterning the Posterior Lateral Line in Teleosts: Evolution of Development

Alain Ghysen, Hironori Wada and Christine Dambly-Chaudière

Abstract The lateral line system of teleost fishes presents large variations of patterns and forms, usually thought of as adaptive. This raises the question of how divergent adult patterns are achieved, and how selective pressures have contributed to this divergence. Our understanding of the development of this sensory system has much improved over the past 10 years, mostly through work on the zebrafish. Because this progress is restricted to a single species, we cannot yet answer questions about the determinism of lateral line evolution, but we can at least propose plausible and testable hypotheses. Here we review the mechanisms that mediate the transition from embryonic to adult pattern in the zebrafish posterior lateral line system (PLL), and we show that the adult pattern is largely determined by developmental events that take place during early larval life. We also show that simple variations in the use of the same mechanisms account for the very different patterns observed in juvenile zebrafish and blue-fin tuna, and could potentially account for many or all of the patterns observed in other adult teleosts. We conclude that, in the case of the lateral line at least, large variations in pattern depend on minor changes in the deployment of conserved developmental programs, with uncertain adaptive value. We propose that organisms neurally adapt to whatever tools they are provided with by their own development, and use them as best as they can, thereby giving the impression that such tools were actually selected for.

Il n'est mouvement qui ne parle.(*There is no such thing as a movement that does not speak*).Michel de Montaigne, Livre II, Chap. XII.

A. Ghysen (✉)
University Montpellier 2 and INSERM U710, Montpellier, France
e-mail: alain.ghysen@univ-montp2.fr

H. Wada
National Institute of Genetics, Mishima, Japan

C. Dambly-Chaudière
University Montpellier 2 and CNRS UMR5235, Montpellier, France

H. Bleckmann et al. (eds.), *Flow Sensing in Air and Water*,   295
DOI: 10.1007/978-3-642-41446-6_12, © Springer-Verlag Berlin Heidelberg 2014

Keywords Neuromast · Hair cell · Plane-polarity · Post-embryonic development ·
Stitch · Budding · Organ migration · *Danio rerio* (zebrafish) · *Thunnus thynnus*
(blue-fin tuna)

12.1 Introduction

12.1.1 Structure of the Lateral Line System

Many reviews have dealt with the basic organization of this sensory system (e.g.,
Coombs et al. 1989, 2014; and this book), and this aspect will only be briefly
summarized here. The lateral line system comprises a number of discrete
peripheral sensory organs, which may be either mechanosensory or electrosensory.
The distribution of organs over the fish body is not exactly reproducible from
individual to individual, but the overall pattern (number and location of the various
lines, density of organs along each line, etc.) is constant within a species.

In most teleost fishes, the lateral line system comprises only mechanosensory
organs, and the present chapter is restricted to a discussion of the mechanosensory
lateral line system. The elementary unit of this system, the neuromast, comprises a
core of mechanosensory cells that resemble very much the hair cells of the vertebrate inner ear. Neuromast hair cells are surrounded by non-sensory support
cells, which secrete a gelatinous cupula in which the sensory hair cell apical
processes (kinocilium, stereocilia) are embedded, and by an outermost rind of
mantle cells.

In many species, a subset of neuromasts sink in the underlying dermal bones of
the head and scales, and become enclosed in canals that communicate with the
outside world through pores. Different sensory stimuli are most effective for
superficial and canal neuromasts, and it is thought that the former measure flow
velocity, whereas the latter measure pressure gradients and flow acceleration, an
aspect of lateral line physiology that has attracted much attention (see e.g.,
Coombs and Montgomery 1999, and this book).

12.1.2 Early Work on Lateral Line Development

The original work on lateral line development was carried on in amphibians, most
notably by Harrison and Stone. Working on the frog *Rana*, Harrison was to our
knowledge the first to suggest that the posterior lateral line system (PLL), which
extends on the trunk and tail, is set up by a cranial primordium that migrates all the
way from the otic region to the tip of the tail (Harrison 1904). During this

migration, the primordium deposits groups of cells, the prospective neuromasts, in its wake. Harrison (1904) also noted that a fiber always connects the primordium to its associated cranial ganglion, and proposed that sensory axons are actually towed by their migrating target cells.

In a number of excellent papers, Stone (1922, 1933, 1937) followed up this pioneering work. He observed migrating primordia in living salamanders (*Ambystoma*), and used grafts of pigmented cells in unpigmented background to demonstrate that both sensory and support cells of the trunk lateral line originate from the primordium. He also observed that a trail of deposited cells extends between neuromasts, and that sheath cells from the grafted placode migrate along the lateral line nerve (1933). Stone further showed that the path followed by the primordium must be defined by some extrinsic cue, but neuromast deposition must be intrinsic to the migrating primordium, as it does not depend on the host tissue. He discovered that each neuromast deposited by the primordium can bud off new neuromasts (which he called accessory neuromasts, or bud-neuromasts) during later development (1937).

Following G. Streisinger's choice of the zebrafish *Danio rerio* as a new "model" system to study the genetics of development, a number of aspects of zebrafish embryology have been described in great detail, including the development of its lateral line system. Metcalfe (1985) showed that, as in amphibians, the zebrafish PLL is laid down by a migrating primordium derived from a postotic placode, and that sensory axons extend into this primordium and accompany it during its migration, as proposed by Harrison (1904) in frogs. He further demonstrated that the sensory neurons that innervate the neuromasts are derived from the same postotic placode as the migrating primordium, and become postmitotic at the very early time of 10 h postfertilization (hpf), during gastrulation (Metcalfe 1983).

12.1.3 Early Work on Lateral Line Evolution

Previous analyses of lateral line evolution have mostly relied (1) on comparative descriptions of adult patterns, or more rarely of larval patterns, with the aim of inferring a putative ancestral pattern, and (2) on the mapping of morphological patterns on independently derived phylogenetic trees, with the aim of revealing evolutionary trends (see e.g., Webb 1989a; Northcutt 1990). Whereas this approach has provided us with a useful catalog of pattern variations, its strictly descriptive nature yields no information on the underlying developmental processes.

Recent progress in our understanding of PLL development has begun to reveal the mechanisms involved in this development, and their molecular bases. Although this progress is mostly limited to zebrafish so far, it paves the way to a re-analysis of the evolution of lateral line development in other species—an analysis looking for changes in the generative mechanisms that cause variation in the final patterns, rather than for pattern variation on its own sake.

In this chapter we summarize the various processes that lead from embryonic to adult PLL patterns in zebrafish, and we examine the implication of similar processes in a few other teleost species, mostly in the blue-fin tuna, *Thunnus thynnus*, where early larval development has been directly compared to that in zebrafish. This comparison is particularly interesting as *Danio* and *Thunnus* belong, respectively, to the Ostariophysi and the Acanthopterygii superorders of teleost fishes, which diverged around 290Myrs ago (Steinke et al. 2006; Hurley et al. 2007), and are therefore as distantly related as any two teleosts can be.

12.1.4 Terminology

Because the same words have been used with different meanings in different contexts (e.g., secondary neuromasts, accessory neuromasts, etc.), we will define the different names used throughout this chapter, mostly based on the recent work in zebrafish, which will be summarized in the next section.

The PLL includes all neuromasts on the body, except those on the head (anterior lateral line system). The neuromasts on the caudal fin are usually considered part of the PLL. Here we retained the name of caudal lateral lines (CLL, Wada et al. 2008) for these neuromasts, as they are a special subset produced by budding from the terminal PLL neuromasts.

Primary PLL neuromasts are those derived from the embryonic primordium, primI, whereas *secondary* neuromasts are those derived from the larval primordia, primII and primD. The distinction between primary and secondary neuromasts is an important one, since it appears that in both the zebrafish, *Danio rerio*, and the blue-fin tuna, *Thunnus thynnus*, hair cells in primary neuromasts are plane-polarized along the anteroposterior axis, whereas those of the secondary neuromasts are polarized along the dorsoventral axis.

Within each group (primary and secondary) some neuromasts are directly deposited by the migrating primordium, but other neuromasts develop later, through the proliferation of interneuromast cells laid down by the migrating primordia. Because those late neuromasts are intercalated between the pre-existing neuromasts deposited by the migrating primordia, they are called *intercalary* neuromasts. Primary (primI-derived) intercalary neuromasts may develop much after secondary neuromasts have been deposited by primII- or primD, and the difference between primary and secondary neuromasts has thus nothing to do with time of appearance.

Neuromasts of all types (primary or secondary, deposited or intercalary) can bud off additional neuromasts during adulthood. Those additional neuromasts are called *accessory*, or *bud-neuromasts*, as originally defined in amphibians by Stone (1937). The words "accessory neuromasts" have unfortunately been given more recently a different meaning in fishes (Coombs et al. 1988), where they are used to name superficial neuromasts associated to canals, although nothing is known of the origin of such superficial neuromasts, and it seems likely that they may actually have different origins in different species (see Sect. 12.3.2.6 below). To avoid unnecessary confusion, therefore, we used "bud-neuromasts" throughout this chapter. Bud-neuromasts remain closely associated to the founder neuromast and end up forming rows of neuromasts that are called *stitches*. The founder neuromast of a stitch is usually impossible to distinguish from its bud-neuromasts, which can themselves bud off new bud-neuromasts (Ledent 2002; Wada et al. 2010).

Pattern will be used throughout with the meaning of spatial distribution of sensory organs on the body surface.

12.2 Development of the Posterior Lateral Line System in zebrafish

12.2.1 Embryonic Development of the Zebrafish PLL

A number of laboratories have joined forces over the past 10 years to get mechanistic insights into the embryonic and early larval development of the zebrafish PLL. As a result, we now have a fairly comprehensive understanding of the embryonic development of the system. This part of lateral line development will be summarized here, to the extent that it contributes to an understanding of evolutionary variation, but only briefly, as there are many reviews published over the past few years that have discussed this process (Ghysen and Dambly-Chaudière 2004, 2007; Lecaudey and Gilmour 2006; Ma and Raible 2009; Aman and Piotrowski 2010; Chitnis et al. 2012).

The embryonic primordium arises as part of the PLL placode, which extends posterior to the otic placode. There is evidence that the PLL placode enters a phase of mitotic quiescence around gastrulation, suggesting a very early determination step (Metcalfe 1983; Laguerre et al. 2005). It seems likely that the placode is determined by interactions between ectoderm and the underlying hindbrain rhombomeres (reviewed in Baker and Bronner-Fraser 2001; Schlosser 2006), but this has not yet been fully elucidated.

Once formed, the embryonic primordium migrates along the horizontal myoseptum to the posterior tip of the body (Fig. 12.1a). This oriented migration is mediated by SDF1/CXCR4 signaling, where the signal (SDF1) is produced by cells along the myoseptum, whereas the receptor (CXCR4) is present in the leading cells of the migrating primordium (David et al. 2002; Knaut et al. 2003; Li et al. 2004). Migration also requires the presence of another SDF1 receptor, CXCR7, in the trailing region of the primordium (Dambly-Chaudière et al. 2007; Valentin et al. 2007). It has been proposed that CXCR7 sequesters SDF1, making it unavailable for CXCR4 signaling (Boldajipour et al. 2008) and thereby producing a gradient of SDF1/CXCR4 signaling within the primordium (Dambly-Chaudière et al. 2007). This gradient was proposed to explain directional migration toward the tail, as has now been demonstrated by Donà et al. (2013).

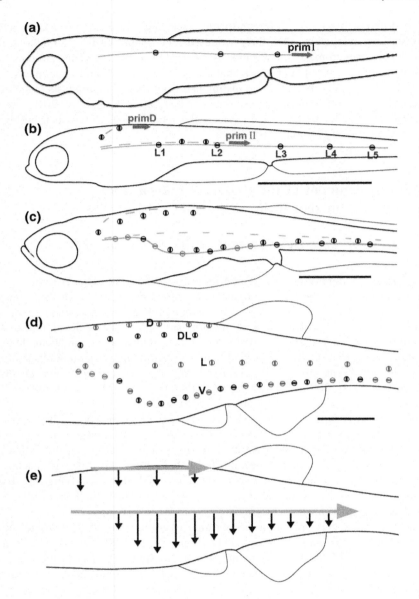

◀ **Fig. 12.1** Development of the PLL in zebrafish. **a** The embryonic primordium primI migrates from the postotic region to the tip of the tail during the second day of life, and deposits five anteroposteriorly polarized neuromasts, L1–L5, as well as a trail of interneuromast cells (*gray line*). **b** Two additional primordia, primII and primD, migrate during larval life and deposit dorsoventrally polarized neuromasts, as well as a discontinuous trail of interneuromast cells (*dashed gray lines*). **c** Over the first half of larval life, primI-derived interneuromast cells proliferate and form intercalary neuromasts (*light gray circles*) with anteroposterior polarity. **d** During larval life, neuromasts of the early lateral line migrate ventrally and end up forming a ventral line (*V line*), whereas the neuromasts of the early dorsal line also migrate ventrally and end up forming a dorsolateral line (*DL line*). At late larval stages, primII- and primD-derived interneuromast cells proliferate in turn, and form two new lines of intercalary neuromasts (*light gray*) with dorsoventral polarity. These new lines are aligned, respectively, along the lateral myoseptum (*L line*) and dorsal midline (*D line*). **e** Scheme to outline the fact that all primordia migrate along the rostrocaudal axis, whereas all neuromasts migrate along the dorsal axis. Scale bar: 1 mm

The anisotropy revealed by the complementary distribution of the two SDF1 receptors in the migrating primordium is mediated by Wnt signaling in the leading region (Aman and Piotrowski 2008), and FGF signaling in the trailing region (Lecaudey et al. 2008; Nechiporuk and Raible 2008). Wnt signaling is responsible for cell proliferation in the PLL placode and primordium (Gamba et al. 2010; Aman et al. 2011; Valdivia et al. 2011), thereby compensating for the loss of cells concomitant to neuromast deposition (Laguerre et al. 2005). FGF signaling is responsible for the mesenchymo-epithelial transition that leads to neuromast deposition (Lecaudey et al. 2008; Nechiporuk and Raible 2008).

Neuromast deposition seems to be an intrinsic property of the PLL primordium (Gompel et al. 2001). It has been proposed that deposition is a simple function of primordium size: proliferation would bring the primordium to some threshold size and trigger the next event of deposition, as suggested by the observation that the rate of deposition is decreased following gain-of-function interference with Wnt signaling (Aman et al. 2011). Although this mechanism is appealingly simple, the control of deposition is likely to be more complicated, however, as a decrease in cell proliferation due to inactivation of the Wnt target, *lef1*, results in a truncation of the line due to eventual lack of cells, rather than to larger spacing between neuromasts (Gamba et al. 2010; Valdivia et al. 2011; McGraw et al. 2011). Our present understanding (and its limitations) of neuromast deposition, self-organization, and differentiation has been excellently covered in Chitnis et al. (2012).

The primordium deposits five clusters of about 25 cells each, at regular intervals, and fragments into 3 terminal clusters upon reaching the tip of the tail. Each cluster differentiates as a neuromast within a few hours after deposition. Besides the periodic deposition of neuromasts, the primordium also deposits a trail of interneuromast cells, which will later form additional (intercalary) neuromasts (see below). Interneuromast cell deposition may be due to strong adhesion between primordium cells, making it almost impossible to break cell continuity. The exact factors, or cell adhesion molecules, involved in the continuity of proneuromasts and interneuromast cells have not been determined.

12.2.2 Postembryonic Development of the Zebrafish PLL

The transformation of the embryonic pattern of five lateral and 2–3 terminal neuromasts into the adult pattern is mediated by five independent mechanisms.

1. Shortly after the embryonic primordium, primI, has begun its journey, a second primordium originates from the same postotic placode that generated primI (Sarrazin et al. 2010) and splits in two groups: primII, which migrates and deposits neuromasts along the horizontal myoseptum as primI did, and primD, which migrates and deposits neuromasts along the dorsal midline until the base of the dorsal fin (Fig. 12.1b; Sapède et al. 2002). Whereas primI-derived neuromasts are polarized along the anteroposterior axis (Fig. 12.2a, b; López-Schier et al. 2004), however, primII- and primD-derived neuromasts are polarized along the dorsoventral axis (Fig. 12.2c, d).

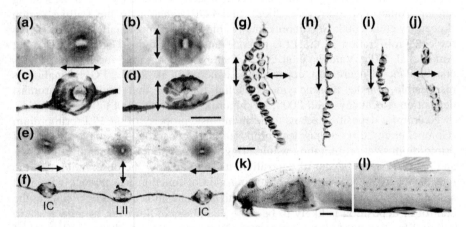

Fig. 12.2 Neuromast polarity and stitches. Anisotropy of a primI-deposited neuromast as revealed by alkaline phosphatase labeling (**a**) or by fluorescence of GFP in the ET20 (Parinov et al. 2004) transgenic line (**c**) parallels the anteroposterior polarization of its hair cells (*double headed arrow*). **b, d** A primII-derived neuromast shows dorsoventral anisotropy, as do its hair cells (*double headed arrow*). Fluorescence displayed in *black*. Scale bars: 20 μm. **e, f** Intercalary neuromasts (*IC*) flanking a primII-derived neuromast (*LII*) retain the anteroposterior anisotropy characteristic of the embryonic neuromasts deposited by primI, as revealed by alkaline phosphatase labeling (**e**) or in the ET20 line (**f**). **g–j** Neuromast polarity is retained by all neuromasts of a stitch. All stitches shown are from the same fish. **g** primII-derived stitch on somite 10, with 13 neuromasts, **h** primII-derived stitch on somite 17, with 9 neuromasts, **i** primII-derived stitch on somite 26 of the same fish, with 6 neuromasts, **j** primI-derived stitch on somite 24 with 6 neuromasts. In **g**, a primI-derived intercalary neuromast was present next to the founder LII neuromast, and has formed a smaller stitch with anteroposterior polarity. Stitches include less and less neuromasts from about somite 10 to the tip of the tail. Scale bar: 200 *μ*. **k, l** PLL stitches in the adult loach *Misgurnus anguillicaudatus* (Cypriniform) in the anterior region (**k**), and at the level of the anal fin (**l**), the size of stitches is similar all along the fish length, as is the fish thickness. Scale bar: 1 mm, size of fish: 36 mm

2. Once neuromasts are deposited, they undergo further migration. In contrast to the primordia, which migrate in anteroposteriorly, differentiated neuromasts migrate dorsoventrally (Fig. 12.1c, d). Due to this migration, the line formed by primI and completed by primII, which is initially located along the horizontal myoseptum, ends up at a much more ventral position (V line, Fig. 12.1d), whereas the line formed by primD, initially located along the dorsal midline, ends up in a more lateral position (DL line, Fig. 12.1d). It has been speculated that the independent control of the anteroposterior dimension through primordium migration, and of the dorsoventral dimension through neuromast migration, could facilitate large changes in the overall PLL pattern among species (Fig. 12.1e; Ghysen and Dambly-Chaudière 2003).

3. During larval development, interneuromast cells proliferate to form additional, "intercalary" neuromasts (Fig. 12.1c, light gray). The delay between deposition and cell proliferation is imposed by the glial cells that accompany the growing neurites and become apposed to the interneuromast cells (Grant et al. 2005; Lopez-Schier and Hudspeth 2005). primI-derived intercalary neuromasts begin to appear at about 8dpf (light gray, Fig. 12.1c), whereas primII- and primD-derived intercalary neuromasts appear around 3 weeks, near the end of larval life. Interestingly, intercalary neuromasts derived from primI-deposited interneuromast cells have hair cells polarized along the anteroposterior axis, as in primI-deposited neuromasts (Fig. 12.2e, f), whereas intercalary neuromasts derived from primII- or primD-derived interneuromast cells have hair cells that are polarized along the dorsoventral axis, as in primII- and primD-deposited neuromasts (Nuñez et al. 2009).

4. Once the juvenile pattern is completed (Fig. 12.1d), an amplification process begins where each juvenile neuromast produces a number of bud-neuromasts (also called "accessory" neuromasts by Stone, who was the first to study the process of budding, in amphibians). In fishes, this process has only been studied to date in the opercular line on the zebrafish head (Wada et al. 2010). All bud-neuromasts of a zebrafish stitch have the same polarity as the founder neuromast (Fig. 12.2g–j), with the exception of the stitches formed by primII-derived intercalary neuromats (Fig. 12.1d, L line). In this case, the founder neuromasts are polarized dorsoventrally, as is fit for primII-derived neuromasts, but some of the bud-neuromasts are polarized along the anteroposterior axis (Ghysen and Dambly-Chaudière 2007). The mechanism underlying this change in polarity is not known, nor is the mechanism whereby neuromast polarity is transmitted from the founder neuromast to its budded progeny. This is most unfortunate, as overall stitch polarity may be a major factor in shaping the information that the brain receives from peripheral sense organs and uses to form sensory perception, and understanding how polarity is determined may help us understand the interplay between developmental constraints and functional selection.

5. A second process that takes place after the juvenile pattern is established, is that some neuromasts sink into the underlying scales on the body, or dermal bones on the head, and become enclosed in "canals." This process is minimal in the PLL of zebrafish, as only the anterior-most three or four primI-derived

intercalary neuromasts become enclosed in a canal (Webb and Shirey 2003; Wada unpublished observations). Formation of a trunk canal is a major PLL feature in many teleost species, however, and the scales that enclose a canal often become morphologically distinct from the other body scales, resulting in the presence of a "lateral line" which is visible to the naked eye, and gave its name to the entire system.

12.3 Evolution of PLL Patterns

12.3.1 Origin of Variations in the Embryonic PLL Pattern

Embryonic PLL patterns are remarkably conserved among those teleost fishes that have been studied, as they always comprise a number of regularly spaced neuromasts aligned along the horizontal myoseptum, and one or a few terminal neuromasts near the tip of the tail (Pichon and Ghysen 2004). There is some variation in the total number of embryonic neuromasts, however. It appears from the small sample of species that we have examined, that neuromast number is correlated with embryo size. For example, blue-fin tuna embryos, in spite of being quite distantly related to zebrafish and living in a totally different environment, are of almost the same size, and have the same number of PLL neuromasts (Ghysen et al. 2012). In contrast, some fishes lay eggs that are much larger than those of *Danio* (e.g., the carp, *Cyprinus carpio*, Fig. 12.3a), and produce longer embryos with a larger number of regularly spaced neuromasts (Fig. 12.3b, c; see also Blaxter and Fuiman 1989; Pichon and Ghysen 2004). The smallest embryo that we examined, of the pygmy filefish *Rudarius ercodes*, also has the smallest number of neuromasts (Fig 12.3d, e). This variation is consistent with the idea that the embryonic pattern reflects a cyclic process of deposition that is intrinsic to the migrating primordium, and essentially conserved in all teleosts.

12.3.2 Origin of Variations in the Adult PLL Pattern

The diversity of PLL patterns observed among adult teleosts (Webb 1989c) cannot be traced back to differences in embryonic patterns, which, as far as we know, are minimal (see Sect. 12.3.1). Diversity must therefore arise during postembryonic development, suggesting that species-specific mechanisms shape the adult pattern during larval development. This aspect of PLL development has been little studied so far. Based on the understanding reached in zebrafish over the past few years, however, it has become at least feasible to propose, and in a very few number of cases, to confirm, explanatory schemes. We will consider successively variations based on each of the five mechanisms for postembryonic development outlined above.

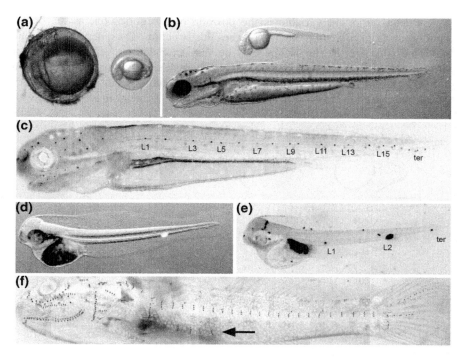

Fig. 12.3 **a–e** Number of embryonic neuromasts correlates with embryo size. **a** Egg of the carp *Cyprinus carpio* (Cypriniform) and egg of the zebrafish *Danio rerio* (Cypriniform), *upper right*; **b** newly hatched larvae of carp (*bottom*, 7.8 mm) and of zebrafish (*top*, 3 mm); **c** PLL of the carp embryo shown in (**b**); **d** embryo of the pygmy filefish *Rudarius ercodes* (Tetraodontiform), 2.1 mm; **e** its lateral line system, and **f** PLL in a young gobi, *Tridentiger trigonocephalus* (Gobioid, Perciform), 17 mm, with a ventral line extending to the anus (*arrow*)

12.3.2.1 PrimII and PrimD

A potential source of variation in the course of primII is the presence of a second stripe of SDF1-producing cells located at a more ventral level. This second stripe serves as a path for the migration of CXCR4-expressing germ cells toward the future gonad (Doitsidou et al. 2002), and extends caudally to the anus. In some zebrafish mutants where the lateral stripe of SDF1 is removed, primI moves ventrally, and follows this alternative pathway (David et al. 2002). We observed that a ventral line is present at the end of embryogenesis in some gobies (Fig. 12.3f). Interestingly, this ventral line stops at the level of the anus, as expected if it were driven by the ventral stripe of *sdf1* expression. We do not know, however, whether this ventral line is formed by a primordium migrating along a ventral path, or results from a ventral migration of neuromasts deposited along the myoseptum.

A second important variable is the position of the dorsal fin, as the dorsal primordium stops migrating when it reaches this structure. Thus, depending on the position of this fin, the dorsal line may be very extended, or very abridged. An extreme case is found in blue-fin tuna larvae, where the development of a dorsal fin in a very rostral position is correlated with the formation of an extremely short, almost abortive, dorsal line (Fig. 12.4c). A dorsal line of neuromasts may be well

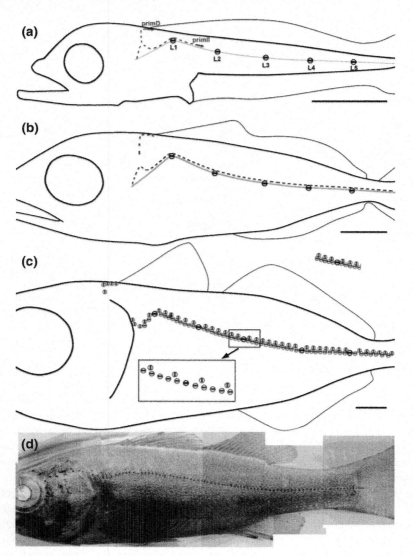

Fig. 12.4 PLL Development in perciforms. **a–c** Three stages of PLL development in the blue-fin tuna, *Thunnus thynnus* (Scombroid, Perciform). **d** Similar pattern in young *Lateolabrax japonicus* (Percoid, Perciform), 48 mm

tuned to respond to surface wavelets produced by fallen insects (reviewed in Bleckmann et al. 1989), and its absence would obviously be of no great concern for tuna.

12.3.2.2 Formation of Intercalary Neuromasts

The density and time of formation of intercalary neuromasts varies considerably between zebrafish and blue-fin tuna. In zebrafish, intercalary neuromasts form progressively during larval development, from anterior to posterior, at a rate of approximately one per day. They form on every intersomitic border except, in general, on those that are occupied by pre-existing neuromasts. In *Thunnus*, primI-derived intercalary neuromasts form synchronously near the end of larval development, at about three weeks of age, and their number is much larger than the number of primII-derived neuromasts. As all PLL neuromasts of *Thunnus* (except those of the diminutive dorsal line) are arranged on a single line, the relative density of AP-polarized, primI-derived and of DV-polarized, primII-derived neuromasts will affect the vectorial selectivity of the line. Very little is known about the development of intercalary neuromasts in other fish species, however, making it impossible to decide whether the proportion of the two types of neuromasts has functional relevance.

12.3.2.3 Neuromast Migration

One early difference between zebrafish and blue-fin tuna is that, in the latter species, the anterior-most embryonic neuromasts migrate dorsally, rather than ventrally as they do in zebrafish (Fig. 12.4a; Ghysen et al. 2012). In both species, primII migrates dorsal to the stripe of primI-derived cells. We do not know if this is due to an inhibitory effect that prevents primII from crossing the path of primI-derived cells, or to physical hindrance (e.g., due to attachment of primI-derived cells to the ectodermal basal lamina) preventing primII from moving across their track. Whatever the case, however, the dorsal versus ventral migration of neuromasts leads to a major difference in the final PLL pattern.

When differentiated neuromasts migrate ventrally in zebrafish, the original lateral and dorsal lines end up in more ventral positions, and the intercalary neuromasts that develop later from primII- and primD-derived interneuromast cells form two new lines, aligned along the horizontal myoseptum and the dorsal midline, respectively, (L and D lines, Fig. 12.1d). The dorsal migration of primI-derived neuromasts and the accompanying line of interneuromast cells in blue-fin tuna has the exactly opposite result: because primII cannot cross this line it migrates along it (Fig. 12.4a, b), thus resulting in a single arched line, dorsal to the horizontal myoseptum, comprising both anteroposteriorly polarized (primI-derived) and dorsoventrally polarized (primII-derived) neuromasts (Fig. 12.4c). This pattern of a single arched line is observed in many teleost species (Webb 1989c; Fig. 12.4d) and

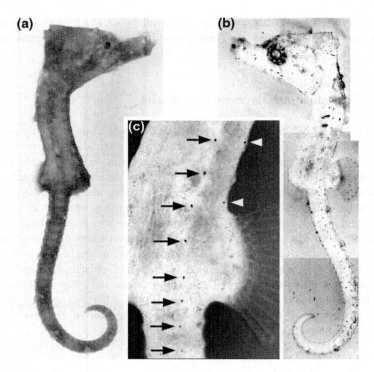

Fig. 12.5 **a** Young sea horse, *Hippocampus mohnikei* (Syngnathiform), 10 mm high, and (**b**) its PLL. **c** Higher magnification showing the neuromasts of the lateral line (*arrows*), and the neuromasts of the dorsal line stopping at the level of the dorsal fin (*arrowheads*)

illustrates how a discrete, and apparently small, change in a conserved mechanism can have drastic consequences on the outcome of later developmental processes.

A second variation arises when differentiated neuromasts do not migrate at all: such is the case of the Japanese seahorse *Hippocampus mohnikei* (locally called tatsunootoshigo, "illegitimate child of dragon"). In this case, one line remains aligned along the horizontal myoseptum, whereas a second line extends along the dorsal midline to the dorsal fin (Fig. 12.5), which is consistent with the basal pattern deduced from the work on zebrafish.

12.3.2.4 Budding and the Formation of Accessory Neuromasts

Adult patterns are derived from juvenile patterns through two processes: the formation of rows of superficial neuromasts (stitches), and the enclosure of neuromasts in canals. In zebrafish, stitches are formed by all juvenile PLL neuromasts (except for the few that become enclosed in canals, see below). Superficial neuromasts keep increasing in number through a budding process, but retain a constant size (Münz 1985), or grow negligibly in relation to canal neuromasts (Webb and Shirey 2003; Janssen et al. 1987). More generally, superficial neuromast size is remarkably similar in teleosts and amphibians (Münz 1985), irrespective of the size of the animal, suggesting either that there is some optimal size for superficial neuromast function, or that there is a conserved mechanism for the determination of neuromast size.

The number of neuromasts within a stitch varies in the ventral-most line (V line, Fig. 12.1d), from more than 40 for the largest stitch, to less than 10 in the smallest one, in 20-month old zebrafish. Stitch size is correlated with body thickness: the largest stitches are found on the belly, and the smallest ones near the caudal peduncle (Fig. 12.2g–j). Because the number and pattern of scales remain constant throughout adult life (Levin et al. 2012), changes in fish size and shape are accommodated by changes in scale morphology, and the largest stitches are found on the largest scales. In contrast, in the pond loach *Misgurnus anguillicaudatus*, where thickness of the trunk is quite uniform along the rostrocaudal axis, stitches have essentially the same size from anterior (Fig. 12.2k) to posterior body positions (Fig 12.2l). Scale growth is not isotropic: ventral scales tend to expand ventrally, and dorsal scales dorsally. Stitch expansion seems to parallel the biased extension of the scales during adult life, as already proposed for the opercular stitch (Wada et al. 2010), where the posteriorward expansion of the stitch parallels the expansion of the opercular bone.

12.3.2.5 Morphogenetic Budding: Patterning of the Terminal System

A special case of budding is found in the terminal system, and is best illustrated by the development of the terminal neuromasts in the blue-fin tuna. At the end of embryogenesis there is only one terminal neuromast (ter1), contrary to the case of zebrafish, where there are on average three terminal neuromasts. Early during larval life, local proliferation of interneuromast cells rostral to ter1 results in the formation of a second terminal neuromast, ter2 (Fig. 12.6a). ter2 then extends a buds in a posterior direction (Fig. 12.6b). This bud becomes a third terminal neuromast, ter3 (Fig. 12.6c), which in turn buds off two additional neuromasts, first ter4 (Fig. 12.6d) and then ter5 (Fig. 12.6e). From ter1 and ter4 additional budding result in the formation of two caudal lines which, themselves, extend by budding, one neuromast at a time (Fig. 12.6f). Thus the fairly complex and highly reproducible pattern of terminal and caudal fin neuromasts in Fig. 12.6e is almost entirely generated by oriented budding events.

The development of the terminal pattern in zebrafish seems quite different, since three terminal neuromasts are already present at the end of embryogenesis (ter1–ter3 numbered from the caudalmost). Interestingly, however, it was established recently that the two species use budding as a means to establish the juvenile terminal/caudal pattern (Wada et al. unpublished). In zebrafish, the rostralmost of the three terminal neuromasts, ter3, is not involved in the formation of the juvenile pattern. The medial neuromast, ter2, undergoes budding much as

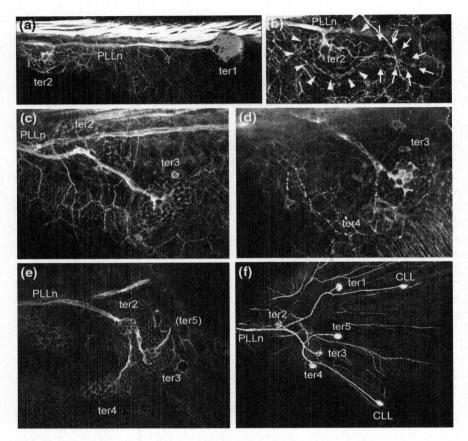

Fig. 12.6 Development of the terminal and caudal PLL in *Thunnus thynnus* (Scombroid, Perciform) as visualized in immunolabeled larvae where actin reveals cells contours, and tubulin reveals neurites and hair cells. **a** An intercalary neuromast, ter2, forms rostral to the single terminal neuromast, ter1. **b** A budding structure (*arrows*) extends from ter2 (*arrowheads*) and (**c**) forms a third neuromast, ter3. **d** ter3 in turn buds off neuromast ter4, and (**e**) ter5, out of focal plane. **f** The juvenile pattern of ter1–ter5 and the onset of the first dorsal and ventral caudal fin lines (CLL). PLLn: PLL nerve

ter2 of tuna, and generates a homolog of tuna's ter3 (called ter2′ in zebrafish). Additional lines then extend on the caudal fin, from ter1 and ter2′, through budding. Thus the principle of completing the terminal pattern and generating the caudal fin lines by budding seems conserved between the two species, even though the details of the intermediate steps vary somewhat.

An interesting variation on the zebrafish/tuna pattern is found in the medaka (*Oryzias latipes*), where a single terminal neuromast is present at the end of embryogenesis, much like in tuna, but remains single at the juvenile stage, unlike in tuna (Wada et al. 2008). Contrary to all PLL neuromasts, which never bud off accessory neuromasts in medaka, the single terminal neuromast becomes a small stitch, but generates no caudal line. It would seem, again, that minute variations in the use of common mechanisms can generate different adult patterns starting from the same set of embryonic terminals, or similar adult patterns starting from different sets of embryonic terminals.

Unfortunately our understanding of terminal and caudal fin patterns is much too limited to draw any firm evolutionary conclusion, which is a pity because the caudal fin lines are obviously distinct from other PLL lines due to the propelling function of the caudal fin. Furthermore, based on the available evidence in zebrafish (Pujol-Marti et al. 2010; Sato et al. 2010; Olszewski et al. 2012), the neurons that innervate the terminal neuromasts play a special role in setting up neural somatotopy (Alexandre and Ghysen 1999), a major aspect of the central connectivity of most sensory systems.

12.3.2.6 Canal Formation

In many fish species, a subset of PLL neuromasts recede into the underlying scales and form a canal that extends rostrocaudally. Canal scales, also called tubular scales, often differ from other body scales in a way that is visible to the naked eye, resulting in the formation of the so-called "lateral line." Canal neuromasts are always polarized along the direction of the canal, i.e., rostrocaudally, consistent with a primI origin. In zebrafish, only the first 3–5 intercalary neuromasts of the primI-derived line become enclosed in canals. Those do not form stitches, in contrast to all other primI-derived intercalary neuromasts. This correlation suggests that the enclosure of a neuromast in a canal prevents its further budding, at the same time as it allows its further growth. On the other hand, these intercalary neuromasts form much later than the embryonic neuromasts (Nuñez et al. 2009), which remain superficial, showing that the first neuromasts to differentiate are not necessarily those that will become canal neuromasts.

Canal neuromasts are often associated with stitches of superficial neuromasts. A simple explanation for this close association could be that stitch neuromasts are bud-progeny formed by the founder neuromast before it sunk and became enclosed in a canal. If this were the case, one would expect that the same branch of the PLL nerve should innervate the canal neuromast and the surrounding superficial neuromasts, as reported in the case of *Champsodon snyderi* (Nakae et al. 2006). We observed the same result through hair cell driven, trans-synaptic labeling of afferent neurons in the sea-bass *Lateolabrax japonicus* (Fig 12.7a).

A switch between budding accessory neuromasts, and sinking into a canal, would provide for a whole range of combinations, i.e., canal neuromasts alone, canal neuromasts accompanied by superficial neuromasts, or superficial stitches not associated to canals. Support for this idea comes from a comparison of two species of the *Pseudorasbora* genus, P. *parva* and P. *pumila*. In the latter, where no canals are formed, stitches are present along the horizontal myoseptum (Fig. 12.7c). In the former, one canal neuromast is present at the center of each stitch (Fig. 12.7b).

Fig. 12.7 Relation between canal and superficial neuromasts in a young adult of *Lateolabrax japonicus* (Percoid, Perciform). **a** DiAsp taken up by the hair cells has been transferred to sensory axons, revealing that the canal neuromast (*center*) and the flanking neuromasts (*arrowheads*) are innervated by branches of the same nerve, suggesting that they belong to a single stitch. **b** canal neuromasts (*arrows*) are accompanied by superficial neuromasts in *Pseudorasbora parva* (Cypriniform). **c** In the closely related species *Pseudorasbora pumila*, all neuromasts of the stitches remain superficial and no canal is formed

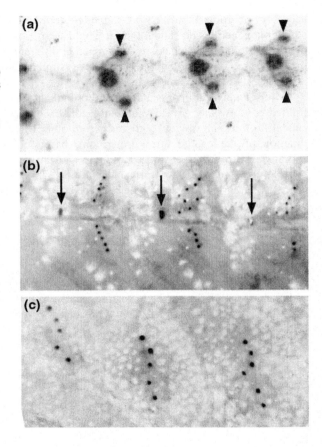

In other cases, a close association between canal neuromast and superficial stitches may derive from a close association of primI-derived and primII-derived neuromasts, as observed in blue-fin tuna larvae (Fig. 12.4c), where the two types of neuromasts are independently innervated (Ghysen et al. 2012 and Ghysen unpublished observations). If primI-derived neuromasts become canal neuromasts, whereas the closely apposed primII-derived neuromasts remain superficial and form stitches, this would account for the observation that in Tilapia (*Sarotherodon niloticus*, a cichlid), different fibers innervate canal and nearby surface neuromasts (Münz 1985). A close apposition of the primI- and primII-derived lines may also explain why canal and superficial neuromasts have orthogonal polarity in at least some cases (Schmitz et al. 2008).

The zebrafish PLL, where only the first 3–5 neuromasts form a short trunk canal, and the *Pseudorasbora* juveniles, where canals do or do not form in species in the same genus, suggest that canal formation is relatively easy to control independently of other PLL features. This would explain the prevalence of the so-called "replacement neuromasts", surface neuromasts found in species that lack canals, and localized at the same positions as canal neuromasts in closely related species (Coombs et al. 1988).

12.4 Behavioral Correlates of Changes in PLL Patterns: Evolution of Development Versus Functional Adaptation

The embryonic PLL pattern is largely conserved among teleost fishes, and the system is used to trigger a fast response in both zebrafish and tuna early larvae. The response is different, however: in zebrafish the response is a C-turn followed by a forward acceleration of the body (escape reaction, Weihs 1973), whereas in tuna, it is a forward movement accompanied by jaw opening (strike reaction, unpublished observations). In yet another species, the hunting archer fish, where escape and strike reactions co-exist, the evidence suggests that the two reactions share a common network of reticulospinal neurons, or elements of it (Wöhl and Schuster 2007).

The difference between the responses of zebrafish and tuna larvae correspond well to the different conditions they encounter: the "escape" reaction of zebrafish may be very useful if lots of hiding places are at close reach, as in the shallow, slow-flowing, well-vegetated waters where zebrafish live (Engeszer et al. 2007; Spence et al. 2008). A similar response would possibly be not very meaningful in open water, because if the predator misses the first time because of an "escape reaction", it is very unlikely to miss a second time. On the other hand, the "strike" reaction of tuna may make all the difference for a larva that has to eat on its first day of larval life, because its yolk sac is exhausted on the next day, contrary to zebrafish larvae that do not need to eat before 3 or 4 days after hatching—a time when both species have developed an excellent visual system.

The behavioral difference between the escape reaction of zebrafish, and the strike reaction of blue-fin tuna, at early larval stages, suggests that the same sensory system can be used to trigger either predator avoidance or prey detection. This is reminiscent of the situation in insects: the fly larva is a worm-like maggot, whereas the grasshopper larva is a diminutive grasshopper, yet the spatial distribution of the various types of mechanosensory organs over the body surface is essentially the same in both larvae (Meier et al. 1991). Thus selection may shape neural circuitry more easily than it does sensory patterns.

The conservation of PLL pattern in early larvae of different species living in entirely different environments, suggests that the pattern of the embryonic PLL is dictated by the way it develops. Functional adaptation may be responsible for other aspects of the PLL, such as the length of the cupula. For example, the cupula is at least five times longer in tuna than in zebrafish (Kawamura et al. 2003; Ghysen et al. 2012), suggesting a very high sensitivity (Coombs and Janssen 1989;

van Netten and Kroese 1989) consistent with an early role in feeding (Mukai et al. 1994).

PLL pattern diversity appears during larval postembryonic development. Based on the scant evidence available, this diversity appears not to be the effect of diverse mechanisms, but on the diverse use of the same five mechanisms that we described in the zebrafish. Developmental mechanisms underlying postembryonic development seem conserved between zebrafish and blue-fin tuna, although they end up producing very different morphologies. Based on this example, it seems easy to imagine how the same five mechanisms could possibly generate *any* PLL pattern.

Part of the developmental variation may depend on differences in timing, as in the time-honored concept of "heterochrony" (discussed in Webb 1989b). Other elements of variation may be related to local (e.g., homeotic genes—dependent) differences in the rate of any of the postembryonic mechanisms outlined above: dorsoventral migration of neuromasts, development of intercalary neuromasts, budding, and formation of canals. Yet other causes for variation may be related to the surrounding tissues: epidermis, underlying muscles and somite boundaries, scales, etc. Clearly we have to know more about the genetic bases of these mechanisms before we can understand how they evolved, and ultimately, what is the molecular basis of pattern variation in the lateral line system.

The conservation of mechanisms governing postembryonic development raises the question of whether adult PLL patterns are adaptive, as commonly thought of, that is, are shaped as a result of a process of natural selection, or whether they are the result of developmental processes that are themselves shaped and constrained by other aspects of fish development. Although this aspect has received little experimental attention so far, partly because it is often taken for granted that all phenotypes are adaptive (i.e., the result of natural selection), at least one study on a monophyletic group of Antarctic fishes has led to the conclusion that "differences in lateral line structures, even large ones, do not necessarily have consequences for function" (Coombs and Montgomery 1994).

We cannot exclude, of course, that some changes in pattern may have adaptive value. We do not have enough comparative data on larval behavior, or on PLL development, to answer this question definitely, but we can at least make an (educated?) guess. Bearing in mind that identical PLL patterns trigger different, and appropriate, behavioral responses in early larvae of zebrafish and tuna, we propose that, in the case of the PLL, behavioral differences (motor outputs to lateral line inputs) reside mostly in neural adaptations (how brain circuits utilize lateral line sensory information) rather than in adaptations of how the sensory organs are organized peripherally. Thus adaptation would reside mostly in neural adjustment to make the best use of sensory systems that are patterned, not by functional constraints, but by developmental ones.

In a penetrating analysis of the early development of insect motor systems, Bate (1998) noted that "the evidence suggests that neurons are born and differentiate in ways that are not conditioned by their future functions as elements of neural circuits. The logic, if there is one, is a developmental one. (...) In contrast to the apparent flexibility with which nervous systems can generate individual variations in behavior, early events such as these, that lay out the foundations of the (neural) network, appear highly stereotyped and the machinery that underlies them is increasingly well understood at a genetic as well as a cellular level." Except for the last line, one would say that the same wording exactly seems to apply to lateral line patterning. The PLL of adult fishes seems determined by "early events" that "lay out the foundations of the pattern" and thereby constrain further development. We propose that its adaptive value depends on how the nervous system makes use and sense of such patterns, and that there is possibly no direct selective pressure on their development. Further elucidation of the "machinery that underlies them" should help clarify this issue.

Acknowledgments We thank Sheryl Coombs for expert editorial assistance and thought-provoking comments on an early draft of this chapter, Hernan Lopez-Schier for excellent critical reading, Jackie Webb for careful editing, Ajay Chitnis for supportive comments, and Nicolas Cubedo for perfect fish handling and help over the past 16 years.

References

Alexandre D, Ghysen A (1999) Somatotopy of the lateral line projection in larval zebrafish. Proc Natl Acad Sci USA 96 13:7558–7562.

Aman A, Piotrowski T (2008) Wnt/beta-catenin and Fgf signaling control collective cell migration by restricting chemokine receptor expression. Dev Cell 15:749–761

Aman A, Piotrowski T (2010) Cell migration during morphogenesis. Dev Biol 341:20–33

Aman A, Nguyen M, Piotrowski T (2011) Wnt/beta-catenin dependent cell proliferation underlies segmented lateral line morphogenesis. Dev Biol 349:470–482

Baker CV, Bronner-Fraser M (2001) Vertebrate cranial placodes I. Embryonic induction. Dev Biol 232:1–61

Bate CM (1998) Making sense of behavior. Int J Dev Biol 42:507–509

Blaxter JHS, Fuiman LA (1989) Function of the free neuromasts of marine teleost larvae. In: Coombs S, Görner P, Münz H (eds) The mechanosensory lateral line: neurobiology and evolution. Springer, New York, pp 481–499

Bleckmann H, Tittel G, Blübaum-Gronau E (1989) The lateral line system of surface-feeding fish, anatomy, physiology, and behavior. In: Coombs S, Görner P, Münz H (eds) The mechanosensory lateral line: neurobiology and evolution. Springer, New York, pp 501–526

Boldajipour B, Mahabaleshwar H, Kardash E, Reichman-Fried M, Blaser H, Minina S, Wilson D, Xu Q, Raz E (2008) Control of chemokine-guided cell migration by ligand sequestration. Cell 132:463–473

Chitnis AB, Nogare DD, Matsuda M (2012) Building the posterior lateral line system in zebrafish. Develop Neurobiol 72:234–255

Coombs S, Bleckmann H, Popper AN, Fay RR (2014) The lateral line system, vol 48, Springer handbook of auditory research. Springer, New York

Coombs S, Janssen J, Webb JF (1988) Diversity of lateral line systems: phylogenetic, and functional considerations. In: Atema J, Fay RR, Popper AN, Tavolga WN (eds) Sensory biology of aquatic animals. Springer, New York, pp 553–593

Coombs S, Görner P, Münz H (1989) The mechanosensory lateral line: neurobiology and evolution. Springer, New York

Coombs S, Janssen J (1989) Peripheral processing by the lateral line system of the mottled sculpin (Cottus bairdi). In: Coombs S, Görner P, Münz H (eds) The mechanosensory lateral line: neurobiology and evolution. Springer, New York, pp 299–319

Coombs S, Montgomery JC (1994) Structural diversity in the lateral line system of antarctic fish: adaptive or non-adaptive? Sensornye Sistemy 8:42–52

Coombs S, Montgomery JC (1999) The enigmatic lateral line system. In: Fay RR, Popper AN (eds) Comparative hearing: fish and amphibians. Springer, New York, pp 319–362

Dambly-Chaudière C, Cubedo N, Ghysen A (2007) Control of cell migration in the development of the posterior lateral line: antagonistic interactions between the chemokine receptors CXCR4 and CXCR7/RDC1. BMC Dev Biol 7:23

David NB, Sapède D, Saint-Etienne L, Thisse C, Thisse B, Dambly-Chaudière C, Rosa F, Ghysen A (2002) Molecular basis of cell migration in the fish lateral line: role of the chemokine receptor CXCR4 and of its ligand, SDF1. Proc Natl Acad Sci USA 99:16297–16302

Dijkgraaf S (1963) The functioning and significance of the lateral line organs. Biol Rev 38:51–105

Doitsidou M, Reichman-Fried M, Stebler J, Köprunner M, Dörries J, Meyer D, Esguerra CV, Leung T, Raz E (2002) Guidance of primordial germ cell migration by the chemokine SDF-1. Cell 111:647–659

Donà E, Barry JD, Valentin G, Quirin C, Khmelinskii A, Kunze A, Durdu S, Newton LR, Fernandez-Minan A, Huber W, Knop M, Gilmour D (2013) Directional tissue migration through a self-generated chemokine gradient. Nature 503:285–289

Engeszer RE, Patterson LB, Rao AA, Parichy DM (2007) Zebrafish in the wild: a review of natural history and new notes from the field. Zebrafish 4:21–40

Gamba L, Cubedo N, Lutfalla G, Ghysen A, Dambly-Chaudiere C (2010) Lef1 controls patterning and proliferation in the posterior lateral line system of zebrafish. Dev Dyn 239:3163–3171

Ghysen A, Dambly-Chaudière C (2003) Le développement du système nerveux: de la mouche au poisson, du poisson à l'homme. médecine-Sciences 19:575–581

Ghysen A, Dambly-Chaudière C (2004) Development of the zebrafish lateral line. Curr Opin Neurobiol 14:67–73

Ghysen A, Dambly-Chaudière C (2007) The lateral line microcosmos. Genes Dev 21:2118–2130

Ghysen A, Dambly-Chaudière C, Coves D, de la Gandara F, Ortega A (2012) Developmental origin of a major difference in sensory patterning between zebrafish and bluefin tuna. Evol Dev 14:204–211

Grant KA, Raible DW, Piotrowski T (2005) Regulation of latent sensory hair cell precursors by glia in the zebrafish lateral line. Neuron 45:69–80

Gompel N, Cubedo N, Thisse C, Thisse B, Dambly-Chaudière C, Ghysen A (2001) Pattern formation in the lateral line of zebrafish. Mech Dev 105:69–77

Harrison RG (1904) Experimentelle Untersuchungen über die Entwicklung der Sinnesorgane der Seitenlinie bei den Amphibian. Arch Mikrosk Anat 63:35–149

Hurley IA, Mueller RL, Dunn KA, Schmidt EJ, Friedmann M, Ho RK, Prince VE, Yang Z, Thomas MG, Coates MI (2007) A new time-scale for ray-finned fish evolution. Proc R Soc B 274:489–498

Jansen J, Coombs S, Hoekstra D, Platt C (1987) Anatomy and differential growth of the lateral line system in the mottled sculpin, Cottus bairdi (Scorpaeniformes: Cottidae). Brain Behav Evol 30:210–229

Kawamura G, Masuma S, Tezuka N, Koiuso M, Jinbo T, Namba K (2003) Morphogenesis of sense organs in the bluefin tuna *Thunnus orientalis*. In: Browman HI, Skitfesvik AB (eds) The big fish bang. Institute of Marine Research, Bergen, pp 123–135

Knaut H, Werz C, Geisler R, Nüsslein-Volhard C (2003) Tübingen 2000 Screen Consortium. A zebrafish homologue of the chemokine receptor Cxcr4 is a germ-cell guidance receptor. Nature 421:279–282

Laguerre L, Soubiran F, Ghysen A, König N, Dambly-Chaudière C (2005) Cell proliferation in the developing lateral line system of zebrafish embryos. Dev Dyn 233:466–472

Lecaudey V, Gilmour D (2006) Organizing moving groups during morphogenesis. Curr Opin Cell Biol 18:102–107

Lecaudey V, Cakan-Akdogan G, Norton WH, Gilmour D (2008) Dynamic FGF signaling couples morphogenesis and migration in the zebrafish lateral line primordium. Development 135:2695–2705

Ledent V (2002) Postembryonic development of the posterior lateral line in zebrafish. Development 129:597–604

Levin BA, Bolotovskiy AA, Levina MA (2012) Body size determines the number of scales in cyprinid fishes as inferred from hormonal manipulation of developmental rate. J Appl Ichtyol 28:393–397

Li Q, Shirabe K, Kuwada J (2004) Chemokine signaling regulates sensory cell migration in zebrafish. Dev Biol 269:123–136

Lopez-Schier H, Hudspeth AJ (2005) Supernumerary neuromasts in the posterior lateral line of zebrafish lacking peripheral glia. Proc Natl Acad Sci USA 102:1496–1501

Lopez-Schier H, Starr CJ, Kappler JA, Kollmar R, Hudspeth AJ (2004) Directional cell migration establishes the axes of planar polarity in the posterior lateral line organ of the zebrafish. Dev Cell 7:401–412

Ma EY, Raible DW (2009) Signaling pathways regulating zebrafish lateral line development. Curr Biol 19:R381–R386

McGraw HF, Drerup CM, Culbertson MD, Linbo T, Raible DW, Nechiporuk AV (2011) Lef1 is required for progenitor cell identity in the zebrafish lateral line primordium. Development 138:3921–3930

Meier T, Chabaud F, Reichert H (1991) Homologous patterns in the embryonic development of the peripheral nervous system in the grasshopper *Schistocerca gregaria* and in the fly *Drosophila melanogaster*. Development 112:241–253

Metcalfe WK (1983) Anatomy and development of the zebrafish posterior lateral line system. Doctoral dissertation, University of Oregon, Eugene

Metcalfe WK (1985) Sensory neuron growth cones comigrate with posterior lateral line primordium cells in zebrafish. J Comp Neurol 238:218–224

Mukai Y, Yoshikawa H, Kobayashi H (1994) The relationship between the length of the cupulae of free neuromasts and feeding ability in larvae of the willow shiner *Gnathopogon elongatus caerulescens* (Teleostei, cyprinidae). J Exp Biol 197:399–403

Münz H (1985) Single unit activity in the peripheral lateral line system of the cichlid fish *Sarotherodon niloticus*. J Comp Physiol A 157:555–568

Nakae M, Asai S, Sasaki K (2006) The lateral line system and its innervation in *Champsodon snyderi* (Champsodontidae): distribution of approximately 1000 neuromasts. Ichtyol Res 53:209–215

Nechiporuk A, Raible DW (2008) FGF-dependent mechanosensory organ patterning in zebrafish. Science 320:1774–1777

Northcutt RG (1990) Ontogeny and phylogeny: a re-evaluation of conceptual relationships and some applications. Brain Behav Evol 36:116–140

Nuñez VA, Sarrazin AF, Cubedo N, Allende ML, Dambly-Chaudière C, Ghysen A (2009) Postembryonic development of the posterior lateral line in the zebrafish. Evol Dev 11:391–404

Olszewski J, Haehnel M, Taguchi M, Liao JC (2012) Zebrafish larvae exhibit rheotaxis and can escape a continuous suction source using their lateral line. PLoS ONE 7:e36661

Parinov S, Kondrichin I, Korzh V, Emelyanov A (2004) Tol2 transposon-mediated enhancer trap to identify developmentally regulated zebrafish genes in vivo. Dev Dyn 231:449–459

Pichon F, Ghysen A (2004) Evolution of posterior lateral line development in fish and amphibians. Evol Dev 3:187–193

Pujol-Martí J, Baudoin JP, Faucherre A, Kawakami K, López-Schier H (2010) Progressive neurogenesis defines lateralis somatotopy. Dev Dyn 239:1919–1930

Sapède D, Gompel N, Dambly-Chaudière C, Ghysen A (2002) Cell migration in the postembryonic development of the fish lateral line. Development 129:605–615

Sarrazin AF, Nuñez VA, Sapède D, Tassin V, Dambly-Chaudière C, Ghysen A (2010) Origin and early development of the posterior lateral line system of zebrafish. J Neurosci 30:8234–8244

Sato A, Koshida S, Takeda H (2010) Single-cell analysis of somatotopic map formation in the zebrafish lateral line system. Dev Dyn 239:2058–2065

Schlosser G (2006) Induction and specification of cranial placodes. Dev Biol 294:303–351

Schmitz A, Bleckmann H, Mogdans J (2008) Organization of the superficial neuromast system in goldfish, *Carassius auratus*. J Morphol 269:751–761

Spence R, Gerlach G, Lawrence C, Smith CH (2008) The behaviour and ecology of the zebrafish, *Danio rerio*. Biol Rev Camb Philos Soc 18:13–34

Steinke D, Salzburger W, Meyer A (2006) Novel relationships among ten fish model species revealed based on a phylogenomic analysis using ESTs. J Mol Evol 62:772–784

Stone LS (1922) Experiments on the development of cranial ganglia and the lateral line sense organs in *Ambystoma punctatum*. J Exp Zool 35:421–496

Stone LS (1933) The development of lateral line sense organs in amphibians observed in living and vital-stained preparations. J Comp Neur 57:507–540

Stone LS (1937) Further experimental studies of the development of lateral line sense organs in amphibians observed in living preparations. J Comp Neur 68:83–115

Valdivia LE, Young RM, Hawkins TA, Stickney HL, Cavodeassi F, Schwarz Q, Pullin LM, Villegas R, Moro E, Argenton F, Allende ML, Wilson SW (2011) Lef1-dependent Wnt/ß-catenin signalling drives the proliferative engine that maintains tissue homeostasis during lateral line development. Development 138:3931–3941

Valentin G, Haas P, Gilmour D (2007) The chemokine SDF1a coordinates tissue migration through the spatially restricted activation of Cxcr7 and Cxcr4b. Curr Biol 17:1026–1031

Van Netten SM, Kroese ABA (1989) Dynamic behavior and micromechanical properties of the cupula. In: Coombs S, Görner P, Münz H (eds) The mechanosensory lateral line: neurobiology and evolution. Springer, New York, pp 247–263

Wada H, Hamaguchi S, Sakaizumi M (2008) Development of diverse lateral line patterns on the teleost caudal fin. Dev Dyn 237:2889–2902

Wada H, Ghysen A, Satou C, Higashijima S, Kawakami K, Hamaguchi S, Sakaizumi M (2010) Dermal morphogenesis controls lateral line patterning during postembryonic development of teleost fish. Dev Biol 340:583–594

Webb JF (1989a) Gross morphology and evolution of the mechanoreceptive lateral line system in teleost fishes. Brain Behav Evol 33:205–222

Webb JF (1989b) In: Coombs S, Görner P, Münz H (eds) The mechanosensory lateral line: neurobiology and evolution. Springer, New York, pp 79–97

Webb JF (1989c) Developmental constraints and evolution of the lateral line system in teleost fishes. In: Coombs S, Görner P, Münz H (eds) The mechanosensory lateral line:neurobiology and evolution. Springer, New York, pp 79–98

Webb JF, Shirey JE (2003) Postembryonic development of the cranial lateral line canals and neuromasts in zebrafish. Dev Dyn 228:370–385

Weihs D (1973) The mechanism of rapid starting of slender fish. Biorheology 10:343–350

Wöhl S, Schuster S (2007) The predictive start of hunting archer fish: a flexible and precise motor pattern performed with the kinematics of an escape C-start. J Exp Biol 210:311–324

Chapter 13
Functional Architecture of Lateral Line Afferent Neurons in Larval Zebrafish

James C. Liao

Abstract Fishes rely on the neuromasts of their lateral line system to detect water flow during behaviors such as predator avoidance and prey localization. While the pattern of neuromast development has been a topic of detailed research, we still do not understand the functional consequences of its organization. Previous work has demonstrated somatotopy in the posterior lateral line, whereby afferent neurons that contact more caudal neuromasts project more dorsally in the hindbrain than those that contact more rostral neuromasts. Recently, patch clamp recordings of posterior lateral line afferent neurons in larval zebrafish (*Danio rerio*) show that larger cells are born earlier, have a lower input resistance, a lower spontaneous firing rate, and tend to contact multiple neuromasts located closer to the tail than smaller neurons, which are born later, have a higher input resistance, a higher spontaneous firing rate, and tend to contact single neuromasts. These data indicate that early-born neurons are poised to detect large stimuli during the initial stages of development. Later-born neurons are more easily driven to fire and thus likely to be more sensitive to local, weaker flows. Afferent projections onto identified glutamatergic regions in the hindbrain suggest a novel mechanism for lateral line somatotopy, where afferent fibers associated with tail neuromasts respond to stronger stimuli and contact dorsal hindbrain regions associated with Mauthner-mediated escape responses and fast, avoidance swimming. The ability to process flow stimuli by circumventing higher order brain centers would ease the task of processing where speed is of critical importance.

Keywords Afferent neurons · Neuromast · Development · Electrophysiology · Flow sensing

J. C. Liao (✉)
The Whitney Laboratory for Marine Bioscience, Department of Biology,
University of Florida, 9505 Ocean Shore Boulevard, St. Augustine FL 32080, USA
e-mail: jliao@whitney.ufl.edu

H. Bleckmann et al. (eds.), *Flow Sensing in Air and Water*,
DOI: 10.1007/978-3-642-41446-6_13, © Springer-Verlag Berlin Heidelberg 2014

13.1 Introduction

The lateral line system consists of discrete clusters of hair cells and support cells that together make up the neuromasts, which in adult fishes either lie directly on the surface of the fish (superficial neuromasts), or associated with fluid-filled canals that are part of the scales (canal neuromasts). Fishes rely on their neuromasts to detect unsteady flows generated by predators, prey and physical obstacles (Coombs 1994; Montgomery et al. 2003; Chagnaud et al. 2006; Liao 2006; McHenry et al. 2009). Neuromasts are responsible for transmitting hydrodynamic information to afferent neurons, which in turn are relayed to the hindbrain (McCormick 1989; Montgomery et al. 1996; Coombs et al. 1998; Nicolson et al. 1998; Bleckmann 2008; Liao 2010). Experimental and modeling work has demonstrated that there are two main classes of lateral line sense organs, superficial and canal neuromasts, which respond to different aspects of the hydrodynamic stimulus, principally—flow velocity (i.e., shear stress) and flow acceleration (i.e., pressure gradients), respectively (Dijkgraaf 1963; Coombs et al. 1989; Montgomery et al. 2000; Engelmann et al. 2002; McHenry et al. 2008). Yet we still lack a framework to understand the overall *functional* organization of the lateral line system. For example, is there a functional significance to how neuromasts are organized along the body axis and how flow information is subsequently routed to the brain? Our ability to understand how this information is processed is greatly enhanced by knowing three things: (1) the number of neuromasts that an afferent fiber contacts, (2) how afferents vary in their response properties as a function of their age and the location of their contacted neuromast(s), and (3) how afferent projections are distributed to identified structures or processing areas in the hindbrain. This level of resolution is not possible in adult fishes, which can possess thousands of neuromasts. In contrast, the posterior lateral line of larval zebrafish (*Danio rerio*) (for which this chapter will focus on) has only a few dozen neuromasts and afferent neurons. Because of their optical transparency and the ease in which one can genetically label neurons with fluorescent proteins, larval zebrafish are emerging as a model system to study the organization of the lateral line during development (Alexandre and Ghysen 1999; Raible and Kruse 2000; Gompel et al. 2001; Ledent 2002; Nagiel et al. 2008; Faucherre et al. 2009; Sarrazin et al. 2010; Sato et al. 2010). The accessibility and tractability of the larval lateral line provides a unique opportunity to now look at a key aspect in lateral line processing that has never been possible with adult fishes; how do afferent physiology and their pattern of connectivity affect information transfer to the hindbrain?

13.2 Using Transgenic Fish to Reveal the Age of Afferent Neurons During Development

As illustrated in the timeline of Fig. 13.1, 5-day post fertilization (dpf) larvae have three distinct populations of neuromasts derived from two distinct placodes (Sarrazin et al. 2010). Placode I develop at 17 h post fertilization (hpf) and give

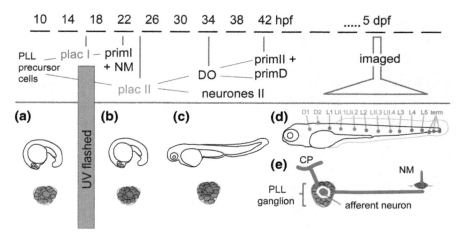

Fig. 13.1 Timeline of afferent neuron photoconversion in HuC:Kaede fish during the development of the posterior lateral line. **a** Initially all afferent neurons are *green* (e.g., at 10 h post fertilization, hpf). *Vertical gray bars* indicate approximate time of development. **b** When larvae are temporarily flashed with UV light at 17 hpf, all afferent neurons are converted to *red*. Neurons that subsequently develop after the UV flash are *green*. **c** When the lateral line ganglion is imaged at 5 days post fertilization (*dpf*), older cells are *red* and younger cells are *green*. **d** Neuromasts are not labeled by HuC:Kaede but are color-coded here to represent the age of the placode that they are derived from. L and terminal neuromasts are derived from placode I, while LII and D neuromasts are derived from placode II. **e** A single bipolar lateral line afferent neuron is highlighted to illustrate how information transduced by the neuromast hair cells are relayed to the hindbrain via central projections (*CP*). *plac* placode, *NM* neuromast, *prim* primordium, *DO* precursor cells. Adapted from Sarrazin et al. (2010)

rise to primordium I and its associated afferent neurons and L neuromasts (red neuromasts in Fig. 13.1d). Placode II develops around 24 hpf and gives rise to primordium II and primordium D, which in turn gives rise to more afferent neurons and the LII and D neuromast, respectively (green neuromast in Fig. 13.1d). In the transgenic line HuC:Kaede, larvae express a UV light-sensitive, photo-convertible protein under the control of a pan-neuronal promoter (Sato et al. 2006). This allows the ability to time-stamp afferent neurons to determine their relative stage of development. To do this, HuC:Kaede larvae are flashed with UV light for 10–20 s at 17 hpf. Under this protocol, all afferent neurons that are born at 17 hpf or before are imaged as red (older) and those that are born after 17 hpf are green (younger). Fish are then raised in the dark until they are imaged at 5 dpf. Note that while placodes give rise to both afferent neurons and neuromast hair cells, HuC:Kaede fish only label neurons and not hair cells. Because the larval lateral line system is experimentally accessible, single-cell labeling techniques can be used in combination with transgenic lines to reveal new information about afferent neurons. For example, old and young afferent cells can be targeted for electroporation and their projections down the body followed to confirm the number and location of neuromasts connected to that cell.

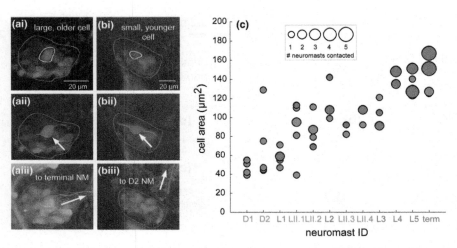

Fig. 13.2 Older and younger afferent neurons in HuC:Kaede larvae can be individually labeled to reveal the location and number of neuromasts contacted. **ai** A large, older cell (*red*) is targeted for labeling in the ganglion (*gray outline*). Dorsal is the *top of the image* and anterior is to the *left*. **aii** Upon successful electroporation the cell (*blue*) is double-labeled (*arrow*). **aiii** The merged image reveals a projection coursing down the body toward the terminal neuromasts of the tail (*arrow*). **bi** A small, younger cell (*green*) is targeted for labeling. **bii** The successfully electroporated cell is double-labeled. **biii** The merged image reveals a projection that rises sharply toward the D2 neuromast (*arrow*). **c** Plot of afferent somata size against the number and body location of the neuromast(s) that it contacts. The size of each data symbol represents the number of neuromasts contacted and the location of the data symbol along the *x*-axis indicates the caudal-most neuromast that the afferent cell contacts. Large, older afferents contact more neuromasts, of which the most caudal tend to be located closer to the tail compared to the neuromasts contacted by smaller, younger afferents. Neuromasts on the *x*-axis *label* are color-coded to symbolize the placode from which they were derived, as in Fig. 13.1e. *Red* indicates a neuromast derived from placode I while *green* indicates a neuromast derived from placode II. A discrepancy between afferent cell and neuromast color indicates that afferent cells are not restricted to contacting neuromasts derived from the same placode

13.3 Afferent Neuron Size and Connectivity is Related to Age

Older afferent neurons are found in the central and dorsal region of the posterior lateral line ganglion, while younger afferent neurons are located in the peripheral and ventral region (Liao 2010; Sato et al. 2010) (Fig. 13.1e). Older cells are significantly larger (Fig. 13.2a, $90.4 \pm 13.7 \ \mu m^2$) than younger cells (Fig. 13.2b, $63.5 \pm 5.8 \ \mu m^2$). More than half of afferent neurons (65 %) contact a single neuromast, and of these, the majority (79 %) consists of younger cells (Fig. 13.2c). Younger cells tend to contact neuromasts that are located closer to the head, although they may contact neuromasts located throughout the body. A smaller percentage of cells (21 %) contact two neuromasts, the majority (67 %) of which consists of older cells. It seems that only older cells contact three or more neuromasts, which often include a terminal neuromast that is located near the tail. Afferent neurons that contact one terminal neuromast typically contact other terminal neuromasts, but can also contact other neuromasts along the body. Interestingly, afferent neurons are not limited to contacting neuromasts derived from the same primordium. Thus, older cells derived from primordium I can contact a neuromast derived from primordium II (e.g., LII neuromasts), and younger cells derived from primordium II can contact a neuromast derived from primordium I. Afferent cell area also differs according to the body location of the neuromast that it contacts. Younger cells that innervate anterior neuromasts are generally smaller than afferents that innervate more posterior neuromasts. For example, an afferent cell that innervates D2 is typically smaller than an afferent cell that innervates terminal neuromasts. In addition, combining labeling experiments with transgenic lines allows cell age to be correlated to its projection patterns to post-synaptic structures in the hindbrain. For example, afferent neurons can be electroporated in Vglut GFP larvae, which mark glutamatergic neurons with GFP (Higashijima et al. 2004), to look at afferent projections in the hindbrain.

13.4 Exploring Hindbrain Targets of Afferent Neuron Projections

Once flow information from the neuromasts is transferred to afferent neurons, it is routed to the hindbrain for further processing. Most of our understanding of the post-synaptic targets of afferent neurons comes from tract-tracing experiments in adult fishes, where lateral line nerve projections were found in at least four different regions: (1) the medial octavolateralis nucleus, (2) the Mauthner neuron, (3) the caudal nucleus, and (4) the eminentia granularis in the cerebellum (McCormick 1989). However, whole-nerve labeling studies in adult fishes do not provide the single-cell resolution needed to reveal relationships between neuromasts, afferent neurons and hindbrain structures that living larval zebrafish provide (Alexandre and Ghysen 1999; Liao 2010; Liao and Haehnel 2012; Pujol-Marti et al. 2012). In 5 dpf larvae, the central projections of the posterior lateral line are likely located in at least three of the regions identified in adults (Mueller and Wullimann 2005). The establishment of these connections in the brain indicates that the lateral line system is a functional and important sensory modality at an early life history stage.

Transgenic zebrafish are poised to make exciting advances in revealing lateral line contacts in the brain. By using transgenic lines that mark neurons expressing difference neurotransmitter phenotypes, the location of putative contacts can be identified. For example, experiments using vglut GFP fish reveal that afferent projections terminate onto the lateral-most glutamatergic hindbrain stripe, contacting the eminentia granularis, medial octavolateralis nucleus (MON), the Mauthner neurons, and possibly the caudal nucleus. (Fig. 13.3a–c). While there seem to be direct contacts onto glutamatergic neuropil corresponding to the MON

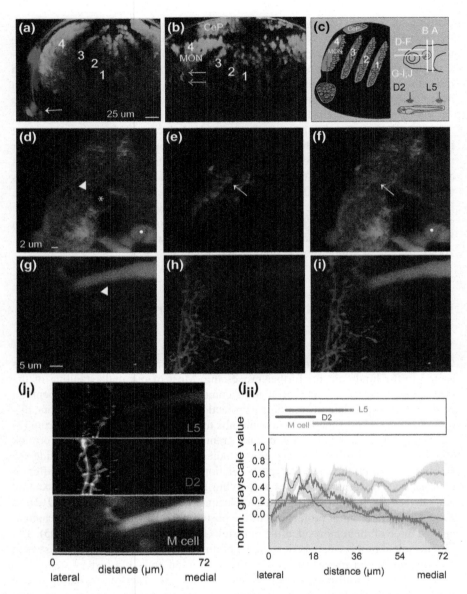

Fig. 13.3 Posterior lateral line afferent neurons contact identified post-synaptic targets in the hindbrain. **a** Afferent soma (*arrow*) labeled by backfilling at the site of a neuromast (D2 *blue*, L5 *red*) in a vglut GFP transgenic fish showing four glutamatergic stripes in one-half of the hindbrain seen in cross-section. **b** A more rostral hindbrain cross-section shows that afferent projections (*arrow*) of the L5 neuromast (*red*) projects more dorsally onto the neuropil of the fourth, lateral-most glutamatergic stripe than afferents associated with the D2 neuromast (*blue*). This region corresponds to the medial octavolateralis nucleus (*MON*) and lies ventral to the cerebellar plate (*CeP*). **c** Schematic of an image stack of (**a**) and (**b**) Medial-lateral position of. Diagram of the

◄ head where cross-sectional images were taken. **d** Glutamatergic neuropil (*arrowhead*) and soma (*dot*) in the MON of a vglut GFP transgenic fish. A cell that is not labeled with GFP is presumably not glutamatergic (*asterisk*). **e** Central projection terminals (*arrow*) of an afferent neuron connected to the L5 neuromast. **f** The merged image illustrates contacts onto glutamatergic neuropil, but not soma. **g** The lateral dendrite of the Mauthner cell. **h** Red projections of afferent neurons connected to the L5 neuromast lie medial to the blue projections associated with the D2 neuromast. **i** The merged image shows putative contacts of L5-associated afferent neurons with the lateral dendrite of the Mauthner cell. **j$_i$** Medial-lateral position of L5 and D2 afferent terminals relative to the Mauthner cell lateral dendrite. **j$_{ii}$** To quantify the spatial distribution for each of the three structures in J$_i$, grayscale pixel intensity was collapsed into a single mean value for each image column of pixels and plotted, along with the standard error, relative to a reference point in the hindbrain. The standard deviation of the intensity distribution was taken as a cut-off value to exclude background noise. Values above the cut-off are plotted in the top panel, which show that L5 hindbrain terminals start and end more medially than D2 terminals and possess more overlap with the Mauthner cell. Note that the cutoff line for the Mauthner cell is obscured by the cutoff line for the L5 projection

(Fig. 13.3d–f), there are no obvious contacts onto cell bodies, similar to general findings in adult fishes (McCormick 1989; Montgomery et al. 1996; New et al. 1996). Furthermore, afferents innervating neuromasts closer to the tail (i.e., L5) contact dorsal glutamatergic neuropil in the MON, while afferents innervating neuromasts closer to the head (i.e., D2) contact ventral neuropil (Fig. 13.3b)(Alexandre and Ghysen 1999; Liao and Haehnel 2012). In addition, afferent projections from L5 neuromasts project more medially than D2 neuromasts, making contact with the lateral dendrite of the Mauthner neuron (Fig. 13.3g–j$_{ii}$).

13.5 Intracellular Recordings of Lateral Line Afferent Neurons

Another strength of the zebrafish system is the ability to perform intracellular recordings in an undissected, behaving preparation, where the physiology is likely the closest to natural as possible. Recordings from afferent neurons show spontaneous spiking activity ranging from 5 to 50 Hz due to the constant release of glutamate from the hair cells of the neuromasts (Trapani and Nicolson 2011; Liao and Haehnel 2012). Whole-cell patch clamp recordings (Fig. 13.4a, b) reveal that afferent neurons possess excitable soma which can generate tonic firing in response to depolarizing current steps, both in the presence of spontaneous activity as well as when the cell is quieted by hyperpolarization (Fig. 13.4c, d).

The same recordings show differences in the intrinsic properties of afferent neurons residing in the same ganglion, suggesting that the sensory information transmitted from the neuromasts is functionally differentiated before it reaches the hindbrain (Liao and Haehnel 2012). Although cells display a continuum of sizes, cells which have an area smaller than $\sim 50 \ \mu m^2$ tend to have a higher spontaneous firing frequency and are more excitable (e.g., have a higher input resistance) than cells larger than $\sim 100 \ \mu m^2$ (Fig. 13.4e–h).

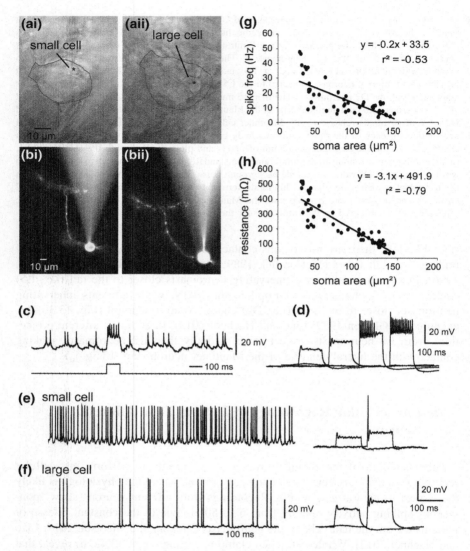

Fig. 13.4 Whole-cell patch clamp recordings of afferent neurons reveal differences in physiology with cell size. **ai–ii** Both small and large cells in the posterior lateral line ganglion can be targeted for recording with standard Nomarski optics. **bi–ii** Dye in the patch pipette labels the cell body and its projections to the neuromasts and hindbrain. **c** Afferent neurons are spontaneously active and increase their firing rate upon depolarization by current injection. **d** Increasing current injection step size elicits tonic firing. **e–f** Smaller cells have a higher spontaneous spike frequency and lower rheobase than larger cells. **g–h** There is an inverse relationship between soma area and spontaneous spike frequency, as well as between soma area and input resistance

The ability to record from single afferent neurons is an important technique that has allowed the first functional interpretation of lateral line somatotopy (Alexandre and Ghysen 1999; Gompel et al. 2001; Fame et al. 2006; Liao and Haehnel 2012). When these data are taken together with other work (Gompel et al. 2001; Pujol-Marti et al. 2010), it is clear that the intrinsic properties and projection patterns of afferent neurons match their birth date. Less excitable, older afferent neurons associated with caudal neuromasts project more dorsally into the hindbrain. These neurons have the largest soma, likely the largest axon diameters (Schellart and Kroese 2002), and have long projections that branch onto multiple neuromasts, resulting in a lower input resistance than later-born neurons. These older neurons are in a unique position to register stronger flow stimuli such as those generated from a predator (McHenry et al. 2009). In contrast, later-developing afferent neurons have smaller diameters, higher spontaneous firing rates, and lower thresholds for firing and are more inclined to contact single neuromasts (Nagiel et al. 2008). Younger afferents are therefore more excitable and may be better suited for detecting weaker flow stimuli in localized regions of the body, maturing perhaps to coincide with the onset of feeding and prey detection. A picture is emerging which suggests that a broad sensory scaffold is initially established by afferents and neuromasts derived from placode I to detect coarse flow, followed by a second wave of cells derived from placode II that would confer the ability to detect flow with greater sensitivity and finer spatial resolution. This architecture would separate different types of flow information at the periphery and likely ease the task of processing inputs at higher order brain centers such as the toris semicircularis (New et al. 1996; Plachta et al. 2003).

13.5.1 Single Neuromast Stimulation can Elicit Motor Behavior

In order to examine how lateral line inputs are correlated to motor output, it is possible to apply a controlled stimulus to a single neuromast while recording ventral root responses in a paralyzed preparation (Liao 2010; Liao and Haehnel 2012). To confirm that a neuromast is stimulated, the intracellular responses of afferent fibers are recorded while images of the deflected neuromast are captured by a high-speed video camera. Stimulation of a single terminal neuromast with a water jet shows that tail neuromasts play an important role in generating avoidances responses such as fast swimming and escapes (Fig. 13.5a). In contrast, there are fewer motor responses when a D2 neuromast (generally contacted by younger afferent neurons) is stimulated (Fig. 13.5b–d). The soma size of afferents contacting D2 and terminal neuromasts did not seem to be correlated to the elicited spike frequency (Fig. 13.5e).

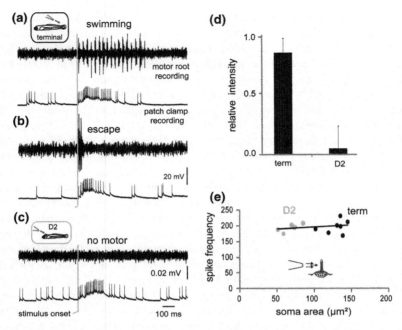

Fig. 13.5 Body location of a stimulated neuromast can determine the probability of a fictive motor response. A water jet directed at a terminal neuromast causes the connected afferent neuron to fire and elicits either a swimming (**a**), or escape response (**b**). **c** The same water jet stimulates the D2 neuromast but does not elicit a reliable motor response. **d** Across individuals there is more motor activity when a terminal neuromast is stimulated compared to a D2 neuromast. **e** A water jet to D2 and terminal neuromasts elicits no difference in afferent neuron spike frequency, suggesting that different projections into the hindbrain rather than spiking activity causes the distinct motor responses observed

13.5.2 Older Afferents are Involved in Fast Avoidance Behaviors

The age-related function we see in the afferent neurons of the lateral line system is part of a broader unifying principle of neuronal organization. This idea has a precedence in motor systems; in the spinal cord, large motor neurons and inter-neurons responsible for powerful movements develop first, while smaller motor neurons and interneurons develop later to facilitate finer motor control (Cope and Sokoloff 1999; Bhatt et al. 2007; McLean et al. 2007). In the hindbrain, neurons form distinct stripes with overlapping neurotransmitter and transcription factor phenotypes and are assembled according to the strength of an associated motor behavior (Kinkhabwala et al. 2010; Koyama et al. 2010). For example, in the medial-most glutamatergic stripe the dorsal neuropil is contacted by older, ventral neurons that are responsible for generating powerful movements such as fast swimming and the escape response. Similarly, in the lateral line system it seems that there is a pattern whereby older afferents that contact multiple neuromasts are also integrated into neuronal stripe organization in the hindbrain. As in the medial glutamatergic stripe for motor systems, the dorsal neuropil in the lateral-most glutamatergic stripe in the lateral line system is also contacted by older cells, which are connected to tail neuromasts (Fig. 13.3b). In this way, the dorsal hindbrain projections of afferent neurons contacting tail neuromasts would be favorably positioned to quickly inform fast motor behaviors, leaving the projections of rostral neuromasts to direct slower motor behaviors.

The central projections of the lateral line have been documented to make extensive contacts to various regions of the brain. The MON is the first processing stage in an ascending lemniscal pathway that includes additional processing stages in the midbrain (torus semicircularis and optic tectum) and forebrain (McCormick 1989). There is also a direct connection between the lateral line and the Mauthner cell that confers an ability to quickly translate flow magnitude and direction into appropriate escape behaviors (Hatta and Korn 1999). The critically important speed of the Mauthner escape response has been shown to benefit from fewer processing stages involving both the anterior (Mirjany and Faber 2011) and posterior (Faber and Korn 1975) lateral line system (Eaton et al. 1977; Faber et al. 1989). Considering the physiology of afferent neurons and their pattern of contact onto hindbrain stripes, this idea suggests that the L5 neuromast–MON circuit could circumvent higher order processing centers in order to (1) make fine adjustments to swimming speed, and (2) provide another mechanism to initiate swimming in addition to the L5 neuromast–Mauthner cell circuit. Stimulation of terminal neuromasts can uniquely influence motor behaviors due to the intrinsic properties and hindbrain projections of their associated afferent neurons, given that there does not seem to be larger numbers of afferent representation for terminal neuromasts (Haehnel et al. 2011). In this way, selective stimulation of the larval lateral line may modulate a range of motor behaviors, from the initiation of escape responses to the modulation of swimming speed.

As zebrafish mature it is likely that the simple organization present at the larval stage becomes obscured through patterns of growth and plasticity (Gaze et al. 1974; Fraser 1983; Chiba et al. 1988). For instance, afferent neurons could selectively change their synaptic strengths with their hair cell partners. This process has been shown to play a part in focusing an originally distributed set of connections, such as in the visual cortex and neuromuscular junction (Bennett and Pettigrew 1974, 1975; Brown et al. 1976; Hubel et al. 1977; Thompson 1985). In addition to the physiological heterogeneity between afferent neurons that may arise from intrinsic channel density and activity (Eatock et al. 2008; Sarrazin et al. 2010; Trapani and Nicolson 2011), this raises the possibility that with age synaptic rearrangement can change the contact strength of older afferent neurons onto multiple neuromasts. This would increase the functional complexity of the system beyond what is documented here in larvae. These findings may turn out to be limited to early stages of development, but it is unlikely that this initial organization would be substantially re-arranged at later stages of growth. Instead, this initial scaffold may be built upon and elaborated, rather than dismantled and reconstructed, according to current challenges that the hydrodynamic environment poses on the organism as it matures into adulthood.

Acknowledgments I would like to thank members of my laboratory who helped with various phases of this work, from data collection and analyses to fish care.
Disclosures This work was supported by NIH RO1DC010809 and NSF IOS 1257150 to James C. Liao. The author declares no conflicting competing interests.

References

Alexandre D, Ghysen A (1999) Somatotopy of the lateral line projection in larval zebrafish. Proc Natl Acad Sci USA 96:7558–7562

Bennett MR, Pettigrew AG (1974) The formation of synapses in striated muscle during development. J Physiol 241:515–545

Bennett MR, Pettigrew AG (1975) The formation of synapses in amphibian striated muscle during development. J Physiol 252:203–239

Bhatt DH, McLean DL, Hale ME, Fetcho JR (2007) Grading movement strength by changes in firing intensity versus recruitment of spinal interneurons. Neuron 53:91–102

Bleckmann H (2008) Peripheral and central processing of lateral line information. J Comp Physiol A 194:145–158

Brown MC, Jansen JK, Van Essen D (1976) Polyneuronal innervation of skeletal muscle in newborn rats and its elimination during maturation. J Physiol 261:387–422

Chagnaud BP, Bleckmann H, Engelmann J (2006) Neural responses of goldfish lateral line afferents to vortex motions. J Exp Biol 209:327–342

Chiba A, Shepherd D, Murphey RK (1988) Synaptic rearrangement during postembryonic development in the cricket. Science 240:901–905

Coombs S (1994) Nearfield detection of dipole sources by the goldfish (*Carassius auratus*) and the mottled sculpin (*Cottus bairdi*). J Exp Biol 190:109–129

Coombs S, Gorner P, Munz H (eds) (1989) The mechanosensory lateral line: neurobiology and evolution. Springer, New York

Coombs S, Mogdans J, Halstead M, Montgomery J (1998) Transformation of peripheral inputs by the first-order lateral line brainstem nucleus. J Comp Physiol A 182:606–626

Cope TC, Sokoloff AJ (1999) Orderly recruitment among motoneurons supplying different muscles. J Physiol Paris 93:81–85

Dijkgraaf S (1963) The functioning and significance of the lateral-line organs. Biol Rev Camb Philos Soc 38:51–105

Eatock RA, Xue J, Kalluri R (2008) Ion channels in mammalian vestibular afferents may set regularity of firing. J Exp Biol 211:1764–1774

Eaton RC, Bombardieri RA, Meyer DL (1977) The Mauthner-initiated startle response in teleost fish. J Exp Biol 66:65–81

Engelmann J, Hanke W, Bleckmann H (2002) Lateral line reception in still- and running water. J Comp Physiol A 188:513–526

Faber DS, Korn H (1975) Inputs from the posterior lateral line nerves upon the goldfish Mauthner cells. II. Evidence that the inhibitory components are mediated by interneurons of the recurrent collateral network. Brain Res 96:349–356

Faber DS, Fetcho JR, Korn H (1989) Neuronal networks underlying the escape response in goldfish. General implications for motor control. Ann NY Acad Sci 563:11–33

Fame RM, Brajon C, Ghysen A (2006) Second-order projection from the posterior lateral line in the early zebrafish brain. Neural Dev 1:4

Faucherre A, Pujol-Marti J, Kawakami K, Lopez-Schier H (2009) Afferent neurons of the zebrafish lateral line are strict selectors of hair-cell orientation. PLoS ONE 4:e4477

Fraser SE (1983) Fiber optic mapping of the Xenopus visual system: shift in the retinotectal projection during development. Dev Biol 95:505–511

Gaze RM, Keating MJ, Chung SH (1974) The evolution of the retinotectal map during development in Xenopus. Proc R Soc Lond B Biol Sci 185:301–330

Gompel N, Dambly-Chaudiere C, Ghysen A (2001) Neuronal differences prefigure somatotopy in the zebrafish lateral line. Development 128:387–393

Haehnel M, Taguchi M, Liao JC (2011) Heterogeneity and dynamics of lateral line afferent innervation during development in zebrafish (Danio rerio). J Comp Neurol 520:1376–1386

Hatta K, Korn H (1999) Tonic inhibition alternates in paired neurons that set direction of fish escape reaction. Proc Natl Acad Sci USA 96:12090–12095

Higashijima S, Mandel G, Fetcho JR (2004) Distribution of prospective glutamatergic, glycinergic, and GABAergic neurons in embryonic and larval zebrafish. J Comp Neurol 480:1–18

Hubel DH, Wiesel TN, LeVay S (1977) Plasticity of ocular dominance columns in monkey striate cortex. Philos Trans R Soc Lond B Biol Sci 278:377–409

Kinkhabwala A, Riley M, Koyama M, Monen J, Satou C, Kimura Y, Higashijima S, Fetcho J (2010) A structural and functional ground plan for neurons in the hindbrain of zebrafish. Proc Natl Acad Sci USA 108:1164–1169

Koyama M, Kinkhabwala A, Satou C, Higashijima S, Fetcho J (2010) Mapping a sensory-motor network onto a structural and functional ground plan in the hindbrain. Proc Natl Acad Sci USA 108:1170–1175

Ledent V (2002) Postembryonic development of the posterior lateral line in zebrafish. Development 129:597–604

Liao JC (2006) The role of the lateral line and vision on body kinematics and hydrodynamic preference of rainbow trout in turbulent flow. J Exp Biol 209:4077–4090

Liao JC (2010) Organization and physiology of posterior lateral line afferent neurons in larval zebrafish. Biol Lett 6:402–405

Liao JC, Haehnel M (2012) Physiology of afferent neurons in larval zebrafish provides a functional framework for lateral line somatotopy. J Neurophysiol 107:2615–2623

McCormick CA (1989) Central lateral line mechanosensory pathways in bony fish. Springer, New York

McHenry MJ, Strother JA, van Netten SM (2008) Mechanical filtering by the boundary layer and fluid-structure interaction in the superficial neuromast of the fish lateral line system. J Comp Physiol A 194:795–810

McHenry MJ, Feitl KE, Strother JA, Van Trump WJ (2009) Larval zebrafish rapidly sense the water flow of a predator's strike. Biol Lett 477–479

McLean DL, Fan J, Higashijima S, Hale ME, Fetcho JR (2007) A topographic map of recruitment in spinal cord. Nature 446:71–75

Mirjany M, Faber DS (2011) Characteristics of the anterior lateral line nerve input to the Mauthner cell. J Exp Biol 214:3368–3377

Montgomery J, Bodznick D, Halstead M (1996) Hindbrain signal processing in the lateral line system of the dwarf scorpionfish Scopeana papillosus. J Exp Biol 199:893–899

Montgomery J, Carton G, Voigt R, Baker C, Diebel C (2000) Sensory processing of water currents by fishes. Philos Trans R Soc Lond B Biol Sci 355:1325–1327

Montgomery JC, McDonald F, Baker CF, Carton AG, Ling N (2003) Sensory integration in the hydrodynamic world of rainbow trout. Proc Roy Soc Lond B 270(Suppl 2):S195–S197

Mueller T, Wullimann MF (2005) Atlas of early zebrafish brain development: a tool for molecular neurogenetics, 1st edn. Elsevier, Amsterdam

Nagiel A, Andor-Ardo D, Hudspeth AJ (2008) Specificity of afferent synapses onto plane-polarized hair cells in the posterior lateral line of the zebrafish. J Neurosci 28:8442–8453

New JG, Coombs S, McCormick CA, Oshel PE (1996) Cytoarchitecture of the medial octavolateralis nucleus in the goldfish, Carassius auratus. J Comp Neurol 366:534–546

Nicolson T, Rusch A, Friedrich RW, Granato M, Ruppersberg JP, Nusslein-Volhard C (1998) Genetic analysis of vertebrate sensory hair cell mechanosensation: the zebrafish circler mutants. Neuron 20:271–283

Plachta DT, Hanke W, Bleckmann H (2003) A hydrodynamic topographic map in the midbrain of goldfish *Carassius auratus*. J Exp Biol 206:3479–3486

Pujol-Marti J, Baudoin JP, Faucherre A, Kawakami K, Lopez-Schier H (2010) Progressive neurogenesis defines lateralis somatotopy. Dev Dyn 239:1919–1930

Pujol-Marti J, Zecca A, Baudoin JP, Faucherre A, Asakawa K, Kawakami K, Lopez-Schier H (2012) Neuronal birth order identifies a dimorphic sensorineural map. J Neurosci 32:2976–2987

Raible DW, Kruse GJ (2000) Organization of the lateral line system in embryonic zebrafish. J Comp Neurol 421:189–198

Sarrazin AF, Nunez VA, Sapede D, Tassin V, Dambly-Chaudiere C, Ghysen A (2010) Origin and early development of the posterior lateral line system in zebrafish. J Neurosci 30:8234–8244

Sato T, Takahoko M, Okamoto H (2006) HuC:Kaede, a useful tool to label neural morphologies in networks in vivo. Genesis 44:136–142

Sato A, Koshida S, Takeda H (2010) Single-cell analysis of somatotopic map formation in the zebrafish lateral line system. Dev Dyn 239:2058–2065

Schellart NA, Kroese AB (2002) Conduction velocity compensation for afferent fiber length in the trunk lateral line of the trout. J Comp Physiol A 188:561–576

Thompson WJ (1985) Activity and synapse elimination at the neuromuscular junction. Cell Mol Neurobiol 5:167–182

Trapani JG, Nicolson T (2011) Mechanism of spontaneous activity in afferent neurons of the zebrafish lateral-line organ. J Neurosci 31:1614–1623

Part IV
Biomechanics and Physiology of Flow Sensors

Chapter 14
Techniques for Studying Neuromast Function in Zebrafish

Primož Pirih, Gaston C. Sendin and Sietse M. van Netten

Abstract The mechano-sensitive hair cells of superficial neuromasts (SNs) of the zebrafish lateral line organ are mechanically coupled to the water motion via gelatinous cupulae. SNs transduce the water motion into electrical signals that can be measured with an extracellular electrode. In this chapter, we review the preparation and measurement techniques for quantifying cupular dynamics and extracellular receptor potentials (ERPs) of SNs. We compare the measuring techniques used in hair cell mechano-physiology and give instructions for building both an intensity-based and an interferometry-based microscope system. We compare the methods used for mechanical excitation of mechanoreceptors, including dipole sources, microfluid jets (FJ) and elastic as well as stiff microprobes. We present the caveats of the measurements of ERPs, especially the crosstalk from the stimulation device. We show that ERPs at twice the stimulation frequency of zebrafish SNs are a reliable measure of mechano-electrical coupling in a restricted range of both stimulus frequency and amplitude. We report the measurements of sub-micrometre motion of SN cupulae using a heterodyne laser interferometer microscope (HLIM) and continuous sinusoidal stimulation with a micro FJ device. Light interference signals were decoded with a phase- and frequency modulation scheme. We compare the robustness of both decoding strategies in terms of accuracy of the measured cupular displacement and velocity.

P. Pirih (✉)
Department of Materials and Metallurgy, Faculty of Natural Sciences and Engineering, University of Ljubljana, Aškerčeva 12 SI-1000, Ljubljana, Slovenia
e-mail: p.pirih@omm.ntf.uni-lj.si

P. Pirih
Department of Biology, Biotechnical Faculty, University of Ljubljana, Ljubljana, Slovenia

G. C. Sendin
Inserm U1051-Institute for Neurosciences of Montpellier, Montpellier, France

S. M. van Netten
Institute of Artificial Intelligence and Cognitive Engineering, University of Groningen, Groningen, The Netherlands

H. Bleckmann et al. (eds.), *Flow Sensing in Air and Water*,
DOI: 10.1007/978-3-642-41446-6_14, © Springer-Verlag Berlin Heidelberg 2014

Both approaches can faithfully monitor cupular movement down to a few nano-
metres, though the velocity decoding technique offered a slightly superior performance and is recommended for higher stimulation frequencies.

Abbreviations

CCD	Charge-coupled device
CMOS	Complementary metal oxide semiconductor
CN	Canal neuromast
DIC	Differential interference contrast
ERP	Extracellular receptor potential
FFT	Fast Fourier transform
FJ	Fluid jet (device)
HLIM	Heterodyning laser interferometer microscopy/microscope
LED	Light emitting diode
MIPO	Microphonic potential
PD	Photodiode
PMT	Photomultiplier tube
PSD	Position sensitive device
RMS	Root mean square
SN	Superficial neuromast
SNR	Signal-to-noise ratio

14.1 Introduction

Aquatic vertebrates (lampreys, cartilaginous and bony fish, amphibia) use a mechanosensory lateral line organ to locate nearby moving underwater sources. This
sense was described as *Ferntastsinn* (remote touch, touch at distance) by Dijkgraaf
(1963). In fish, the lateral line organ consists of an array of neuromasts distributed
along the trunk and the head. Distance touch is important in prey detection,
predator avoidance and schooling behavior and is a very important sense in dark
and turbid environments.

There are two types of neuromasts: superficial neuromasts (SNs) are located
directly on the skin while canal neuromast (CNs) are embedded in subcutaneous
canals. A neuromast consists of tens to hundreds of sensory hair cells which have
their ciliary bundles embedded in a gelatinous cupula. The cupula is driven by the
local fluid flow, either outside the fish (SNs) or inside the canal (CNs). Cupular
motion is transferred to mechanosensitive bundles of hair cells, which transduce
motion into electrical signals (Howard and Hudspeth 1988). Adult teleost fish
usually have both SNs and CNs while fish in the larval stage have only SNs (Metcalfe et al. 1985). Amphibian lateral line contains only SNs. Lampreys have lateral
line receptors which are functional but do not have a cupula (Gelman et al. 2007).

The mechanosensory lateral line has come into focus for its potential to create biomimetic sensor arrays for near-field detection (Yang et al. 2006; Ćurčić-Blake and van Netten 2006). From the 1990s on, zebrafish (*Danio rerio*) has been established as a vertebrate genetic model. The lateral line of zebrafish became a general model to study lateral line detection (McHenry et al. 2009), neural development and cell targeting (Lopéz-Schier and Hudspeth 2005) and hearing disorders. To that end, the measurement of extracellular receptor potentials (ERPs) made it possible to confirm results in behavioral screening tests indicating hair cell dysfunction in a number of mutant strains (Nicolson et al. 1998). Recently, electrophysiological recordings from afferent neurons have been linked to motor neuron responses (Liao and Haehnel 2012).

Studies of zebrafish SN mechanics have revealed that the gelatinous upper part of the cupula is very compliant, while the lower part of the cupula is dominated by the stiffness of the cilia (McHenry and van Netten 2007). A two-beam mechanical model (McHenry et al. 2008) predicts that the mechanical tuning of zebrafish SNs is that of a band-pass filter, when the displacement of cilia is referred to a free-stream velocity. This is due to a combined effect of transfer functions of the cupula (low-pass filter) and the boundary layer (high-pass filter). On the other hand, CNs, which are generally bigger and have a stiffer cupula than the SNs, effectively detect pressure gradients produced by underwater sources (Denton and Gray 1983; van Netten 2006). A detailed comparison of SNs and CNs mechanosensitivity has been reported recently (van Netten and McHenry 2013).

The tuning characteristics of CNs have been reported in terms of mechanical, ERP and afferent neuron spiking tuning curves for several fish species (ruffe, African knife fish, trout, goldfish), while data on SNs are scarce. A mechanical tuning curve of the large SNs (400 µm height) of a pædomorphic salamander, the mudpuppy *Necturus maculosus,* has been measured by means of a stroboscopic method and complemented with a mechanical model (Liff and Shamres 1972). The tuning of amphibian SNs has been characterised electrophysiologically by ERP recordings and analysis of spike trains of afferent nerve fibres (Kroese et al. 1978; Kroese and Schellart 1992). Mechanical tuning curves of zebrafish SNs have been measured (Dinklo 2005) and will be expanded into a more complete study, linking mechanical tuning curve, ERP tuning curve and a cupular mechanical model (Sendin et al. in preparation).

This chapter is written both as a review of existing techniques (visualisation, mechanical stimulation, optical measurements) and as a collection of experimental protocols (fish husbandry, experimental chamber design, analysis of tuning curves). We compare light intensity-based methods to laser-interferometric methods for optical measurements of SNs mechanics. We briefly review the methods for mechanical stimulation [flexible and stiff probe, stimulus sphere, fluid jet (FJ)]. We give hints for adjusting standard microscopy equipment to visualise SNs. We describe the details of recording and analysis techniques used in measuring tuning curves with sinusoidal stimulation. We illustrate the heterodyning laser interferometric microscopy (HLIM) and ERP techniques with measurement examples. This chapter complements a recently published article focusing on measurements of zebrafish ERPs and action currents from afferent neurons (Trapani and Nicolson 2010).

14.2 Fish Husbandry

14.2.1 Starting the Husbandry

First, we give a brief recount on our experience with handling a zebrafish stock. We generally followed instructions from the practical zebrafish manual (Nusslein-Volhard and Dahm 2002). The zebrafish culture was started with bleached embryos (AB and TL strains) ordered from a stock centre (e.g. ZIRC, University of Oregon, USA) or obtained from another academic institution. The alternative approach of buying adults from a pet shop is less recommended due to adaptability problems in the new environment or infections, which could result in lesser or nil fecundity. A school of about 50 adults was kept in a 50-L tank with environmental enhancements (live and plastic plants, pebbles). We used smaller 4-L tanks for breeding. Tanks had to be covered with a lid in order to prevent evaporation of water and jumping of fish. Zebrafish were kept on a 14 h: 10 h light-dark cycle at 27 °C. The temperature was achieved by controlling the culture room temperature with an electric radiator heater coupled to a thermostat. Water in all the tanks was aerated with air diffusors.

Water in the big tank was recycled through a series of external mechanical and biological filters (Eheim, Germany). Food debris was cleaned with a small battery-operated mechanical cleaner. Water conductivity and pH were monitored with a commercial meter (Hanna Instruments, Germany). Target conductivity and acidity were 200–300 μS/m and pH 6.5–8.5, respectively. A more extensive assessment of water quality including total hardness, carbonate hardness, NO_3 and NO_2 concentrations was performed with colorimetric dip strips (Merck, Germany). Once a week, or if the pH got too low, up to three quarters of the tank water were exchanged. We prepared the water by adding 0.1 g/L of sea salt (e.g. Hobby Marine, Germany) to reverse osmosis (RO) water. The pH was set to 7.5 by titration with drops of HCl (1 mol/L) or NaOH (1 mol/L). Water was then inoculated with about one tenth volume of used tank water and aerated for at least 1 day before being added to the tanks.

Fish tanks were monitored on a daily basis for general hygiene and fish behavior. Lethargy, separation of a fish from the school or swimming close to the surface can be signs of e.g. physical injury or fungal infections. We quarantined the fish exhibiting such behavior to an isolation tank. If they did not recover in a week, they were decapitated and disposed of.

14.2.2 Feeding Adult Fish

Adult fish were fed twice a day. Fry food of bigger size (ZM200 and ZM400; ZM Systems, UK) was the easy solution but zebrafish seemed to enjoy more live brine shrimp larva. The additional advantage of using brine shrimp was that usually less food debris was left in the tank. Brine shrimps were prepared by putting a small amount of cysts (Silver Star Artemia, Salt Lake, USA) into a cone-shaped tank containing bubbled salty water (25 g/L) that was prepared by diluting a stock solution (100 g/L). After 24 h, bubbling was stopped and the tank top was covered with a screen. After a few minutes, empty eggshells settled on the bottom while live shrimps aggregated above the sediment. Upon opening of the valve at the bottom of the tank, the sediment fraction was discarded. The fraction with live shrimps was collected onto a sieve with a very fine mesh, rinsed several times with RO water and put into a glass. The harvest may be kept in the fridge for up to 2 days, but one should keep in mind that larvae do not survive for very long in the hypo-osmotic RO water. It is a good practice to repeat the harvest every 2–3 days.

14.2.3 Spawning, Egg Harvesting and Breeding of Larvae

In order to get a new batch of zebrafish larvae, several parallel batches of four fish, (two females and two males; females have more pronounced bellies while males keep a torpedo-like shape) were transferred into aerated 4-L tanks with fake plants and a net preventing the adults from reaching the bottom and eating the eggs. Spawning took place when the lights went on. If a batch of fish did not spawn after being in the tank for 2 days, we returned it to the big tank and took a new batch. If the selection of breeder fish was productive, it was kept together in the small tank for longer periods. Our observation was that some females were able to produce eggs every day even when they were kept in the smaller tanks for weeks. We note that according to video instructions on large-scale zebrafish breeding (available on Internet), egg deposition may be induced by reducing swimming water depth, e.g. by raising the net or a perforated barrier in the water tank.

The eggs were harvested with a plastic Pasteur pipette and transferred into a Petri dish containing E3 medium. The first medium was lightly coloured with a small drop of Methylene blue. Healthy eggs were selected on the basis of their exclusion of the blue dye and then kept in clean E3 until hatching, which generally occurs after 24–48 h incubation. Newly hatched zebrafish initially do not need to be fed as they use their yolk sac reserves. From 5 days post-fertilisation (P5), we fed the larvae with granulated fry food (ZM000, ZM Systems, UK) up to P15, and further with adult food.

14.3 Experimental Preparation

14.3.1 Bath Solutions

Zebrafish larvae were transferred with a broad tip plastic Pasteur pipette to another Petri dish filled with Modified Barth's solution (MBS) (Dinklo 2005) or 10 % Rüsch solution (RS) (adapted from Nicolson et al. 1998). The lateral line electrophysiology of zebrafish larvae was successfully recorded before in full extracellular solution (RS) with osmolarity 260 mOsm/L (Nicolson et al. 1998), while the larvae normally develop in E3 (20 mOsm/L). Given the osmolarity range, it seems that the selection of bath fluid is not critical for viability of whole-larva preparation and that 5–120 mmol/L of sodium, 0.2–2 mmol/L of potassium and up to 2 mmol/L of calcium seems to be adequate. If experiments are to be performed on isolated tails, RS seems to be the natural choice. We note that extremely low calcium concentration would probably destroy tip links of hair cells (Zhao et al. 1996).

E3   NaCl 5 mM, KCl 0.17 mM, $CaCl_2$ 0.33 mM, $MgSO_4$ 0.33 mM;
RS   NaCl 120 mM, KCl 2 mM, $CaCl_2$ 2 mM, HEPES 10 mM; pH 7.3;
MBS   NaCl 8.8 mM, KCl 1 mM, $NaHCO_3$ 2.4 mM, $MgSO_4$ 0.82 mM, $Ca(NO_3)_2$ 0.33 mM, $CaCl_2$ 0.41 mM, HEPES 1 mM; pH 7.4.

14.3.2 Anaesthesia

We used the standard fish anaesthetic tricaine methanesulfonate (tricaine ethyl 3-aminobenzoate methanesulfonic acid, also called MS-222 or MESAB, from Sigma-Aldrich). The stock solution of MS-222 (4 mg/mL) was buffered with sodium bicarbonate to pH 7–8. We used MS-222 diluted to a final concentration of 0.16 mg/mL (Nusslein-Volhard and Dahm 2002; McHenry and van Netten 2007). The state of anaesthesia was monitored under the dissecting microscope by evaluating the response of the fish to soft tactile stimulation with an eyelash, and by monitoring the rate of opercular and fin movements.

We must emphasise that MS-222 unfortunately works as a partial blocker of mechanotransduction in a dose-dependent manner (Palmer and Mensinger 2004). If the experiments require an unhampered electrophysiological response of the lateral line, the block should be removed. MS-222 can be used in the first stage of the preparation, which is then followed by either injecting alpha-bungarotoxin into the heart of the larva (Nicolson et al. 1998; Trapani and Nicolson 2010) or by adding tetrodotoxin (100–200 nM, Nicolson et al. 1998) to the bath to cause muscle paralysis, followed by a dose reduction or complete washout of MS-222. Given the complexity of this procedure, alternative means of anaesthesia for mechanosensory research would be most welcome.

Neurosteroid anaesthetics seem to be a viable alternative to MS-222 in lateral line research as they do not have the blocking effect on the mechanosensory and electrosensory receptors (Peters et al. 2001). Neurosteroid alphaxalone is an agonist of $GABA_A$ (gamma aminobutyric acid) chloride channels and prevents nerve impulse conduction through hyperpolarisation, causing reversible central anaesthesia in mammals (Muir et al. 2008). The original mix of alphaxalone and alphadolone (Saffan) was administered either intraperitoneally in bigger fish (Peters et al. 2001; Ćurčić-Blake and van Netten 2006) or through bath water in smaller fish, e.g. in the elephantnose fish *Gnathosthomus petersii*, using induction concentration 2 mg/L and the maintenance concentration 1.5 mg/L (Peters and Dénizot 2004). While Saffan was discontinued because its solubilising agent was causing allergic reactions in pets, a reformulation of alphaxalone with cyclodextrane (Alfaxan-CD RTU) is being routinely used as an intravenous anaesthetic for cats and dogs in some countries (e.g. Australia, United Kingdom, France, Germany). In fish, it has already been used through bath-water administration (5 mg/L) to achieve operational anaesthesia of trout cods *Maccullochella macquariensis* (Ebner et al. 2007) and goldfish *Carassius auratus* (O'Hagan 2006) but, to our knowledge, it has not been used in zebrafish experiments yet.

A potentially useful anaesthetic being routinely used on aquarium fish is clove oil. Its active component, eugenol, has been shown to block sodium voltage-gated channels and to activate TRPV1 channels in rats (Park et al. 2009). A comparative pharmacological study on juvenile zebrafish has shown that eugenol has a higher safety margin than MS-222, with the induction dose for full anaesthesia at 60–100 ppm (Grush et al. 2004). To our knowledge, the effect of eugenol on the lateral line mechanosensitivity has not been tested yet. Another fish anaesthetic is 2-phenoxy ethanol (Mylonas et al. 2005). It has already been used in auditory and lateral line experiments on the goldfish (dose 2 µmol/L; Higgs and Radford 2013), but not on the zebrafish.

14.4 Experimental Chamber

The experimental chamber should be designed having in mind mechanical stability, accessibility for pipettes and visibility of the preparation. Another point to consider is that the trunk neuromasts that we investigated were oriented along the longitudinal axis of the fish and, therefore, we aligned the fluid jet stream accordingly. In the case of head neuromasts, a different orientation may be optimal for recording of ERPs.

We immobilised the larvae between slabs of cured Sylgard 184 (Dow Corning) of different thicknesses. The advantage of Sylgard is that it is optically clear and therefore allows good visualisation in the transillumination mode. The experimental chamber (Fig. 14.1) was built on a plastic tissue slide (75×25 mm^2, thickness 2 mm) with a centre hole into which a round coverslip window (Ø 22 mm, Chance Propper) was glued with silicone glue (Bison). We placed a

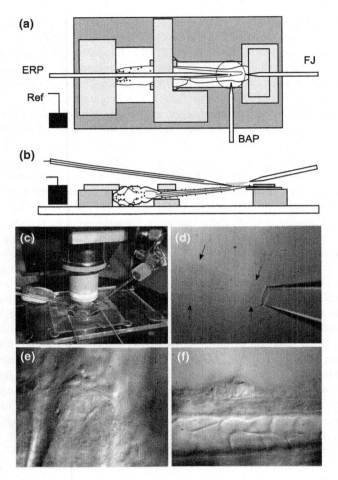

Fig. 14.1 Design of the experimental chamber used for the recording of extracellular receptor potentials (*ERP*) and mechanical measurements of cupular motion in superficial neuromasts (*SN*) of the zebrafish lateral line organ. **a** *Top view*. The fish is placed in a niche (*white area*) carved out from a 1 mm thick Sylgard bed (*dark grey*). Additional thin slabs of cured Sylgard were used in order to mechanically stabilize the fish body (*light grey*). FJ stands for the fluid jet pipette. Beads were gently delivered to the cupula by a gravity-fed thick pipette (*BAP*). Once a few beads were stably attached to the cupula, the pipette was withdrawn. The ERP pipette was placed at an angle of about 20° with respect to the fish body surface (*ERP*) and along its longitudinal axis. A reference electrode (*Ref*) was put into the side of the bath. **b** *Side view* of the experimental chamber with the larva lying on its right side, showing the angle of both pipettes (*ERP, FJ*). The water flow exciting a SN is shown in *light grey*. **c** *Close-up* of the chamber showing the 40x water objective, the reference and recording pipettes (*left*) and the fluid jet pipette (*right*). **d** Photograph showing the fluid jet pipette (diameter 10 μm) and the water stream coming out of its tip. Note that the water flow shape (*black arrows*) has a divergence similar to that of the pipette tip. **e** *Top view* of a SN from the tail of a juvenile zebrafish larva obtained at a deep focal plane using the 40 × 0.80NA objective and Nomarski optics. The picture shows the *rosetta* of hair cell bodies (post-natal day 30). **f** *Side view* of the SN in the most caudal portion of a zebrafish body, showing the kinocilia (post-natal day P5)

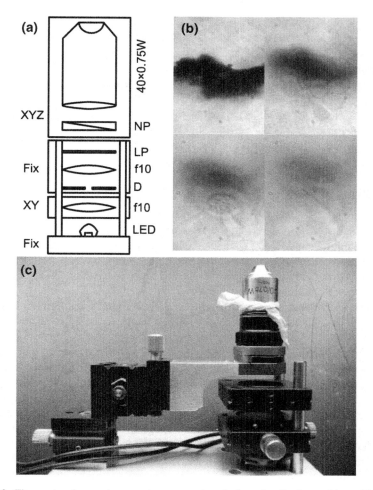

Fig. 14.2 The water immersion condenser employed for visualisation without Nomarski contrast. **a** Schematic representation. The condenser was built around a 40 W × 0.75NA objective. The light from a LED was focused with an f10 lens onto a pinhole (D), which acted as a secondary light source. The beam was collimated with a second f10 lens and sent through a linear polarizer (*LP*) and optionally through a Nomarski prism (*NP*). The pinhole and the objective were mounted on translator stages. **b** Pictures of a SN from a juvenile zebrafish (post-natal day 30) at different optical planes, taken with a Leitz 25 W × 0.50NA objective with crossed polarizers but without the Nomarski prism. *Above left* Deep focal plane showing the hair cell bodies. *Below right* Focusing on the hair bundles. **c** *Side view* picture of the condenser with the 40 W × 0.75NA objective and the optical elements described above

10 × 10 mm slab of 1 mm thick Sylgard to the coverslip and carved out a piston-shaped opening. A slab of 0.5 mm thick Sylgard was inserted in between, into the central part of the opening. The head of the larva was placed into the broader part of the opening and the trunk was lying over the centre slab. A rectangular slab was

placed over the rostrum, gently pressing it down. The trunk of the larva, caudally from the gas bubble, was sandwiched between the centre slab and an L-shaped slab of the same thickness. The caudal fin was sandwiched between two thin (<0.2 mm) slabs. In order to increase the sticking of the slabs to the coverslip and among each other, we used minute amounts of Vaseline. This way, the larva was well immobilized and lying on its side, with a large part of the lateral line lying directly on the coverslip and accessible to pipettes. The blood flow was usually maintained for at least 3 h. We note that the mechanical stability of older larvae was compromised by peristaltic movements of the gut.

14.5 Visualisation of Neuromasts

Physiological studies of neuromasts are usually conducted in a setup built around an upright microscope with a water immersion objective. The larvae can be put on their side, so the neuromasts are facing the objective, or on their belly (as used in inner ear microphonic recordings, see for instance Corey et al. 2004), so that the neuromasts side is facing the objective.

In our experiments, we used the side preparation and transillumination. Since neuromasts are objects with a low intensity contrast, better visualisation was achieved with differential interference contrast (DIC) microscopy. For this, the condenser and the objective need to be equipped with linear polarizers and Nomarski-modified Wollaston prisms. For visualisation in electrophysiological experiments, we used a Zeiss Achroplan 40 × 0.80NA/∞ objective with working distance 1.6 mm, and an air condenser (0.90NA). Due to scattering, the quality of DIC deteriorates with the thickness of the underlying substratum and fish tissue, which limits its use to the tail neuromasts and relatively small larvae (<P10). We found that a combination of Leitz 25 × 0.50NA/160 water objective and a custom built water immersion condenser with a Zeiss 40 × 0.75NA/160 objective enabled us to visualise the neuromasts in transillumination without DIC, which in turn allowed experiments in thicker larvae up to day P30 (Fig. 14.2). Contrast was optimised by adjusting rotatable polarizers.

14.6 Optical Measurements of Neuromast Mechanics

Optical methods for measuring mechanical tuning curves on microscopic objects in the nano- and sub-micrometre range are customarily based on a compound microscope that is mechanically stabilised on a vibration isolation table. There is a fair amount of additional electronics (amplifiers, filters) and custom optics needed. Although off-the-shelf equipment with the required resolution is steadily becoming available, considerable time and experience is necessary for integrating the components into a working rig. An example of the microscope modification is presented in Figs. 14.3a and 14.4a, b.

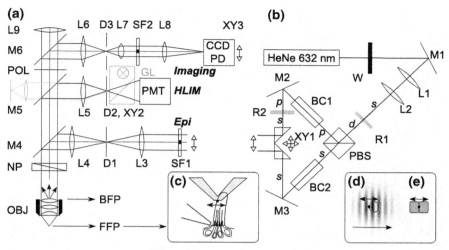

Fig. 14.3 Scheme of the epi-illumination microscope used for intensity-based and interferometry-based measurements of neuromast mechanics. **a** The microscope. The epi-illumination pathway contains a spatial filter (*SF1*) that is conjugate with the back-focal plane of the objective (Obj, Zeiss 40 W × 0.80NA, f4), two telescopic lenses (*L3, L4*) and a field diaphragm [Linos photonics (Qioptiq) microbench]. The spatial filter *SF1* was not needed for HLIM measurements as the two laser beams were entering the side of the objective back lens. By slightly offsetting *L3*, the crossing plane of the laser beams could be adjusted. A half mirror (*M4*) delivers the light down to the main pathway. Nomarski-modified Wollaston prism (*NP*) was used for Nomarski DIC imaging together with a standard Nomarski transillumination air condenser (*not shown*). For HLIM, the back-scattered light from a bead is imaged via mirror *M5* and through a lens *L5* onto a translatable pinhole *D2/XY2* and interfered on the photomultiplier tube (*PMT*). The pinhole position was adjusted with the help of the guiding light (*GL*) and a back-reflecting lens/mirror system (*greyed*). For intensity-based measurements, the image of the bead is reflected via mirror M6 and imaged by the lens *L6* (f80) onto a diaphragm D3. The image is relayed with the lenses *L7* (f20) and *L8* (f80) onto a CCD camera or a differential photodiode (*PD*). With the described lenses, the lateral magnification of the lens system is (80/4) × (60/20) = 60. The calibration of the PD is achieved by moving the stage with the translator *XY3*. The position of the spatial filter *SF2* can be imaged on the camera by exchanging *L8* for f20. *L9* is the standard tube lens of the microscope or an Optovar lens. **b** The heterodyning laser interferometer source. A laser beam generated by a He–Ne laser (632 nm) is expanded with a telescope pair *L1&L2*, its polarisation plane rotated for diagonally with a half-lambda plate R1 and split using a polarising beam-splitter (*PBS*). The two beams are directed through Bragg cells (*BC1* and *BC2*), driven at 40 and 40.4 MHz, respectively. Letters *p, s, d* denote parallel, normal and diagonal polarisation, respectively, using the optical table as the reference plane. The polarisation of the *p* beam is rotated to *s* via the second half-lambda plate *R2*. Rotatable mirrors *M2* and *M3* are used to parallelise the two beams which are then reflected by two reflected prisms that are sitting on a translator *XY1*. By adjusting the translator, the distance and position of the two beams is controlled, eventually determining the angles of the beam incidence on the neuromast and thus the calibration of the heterodyning signal. **c** Scheme of a SN and two laser beams (*shown in grey*) converging on a small bead attached on the cupula. Their angles are determined by the translator *XY1*. *The arrow* indicates the direction of the cupular displacements. **d** A scheme of the fringe pattern with the silhouette of a cupula seen from above (*black open circle*) and an attached bead (*small full black circle*). **e** A light-scattering object (a bead) imaged onto a pair of differential photodiodes, a similar experimental arrangement as previously used in the measurements of hair bundle displacements (Crawford and Fettiplace 1985)

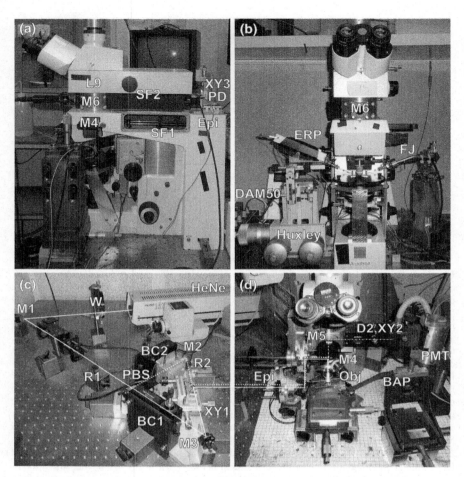

Fig. 14.4 Photos of the two experimental setups. **a** *Side view* of the Axiotron microscope, showing the epi-illumination pathway (*Epi*) and the position of its spatial filter (*SF1*). The image detection pathway contains the second spatial filter *SF2* and a photodiode sensor (*PD*). The tube lens (L9) in this case was an Optovar. **b** Front view of the microscope showing the Huxley micromanipulator being used to position the ERP pipette in the chamber. The fluid jet pipette (*FJ*) is on the right. Its position was controlled by a custom-made translator. **c** Picture of the HLIM illuminator, showing the He–Ne laser, the beam-splitter (*PBS*), both Bragg cells (*BC1* and *BC2*) and additional translatable elements necessary to direct the beams (*R1, R2* and *XY1*). **d** Front view of the HLIM microscope showing the manipulator with the bead applying pipette (*BAP*), the detection pathway with the translatable pinhole (*D2, XY2*) and the photomultiplier (*PMT*)

Methods for optical detection of movement can be broadly classified into light intensity-based methods and interference-based methods. In all cases, photonic detectors [photodiodes (PDs), photomultiplier tubes (PMTs) and CCD/CMOS cameras] are essentially detecting changes in light intensity falling upon them. When choosing the detector, one should consider quantum efficiency and the detector settling time, which will determine the detection gain-bandwidth product and consequently the measurement frequency range. Due to the stochastic nature of light, the accuracy of measurement benefits both from an increase of the signal and from a reduction of background. In the case a laser is used for the measurement, adding a sharp wavelength band-pass filter in the detection pathway may significantly improve the signal-to-noise ratio (SNR). Averaging of sweeps with repeated stimulation may be necessary in order to further increase the measurement accuracy. The implementation of averaging will be relatively unproblematic in the case a photonic detector is being used with an acquisition unit with AD-DA synchronisation and continuous recording capability for both the stimulation and the measurement. Averaging may become cumbersome with imaging sensors which usually have their own clocking.

We briefly review intensity-based methods and describe in more detail an interference-based method (heterodyning interferometry, HLIM). We note that intensity-based methods are capable of measuring displacement, while the interference-based method, essentially employing the Doppler shift effect, is capable of directly measuring both displacement and velocity, which makes it particularly suitable for measurements at higher mechanical stimulation frequencies. The details of the methods are summarised in Table 14.1.

14.6.1 Intensity-Based Methods

14.6.1.1 Differential Photodiode Measurements

Intensity-based methods detect the displacement of the object. Commonly, differential silicon PDs (two and four-segment PDs) are used as detectors. The moving object casts a shadow or a bright spot over the PD segments (Fig. 14.3e). The photocurrents are amplified and converted to voltages. In the first stage of processing, the difference signal is normalised to the sum signal to correct for the drift in light intensity. A two-segment PD produces a displacement signal in one dimension, while a four-segment PD may be wired to produce a two-dimensional signal. An important point to keep in mind is that absolute calibration in terms of object displacement versus signal output is needed. Calibration is normally done by moving the detector head with piezo actuators, taking into account the microscope system magnification. One should keep in mind that the calibration curve obtained with this method may be non-linear and that the calibrated range will be limited to the condition when the moving image is actually falling on both sensors.

In some of our experiments, we used a position-sensitive detector (PSD, Ontrak Photonics OT-301 with PSM 2–10 module). PSD is a special type of a PD sensor, a monolithic PD with two anodes and two cathodes. The amplifier outputs the coordinates of the beam falling on the detector in terms of the centre of the mass. This detector type would ideally be giving a linear signal over a larger

Table 14.1 Comparison of techniques for mechanical stimulation and optical measurements of hair cell and neuromast mechanics

Author	Liff and Shamres	Crawford and Fetiplace	Van Netten	Denk and Webb	Geleoc et al.	McHenry and van Netten	Dinklo et al.	Sendin et al.
Year	1972	1985	1988	1989	1998	2007	2007	In prep.
Experiment	Mechanical tuning curve	Mechano-electrical transduction	Mechanical tuning curve	Hair bundle Brownian motion	Mechano-electrical transduction	Static flexural stiffness of SN cupula	Calibration of fluid jet device	Mechanical and ERP tuning curve
Object	Mudpuppy SN	Turtle HC	Ruffe CN	Frog sacculus HC	Mouse OHC & VHC	Zebrafish SN	Flexible fibre with a glue ball	Zebrafish SN
Measured mechanical quantity	Displacement	Displacement, force	Displacement, velocity	Displacement	Displacement	Displacement, force	Displacement, velocity	Displacement, velocity
Measurement method	Stroboscopic	Intensity	Interferometry	Intensity	Intensity	Imaging	(1) Interferometry, (2) Intensity	Interferometry
Measurement contrast	Ink particles	Glass fibre	Polystyrene beads (1 μm) or none	DIC	DIC	Polystyrene beads (0.1 μm) and glass fibre	glue ball	Polystyrene beads (1 μm)
Detector	Eye	Differential PD	PMT, single PD	Quadrant PD	Differential PD	CCD camera	(1) PMT, (2) PSD	PMT
Calibration	Ruling	Object stage scan	Intrinsic: beam angle	Object stage oscillations	Combination of laser and detector scan	Intrinsic: pixel size/optical magnification	(1) Intrinsic: beam angle, (2) detector head	Intrinsic: beam angle
Stimulus device	Stimulus ball	Glass fibre	Stimulus ball	None (thermal motion)	Fluid jet	Glass fibre	Stimulus ball, fluid jet	Fluid jet
Stimulus type	Sinusoidal	Step	Sinusoidal	None	Step	Ramp	Sinusoidal	Sinusoidal
Visualisation contrast	?	DIC	XPOL	DIC	DIC	DIC	DIC	XPOL
Measurement illumination	? Epi	Trans	Epi	Trans	Trans	Trans	Trans	Epi
Visualisation illumination	? Epi	Trans	Epi	Trans	Trans	Trans	Trans	Trans
Measurement source	Stroboscope	Incandescent	He–Ne Laser	He–Ne Laser	Diode laser	Incandescent	He–Ne Laser	He–Ne Laser
Visualisation source	?	Incandescent	Xenon	Incandescent	Monochrome	Incandescent	Monochrome	White LED
Main objective	?	W40 × 0.75NA	W40 × 0.75NA	W40 × 0.75NA	W40 × 0.75NA	W40 × 0.80NA	W40 × 0.80NA	W25 × 0.50NA W40 × 0.80NA
Transillumination objective	?	Air condenser	Not used	W40 × 0.75NA	Air condenser	Air condenser	Air condenser	W40 × 0.75NA

? not reported in the original work

displacement range than the differential detector. However, we experienced that the PSD is quite sensitive to stray light and has a somewhat low gain-bandwidth product, limiting its use at higher frequencies. If the detection system is used with epi-illumination, there may be unwanted sources of stray light arising from the objective lens reflections. These may be blocked by placing spatial filters in the conjugate planes of the illumination or detection pathways (Fig. 14.3a).

14.6.1.2 Imaging Sensors

Contemporary area-scan and line-scan cameras are an attractive choice of an imaging detector, especially because the absolute calibration will be given directly by the magnification factor of the imaging system. With 10 μm camera pixels and an overall magnification of 400×, a camera pixel will correspond to 25 nm. We note that the accuracy of tracking a visible particle of a known size (e.g. a reflecting bead with diameter 1 μm) when used with appropriate image detection and analysis method, is not limited by the Rayleigh criterion; the accuracy may well be in the 10–100 nm range.

As cameras use internal digitisers, synchronisation with the stimulus may be problematic. Sampling rate, minimal exposure time, external synchronisation capabilities and temporal jitter performance of an imaging sensor should be carefully considered when choosing the detector. In order to measure the mechanical response to an oscillatory (e.g. sinusoidal) stimulus both in terms of amplitude and phase, including several harmonics, it is practical to acquire at least 16 points per cycle. In the case of a fast area-scan camera with 400 frames per second, the highest stimulation frequency thus becomes meagre 25 Hz. Line-scan cameras normally provide scan rates in the range 1–100 kHz, so the limitation is way less severe. In order to overcome the sampling rate limitation, one could use a phase-locked stroboscope method (Liff and Shamres 1972). There, at each stimulation frequency, the phase delay between the stimulus period and the flash is adjusted in a number (e.g. 16) of steps. A gated super luminescent LED seems to be the contemporary illumination technology of choice for this purpose. An additional advantage of using gated illumination is that there will be less motion blur if short, intense light pulses are used.

14.6.1.3 Nomarski DIC

The attractiveness of using DIC is that it does not require any tissue contrasting. This imaging method uses microscope transillumination mode and is suitable for thin objects. The Nomarski-modified Wollaston prism in the illuminator separates the focal points of the two polarisation planes. The light of the two polarisation planes travels slightly different optical paths through the tissue and is then collected by the objective and combined back with the second prism. Due to light interference, objects appear as white and black shadows upon a grey background.

We emphasise that Nomarski DIC contrast will be visible only when there is little light-scattering tissue in the background of the object. In our experience with the SNs, stereocilia (or correctly, *stereovilli*) are visible with enough contrast while the cupula is not, so it should be additionally contrasted, e.g. with small diameter (0.1 μm) styrene beads that form a visible coating of the cupula (McHenry and van Netten 2007).

In a study of fluid jet stimulation of cultured mouse hair cells (Géléoc et al. 1997), laser light was used as the transillumination source and a differential PD as the detector. The ciliary bundles were oriented along right angles of the objective axis. Detector calibration was implemented by moving the laser and the detector head in concert. The method with the highest reported accuracy of a measured displacement spectrum (1 pm/$\sqrt{Hz}$, using an averaging interval of several minutes) was used in a study of thermal motion of stereovilli of frog sacculus hair cells (Denk and Webb 1992). The setup was built from an inverse and an upright microscope, with a He–Ne fibre laser as the source. The tallest stereovillus in the hair bundle was aligned with the focal plane, between the two focal points. Displacement of the stereocilium was detected with a quadrant PD. A continuous calibration signal was provided by piezo actuators oscillating the stage with the preparation.

14.6.1.4 Stimulation and Contrasting with a Fibre Probe

Displacement and stiffness of an object can be the detected through the bending of a glass or carbon fibre probe. In a study of turtle hair cells, the image of a fibre was projected to a differential PD (Crawford and Fettiplace 1985). The microscope operated in transillumination mode with a halogen lamp light source. Calibration of the PD signal was performed by physically displacing the sensor head with piezo actuators, taking into account the magnification of the system. The disadvantage of such a method is that the measured object (e.g. the cupula) is mechanically constrained with the fibre. The accuracy of this method is not directly reported, but we may estimate the noise level in the order of nm, within the bandwidth of 0–100 Hz (Martin and Hudspeth 1999).

In zebrafish, a combination of a (carbon) fibre probe and a camera system was used in a study of static stiffness of SN cupulae. The stiffness of carbon fibres was calibrated with micro weights. The cupulae were visualised by deposition of small (0.1 μm) styrene beads. A digital camera was used as the imaging device (McHenry and van Netten 2007).

14.6.1.5 Contrasting with Styrene Beads

Micrometre-sized polystyrene beads are convenient contrasting agents because they have a high refractive index (~ 1.6) which renders them visible in the water medium and because they stick to the cupulae of SNs. They are superior to vital dyes because they do not penetrate the cupula. In a study of flexural stiffness of SNs (McHenry and van Netten 2007), styrene beads with 0.1 µm diameter, applied over the whole surface of the neuromast have provided an outline image of the cupula under DIC contrast. This contrasting approach could be very convenient in combination with an area-imaging detector. On the other hand, single styrene beads of 1 µm diameter are visible with or without DIC contrast and can be used with any differential or imaging detector, or with the interferometric method.

14.6.2 Heterodyning Laser Interferometry

This method is employing the laser Doppler effect and is essentially capable of measuring both displacement and velocity of an object. A beating fringe pattern is produced by modulating two laser half-beams with Bragg cells. The two beams cross in the focal plane of the objective and produce a moving fringe pattern. The light reflected from the moving object is detected by a single PMT or by a fast PD (van Netten 1988; Ćurčić-Blake 2006).

As opposed to the intensity-based methods, the information on the position of the reflecting object in the fringe pattern is contained in the phase of the detector signal and not in its intensity (Fig. 14.3d). The phase information is extracted from the beating signal with an FM demodulator, which provides a signal proportional to the velocity of a moving object and is, therefore, most accurate at higher oscillation frequencies. For the measurement of displacement, suitable for DC and low frequency displacements, phase demodulation based on fringe counting can be used. An important advantage of this method is that it is intrinsically calibrated, i.e. the conversion factor depends solely on the angle of the two beams (Fig. 14.3c) and on the angle between the object movement and the beam plane. The method is to a large extent insensitive to the fluctuations in light intensity (van Netten 1988; Ćurčić-Blake 2006). The accuracy of this method is in the range of 100 pm/$\sqrt{\text{Hz}}$.

In our measurements of SN submicrometer displacements, we used a custom HLIM system built on an optical table (Figs. 14.3b, 14.4c, d) in combination with a He–Ne laser and two Bragg cells oscillating at 40 and 40.4 MHz, respectively. The two beams were sent through a water immersion objective (40 × 0.80 or 25 × 0.50NA) and crossed in the focal plane, creating a fringe pattern beating at 400 kHz (Fig. 14.3c, d). Prior to the measurement, a flow of bath solution with a low concentration of polystyrene beads (diameter 1 µm) was applied close to the neuromast until one or a few beads attached to the cupula. A part of the light backscattered from the polystyrene bead that attached to the cupula and was collected back with the objective, sent through a pinhole, detected with a PMT and the photocurrent was converted to voltages with a fast current-to-voltage amplifier. The 400 kHz carrier frequency was filtered with a narrow band-pass amplifier and sent to the analogue FM demodulator and to the fringe counting demodulator of an off-the-shelf decoder (Polytec 3000). Because the decoder was factory-tuned to 40 MHz, the detected modulated 400 kHz carrier signal was added to the driving frequency of the Bragg cell (40.4 MHz) prior to decoding. We note that we were unable to find a contemporary turn-key solution for a HLIM setup, but it seems possible that a double fibre-optic interferometer (OFV-552, Polytec) could be used as the illuminator system that could be coupled to the microscope, and optionally complemented with a separate detector. In case a single Bragg cell is used for the heterodyning system, its heterodyning frequency is 40 MHz and care should be taken that the detector and amplifier electronics have an appropriate bandwidth.

14.7 Methods for Stimulating Mechanosensory Organs

The kick-off event in sensory encoding by the lateral line organ is the activation of mechano-transducer channels, which are present at the tips of the hair bundles. These channels are activated by force, similar to the hair cells in the inner ear of amphibians and mammals. The hair bundles of the lateral line system are covered by a cupula that is freely exposed to the surrounding aquatic environment. Hair bundles extend only up to 5 μm of the total length of a neuromast, which is around 45 μm long in the zebrafish (Dinklo 2005). Since the cupula mechanically couples to the underlying hair bundles, deflections of the cupula, leading to forces on the tip links, raise the opening probability of mechano-transducer channels. An essential step in many physiological studies of neuromast function is to ensure proper excitation of hair bundles. Historically, a variety of experimental approaches have been employed to achieve this goal. We will briefly sum up the main characteristics of these methods applied in hair cell and lateral line physiology, trying to emphasise on their advantages and limitations.

14.7.1 Direct Stimulation Methods

14.7.1.1 Magnetic Stimulation

The 1950s heralded the first attempts at studying the mechanics of lateral line hair cells. Kuiper decorated the cupulae of SNs of ruffe (*Gymnocephalus cernuus*) with magnetic particles, which could be attracted by the magnetic fields generated by coils that were conveniently positioned around the fish (Kuiper 1956).

14.7.1.2 Stiff Probe

A method that involved direct physical contact between a fibre probe driven by a piezoelectric actuator and the hair bundle of vestibular hair cells used to be a common approach (Strelioff and Flock 1984; Crawford and Fettiplace 1985; Howard and Ashmore 1986; Howard and Hudspeth 1987). In this method, the fibre sticks to the cell membrane of the kinocilium or tallest hair bundle and transmits the force applied by the actuator, directly deflecting the hair bundle. Rigid borosilicate glass fibres with tips polished to 1–3 μm were successfully used to study transducer channel kinetics (Ricci and Fettiplace 1997; Kennedy et al. 2003).

14.7.1.3 Flexible Probe

If the intrinsic stiffness of the stimulus probe is substantially larger than the hair bundle stiffness, it hinders the characterisation of hair bundle intrinsic mechanical properties. A variation of the probe method involves the use of a fibre whose intrinsic stiffness is on the order of that of the hair bundle. A fibre suitable for monitoring hair bundle stiffness should be able to bend during the application of a small force; this allows measuring the exerted force via bending of the fibre, as well as displacement of the hair bundle in the same experiment.

In order to measure bundle characteristics under dynamic conditions, the fibre should have as low viscous drag as possible, ideally less than the bundle. Since compliance increases with increasing fibre length and decreasing diameter while viscous drag is reduced by decreasing both, the best solution in this case would be to use a sufficiently short fibre with small diameter that ensures effective hair bundle displacement. Viscous forces, however, will attenuate high frequency stimuli, effectively limiting these compliant fibres to low frequencies (Howard and Hudspeth 1987). Typical quartz fibre diameters used in these studies had the diameter in the range of 0.5–4 μm and the length spanning from 50 μm to a few hundred micrometres. Long fibres (>100 μm as in Crawford et al. 1985) seemed to exhibit mechanical filtering due to viscous drag so it is generally advisable to keep them short (e.g. 30 μm as in Kennedy et al. 2005). Fibres were usually cleaned with chromic acid or 4 % hydrofluoric acid and prepared freshly before each experiment to guarantee adherence to the sensory hairs in order to facilitate pulling the hair bundle (Howard and Ashmore 1986; Ricci et al. 2000; Kennedy et al. 2003).

Determination of the fibre stiffness is commonly performed by measuring with a horizontal microscope the deflection caused by a mass of known weight (polymethyl methacrylate beads or small copper wire pieces, see for instance Howard and Ashmore 1986; Kros et al. 1992; McHenry and van Netten 2007). We note that in order to study the mechanics of SN cupulae, extremely compliant (e.g. carbon) fibre probes should be used.

14.7.2 Flow-Coupled Stimulation Methods

All the methods discussed so far make use of a direct contact between the stimulus and the hair bundle. A different way of stimulating hair bundles or cupulae exploits the fact that, in the animal's natural habitats, they are moved by water currents. Fluid-coupled stimulation strategies can be broadly put in one of these two categories: those that rely on a way of directing a water flow towards an individual neuromast or hair bundle (micro FJ, water jet) and those that are able to generate water movements with a vibrating stimulus sphere (dipole stimulus).

14.7.2.1 Fluid Jet

The first attempts to generate a well-defined fluid jet stimulation method date back to the early 1950s, when a fluid jet producing device was applied to drive the cupula of a fish lateral line neuromast (Jielof et al. 1952). The pipette used in that study was wide ($\phi = 4$ mm) in comparison with later efforts that took special care in narrowing the end in order to displace individual hair bundles (Flock and Orman 1983). A potential disadvantage of these devices is that the force exerted by the FJ flow on the sensory hair bundle cannot be precisely calculated (Kros et al. 1992). Other drawbacks include inaccurate determination of the angle formed by the preparation and the fluid stream, viscous drag near the surface and distance from the FJ tip to the hair bundle. An estimation of the force exerted by the FJ can be obtained by pointing the water jet towards a glass fibre of known stiffness, and measuring the deflection of its tip (Kros et al. 1992; Géléoc et al. 1997). If the cross-sectional area of the fluid jet stream is smaller than the cross-sectional area of the fibre, the force generated by the FJ per volt applied to the driver and per unit of stream area can be estimated.

Dinklo and colleagues described a similar piezoelectric-driven FJ device and its properties. In that study, a compliant glass fibre with a resin sphere attached to its tip was used to calibrate the displacement output, which was monitored by means of a heterodyne laser interferometer coupled through the condenser of a trans-mitted light microscope. This procedure allows for the determination of the fre-quency-dependent characteristics of the fluid jet. The pipette's tip diameter turned out to be the key parameter governing the amplitude and phase behavior (Dinklo et al. 2007). Tip sizes in the order of 10 µm produced a constant fluid velocity with a flat frequency response in phase with the applied voltage, over a broad range of frequencies (1–1,000 Hz). Knowing frequency response characteristics may not be so relevant when hair bundles are stimulated at only one frequency, but it becomes essential when one wishes to obtain mechanical or electrical tuning curves.

The use of fluid jet stimulation might be advantageous over the fibre stimulation when trying to stimulate short hair bundles. Kros et al. (1992) obtained more stable transducer currents when using this device to drive the motion of mouse OHC hair bundles, which can be 3 µm long at the apex and just about 1 µm long at the base. Fluid jets can be used to deflect hair bundles or neuromast cupulae in a step-like fashion (Ćurčić and van Netten 2005; Dinklo et al. 2007) or to generate a vibra-tional response, e.g. with (double) sine stimulation (Dinklo 2005). As the FJ tip can be placed close (100–300 µm) to the fish surface, it will essentially reside in the boundary layer and will thus be more effective at low stimulation frequencies (Sendin et al. in preparation).

14.7.2.2 Fast Pressure Clamp

A fast pressure clamp technique originally designed for the study of mechanically gated channels (McBride and Hamill 1995) has been successfully integrated into a fluid jet generating device in order to stimulate hair bundles (Denk and Webb 1992). This method is based on the use of a compact mixing chamber with three functional ports: one is connected to the suction port of a pipette holder, and the remaining ones lead to two piezoelectric valves which are connected to a vacuum pump and to a nitrogen gas tank. Both valves are controlled via a feedback loop with a small transducer integrated into one of the walls of the mixing chamber. The desired pressure is achieved by balancing the pressure and suction and can be attained with sub-millisecond kinetics (McBride and Hamill 1995). This design has been recently improved to incorporate a newly designed valve with a single piezoelectric element controlling both pressure and suction. A further reduction in the internal fluid volume of the valves and interconnecting tubing resulted in a faster response time (Besch et al. 2002). A similar operating principle lies at the core of water jet delivering systems currently being used in lateral line research to record ERPs (*microphonic potentials*) of neuromasts and extracellular action potentials (*action currents*, when measured in voltage-clamp mode) of afferent neurons (Obholzer et al. 2008; reviewed in Trapani and Nicolson 2010). We note that the transfer properties of the feedback system may be limiting the frequency response of such a stimulus device to a few tens of Hertz. In practice, the tubing between the pipette and the clamp device should be as stiff and as short as possible. The actual transfer characteristic of the particular system should be assessed as in the case of the FJ (Dinklo et al. 2007).

14.7.2.3 Microinjector

A microinjector ("Pico-spritzer") apparatus delivers a standardised pressure pulse of air to a standing column of fluid, thanks to a high-speed valve. Originally designed to inject nanolitres of fluid into embryos, it has been readapted to generate pulsed and sustained micro-FJ streams that are suitable for the direct current excitation of SNs (Liao 2010).

14.7.2.4 Stimulus Sphere

Well-defined fluid displacements can also be obtained using a stimulus sphere attached to a piezoelectric element. It generates a water displacement output as a function of applied voltage across the piezoelectric material that is constant over a broad range of frequencies up to several hundred Hertz. However, its effectiveness diminishes at larger distances, falling in stimulus strength with the third power of the distance to the stimulus sphere centre. In this respect, the FJ might be advantageous since its effective range of action is extended due to the fact that its output strength falls with the second power of distance from the pipette's tip. It is possible to counteract this effect by simply increasing the sphere diameter, which in turns requires stimuli to be delivered in larger volumes. We note that the stimulus sphere is big in comparison with micropipettes and may, therefore, cause trouble with fitting under a water immersion microscope.

This stimulation strategy has been successfully used to measure the frequency response of the lateral line organ of *Xenopus laevis* (Kroese et al. 1978), the frequency response and receptive fields of the lateral line of the ruffe *G. cernuus* (*syn. Acerina cernua*; van Netten and Kroese 1987; Ćurčić-Blake and van Netten 2006), as well as central unit responses to water motion in the goldfish (e.g. Coombs et al. 1998) and behavioral thresholds for feeding response in the mottled sculpin (Coombs and Janssen 1990). This stimulus has also been combined with methods to apply DC-fluid velocity to the lateral line system via a pipe system and a pump that produces a uniform water flow via sieve-like grids placed in the pipe's cross-section (Engelmann et al. 2000).

14.8 Recording of Extracellular Receptor Potentials

The ERPs, also called microphonic potentials, or colloquially microphonics (*mipos*), are electrical signals arising from the flow of electric current in a closed loop. In the first part of the loop, the ions are entering into the hair cells through mechanically activated transduction channels in the stereovilli, and are leaving them through voltage-gated channels in the basolateral membrane. Recently, patch-clamp recordings from hair cells of SNs of zebrafish revealed K^+ currents and Ca^{2+} currents which are similar to those present in the hair cells of birds, frogs and turtles (Ricci et al. 2013).

The mechanosensory channel is a large-conductance channel (100–300 pS), preferentially permeable to divalent cations (permeability ratio Ca^{2+}: $Na^+ \approx 5$; Fettiplace 2009). Given the ionic composition of fresh water, it is likely that the majority of inward mechanosensory current in SN hair cells is carried by calcium ions.

The relevant component of the ERP is produced at twice the stimulus frequency, due to combined activation of two oppositely oriented populations of hair cells, each with a rectifying displacement-response curve (Kuiper 1956; Flock and Wersäll 1962; Flock 1971; Faucherre et al. 2009). This response at twice the stimulation frequency has been observed in all vertebrate lateral line systems and otolith organs but it is absent from semicircular canals and the mammalian cochlea, where the hair cells bundles share only one orientation and, therefore, respond at the fundamental stimulus frequency only. In zebrafish, the ERP signals can be recorded from the neuromasts of the lateral line system and also from hair cells in the inner ear. Since ERPs are vulnerable to manipulations that abolish mechanotransduction, they have been widely used as diagnostic tools in the screening of genetic mutations leading to mechanosensory impairment (Nicolson et al. 1998; Corey et al. 2004; Kappler et al. 2004; Obholzer et al. 2008).

It is problematic to determine the cellular components that contribute to the generation of the ERP. In the bullfrog (*Rana catesbeiana*) sacculus, Corey and Hudspeth

(1983) established that the microphonic current is primarily generated by changes in the mechano-transduction channel conductance, but not exclusively: as the receptor potential changes the membrane potential, voltage-gated ionic channels can in turn be activated and add to the microphonic current. In this system, at least four parameters seem to dictate the shape of the ERP: the tissue impedance with its capacitive and resistive branches, the saturating displacement-response curve, a mechanical adaptation mechanism and a large K^+ outward conductance (Corey and Hudspeth 1983).

In the lateral line organ of *X. laevis,* it is possible to record ERPs and single fibre afferent activity from the same neuromast. In the study by Kroese et al. (1980), the neuromast was stimulated with a small glass bead ($d = 300$ μm) attached to a loudspeaker that was driven by a sinusoidal wave generator. Exploiting this preparation, Kroese and colleagues showed that the ERP amplitude increased linearly as a function of the stimulus amplitude in a range of 20 dB. Similar to the experiments of Corey and Hudspeth (1983), the plot of ERP amplitude versus constant displacement amplitude showed a maximum which was about 20 Hz for *Xenopus* and 100 Hz for bullfrog sacculus. Moreover, at least in the *Xenopus* lateral line, the frequency response of afferent nerve activity had a maximum that closely matched this value. At low frequencies both in the sacculus and in the lateral line organ, the phase of ERPs was first leading the stimulus but decreased with increasing stimulation frequency, eventually approaching the zero value (*Xenopus* lateral line, Kroese et al. 1980) or even lagging behind the stimulus at high frequencies (bullfrog sacculus, Corey and Hudspeth 1983).

14.8.1 Practical Approaches for the Recording of ERPs in Zebrafish

For lateral line ERP measurements, a recording electrode is advanced towards the base of the neuromast, as close as possible to the site of generation of the biological signal (hair cells) but without hindering the cupula movement, as described previously (Nicolson et al. 1998). The extracellular signals from zebrafish SNs have amplitudes of a few tens of μV and decay exponentially with distance from the hair cell epithelium with a length constant of about 5 μm (Nicolson et al. 1998) The characteristic second harmonic component of the ERP response disappears when the recording electrode or the neuromast are displaced (Fig. 14.5) or when dihydrostreptomycin (300 μM) is added. Both procedures are generally used to confirm the genuine character of the recorded potentials. For a description of the standard experimental procedures used for studying ERPs in neuromasts, we recommend the chapter of Trapani and Nicolson (2010).

14.8.1.1 Preamplification and Conditioning

In our recording of ERPs from SNs of the zebrafish, we used a DAM-50 preamplifier as the head stage. This relatively cheap battery-operated preamplifier is

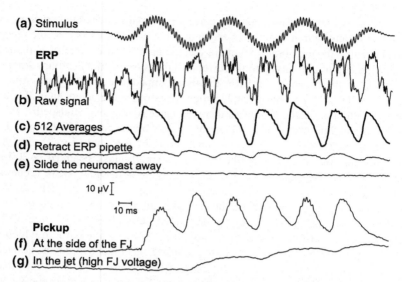

Fig. 14.5 The ERP waveforms evoked by windowed double sinusoidal stimulation sweeps delivered to an SN. **a** Stimulus waveform. **b** One presentation of the stimulus was enough to obtain a signal with a second harmonic component. **c** Averaging repeated presentations of the same stimulus helped to improve the signal-to-noise ratio. **d** Slight retraction of the recording ERP pipette (10 μm) attenuated the amplitude but did not abolish the signal. **e** After sliding the SN away from the fluid jet, ERP could not be recorded. **f** A spurious second harmonic component can appear when the ERP pipette tip is at the opening of the FJ pipette tip. **g** Mechanical pickup can appear when a strong water jet is generated by a high piezo voltage

convenient because it has accessible head stage connections and selectable input mode (unipolar or differential). As the secondary stage, we were using a conditioning module (AM 502, Tektronix, USA). We used the preamplifier in the AC mode with the pass band from 0.1 Hz to 1 kHz and an overall amplification of 10^5. In this configuration, using 1 MΩ electrodes, we were usually achieving 20 μV p/p noise (referred to input), close to the theoretical value of Johnston-Nyquist thermal noise (see below). We also tried to record ERPs using a patch clamp amplifier in the current-clamp mode, (Axopatch 200B, Axon Instruments, UK), as previously reported (Dinklo 2005; Nicolson et al. 1998). However, the patch clamp amplifier turned out to be somewhat noisier (typically 50 μV p/p noise) than the DAM-50 preamplifier.

14.8.1.2 Pipettes and Reference Electrodes

The microelectrodes were pulled on a horizontal puller (P97, Sutter, USA). We used borosilicate glass (outer diameter 1.5 mm, inner diameter 0.86 mm, with filament; GC150F-10, Harvard Apparatus, USA). The electrode was pulled with a square filament in five pulls. The tip sizes were 2–5 μm. Under visual control of the microscope, the micropipette's tip was advanced as close as possible to the base of the neuromast but without touching the cupula. We utilised a small Ag/AgCl pellet (WPI Instruments) as the reference bath electrode. Alternatively, a chlorinated silver wire was placed under the fish. In order to reduce the crosstalk of the FJ signal, we also tried to introduce a circumferential grounding wire on the perimeter of the bath, but this did not seem to make much difference; in practice, some stimulus crosstalk always remained.

Because the ERP's are typically in the range of a few microvolts, utmost care had to be taken to ensure minimal line hum and noise levels. In the ideal case, the biggest contribution to the voltage noise is thermal (Johnston-Nyquist) noise $V_{RMS} = (4k_B TR\Delta f)^{1/2}$, where V_{RMS} is the root mean square of the voltage noise, k_B is Boltzmann's constant, T is the temperature, R is the pipette resistance, and Δf is the frequency bandwidth, which was set to 1 kHz for the ERP measurements. For a microelectrode with the resistance $R = 1$ MΩ, at 25 °C, Johnston-Nyquist noise RMS will, therefore, be $V_{RMS} = 4$ μV, and for $R = 10$ MΩ, 12 μV (the peak-to-peak noise is roughly 3–5 times RMS). It is, therefore, desirable to use electrodes with reasonably low resistance, which can be achieved either by increasing the tip diameter or by increasing the conductivity of the electrode fluid. When the pipettes were filled with 1 M NaCl or 1 M KCl, their resistance was in the acceptable range (2–10 MΩ). The different pipette and bath solutions bring about a liquid junction potential that may be hard to compensate in the DC mode; however, this is not an issue if AC coupling (e.g. 0.1 Hz) is used.

14.8.1.3 Grounding, Hum and Crosstalk

Line hum and crosstalk of the FJ driving voltage could compromise the recordings when the grounding is not optimal, so great care should be taken to avoid these sources of contamination. In our case, the head stage ground of the DAM-50 was connected to the isolated electrode holder on the Huxley-type micromanipulator. The lower part of the manipulator, the microscope, the FJ-manipulator and the Faraday cage, which was built from old computer cases, were connected to the common ground point of the mechanically isolated table top (Fig. 14.4b). We also noticed that water leak to the microscope platform caused an increase of the hum. If a ground loop between the preamplifier and the main rig becomes a source of hum, it may be handy to use coax cables with shields interrupted at one connector.

In our experimental setup, the differential mode had the least FJ crosstalk. After careful optimisation, the line hum was buried in the thermal noise and the head-stage was almost insensitive to hum from hand manipulation. In order to additionally protect the electrodes from hum, the front of the microscope stage with the preparation was covered with a grounded metal plate. We checked that the microelectrode was picking up signals correctly by tapping on the stage, which resulted in a pickup similar to that produced by tapping on a microphone. Under these conditions, we were able to characterize the crosstalk from the FJ device

Fig. 14.6 Analysis of ERP over a range of fluid jet stimulation voltages. **a** Averaged ERP responses at stimulation voltages from 0.01 to 1 V. The fundamental component prevailed above 0.5 V. The FT analysis window is shown on the stimulation trace. **b** The ERP amplitude of the fundamental (*open circles, dashed line*) and the second harmonic (*full circles, solid line*). The fundamental amplitude becomes very strong with stimulation voltages above 0.6 V, at least partially due to crosstalk. The second harmonic increases approximately linearly with the stimulation voltage. **c** The phase of the fundamental starts changing above 0.3 V stimulation, while the phase of the second harmonic is changing below this voltage

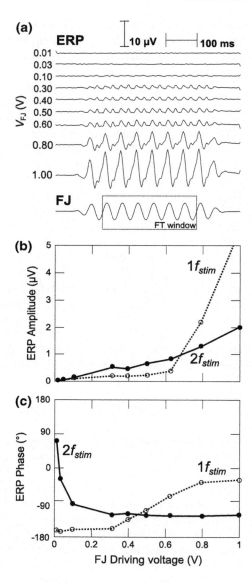

(Fig. 14.5). At moderate FJ stimulation voltages, the crosstalk will be mostly kept at the fundamental frequency and will rise proportionally to the stimulation frequency (not shown) and voltage (Fig. 14.6a, b). At extremely high stimulation voltages, however, a spurious second harmonic may appear, probably due to the distortion of the electric signal, or due to a mechanical pickup (Fig. 14.5f, g).

14.8.2 Interpretation of ERP Recordings

Although ERP measurements are relatively easy to obtain, their interpretation is not as simple. In zebrafish, the ERP is usually recorded with one electrode in the vicinity of the neuromast and with the other electrode somewhere in the medium. In this configuration, the major contribution to the ERP is desirably the potential difference sensed between the two electrodes due to the minute currents flowing towards the neuromast and into the hair cells. On the other hand, a part of the ERP may be due to an antenna effect of the changing electric field in the hair cell membrane. As the two contributions may have different amplitudes and phases over a range of frequencies, the interpretation of the ERP recording (especially its phase in relation to the stimulus phase) may not be as straightforward as desired. A classical treatise on the complexity of extracellular detection of bioelectrical signals was given by Geddes (1972).

In the case shown in Fig. 14.6, the first harmonic component shows crosstalk at higher stimulus amplitudes (>0.6 V), with a concurrent phase change. The secondary harmonic component of ERP amplitude is rising approximately linearly with the stimulation voltage (Fig. 14.6b), while the ERP phase was changing in the lower range of stimulation voltages and was constant in the upper range (Fig. 14.6c).

14.9 Stimulus Generation and Data Acquisition

14.9.1 Stimulation Functions

Mechanical and electrical properties of mechanoreceptors are most often assessed with step, ramp and sinusoidal stimulation protocols. In the case of step stimulation, care must be taken to account for mechanical resonances of the stimulus apparatus, which may cause ringing (Ćurčić-Blake and van Netten 2006; Dinklo et al. 2007). Sinusoidal stimulation can be implemented with a continuous or a windowed signal. The stimulation function may be a sum of two sinusoids or can have a ramp or DC offset component. A windowed sine may provide insight into transient and steady-state components of the response (Figs. 14.5, 14.6). An offset sine stimulation (Nicolson et al. 1998) or double sine stimulation (van Netten et al. 2003) may provide an insight into the non-linearity of the studied system. While offsetting the displacement of a stiff probe is straightforward to implement, a sustained offset of the water flow velocity is possible with the pressure clamp device, but not with the stimulus ball. The FJ device is able to sustain the DC flow for a limited amount of time of a few seconds, depending on the dimensions of the FJ device, and could thus be used with a DC offset, ramp or slow sine to approximate the directional flow. Proper calibration of the FJ stimulus (Dinklo et al. 2007) is in this case of utmost importance.

14.9.2 Laboratory Interfaces

It is advisable to use a laboratory interface that is capable of continuous and synchronous signal generation and acquisition. The unfiltered signal output contains steps between samples and should be smoothed with a low-pass filter set to, e.g. one quarter of the sampling frequency. Although a 16-bit DA interface offers enough bit depth to avoid trouble with quantisation round-off, it may be still a good idea to use a computer-controlled or a manual attenuator of the analogue stimulus signal, especially if a large stimulus dynamic range must be tested. An attenuator is a must if the DA interface has only 12 bits resolution. For the recording pathway, the anti-aliasing filter should be set at least at eight times the fundamental stimulation frequency and at most at half the sampling frequency.

14.9.3 Frequency Domain Analysis

Synchronisation of stimulation and acquisition will be very useful when frequency domain analysis is being used (see also van Netten and McHenry 2013). Laboratory interfaces with high sampling rates (100 kHz per channel) and synchronised input–output are easily obtainable, and since data memory is not a problem nowadays, the most practical approach is to design a measurement system which generates and records the signals at a constant (high) sample rate. This way, the output and input anti-aliasing filters may remain set at a fixed frequency, simplifying the filter electronics and measurement system transfer function calibration. The signal bouts can then be selected and analysed off-line. We note that in the case of using a digital imaging system, synchronisation of imaging with stimulation and electrophysiological signals may require quite some effort.

For sinusoidal stimuli, it is advisable to choose stimulation frequencies f_{stim} which are an integer multiple of the inverse of the stimulus bout length ($1/T$). In practice, it is a good idea to have one cycle constructed from, e.g. at least 16 points, so the maximum stimulation frequency should be ideally not more than 1/16 of the sampling frequency. The stimulus bout should best contain a binary power (2^n) of samples, so that the most efficient Fourier transform (FFT) algorithm can be used. If these conditions are met, the whole signal energy at the fundamental f_{stim} and the relevant higher harmonics will be kept in discrete frequency components, greatly simplifying the frequency analysis. The local noise at each frequency can be in this case easily estimated from the neighbouring non-harmonic components (Fig. 14.7c).

With modern computers speed and memory capacity, the FFT of 2^{20} (1 MB) samples takes a fraction of a second, offering a possibility of using a constant sampling frequency for the analysis of 10 s signal bouts with the stimulation frequencies ranging from about 0.1 to 5,000 Hz, thus exceeding the frequency limitations of both the stimulator (e.g. the FJ) and the biological system studied.

14.9.4 Measuring Tuning Curves with Sinusoidal Stimuli

Mechanical and electrical tuning curves are easily obtained with continuous sinusoidal stimulation (van Netten 1988; Dinklo 2005). In order to reliably estimate the tuning curve of the studied mechanosensory system, the transfer characteristics of the stimulation and the measurement (detection, signal conditioning and acquisition) systems must be calibrated. A thorough description of mechanical calibration of the stimulus ball and the FJ device was described by Dinklo et al. (2007). We note that in order to calibrate the acquisition system for the ERP measurements, an electronic model neuromast ("fake-fish") that will provide the second harmonic component can be built from two semiconductor diodes and a few resistors and capacitors.

As an example, we present measurements of the mechanical response of a neuromast, stimulated with an FJ device, obtained with the interferometric setup (HLIM) operating with both velocity and displacement decoders (Fig. 14.7). We note that the decoder outputs shown in this figure have not been calibrated for the transfer characteristics of the stimulus device and the measurement system.

Approximately one second of the initial part of the recording at each frequency was discarded due to possible transient effects. The continuous signal at each stimulation frequency was filtered at eight times the stimulation frequency and was sliced into eight 1,024-sample bouts containing 16 full sinusoidal cycles which were averaged online. The absolute magnitude of the averaged displacement signal is bigger at low stimulation frequencies, while for higher stimulation frequencies the velocity signal is bigger (Fig. 14.7a). The averaged signals were transformed with FFT. Due to the fact that we used an integer number of cycles per bout, the signal energy was kept in single harmonic components (Fig. 14.7b). The local noise of each harmonic component was estimated from the average of four neighbouring non-harmonic components (Fig. 14.7c). The SNR achieved with the averaging of 8 signal bouts (record time 1.3 s at 6 Hz and 0.1 s at 95 Hz) was between 100 and 10 for both decoding schemes (Fig. 14.7d) and on average slightly better for the velocity decoder (Fig. 14.7d, *inset*). We note that due to quantization round-off, the velocity decoder will provide better measurements at high frequencies even though its absolute noise levels are higher there.

The theoretical SNR of a signal contaminated with pure white noise is increasing with the square root of sampling time. In the used averaging scheme, the SNR curve should follow the power slope of $-1/2$ because the sampling time was inversely proportional to the stimulation frequency. The experimental SNR curve follows this slope in the mid frequency part (10–100 Hz). The SNR in the low frequency part is worse due to thermal drift; for the same reason, increasing the recording time will be timely and less effective at low frequencies. At high frequencies, the SNR is somewhat worse due to the increase of filter cut-off frequency. In this frequency range, the SNR will benefit from increasing the averaging cycle number at a very small expense of extra recording time. As a rule of thumb, the time spent at each frequency should be approximately the same in

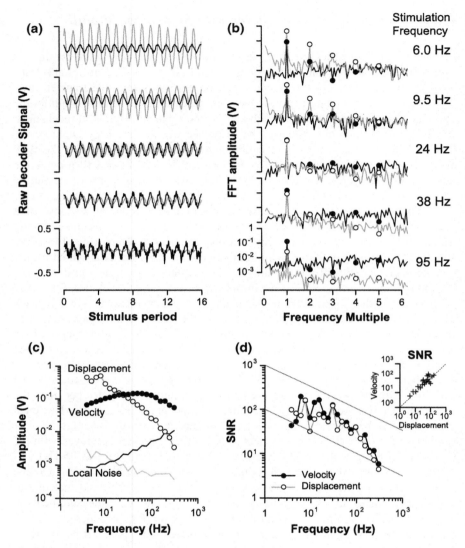

Fig. 14.7 Comparison of displacement and velocity measurements of cupular motion. **a** A series of recordings of cupular motion obtained from a bead located at 36 μm above the skin of the zebrafish. The amplitude of the decoder signals is shown here for 16 periods of the stimulus for both velocity (*black*) and displacement (*grey*) channels. All the traces included here represent the average of eight recordings. The conversion factor between the two demodulators was as follows: 1 V = 830 nm for the phase demodulator (displacement) and 1 V = 130 μm s^{-1} for the frequency demodulator (velocity). **b** The FFT amplitude of the waveforms shown in A. The signal energy was contained in discrete Fourier components, shown here for the velocity (*black full circles*) and the displacement (*open circles*) demodulators. The stimulation frequencies used here are given on the right. **c** The signal component at the stimulus frequency (*circles*) is shown for the velocity (*full black circles*) and displacement channels (*open circles*) together with the local noise of each harmonic (*black line* for velocity and *grey line* for displacement detectors). We note that these raw measurements are not corrected for the transfer characteristics of the

◀ measurement system. **d** Signal-to-noise ratio (*SNR*) for both channels as a function of the stimulation frequency. Note that the SNR diminishes with stimulation frequency due to the fact that the averaging time was inversely proportional to frequency. In theory, this averaging scheme for a signal contaminated with a pure white noise would result in the SNR being proportional to $1/\sqrt{f}$ (*dotted lines*). *Inset* shows the SNR of velocity measurements versus the SNR of displacement measurements. Velocity measurements mostly have a slightly better SNR

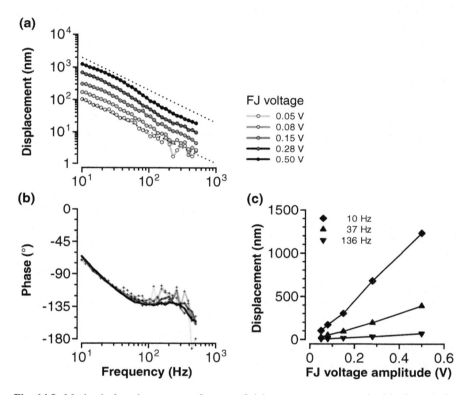

Fig. 14.8 Mechanical tuning curves of a superficial neuromast, measured with the velocity decoder of the HLIM setup. Stimulation was applied with a fluid jet (*FJ*) at five different voltage amplitudes, producing approximately constant jet velocities over the tested frequency range. The contrasting bead was attached to the cupula at about 35 μm above the skin, the fluid jet tip was about 100 μm above the skin and about 100 μm to the side of the neuromast. **a** Displacement amplitude tuning curves. The *dotted lines* represent a 1/*f* proportionality. **b** Displacement phase tuning curves. **c** The displacement amplitude as a function of the FJ driving voltage

order to obtain constant SNR over the whole frequency range. The SNR of the signal amplitude also affects the reliability of phase estimation. For SNR > 10, the error of the phase can be approximated with $\Delta\varphi = 360°/(2\pi * SNR)$. At SNR = 10, the phase angle error $\Delta\varphi$ is in the range of about 5°.

In Fig. 14.8, we show a family of mechanical tuning curves of an SN at a range of flow velocities and frequencies. The tuning curves were obtained with the velocity decoder of the HLIM setup and were calibrated and corrected for the transfer characteristics of the conditioning and acquisition systems. Stimulation was applied with an FJ which produced approximately constant jet velocities over the tested frequency range. This particular displacement amplitude tuning curve shows a low-pass transfer characteristic with a power slope of about -1 (Fig. 14.8a). The displacement phase curves are independent of the driving voltage (Fig. 14.8b). The displacement was linear with the driving voltage (Fig. 14.8c), implying that the neuromast was stimulated in the non-saturating range.

14.10 Conclusions

In this chapter, we have reviewed the techniques used for visualisation and mechanical stimulation of neuromasts, for contact and non-contact measurements of their mechanics and for extracellular measurements of their electrical activity. We also gave practical advice for frequency domain analysis of electrical or mechanical tuning properties of neuromasts (see also van Netten and McHenry 2013). We believe that some of the described methods will continue to play an important role in the research of the lateral line, but also in the research of other mechanoreceptor organs. We would like to deprecate the use of the anaesthetic MS-222, which acts as a partial blocker of mechanotransduction (Palmer and Mensinger 2004). There seems to be a number of less messy alternatives (eugenol: Grush et al. 2004; 2-epoxy ethanol: Higgs and Radford 2013; alphaxalone: O'Hagan and Raidal 2006); the use of MS-222 as the standard anaesthetic for mechanosensory research on zebrafish should thus be seriously reconsidered.

In the research of the zebrafish lateral line physiology, probably the biggest setback is that the first intracellular recordings from the neuromast hair cells have only been published at the time of this chapter revision (Ricci et al. 2013) and it may be that the patch-clamp technique on neuromast hair cells will have a yield too low to be routinely used for lateral line research. On the other hand, a combination of electrophysiological recordings from afferent nerves (Liao 2010) or neuronal activity imaging (Muto et al. 2013) with biologically relevant and calibrated mechanical stimulation of single neuromasts or of neuromast arrays may be a promising direction. We hope that the description of mechanical stimulation techniques and optical and electrophysiological recording techniques presented in this chapter will be of help in the development of new experimental and analytical tools for this exciting field of sensory physiology.

Acknowledgments The authors wish to thank the reviewers and the editors for their insightful comments. The work presented in this chapter was chiefly funded by the EU FP6 Project Cilia.

References

Besch SR, Suchyna T, Sachs F (2002) High-speed pressure clamp. Pflügers Arch 445:161–166

Coombs S, Janssen J (1990) Behavioral and neurophysiological assessment of lateral line sensitivity in the mottled sculpin, *Cottus bairdi*. J Comp Physiol A 167:557–567

Coombs S, Mogdans J, Halstead M, Montgomery J (1998) Transformation of peripheral inputs by the first-order lateral line brainstem nucleus. J Comp Physiol A 182:606–626

Corey DP, Hudspeth AJ (1983) Analysis of the microphonic potential of the bullfrog's sacculus. J Neurosci 3:942–961

Corey DP, Garcia-Anoveros J, Holt JR, Kwan KY, Lin SY, Vollrath MA, Amalfitano A, Cheung EL, Derfler BH (2004) TRPA1 is a candidate for the mechanosensitive transduction channel of vertebrate hair cells. Nature 432:723–730

Crawford AC, Fettiplace R (1985) The mechanical properties of ciliary bundles of turtle cochlear hair cells. J Physiol 364:359–379

Ćurčić-Blake B (2006) Spatial and temporal characteristics of fish lateral line detection. PhD thesis, University of Groningen, The Netherlands. Available via University of Groningen PhD Thesis repository. http://irs.ub.rug.nl/ppn/296009164. Accessed 13 Jan 2013

Ćurčić-Blake B, van Netten SM (2005) Rapid responses of the cupula in the lateral line of ruffe (*Gymnocephalus cernuus*). J Comp Physiol A 19:393–401

Ćurčić-Blake B, van Netten SM (2006) Source location encoding in the fish lateral line canal. J Exp Biol 209:1548–1559

Denk W, Webb WW (1992) Forward and reverse transduction at the limit of sensitivity studied by correlating electrical and mechanical fluctuations in frog saccular hair cells. Hearing Res 60:89–102

Denton EJ, Gray JAB (1983) Mechanical factors in the excitation of clupeid lateral lines. Proc R Soc Lond B 218:1–26

Dijkgraaf S (1963) The functioning and significance of the lateral-line organs. Biol Rev 38:51–105

Dinklo T (2005) Mechano- and electrophysiological studies on cochlear hair cells and superficial lateral line cupulae. PhD thesis, University of Groningen, The Netherlands. Available via University of Groningen PhD Thesis repository. http://irs.ub.rug.nl/ppn/271278285. Accessed 13 Jan 2013

Dinklo T, Meulenberg CJ, van Netten SM (2007) Frequency-dependent properties of a fluid jet stimulus: calibration, modeling, and application to cochlear hair cell bundles. J Assoc Res Otolaryngol 8:167–182

Ebner BC, Thiem JD, Lintermans M (2007) Fate of 2 year-old, hatchery-reared trout cod *Maccullochella macquariensis* (Percichthyidae) stocked into two upland rivers. J Fish Biol 71:182–199

Engelmann J, Hanke W, Mogdans J, Bleckmann H (2000) Hydrodynamic stimuli and the fish lateral line. Nature 408:51–52

Faucherre A, Pujol-Martí J, Kawakami K, López-Schier H (2009) Afferent neurons of the zebrafish lateral line are strict selectors of hair-cell orientation. PLoS ONE 4:1–12

Fettiplace R (2009) Defining features of the hair cell mechanoelectrical transducer channel. Pflugers Arch Eur J Physiol 458:1115–1123

Flock Å (1971) Sensory transduction in hair cells. In: Loewenstein WR (ed) Handbook of sensory physiology, vol I. Springer, Heidelberg, pp 396–441

Flock Å, Orman S (1983) Micromechanical properties of sensory hairs on receptor cells of the inner ear. Hearing Res 11:249–260

Flock Å, Wersäll JMD (1962) A study of the orientation of the sensory hairs of the receptor cells in the lateral line organ of fish with special reference to the function of receptors. J Cell Biol 15:19–27

Geddes LA (1972) Electrodes and the measurement of bioelectric events. Wiley, New York

Géléoc GSG, Lennan GWT, Richardson GP, Kros CJ (1997) A quantitative comparison of mechanoelectrical transduction in vestibular and auditory hair cells of neonatal mice. Proc R Soc Lond B 264:611–621

Gelman S, Ayali A, Tytell ED, Cohen AH (2007) Larval lampreys possess a functional lateral line system. J Comp Physiol A 193:271–277

Grush J, Noakes DLG, Moccia RD (2004) The efficacy of clove oil as an anesthetic for the zebrafish, *Danio rerio* (Hamilton). Zebrafish 1:46–53

Higgs DM, Radford CA (2013) The contribution of the lateral line to 'hearing' in fish. J Exp Biol 216:1484–1490

Howard J, Ashmore JH (1986) Stiffness of sensory hair bundles in the sacculus of the frog. Hearing Res 23:93–104

Howard J, Hudspeth AJ (1987) Mechanical relaxation of the hair bundle mediates adaptation in mechanoelectrical transduction by the bullfrog's saccular hair cell. Proc Natl Acad Sci USA 84:3064–3068

Howard J, Hudspeth AJ (1988) Compliance of the hair bundle associated with gating of mechanoelectrical transduction channels in the bullfrog's saccular hair cell. Neuron 1:189–199

Jielof R, Spoor A, de Vries H (1952) The microphonic activity of the lateral line. J Physiol 116:137–157

Kappler JA, Starr CJ, Chan DK, Kollmar R, Hudspeth AJ (2004) A nonsense mutation in the gene encoding a zebrafish myosin VI isoform causes defects in hair-cell mechanotransduction. Proc Natl Acad Sci USA 101:13056–13061

Kennedy HJ, Evans MG, Crawford AC, Fettiplace R (2003) Fast adaptation of mechanoelectrical transducer channels in mammalian cochlear hair cells. Nat Neurosci 6:832–836

Kennedy HJ, Evans MG, Crawford AC, Fettiplace R (2005) Force generation by mammalian hair bundles supports a role in cochlear amplification. Nature 433:880–883

Kroese ABA, Schellart NAM (1992) Velocity- and acceleration sensitive units in the trunk lateral line of the trout. J Neurophysiol 68:2212–2221

Kroese ABA, van der Zalm JM, van den Bercken J (1978) Frequency response of the lateral line organ of *Xenopus laevis*. Pflügers Arch 375:167–175

Kroese ABA, van der Zalm JM, van den Bercken J (1980) Extracellular receptor potentials from the lateral-line organ of *Xenopus laevis*. J Exp Biol 86:63–77

Kros CJ, Rusch A, Richardson GP (1992) Mechano-electrical transducer currents in hair cells of the cultured neonatal mouse cochlea. Proc Biol Sci 249:185–193

Kuiper JW (1956) The microphonic effect of the lateral line organ. PhD Thesis, University of Groningen, The Netherlands

Liao JC (2010) Organization and physiology of posterior lateral line afferent neurons in larval zebrafish. Biol Lett 6:402–405

Liao JC, Haehnel M (2012) Physiology of afferent neurons in larval zebrafish provides a functional framework for lateral line somatotopy. J Neurophysiol 107:2615–2623

Liff HJ, Shamres S (1972) Structure and motion of cupulae of lateral line organs in *Necturus maculosus* III. A technique for measuring the motion of free-standing lateral line cupulae. Q Progr Rep Res Lab Electronics MIT 104:332–336

Lopéz-Schier H, Hudspeth AJ (2005) Supernumerary neuromasts in the posterior lateral line of zebrafish lacking peripheral glia. Proc Natl Acad Sci USA 102:1496–1501

Martin P, Hudspeth AJ (1999) Active hair-bundle movements can amplify a hair cell's response to oscillatory mechanical stimuli. Proc Natl Acad Sci USA 96:14306–14311

McBride DW Jr, Hamill OP (1995) A fast pressure-clamp technique for studying mechanogated channels. In: Neher E, Sakmann B (eds) Single channel recording, 2nd edn. Plenum Press, New York

McHenry MJ, van Netten SM (2007) The flexural stiffness of superficial neuromasts in the zebrafish (*Danio rerio*) lateral line. J Exp Biol 210:4244–4253

McHenry MJ, Strother JA, van Netten SM (2008) Mechanical filtering by the boundary layer and fluid–structure interaction in the superficial neuromast of the fish lateral line system. J Comp Physiol A 194:795–810

McHenry MJ, Feitl KE, Strother JA, Van Trump WJ (2009) Larval zebrafish rapidly sense the water flow of a predator's strike. Biol Lett 5:477–497

Metcalfe WK, Kimmel CB, Schabtach E (1985) Anatomy of the posterior lateral line system in young larvae of the zebrafish. J Comp Neurol 233:377–389

Muir W, Lerche P, Wiese A, Nelson L, Pasloske K, Whittem T (2008) Cardiorespiratory and anesthetic effects of clinical and supraclinical doses of alfaxalone in dogs. Vet Anaesth Analg 35:451–462

Muto A, Ohkura M, Gembu A, Nakai J, Kawakami K (2013) Real-time visualization of neuronal activity during perception. Curr Biol 23:307–311

Mylonas CC, Cardinaletti G, Sigelaki I, Polzonetti-Magni A (2005) Comparative efficacy of clove oil and 2-phenoxyethanol as anaesthetics in the aquaculture of European sea bass (*Dicentrarchus labrax*) and gilthead sea bream (*Sparus aurata*) at different temperatures. Aquaculture 246:467–481

Nicolson T, Rüsch A, Friedrich RW, Granato M, Ruppersberg JP, Nusslein-Volhard C (1998) Genetic analysis of vertebrate sensory hair cell mechanosensation: the zebrafish circler mutants. Neuron 20:271–283

Nusslein-Vollhard C, Dahm R (2002) Zebrafish: a practical approach. Oxford University Press, Oxford

O'Hagan BJ, Raidal SR (2006) Surgical removal of retrobulbar hemangioma in a goldfish (*Carassius auratus*). Vet Clin Exot Anim 9:729–733

Obholzer N, Wolfson S, Trapani JG, Mo W, Nechiporuk A, Busch-Nentwich E, Seiler C, Sidi S, Söllner C, Duncan RN, Boehland A, Nicolson T (2008) Vesicular glutamate transporter 3 is required for synaptic transmission in zebrafish hair cells. J Neurosci 28:2110–2118

Palmer LM, Mensinger AF (2004) Effect of the Anesthetic Tricaine (MS-222) on Nerve Activity in the Anterior Lateral Line of the Oyster Toadfish, *Opsanus tau*. J Neurophysiol 92:1034–1041

Park CK, Kim K, Jung SJ, Kim MJ, Ahn DK, Hong SD, Kim JS, Oh SB (2009) Molecular mechanism for local anesthetic action of eugenol in the rat trigeminal system. Pain 144:84–94

Peters RC, Dénizot J-P (2004) Miscellaneous features of electroreceptors in *Gnathonemus petersii* (Günther, 1862) (Pisces, Teleostei, Mormyriformes). Belg J Zool 134:61–66

Peters RC, Van den Hoek B, Bretschneider F, Struik ML (2001) Saffan: a review and some examples of its use in fishes (Pisces: Teleostei). Neth J Zool 51:421–437

Ricci AJ, Fettiplace R (1997) The effects of calcium buffering and cyclic AMP on mechano-electrical transduction in turtle auditory hair cells. J Physiol 501:111–124

Ricci AJ, Crawford AC, Fettiplace R (2000) Active hair bundle motion linked to fast transducer adaptation in auditory hair cells. J Neurosci 20:7131–7142

Ricci AJ, Bai J-P, Song L, Lv C, Zenisek D, Santos-Sacchi J (2013) Patch-clamp recordings from lateral line neuromast hair cells of the living zebrafish. J Neurosci 33:3131–3134

Sendin GC, Dinklo T, Pirih P, van Netten SM (in preparation) Mechanical and electrical filter characteristics of superficial neuromasts in the zebrafish lateral line

Strelioff D, Flock A (1984) Stiffness of sensory-cell hair bundles in the isolated guinea pig cochlea. Hearing Res 15:19–28

Trapani JG, Nicolson T (2010) Physiological recordings from zebrafish lateral-line hair cells and afferent neurons. Methods Cell Biol 100:219–231

van Netten SM (1988) Laser interferometer microscope for the measurement of nanometer vibrational displacements of a light scattering microscopic object. J Acoust Soc Am 83:1667–1674

van Netten SM (2006) Hydrodynamic detection by cupulae in a lateral line canal: functional relations between physics and physiology. Biol Cyber 94:67–85

van Netten SM, Kroese ABA (1987) Laser interferometric measurements on the dynamic behaviour of the cupula in the fish lateral line. Hearing Res 29:55–61 van Netten SM, McHenry MJ (2013) The biophysics of the fish lateral line. In: Popper A (ed) Springer handbook of auditory research: the lateral line system (in press)

van Netten SM, Dinklo T, Marcotti W, Kros CJ (2003) Channel gating forces govern accuracy of mechano-electrical transduction in hair cells. Proc Natl Acad Sci USA 100:15510–15515

Yang Y, Chen J, Engel J, Pandya S, Chen S, Tucker C, Coombs S, Jones DL, Liu C (2006) Distant touch hydrodynamic imaging with an artificial lateral line. Proc Natl Acad Sci USA 103:18891–18895

Zhao YD, Yamoah EN, Gillespie PG (1996) Regeneration of broken tip links and restoration of mechanical transduction in hair cells. Proc Natl Acad Sci USA 93:15469–15474

Chapter 15
Neuronal Basis of Source Localisation and the Processing of Bulk Water Flow with the Fish Lateral Line

Horst Bleckmann and Joachim Mogdans

Abstract Fish perceive local water motions and local pressure gradients with their mechanosensory lateral line. The sensory units of the lateral line are the neuromasts that are distributed across the surface of the animal. Water motions are received and transduced into neuronal signals by neuromast hair cells. The information is then conveyed by afferent nerve fibers to the fish brain and processed by lateral line neurons in distinct nuclei of the brainstem, cerebellum, midbrain, and forebrain. The present review introduces the peripheral morphology of the lateral line and describes physiological work, thereby focussing on recent studies that have investigated what kind of sensory information is provided by dipole sources and bulk water flow, and how fish use and process this flow information.

Keywords Mechanosensory · Neuromast · Dipole · Kármán vortex street · Neural coding

Abbreviations

ALLN   Anterior lateral line nerve
CN   Canal neuromast
MON   Medial octavolateralis nucleus
PLLN   Posterior lateral line nerve
PIV   Particle image velocimetry
PSTH   Peri stimulus time histogram
SN   Superficial neuromast
RF   Receptive field
TS   Torus semicircularis

H. Bleckmann (✉) · J. Mogdans
Institute for Zoology, University of Bonn, 53115 Bonn, Germany
e-mail: bleckmann@uni-bonn.de

H. Bleckmann et al. (eds.), *Flow Sensing in Air and Water*,
DOI: 10.1007/978-3-642-41446-6_15, © Springer-Verlag Berlin Heidelberg 2014

15.1 Introduction

Fishes and aquatic amphibians detect minute water motions with their mechano-sensory lateral line. The lateral line plays an essential role in many behaviors, including intraspecific communication (Kaus and Schwartz 1986; Satou et al. 1991; Walkowiak and Münz 1985), detection of predators and prey (Bleckmann et al. 1989a; Janssen 1997; Janssen and Corcoran 1993; Montgomery and Macdonald 1987), detection, discrimination, and avoidance of inanimate objects (Campenhausen et al. 1981; Weissert and Campenhausen 1981), and orientation to ambient water flows (Baker and Montgomery 1999a, b). Furthermore, the lateral line of rheophilic fish provides some information for station holding (Bleckmann et al. 2012; Liao 2007). Like other sensory systems, the lateral line must extract biologically relevant information in the presence of self-generated and/or external background noise, encode sensory information in the ascending central pathway and decode this information for adequate guidance of behavior. A number of studies have described the physiology of lateral line hair cells (Harris et al. 1970), primary afferent nerve fibers (Münz 1989), and central lateral line units (Bleckmann 2007) to hydrodynamic stimuli. These studies have revealed a wealth of information about the encoding of water motions generated by a stationary dipole (vibrating sphere), moving objects, and bulk water flow (Bleckmann 2007). This review is confined to recent data on the processing of dipole stimuli in still- and running water and the processing of bulk water flow by the ascending lateral line pathway. We ask how the fish brain may use the information provided by the lateral line periphery for source identification and source localization, and how the information provided by bulk water flow is encoded. In addition, this review points out the major gaps we still have in understanding the central processing of lateral line information.

15.2 The Lateral Line Periphery

The lateral line system of fish comprises anywhere from less than 100 to several thousand sensory endorgans called neuromasts (Coombs et al. 1988). A neuromast may contain only a few or up to several thousand hair cells. Lateral line hair cells have up to 150 stereovilli and a single kinocilium that occurs eccentrically at the tall edge of the bundle. The mechanical linkage between the hair cell's ciliary bundle and the surrounding water is provided by a gelatinous cupula (Münz 1989). In fishes, usually two types of neuromasts can be distinguished: superficial neuromasts (SNs) which occur freestanding on the skin, and canal neuromasts (CNs) which are located in subepidermal canals that are in contact with the water through canal pores. Typically, CNs are located between adjacent canal pores. A single lateral line afferent may innervate several neighboring SNs, but never a combination of SNs and CNs (Münz 1985). Thus, the lateral line periphery begins with at least two, independent pathways that carry hydrodynamic information to the brain.

With the exception of the neuromasts of Pacific hagfishes (Wullimann and Grothe 2013), lateral line neuromasts contain two populations of hair cells with oppositely oriented ciliary bundles (Flock and Wersäll 1962). Since single lateral line afferents innervate only hair cells that are aligned in the same direction (Görner 1963) they either increase or decrease their ongoing discharge rates if the cupula is displaced (Flock 1965; van Netten and Kroese 1989). Thus, both the SN pathway and the CN pathway can be further subdivided into pathways that encode the direction of cupula motion. Besides an afferent innervation, lateral line neuromasts may receive an efferent innervation.

SNs are primarily driven by viscous drag forces that are proportional to the velocity of water movements (Kalmijn 1989). Thus, in their frequency range of operation (>0 Hz up to about 80 Hz) SNs function as velocity detectors. Although the frequency range of SNs and CNs overlap, that of CNs extends to higher frequencies to about 150 Hz. Moreover, CNs function as pressure gradient detectors because fluid flow inside a lateral line canal occurs only if there is a pressure gradient between canal pores. Outside the canal, the pressure gradient is proportional to the acceleration of the water particles. Thus, in their operating range CNs can also be thought of as acceleration detectors (Kalmijn 1989).

15.3 The Ascending Lateral Line Pathway

The central neuroanatomy of the mechanosensory lateral line has been studied in various vertebrate taxa, but to a different degree (for a comprehensive description of the central lateral line pathways of agnathans, cartilaginous, and ray-finned fishes as well as electrosensory and non-electrosensory teleosts see Wullimann and Grothe (2013)). In the present review, only the central lateral line pathway of cyprinids (carps and relatives) will be described. This description serves as an example for all non-electroreceptive ray-finned fishes (actinopterygians). The major reason for this choice is, that the central lateral line physiology of cyprinids has been far better investigated than the central lateral line physiology in all other fish taxa. Different fish taxa can show, however, marked differences in their central lateral line pathways and thus probably also in the physiology of these pathways. In all fish, primary afferents that innervate lateral line neuromasts travel to the brain in separate cranial nerves (e.g., Northcutt 1989). The most prominent nerves are the anterior lateral line nerve (ALLN) that innervates cephalic neuromasts, and the posterior lateral line nerve (PLLN) that innervates neuromasts on the trunk. In cyprinids, afferent fibers from both nerves terminate in the medullary ipsilateral medial (MON) and caudal (CON) octavolateralis nucleus (e.g., McCormick and Hernandez 1996; New et al. 1996). Additional fibers may reach the cerebellar granular eminence, the corpus cerebellum (Puzdrowski 1989), the Mauthner cells, which initiate escape behavior (Mirjany and Faber 2011), and the magnocellular nucleus. Upon reaching the medulla, many single lateral line afferents of cyprinids bifurcate into an anterior and posterior branch within the MON. The posterior branch continues to the CON. The termination sites of anterior and posterior lateral line nerve fibers are segregated in the MON, with ALLN fibers lying anteroventrally and PLLN fibers lying more posterodorsally (Wullimann and Grothe 2013). Output neurons of the MON (crest cells or Purkinye-like cells) give rise to commissural projections and to fibers that terminate in the contralateral midbrain torus semicircularis (e.g., McCormick and Hernandez 1996). There are also reciprocal connections with the nucleus preeminentialis (McCormick and Hernandez 1996). Some weak projections from the MON to the ipsilateral torus semicircularis, the contralateral optic tectum and the principal sensory trigeminal nucleus also exist (McCormick and Hernandez 1996). The torus semicircularis has strong ipsilateral projections to the dorsal division of the diencephalic lateral preglomerulosus nucleus (Echteler 1984; McCormick and Hernandez 1996). Efferent fibers of this nucleus reach the dorsal pallium, in particular the dorsal part of the lateral zone, but also project weakly to the dorsal part of the medial and the central zone (for review see Wullimann and Grothe 2013).

15.4 Identification and Localization of Dipole Sources

To understand how fish identify and localize the source of a hydrodynamic stimulus with the lateral line, researchers have conducted experiments in which a mechanical dipole, i.e., a stationary, sinusoidally vibrating sphere, was used as a stimulus source. In most of these experiments the dipole was generating a monofrequency stimulus. The advantage of such a stimulus is that it can easily be manipulated in amplitude and frequency. A further advantage is, that the flow field and the pressure distribution around a dipole can be calculated using simple mathematical equations, thus allowing predictions of the stimulus received by a neuromast at any location of the animal's surface (Kalmijn 1988).

15.4.1 Neuronal Representation of Spectral and Temporal Aspects of a Dipole Stimulus by Afferent Nerve Fibers

Primary lateral line afferents show ongoing activity. They respond to a suprathreshold monofrequency dipole stimulus with phase-coupled action potentials (Fig. 15.1a, b). With increasing displacement of the dipole both the degree of phase-coupling and the discharge rate increase (Fig. 15.1c). In terms of frequency, two types of response functions can be found, those with a slope of about 6 dB and those with a slope of about 12 dB per frequency doubling, at least up to about 100 Hz (Fig. 15.1d). A gain of 6 dB/octave is indicative of a velocity detector, whereas a gain of 12 dB/octave is indicative of an acceleration detector. In other words, the slope of the frequency-response function indicates whether the recorded fiber was

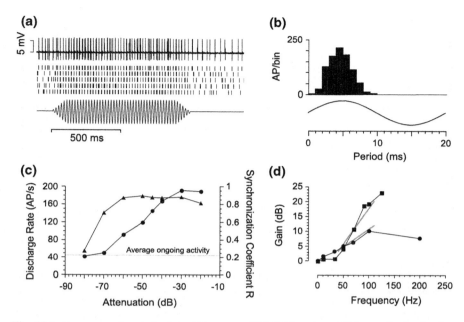

Fig. 15.1 Responses of posterior lateral line nerve (*PLLN*) fibers in goldfish to a dipole stimulus. **a** Original recording (*top*), raster plot (each marker represents one action potential) across five stimulus representation (*middle*) and the applied 50 Hz stimulus (peak-to-peak vibration amplitude 4 μm) (*bottom*). **b** Period histogram of the distribution of spikes within each cycle of the 50 Hz stimulus. **c** Level-response function of a PLLN fiber. Discharge rate (*line* connecting *circles* re: *left* hand axis) and synchronization coefficient *R* as a measure of the strength of phase-coupling (*line* connecting *triangles* re: *right* hand axis) are plotted as function of stimulus level (in dB). An attenuation of −20 dB corresponds to a peak-to-peak sphere displacement of 425 μm. **d** Gain of two PLLN fibers plotted as function of stimulus frequency. One fiber (*circles*) had a gain of about 6 dB per octave, the other fiber (*squares*) had a gain of about 12 dB per octave. *Dotted lines* represent slopes of 6 and 12 dB per octave, respectively. Adapted from Bleckmann et al. (2003)

innervating a SN or a CN (Kroese and Schellart 1992). If the lateral line is stimulated with amplitude-modulated sinusoidal water movements, primary afferents phase-lock to both the carrier frequency and the amplitude modulation frequency (Mogdans and Bleckmann 1999). Thus, primary lateral line afferents encode both, spectral and temporal aspects of a dipole stimulus. This provides information about a stimulus that the brain may use to determine source identity.

15.4.2 Neuronal Representation of Spatial Information of a Dipole Stimulus by Afferent Nerve Fibers

To localize a dipole source the lateral line must extract spatial information about the source, like azimuth, elevation, and distance, from the stimulus field that is provided by the source. Since the stimulus field not only changes with the

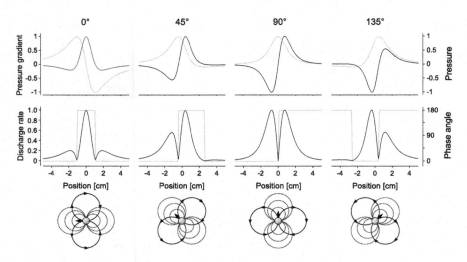

Fig. 15.2 Modeled pressure gradient patterns generated by a dipole source, i.e., a sinusoidally vibrating sphere, across a linear array of lateral line receptors and predicted spatial excitation patterns. *Upper*: Local pressure (*grey*) and pressure gradients (*black*) for different angles of sphere vibration (0°, parallel to the array, 90°, perpendicular, i.e., toward the array, 45° and 135° intermediate). *X*-axis indicates the position of the dipole source. *Middle*: Spatial excitation patterns of primary lateral line afferents (*black*) and phase angles of their responses (*gray*). *Lower*: Schematic representation of iso-pressure contours (*dashed lines*) and flow lines (*solid lines* with *arrows*) around a dipole source (*gray circle*). Iso-pressure contours are plotted for a single plane through the axis of sphere oscillation (*large arrow head* close to the source). *Flow lines* indicate the movement of water around the dipole for one half cycle of sphere oscillation. Adapted from Meyer et al. (2012)

displacement amplitude of the dipole but also with the distance from the source, its location cannot be encoded unequivocally by the activity of a single neuromast. For example, a nearby dipole vibrating with a small displacement amplitude may elicit the same neuronal response in a single afferent fiber as a dipole that is further away but vibrating with a larger displacement. However, source location can be determined from the activity across an array of neuromasts. This is due to the fact that the stimulus field generated by a dipole changes in a predictable way across the surface of a fish depending on azimuth, elevation, distance, and orientation (vibration axis) of the dipole with respect to the fish (Coombs and Conley 1997; Coombs et al. 1996; Goulet et al. 2008; Sand 1981; Curcic-Blake and van Netten 2006).

A dipole generates a distinct pressure pattern that can be measured by moving a vibrating sphere passed a pressure receiver, e.g., a hydrophone. Since neuromasts function as velocity (SNs) or pressure gradient (CNs) receptors, the pressure field must be converted into the corresponding velocity or pressure gradient fields. Analytical analysis demonstrated that the velocity field and pressure difference field generated by a dipole are proportional to each other. In other words, the pressure difference distribution along a lateral line canal has a form identical to the velocity field (Goulet et al. 2008). If the sphere is vibrating parallel to the longitudinal axis of the fish (Fig. 15.2, direction of vibration 0°), the pressure pattern is biphasic, i.e., it consists of two peaks with opposite sign. The corresponding pressure gradient pattern is triphasic, consisting of a large positive peak bordered by two smaller negative sidepeaks. In contrast, a sphere that is vibrating orthogonal to the longitudinal axis of the fish, i.e., towards the fish (Fig. 15.2, 90°), generates a single symmetrical pressure peak. The corresponding pressure gradient pattern is biphasic, consisting of a negative peak followed by a positive peak of identical absolute height. A sphere that vibrates at an intermediate angle toward the fish (Fig. 15.2, 45° or 135°), also generates a single pressure peak and, consequently, a biphasic pressure gradient pattern. However, the negative and positive pressure gradient peaks differ in absolute height, i.e., a smaller peak is followed by a larger one (45° vibration angle) or vice versa (135° vibration angle), and the location of the trough between the peaks is shifted in one (45°) or the other (135°) direction.

If source distance is increased, the pressure gradient pattern is affected in a distinct way. First, since stimulus strength decreases with increasing source distance, the height of the pressure gradient peaks decreases. Second, and more importantly, the spatial separation of the troughs between the response peaks, and consequently, the spatial separation between response peaks, increases if source distance is increased. Since the spatial separation between response troughs (peaks) is linearly related to source distance, it is a direct measure for the distance between fish and source (Coombs et al. 1996; Goulet et al. 2008; Curcic-Blake and van Netten 2006). In contrast to changes in source distance, changes in size and vibration amplitude of a dipole will affect the height of a response peak but not the spatial separation between troughs (peaks) (Coombs et al. 1996; Goulet et al. 2008).

Afferent nerve fibers can respond with an increase or decrease in spike rate to a pressure gradient peak, depending on hair cell orientation and depending on whether the peak is negative or positive. However, the phase angle to which the fiber's response is locked will shift by 180° if the pressure gradient changes from negative to positive (Fig. 15.2). Consequently, if sphere vibration direction is 0°, the predicted excitation pattern across an array of linearly arranged lateral line receptors will be triphasic, consisting of a central peak bordered by two smaller peaks of excitation and the peaks are separated by two distinct changes in phase angle of 180° (Fig. 15.2, middle). If, in contrast, sphere vibration direction is 90°, the excitation pattern will exhibit two peaks of identical discharge rate separated by a single 180° change in phase angle. If sphere vibration is intermediate (45° or 135°), the excitation pattern will also exhibit two peaks of discharge rate separated by a single 180° change in phase angle. However, the peaks will be of different height and the changes in phase angle will be in opposite directions (Fig. 15.2, middle).

Physiological studies confirm the theoretically predicted activity both in terms of discharge rate and phase angle (Coombs and Conley 1997; Coombs et al. 1996; Goulet et al. 2008; Curcic-Blake and van Netten 2006). In other words, the spatial receptive fields of primary lateral line afferents match the spatial velocity/pressure gradient fields generated by a dipole source (Fig. 15.3). This means that the spatial information that is contained in the velocity/pressure gradient field around a dipole is maintained in the neuronal activity across the population of primary afferent

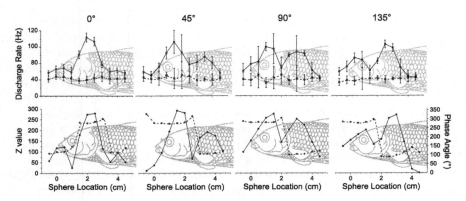

Fig. 15.3 Responses of a primary lateral line nerve fiber recorded in the posterior lateral line nerve of goldfish, *Carassius auratus*, to a sinusoidally (50 Hz) vibrating sphere placed at various locations alongside the fish and vibrating at different angles. This figure and Figs. 15.4 and 15.5 are organized as follows. *Upper graphs*: Ongoing discharge rates (*black lines*) and evoked rates (*dotted lines*) as a function of sphere location along the side of the fish. *Vertical* bars represent one standard deviation. *Lower graphs*: Z-values (*black lines*, *left Y*-axis) and mean phase angles (*dotted lines*, *right Y*-axis) as a function of sphere location. The fish drawing is scaled to the fish's actual body size. Adapted from (Künzel et al. 2011)

nerve fibers, i.e., across the array of lateral line neuromasts. Consequently, the brain receives the information that is necessary and sufficient to reconstruct the spatial location of a dipole source. This raises the question how the brain extracts specific aspects of a wave source from the activity across the neuromast array and how these aspects are encoded by the central nerve cells. Are there single cells tuned, for instance, to specific spatial locations (including distance) or vibration angles of a dipole, and if so, are these cells systematically organized in central maps? To address these questions, responses of single units in the MON and TS to dipole stimuli have been recorded.

15.4.3 Processing of Spectral and Temporal Aspects of a Dipole Stimulus by Central Neurons

Like primary lateral line afferents, many central lateral line units exhibit ongoing activity. However, the spike rates of both ongoing and evoked activity, on average, decrease along the ascending lateral line pathway (Bleckmann and Bullock 1989). In the MON, many units do not respond to a dipole stimulus even if displacement amplitudes of up to 800 μm are applied. However, these units respond readily to the complex water motions generated by a moving object (Mogdans and Goenechea 2000). Thus, MON units can be selective. Units that do respond to a dipole fall into several response categories. Some exhibit primary-like responses with robust phase-locking and sustained increases in discharge rate, as observed in primary afferent fibers, whereas others exhibit non-primary-like responses with poor phase-locking and variable response patterns, including increases or decreases in discharge rate, sustained (tonic) or phasic responses, and intermittent or build-up responses (Ali et al. 2010; Coombs et al. 1998; Künzel et al. 2011; Mogdans and Kröther 2001). In contrast to MON units, most midbrain units exhibit phasic responses to constant-amplitude dipole stimuli (Meyer et al. 2012; Plachta et al. 1999).

Based on responses to equal displacements of the dipole at different frequencies (isodisplacement response curves), both MON and TS units preferred either low (33 Hz), mid (50 and 100 Hz) or high (200 Hz) frequencies, i.e., units exhibited low-pass, band-pass, or high-pass characteristics (Plachta et al. 1999; Ali et al. 2010). If stimulated with amplitude-modulated water motions, both MON and TS units responded with a burst of discharge to each modulation cycle; that is, units exhibited strong phase-locking to the amplitude modulation frequency, but weak phase-locking to the carrier frequency (Plachta et al. 1999; Ali et al. 2010). The observation that responses of MON units to amplitude-modulated stimuli are in many respects comparable to those of TS units, suggests that many response properties emerge first in the lateral line brainstem. Moreover, these findings indicate that the central lateral line is more adapted for the processing of natural, amplitude, and/or frequency-modulated stimuli than for the processing of mono-frequent stimuli, which are rare or may not even occur at all in a natural environment.

15.4.4 Processing of Spatial Information of a Dipole Stimulus by Central Neurons

The spatial receptive fields of MON and TS units have been characterized by recording their responses to a dipole that slowly moved alongside the fish (Coombs et al. 1998) or that was placed at various locations along the side of the fish (Mogdans and Kröther 2001; Künzel et al. 2011; Meyer et al. 2012). In the MON, units with primary-like receptive fields were recorded, but the majority of MON units have non-primary-like receptive fields of different shapes and sizes that cannot directly be related to pressure gradient patterns generated by a dipole source (Coombs et al. 1998; Künzel et al. 2011). Some, but not all of the primary-like receptive fields could be modeled with excitatory center/inhibitory surround and inhibitory center/excitatory surround organizations (Coombs et al. 1998). Non-primary-like receptive fields may consist of (1) areas from which stimulation with the dipole causes an increase or, in other units, a decrease in discharge rate, (2) two or more areas from which the dipole causes an increase in discharge rate, (3) one area from which the dipole causes an increase and another area from which it causes a decrease in discharge rate (Fig. 15.4). Many of the non-primary-like receptive fields extended across most of the fish surface, indicating that many MON units receive information from afferent fibers that innervate large parts of the lateral line periphery (see also Caird 1978). Thus, there is substantial convergence

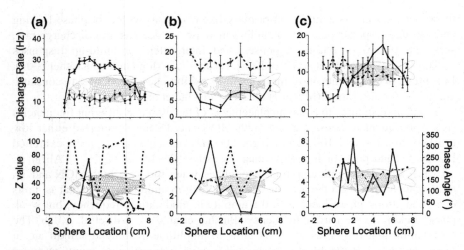

Fig. 15.4 Responses of MON units to a 50 Hz vibrating sphere placed at various locations alongside the fish. Angle of vibration was parallel to the longitudinal axis of the fish (0°). *Solid lines*: evoked spike rates, *stippled lines*: ongoing activity. Z-values ≥4.6 indicate significant phase-locking. **a** Example of a unit that responded with increased discharge rates to stimulation from nearly all locations along the side of the fish. **b** Example of a unit that responded with decreased discharge rates. **c** Example of a unit that responded with decreased discharge rates to stimulation near the head and increased discharge rates to stimulation near the trunk. Adapted from (Künzel et al. 2011)

of sensory information from the neuromast array onto MON cells. MON units tuned to specific spatial aspects of a dipole source, i.e., units responding exclusively when the dipole was at a distinct spatial location or vibrating at a distinct angle with respect to the surface of the fish have not been found.

Three studies have investigated the receptive fields of TS units (Engelmann and Bleckmann 2004; Meyer et al. 2012; Voges and Bleckmann 2011). The few single units recorded to date had non-primary-like receptive fields that were either broad, extending across large parts of the body surface, or consisted of two or more areas from which stimulation with the dipole caused an increase in discharge rate (for example see Fig. 15.5). These areas were located at specific sites with respect to the rostrocaudal and the dorso-ventral axis of the fish (Voges and Bleckmann 2011). Thus, the receptive fields of TS units are similar to those of many MON units, indicating that these units integrate across large parts of the sensory periphery. In contrast to the MON, TS units with primary-like receptive fields were not found. This suggests that the information that is provided by primary afferents to the fish brain about location and vibration direction of a vibrating sphere is encoded in a different way by TS units. Position and vibration direction of a stationary sphere might be determined unequivocally if the responses (spike rates and phase angles) of several units were taken into account. Thus, an alternative hypothesis is that at the level of the TS the lateral line uses a population code to determine the position of an object. However, considering the rather low number

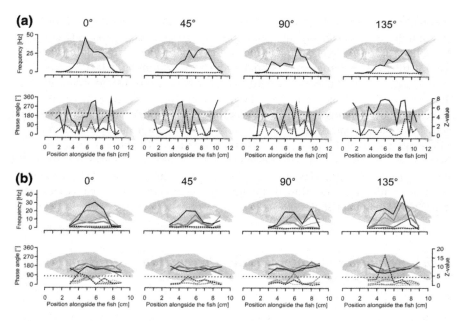

Fig. 15.5 Responses of toral lateral line units of the goldfish, *C. auratus*, to a 50 Hz vibrating sphere placed at various locations alongside the fish and vibrating in different directions. *Solid lines*: evoked spike rates, *stippled lines*: ongoing activity. **a** Example of a unit that responded with increased discharge rates to stimulation from nearly all locations along the side of the fish. Degree of phase-locking and phase angles was unpredictable across stimulus locations. **b** Example of a unit that responded with increased discharge rates to stimulation from a more restricted area in space. Different lines represent different distances of the sphere (black 10 mm, red 20 mm, blue 30 mm, brown 40 mm). Note that the area from which stimulation caused a rate increase was independent of sphere distance and that the degree of phase-locking and phase angle was constant across all stimulus locations. Adapted from Meyer et al. (2012)

of TS recordings obtained, the existence of space sensitive lateral line units cannot be ruled out. They may simply not have been found or exist at the next higher level of central integration, e.g., the optic tectum.

15.5 Processing of Flow Information

In nearly all physiological experiments performed to study lateral line function fish were fixed in a holder. Moreover, experiments were done in still water so that the fish were never exposed to unidirectional water flow or to hydrodynamic turbulences. In their natural habitat, fish rarely face still water conditions. Along the ocean shoreline as well as in rivers and creeks, fish are constantly exposed to gross water motions. And even in the deep sea or in ponds and lakes, most fish move around and thus are exposed to net water motions. The flow in rivers and creeks is often unsteady and contains vortices caused by rocks, roots, or boulders (Sutterlin and Waddy 1975).

Rheophilic fish use vortices and unsteady flow to save locomotory energy while station holding and probably also while migrating up- or downstream (Liao 2007; Przybilla et al. 2010). Wakes containing vortices not only occur in running water but also in still water environments behind undulatory swimming fish (Blickhan et al. 1992; Rosen 1959). Wakes can be used by seals and sea lions (Dehnhardt et al. 2001; Gläser et al. 2011; Hanke et al. 2000; Pohlmann et al. 2001; Wieskotten et al. 2011) (see chapter by Dehnhardt et al.) and, possibly, also by piscivorous fish (Pohlmann et al. 2004; Steiner and Bleckmann 2012) for prey detection and hydrodynamic trail following. To study the physiology of the lateral line in a more natural situation, researchers have stimulated fish with single vortex rings (Chagnaud et al. 2006), with bulk water flow (Chagnaud et al. 2007, 2008a, b; Engelmann et al. 2000; Voigt et al. 2000) and with bulk water flow that contained a Kármán vortex street (Bleckmann et al. 2012; Chagnaud et al. 2006).

15.5.1 Responses of Lateral Line Afferent Fibers to Bulk Water Flow

When fish (*Carassius auratus* or *Oncorhynchus mykiss*) are exposed to unidirectional bulk water flow (≤ 10 cm s^{-1}) lateral line afferents may (type I) or may not (type II) increase their discharge rate for as long as the flow continues (Engelmann et al. 2000; Voigt et al. 2000). This is true even if flow direction (anterior to posterior or posterior to anterior) is reversed (Chagnaud et al. 2008b) (see Fig. 15.6). Type I afferents most likely innervate SNs and type II afferents CNs (Engelmann et al. 2000). Since the two antagonistically arranged populations of hair cells within a lateral line neuromast are innervated by different afferent fibers (see above), this result was unexpected (Engelmann et al. 2000; Voigt et al. 2000). In any case, these data show that a single lateral line afferent cannot encode the direction of bulk water flow. Particle image velocimetry (PIV) revealed that even seemingly laminar water flow is superimposed by minute flow fluctuations and that the amplitude of these fluctuations increases with increasing bulk flow velocity (Chagnaud et al. 2008b) (See Fig. 15.6). Maximal spectral amplitudes of the flow fluctuations were below 5 Hz (bulk flow velocity 4–15 cm s^{-1}). The frequency spectra of the firing rates of lateral line afferents also showed an increase in amplitude when fish were exposed to running water. The maximal spectral amplitudes of the neuronal data were in the frequency range 3–8 Hz. This suggests that lateral line afferents do not respond to the DC-component of bulk water flow but only to the flow fluctuations (Chagnaud et al. 2008b). Since the frequency and amplitude content of these flow fluctuations not only depends on bulk flow velocity, but also on the size, shape, and distance of upstream objects, a single lateral line afferent cannot unequivocally encode bulk flow velocity.

From behavioral studies we do know that fish can determine bulk flow direction and bulk flow velocity (Baker and Montgomery 1999a; Montgomery et al. 1997).

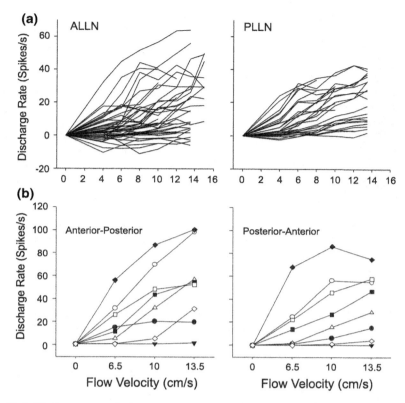

Fig. 15.6 Discharge rates of primary lateral line afferents to bulk water flow. **a** Discharge rates of anterior (*left*; $n = 42$) and posterior lateral line nerve fibers (*right*; $n = 29$) in reaction to bulk water flow. Spike rates (ongoing rates subtracted) are plotted as function of bulk flow velocity. **b** Discharge rates of anterior lateral line nerve fibers ($n = 8$) to water flow from anterior to posterior (*left*) and posterior to anterior (*right*). Each symbol refers to a single unit that was tested with flow in both directions. Figure modified after Chagnaud et al. (2008b)

If single lateral line afferents do not provide the necessary information, what allows fish to obtain this information? PIV-measurements have shown that flow fluctuations move downstream with bulk flow velocity. Since primary lateral line afferents are highly sensitive to flow fluctuations, nearly any flow fluctuation will be picked up by consecutively arranged SNs (Chagnaud et al. 2008a). Theoretically, two SNs separated in a downstream (rostrocaudal) direction are sufficient to determine bulk flow velocity by cross-correlating their responses (Chagnaud et al. 2008a) (for an example see Fig. 15.7). However, at a bulk flow velocity of 10 cm s^{-1} a cross correlation requires a time difference of 20 ms if two neuromasts are separated by a distance of 2 mm. At lower flow velocities, the necessary time difference is even larger. Such a large time difference cannot easily be obtained with a delay line.

Fig. 15.7 Cross-correlation functions of the responses of two lateral line afferents recorded simultaneously. Gross water flow direction was from rostral to caudal. Bulk flow velocity varied between 0 and 13.5 cm s^{-1}. Adapted from Chagnaud et al. (2008a). Note that cross-cerrelation peaks shift systematically with increasing flow velocity, corresponding to a distance between the neuromasts that were innervated by the two fibers of about 1.2 cm

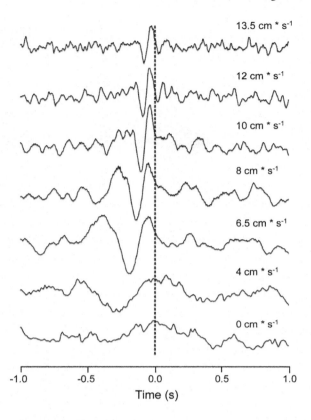

15.5.2 Responses of Central Lateral Line Units to Bulk Water Flow

Like primary lateral line afferents, many central units are also flow-sensitive, i.e., they may show sustained increases or decreases in discharge rate under flow conditions (up to 15 cms^{-1}) (Kröther et al. 2002). However, unlike primary afferents, the discharge rates of flow-sensitive MON and TS units either increase or decrease in response to bulk water flow (Fig. 15.8) (Engelmann and Bleckmann 2004).

15.5.3 Response Masking

When the lateral line is stimulated in still water with a stationary vibrating sphere, both type I and type II afferents respond with sustained, phase-locked discharges. If exposed to bulk water flow (<15 cm s^{-1}), the responses of type I fibers to a

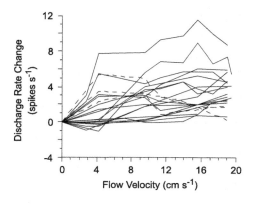

Fig. 15.8 Sensitivity of MON units to running water. Change in discharge rate is plotted as function of flow velocity. Data are shown from flow-sensitive units that responded with increasing ongoing discharge rates to increasing flow velocities. *Broken lines* represent data from units in which discharge rates decreased at flow velocities above approximately 10 cm s^{-1}. Modified after Kröther et al. (2002)

vibrating sphere stimulus are masked (Engelmann et al. 2000). In contrast, the responses of type II fibers are not masked; thus, trunk lateral line canals act as high-pass filters.

The responses of flow-sensitive MON units to a dipole stimulus are masked in running water, whereas the responses of flow-insensitive MON units are not. A third type of MON unit is not flow-sensitive, nevertheless the responses of these units are masked if the fish is exposed to bulk water flow (Kröther et al. 2002). These data again indicate that the functional subdivision of the lateral line periphery in an SN-system and a CN-system is maintained to a large degree at the level of the MON, although interactions between these two systems do occur.

15.5.4 Responses to Vortex Streets

At Reynolds numbers >140, flow behind a bluff 2-D body, such as a cylinder, generates a staggered array of discrete, periodically shed, columnar vortices of alternating sign. This flow pattern is called a Kármán vortex street (Vogel 1983). By changing flow velocity and/or cylinder diameter, the vortex shedding frequency and/or the wake wavelength, i.e., the spacing between successive vortices, systematically can be changed (Liao et al. 2003). If exposed to a Kármán vortex street, the average discharge rate of primary lateral line afferents hardly changes. However, in many afferents, the temporal pattern of discharge follows the vortex shedding frequency, as evidenced by a spectral analysis of the spike train (Chagnaud et al. 2007) (Fig. 15.9), especially if the fish intercepts the edge of the Kármán vortex street.

If exposed to a Kármán vortex street, MON units may respond with bursts of spikes, the burst frequency of which matches the vortex shedding frequency (for an example see Fig. 15.10). Thus, like primary lateral line afferents (see above) some

Fig. 15.9 Averaged frequency spectra of the neuronal activity of ALLN fibers (*upper curves, left y*-axis) and frequency spectra of the corresponding PIV traces (*lower curves, right y*-axis). From *top* to *bottom*: still water ($n = 85$), running water (flow velocity 10 cms^{-1}, $n = 58$), and Kármán vortex street condition (flow velocity 10 cms^{-1}, cylinder diameter 2.5 cm, $n = 46$). *Arrow* indicates the frequency of the calculated vortex shedding frequency. Adapted from Chagnaud et al. (2006)

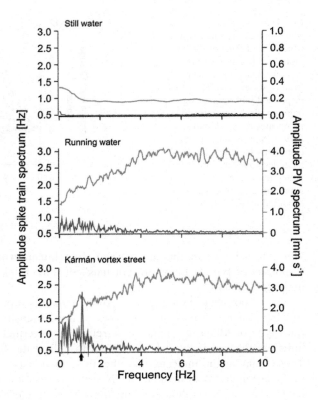

MON units encode certain aspects of a vortex street. The representation of the vortex shedding frequency is even better in the MON than in primary lateral line afferents (Bleckmann et al. 2012), i.e., central mechanisms lead to an improved signal-to-noise ratio.

As already mentioned, a cross correlation of at least two flow-sensitive afferent responses, one of which derives from an upstream and one from a downstream neuromast, is a possible mechanism to determine bulk flow velocity (Chagnaud et al. 2008b). To implement such a mechanism, the nervous system would likely use a delay line as found in the visual system of invertebrates, the electrosensory system of weakly electric fish and in the auditory system of barn owls (Carr 1993; Carr and Konishi 1988; Egelhaaf et al. 1993; Joris et al. 1998). In order for fish to use a cross-correlation mechanism, there should be a subpopulation of flow-sensitive central lateral line units that are (1) highly directionally sensitive, (2) tuned to a particular bulk flow velocity, and (3) insensitive to flow fluctuations. Although flow-sensitive MON and TS units may be highly sensitive to gross flow direction (Fig. 15.10) neurons which satisfy all three criteria have not been found so far (Hellinger and Bleckmann, unpublished), and thus the current physiological data do not support the idea of a cross-correlation mechanism (Fig. 15.11).

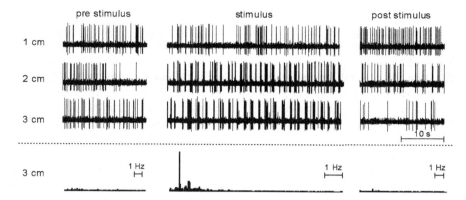

Fig. 15.10 Responses of a MON unit to uniform flow (*left* and *right*) and to the vortex motions caused by a cylinder of 1, 2, and 3 cm diameter (*middle*). Flow velocity varied between 6 and 7 cms^{-1}. Note that the activity of the unit was fairly regular in uniform flow. *Bottom*: Spectral composition of the neuronal responses. The peak in the *middle* spectrum at 0.56 Hz was close to the measured vortex shedding frequency (0.63 Hz). From Bleckmann et al. (2012)

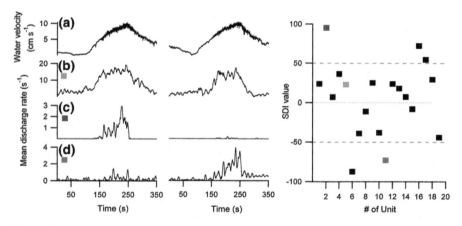

Fig. 15.11 *Left* (**a–d**): Anemometer trace showing flow stimuli (**a**). Response of a unit without directional preference (**b**). Directionally selective responses preferring anterior/posterior flow (**c**) or posterior/anterior flow (**d**). *Right*: Signed directionality index (*SDI*) for 18 flow-sensitive midbrain units. Positive SDI values indicate a preference for anterior–posterior, negative SDI values for posterior–anterior flow. An SDI of 100 (−100) indicates that the unit responds exclusively to one flow direction. Note the wide range of directional selectivity. *Colors* refer to the example units shown in **b**, **c**, and **d**. The figure was generously provided by V. Hofmann

15.6 Conclusions

Since the lateral line conference that took place in Bielefeld in 1989, our knowledge about the central processing of lateral line information in fish has increased considerably. Nevertheless, many questions remain unanswered.

First, our knowledge about the temporal and spatial characteristics of biologically meaningful hydrodynamic stimuli is still sparse, although some progress has been made in recent years (Bleckmann et al. 1991; Hanke and Bleckmann 2004; Hanke et al. 2000; Montgomery and Macdonald 1987). Second, under natural conditions, the lateral line system has to cope with biological meaningful stimuli but also with self-generated and externally generated hydrodynamic noise. Physiological studies (Engelmann et al. 2000) and studies using artificial lateral lines (Denton and Gray 1983, 1988; Klein et al. 2013) have shown that mechanical filters are implemented in the peripheral lateral line, but physiological filters may also be involved (Montgomery and Macdonald 1987; Weeg and Bass 2002). To uncover the filter properties of the lateral line, more studies are needed in which the animals are exposed to quasi-natural noise conditions. Recordings from peripheral and central lateral line units of freely swimming fish may be particularly useful to uncover hydrodynamic filters that deal with self-generated noise. Natural noise conditions and biologically meaningful stimuli may be highly species specific, depending on the lifestyle and natural hydrodynamic environment of the respective species.

In both aquatic amphibians and fish, primary lateral line afferents bifurcate in the MON and form ascending and descending tracts (Bell 1981; Blübaum-Gronau and Münz 1987; Claas and Münz 1981; Fritzsch 1981; Puzdrowski 1989; Song and Northcutt 1991; Will et al. 1985). The functional significance of this bifurcation is still not known. Anatomical studies have identified at least five cell types in the MON of goldfish (Caird 1978; New et al. 1996). Although we have some knowledge about the physiology of MON cells (see above), we know little about their functions and interconnections. Intracellular recording techniques with subsequent staining of the cells—as first done by Caird (1978)—will help to reveal this relationship. Knowing about this relationship is a prerequisite to developing circuit diagrams of any kind. To date, the only comprehensive lateral line circuit diagram is that given by (Montgomery and Bodznick 1994). This still hypothetical diagram suggests that one function of the MON is the cancelation of self-generated noise. We also do not know the functional significance of the CON.

Like primary lateral line afferents, many central lateral line units phase-lock to sinusoidal water motions, only some of which phase-lock to both halves of a full wave cycle, indicating that they do not integrate inputs with opposing directional sensitivities (e.g., headward or tailward). Some central neurons do not phase-lock at all, whereas others phase-lock to both halves of a full wave cycle. Obviously, these neurons receive input from hair cells that are antagonistically aligned. While phase-locking is one way to encode the carrier and/or modulation frequency of a sinusoidal stimulus, the functional significance of this convergence is not known. The absence of phase-locking in some central units is also not well understood, although it could reflect a wide integration of inputs from differently oriented neuromasts on different body regions.

In gnathostomes, a part of the primary lateralis projection to the MON commonly continues into the eminentia granularis of the vestibulolateral cerebellum (Claas and Münz 1981; New and Northcutt 1984). Nothing is known about the sensory information that is transmitted through these fibers. Some granular cells of the corpus cerebellum of the thornback guitarfish, *Plathyrinoides triseriata* (Fiebig 1988), but also in the cerebellum of teleosts (Lee and Bullock 1984) respond to lateral line stimuli. These cells my be involved in noise cancelation (Bell et al. 1997).

A fairly large number of studies investigated the response properties of toral lateral line units (Engelmann and Bleckmann 2004; Plachta et al. 2003; Wojtenek et al. 1998; Zittlau et al. 1986). According to these studies, the response properties of many toral lateral line units, if stimulated with sinusoidal water motions, are similar to the response properties of MON units. Toral lateral line units may be highly directional if stimulated with gross water flow or an object that passes the fish laterally. Despite a fairly large number of physiological studies, not much is known about the toral circuit diagrams that lead to this directionality and the information that enters and leaves the torus and the specific processing of information in the TS. Some toral lateral line units project to the ipsilateral tectum opticum, but nothing is known about the processing of lateral line information in the tectum of fish. This is especially unfortunate since we know from studies in amphibians that tectal lateral line maps do exist and that these maps encode the direction from which a surface wave train impinges on the animal (Bartels et al. 1990; Zittlau et al. 1986). With these exceptions, very few physiological studies have uncovered lateral line directional maps (Bleckmann et al. 1989b; Plachta et al. 2003).

Another huge gap in our knowledge refers to the forebrain of fishes. Both the diencephalon and telencephalon receive lateral line information (Bleckmann et al. 1987, 1989b; Prechtl et al. 1998). In the diencephalon, the posterior tuberculum (preglomerular complex in teleosts), the dorsal thalamus and, in some fish, hypothalamic centers (e.g., the anterior tuberal nucleus) are involved in lateral line processing. Projections to the telencephalic pallium arise from the posterior tubercular and/or dorsal thalamic lateral line nuclei. Lateral line units (and evoked potentials) recorded from forebrain lateral line areas require inter-stimulus intervals of up to 50 s in order to respond (Bleckmann et al. 1987, 1989a; Kirsch et al. 2002). This makes forebrain studies in fishes very difficult and is probably one of the reasons why so little is known about the physiology of these brain areas. Also poorly understood is the functional significance of an apparent decrement in the magnitude of neural responses to, e.g., a dipole along the ascending lateral line pathway. One intriguing question is whether the responses of forebrain units can be dishabituated with novel stimuli. If so, the discrimination abilities of the central lateral line pathway could easily be studied.

The study of the visual, acoustic, and electrosensory systems has revealed highly selective central neurons in mammals (Hubel and Wiesel 1962), amphibians (Maturana et al. 1960), crustaceans (Waterman and Wiersma 1963), birds (Konishi 1983), weakly electric fish (Heiligenberg and Bastian 1984), and insects (Pollack 1998). Although highly selective central lateral line units do exist (see above), most central lateral line units respond to a variety of hydrodynamic stimuli (e.g., Bleckmann et al. 1989b). One reason for the rarity of selective central lateral line units is that we may have simply failed to find the right stimuli or combinations of features for evoking selectivity.

Another huge gap in our knowledge refers to the function of the efferent lateral line pathway. In the electrosensory system, the praeeminential nucleus (the nucleus B in cartilaginous fishes may be its homolog) is involved in gain control and attention (Bastian 1986), a finding that may also apply to the mechanosensory lateral line and the auditory system. Although in fish—like in other vertebrates—several senses are involved in the identification and localization of inanimate and animate objects we have still not enough knowledge about where the information from different sensory modalities converge and how they are integrated by central units (Bleckmann et al. 1987; Kirsch et al. 2002).

Besides asking more and more refined questions about the physiology of the lateral line of a "standard" mid-water teleost, e.g., the goldfish, we should also investigate taxa that are highly specialized with respect to this sensory modality—e.g., fish that have a close peripheral contact between the lateralis and the otic system (like some clupeids, mormyrids, and catfish), blind cave fish that highly depend on lateral line information while swimming around, or fish with multiple lateral lines like *Xiphister atropurpureus*. Most lateral line studies were done with teleosts. Nothing (or very little) is known about the peripheral and central processing of lateral line information in myxinoids, sharks, in non-teleost ray-finned fishes (polypteriformes and chondrosteans), and in teleosts with electroreception (gymnotids, mormyrids, silurids). We also should look for commonalities or differences between more recent and more ancient taxa.

Acknowledgments We are indebted to Sheryl Coombs for carefully reading and commenting on the manuscript. The original research of the authors was generously supported by the DFG, the BMBF, DARPA, the BfG, DAAD, and the EU.

References

Ali R, Mogdans J, Bleckmann H (2010) Responses of medullary lateral line units of the goldfish, *Carassius auratus*, to amplitude-modulated sinusoidal wave stimuli. Intern J Zool 2010:1–14

Baker CF, Montgomery JC (1999a) Lateral line mediated rheotaxis in the Antarctic fish *Pagnothenia borchgrevinki*. Polar Biol 21:305–309

Baker CF, Montgomery JC (1999b) The sensory basis of rheotaxis in the blind Mexican cave fish, *Astyanax fasciatus*. J Comp Physiol A 184:519–527

Bartels M, Münz H, Claas B (1990) Representation of lateral line and electrosensory systems in the midbrain of the axolotl, *Ambystoma mexicanum*. J Comp Physiol A 167:347–356

Bastian J (1986) Gain control in the electrosensory system: a role for the descending projections to the electrosensory lateral line lobe. J Comp Physiol A 158:505–515

Bell CC (1981) Central distribution of octavolateral afferents and efferents in a teleost (Mormyridae). J Comp Neurol 195:391–414

Bell C, Bodznick D, Montgomery J, Bastian J (1997) The generation and subtraction of sensory expectations within cerebellum-like structures. Brain Behav Evol 50:17–31

Bleckmann H (2007) Peripheral and central processing of lateral line information. J Comp Physiol A 194:145–158

Bleckmann H, Bullock TH (1989) Central nervous physiology of the lateral line, with special reference to cartilaginous fishes. In: Coombs S, Görner P, Münz H (eds) The mechanosensory lateral line. Neurobiology and evolution. Springer, New York, pp 387–408

Bleckmann H, Breithaupt T, Blickhan R, Tautz J (1991) The time course and frequency content of hydrodynamic events caused by moving fish, frogs, and crustaceans. J Comp Physiol A 168:749–757

Bleckmann H, Bullock TH, Jørgensen JM (1987) The lateral line mechanoreceptive mesencephalic, diencephalic, and telencephalic regions in the thornback ray, *Platyrhinoidis triseriata* (Elasmobranchii). J Comp Physiol A 161:67–84

Bleckmann H, Mogdans J, Dehnhardt G (2003) Processing of dipole and more complex hydrodynamic stimuli under still- and running-water conditions. In: Collin SP, Marshall NJ (eds) Sensory processing in aquatic environments. Springer, New York, pp 108–121

Bleckmann H, Przybilla A, Klein A, Schmitz A, Kunze S, Brücker C (2012) Station holding of trout: behavior, physiology and hydrodynamics. In: Tropea C, Bleckmann H (eds) Nature-inspired fluid mechanics. Notes on numerical fluid mechanics and multidisciplinary design. Springer, Berlin, pp 161–187

Bleckmann H, Tittel G, Blübaum-Gronau E (1989a) The lateral line system of surface-feeding fish: Anatomy, physiology, and behavior. In: Coombs S, Görner P, Münz H (eds) The mechanosensory lateral line. Neurobiology and evolution. Springer, New York, pp 501–526

Bleckmann H, Weiss O, Bullock TH (1989b) Physiology of lateral line mechanoreceptive regions in the elasmobranch brain. J Comp Physiol A 164:459–474

Blickhan R, Krick C, Breithaupt T, Zehren D, Nachtigall W (1992) Generation of a vortex-chain in the wake of a subundulatory swimmer. Naturwi 79:220–221

Blübaum-Gronau E, Münz H (1987) Topological representation of primary afferents in various segments of the lateral line system in the butterflyfish, *Pantodon buchholzi*. Verhandlungen der Deutschen Zoologischen Gesellschaft. Gustav Fischer, Stuttgart, pp 268–269

Caird DM (1978) A simple cerebellar system: the lateral line lobe of the goldfish. J Comp Physiol A 127:61–74

Carr CE (1993) Processing of temporal information in the brain. Ann Rev Neurosci 16:223–243

Carr CE, Konishi M (1988) Axonal delay lines for time measurement in the owl's brainstem. Proc Natl Acad Sci USA 85:8311–8315

Von Campenhausen C, Riess I, Weissert R (1981) Detection of stationary objects in the blind cave fish *Anoptichthys jordani* (Characidae). J Comp Physiol A 143:369–374

Chagnaud BP, Bleckmann H, Engelmann J (2006) Neural responses of goldfish lateral line afferents to vortex motions. J Exp Biol 209:327–342

Chagnaud BP, Bleckmann H, Hofmann M (2007) Kármán vortex street detection by the lateral line. J Comp Physiol A 193:753–763

Chagnaud B, Brücker C, Hofmann MH, Bleckmann H (2008a) Measuring flow velocity and flow direction by spatial and temporal analysis of flow fluctuations. J Neurosci 28:4479–4487

Chagnaud BP, Bleckmann H, Hofmann MH (2008b) Lateral line nerve fibers do not respond to bulk water flow direction. Zoology 111:204–207

Claas B, Münz H (1981) Projection of lateral line afferents in a teleost brain. Neurosci Lett 23:287–290

Coombs S, Conley RA (1997) Dipole source localization by mottled sculpin II. The role of lateral line excitation patterns. J Comp Physiol A 180:401–416

Coombs S, Hastings M, Finneran J (1996) Modeling and measuring lateral line excitation patterns to changing dipole source locations. J Comp Physiol A 178:359–371

Coombs S, Janssen J, Webb JF (1988) Diversity of lateral line systems: evolutionary and functional considerations. In: Atema J, Fay RR, Popper AN, Tavolga WN (eds) Sensory biology of aquatic animals. Springer, New York, pp 553–593

Coombs S, Mogdans J, Halstead M, Montgomery J (1998) Transformation of peripheral inputs by the first-order lateral line brainstem nucleus. J Comp Physiol A 182:609–626

Curcic-Blake B, van Netten SM (2006) Source localization encoding in the fish lateral line. J Exp Biol 209:1548–1559

Dehnhardt G, Mauck B, Hanke W, Bleckmann H (2001) Hydrodynamic trail-following in harbor seals (*Phoca vitulina*). Science 293:102–104

Denton EJ, Gray JAB (1983) Mechanical factors in the excitation of clupeid lateral lines. Proc R Soc London B 218:1–26

Denton EJ, Gray JAB (1988) Mechanical factors in the excitation of lateral line canals. In: Atema J, Fay RR, Popper AN, Tavolga WN (eds) Sensory biology of aquatic animals. Springer, New York, pp 595–617

Echteler SM (1984) Connections of the auditory midbrain in a teleost fish, *Cyprinus carpio*. J Comp Neurol 230:536–551

Egelhaaf M, Borst A, Reichardt W (1993) Computational structure of a biological motion-detection system as revealed by local detector analysis in the fly's nervous system. J Opt Soc Am A 6: 1070–1087

Engelmann J, Bleckmann H (2004) Coding of lateral line stimuli in the goldfish midbrain in still-and running water. Zoology 107:135–151

Engelmann J, Hanke W, Mogdans J, Bleckmann H (2000) Hydrodynamic stimuli and the fish lateral line. Nature 408:51–52

Fiebig E (1988) Connections of the corpus cerebelli in the thornback guitarfish, *Platyrhinoidis triseriata* (Elasmobranchii): a study with WGA-HRP and extracellular granule cell recording. J Comp Neurol 268:567–583

Flock A (1965) Electronmicroscopic and electrophysiological studies on the lateral line canal organ. Acta Otolaryngol 199:1–90

Flock A, Wersäll J (1962) A study of the orientation of sensory hairs of the receptor cells in the lateral line organ of a fish with special reference to the function of the receptors. J Cell Biol 15:19–27

Fritzsch B (1981) The pattern of lateral-line afferents in urodeles: a horseradish-peroxidase study. Cell Tiss Res 218:581–594

Gläser N, Otter C, Dehnhardt G, Hanke W (2011) Hydrodynamic trail following in a California sea lion (*Zalophus californianus*). J Comp Physiol A 197:141–151

Görner P (1963) Untersuchungen zur Morphologie und Elektrophysiologie des Seitenlinienorgans vom Krallenfrosch (*Xenopus laevis* Daudin). J Comp Physiol A 47:316–338

Goulet J, Engelmann J, Chagnaud B, Franosch J-MP, Suttner MD, van Hemmen JL (2008) Object localization through the lateral line system of fish: theory and experiment. J Comp Physiol A 194:1–17

Hanke W, Bleckmann H (2004) The hydrodynamic trails of *Lepomis gibbosus* (Centrarchidae), *Colomesus psittacus* (Tetraodontidae) and *Thysochromis ansorgii* (Cichlidae) investigated with scanning particle image velocimetry. J Exp Biol 207:1585–1596

Hanke W, Brücker C, Bleckmann H (2000) The ageing of the low frequency water disturbances caused by swimming goldfish and its possible relevance to prey detection. J Exp Biol 203:1193–1200

Harris GG, Frishkopf LS, Flock Å (1970) Receptor potentials from hair cells of the lateral line. Science 167:76–79

Heiligenberg W, Bastian J (1984) The electric sense of weakly electric fish. Ann Rev Psychol 46:561–583

Hubel DH, Wiesel TN (1962) Receptive fields, binocular interaction and functional architecture of the cat's visual cortex. J Physiol 160:106–154

Janssen J (1997) Comparison of response distance to prey via the lateral line in the ruffe and the yellow perch. J Fish Biol 51:921–930

Janssen J, Corcoran J (1993) Lateral line stimuli can override vision to determine sun fish strike trajectory. J Exp Biol 176:299–305

Joris PX, Smith PH, Yin TCT (1998) Coincidence detection in the auditory system: 50 years after Jeffress. Neuron 21:1235–1238

Kalmijn AJ (1988) Hydrodynamic and acoustic field detection. In: Atema J, Fay RR, Popper AN, Tavolga WN (eds) Sensory biology of aquatic animals. Springer, New York, pp 83–130

Kalmijn AJ (1989) Functional evolution of lateral line and inner ear sensory systems. In: Coombs S, Görner P, Münz H (eds) The mechanosensory lateral line. Neurobiology and evolution. Springer, New York, pp 187–216

Kaus S, Schwartz E (1986) Reaction of young *Betta splendens* to surface waves of the water In: Barth FG, Seyfarth EA (eds) Verhandlungen der Deutschen Zoologischen Gesellschaft. Gustav Fischer, Stuttgart, pp 218–219

Kirsch JA, Hofmann MH, Mogdans J, Bleckmann H (2002) Response properties of diencephalic neurons to visual, acoustic and hydrodynamic stimulation in the goldfish, *Carassius auratus.* Zoology 105:61–70

Klein A, Münz H, Bleckmann H (2013) The functional significance of lateral line canal morphology on the trunk of the marine teleost *Xiphister atropurpureus* (Stichaeidae). J Compe Physiol A 199:735–749

Konishi M (1983) Neuroethology of acoustic prey localization in the barn owl. In: Huber F, Markl H (eds) Neuroethology and behavioral physiology. Springer, Berlin, pp 304–317

Kroese ABA, Schellart NAM (1992) Velocity- and acceleration-sensitive units in the trunk lateral line of the trout. J Neurophysiol 68:2212–2221

Kröther S, Mogdans J, Bleckmann H (2002) Brainstem lateral line responses to sinusoidal wave stimuli in still- and running water. J Exp Biol 205:1471–1484

Künzel S, Bleckmann H, Mogdans J (2011) Responses of brainstem lateral line units to different stimulus source locations and vibration directions. J Comp Physiol A 197:773–787

Lee LT, Bullock TH (1984) Sensory representation in the cerebellum of the catfish. J Comp Physiol A 13:157–169

Liao JC (2007) A review of fish swimming mechanics and behaviour in altered flows. Phil Trans R Soc B 362:1973–1993

Liao JC, Beal DN, Lauder GV, Triantafyllou MS (2003) The Kármán gait: novel body kinematics of rainbow trout swimming in a vortex street. J Exp Biol 206:1059–1073

Maturana HR, Lettvin JY, McCulloch WS, Pitts WH (1960) Anatomy and physiology of vision in the frog (*Rana pipiens*). J Gen Physiol 43:129–175

McCormick CA, Hernandez DV (1996) Connections of the octaval and lateral line nuclei of the medulla in the goldfish, including the cytoarchitecture of the secondary octaval population in goldfish and catfish. Brain Behav Evol 47:113–138

Meyer G, Klein A, Mogdans J, Bleckmann H (2012) Toral lateral line units of goldfish, *Carassius auratus,* are sensitive to the position and vibration direction of a vibrating sphere. J Comp Physiol A 198:639–653

Mirjany M, Faber DS (2011) Characteristics of the anterior lateral line nerve input to the Mauthner cell. J Exp Biol 214:3368–3377

Mogdans J, Bleckmann H (1999) Peripheral lateral line responses to amplitude modulated hydrodynamic stimuli. J Comp Physiol A 185:173–180

Mogdans J, Goenechea L (2000) Responses of medullary lateral line units in the goldfish, *Carassius auratus,* to sinusoidal and complex wave stimuli. Zoology 102:227–237

Mogdans J, Kröther S (2001) Brainstem lateral line responses to sinusoidal wave stimuli in the goldfish, *Carassius auratus.* Zoology 104:153–166

Montgomery JC, Baker CF, Carton AG (1997) The lateral line can mediate rheotaxis in fish. Nature 389:960–963

Montgomery JC, Bodznick D (1994) An adaptive filter that cancels self-induced noise in the electrosensory and lateral line mechanosensory systems of fish. Neurosci Lett 174:145–148

Montgomery JC, Macdonald JA (1987) Sensory tuning of lateral line receptors in Antarctic fish to the movements of planctonic prey. Science 235:195–196

Münz H (1985) Single unit activity in the peripheral lateral line system of the cichlid fish *Sarotherodon niloticus L.* J Comp Physiol A 157:555–568

Münz H (1989) Functional organization of the lateral line periphery. In: Coombs S, Görner P, Münz H (eds) The mechanosensory lateral line. Neurobiology and evolution. Springer, New York, pp 285–298

New JG, Coombs S, McCormick CA, Oshel PE (1996) Cytoarchitecture of the medial octavolateralis nucleus in the goldfish, *Carassius auratus*. J Comp Neurol 366:534–546

New JG, Northcutt RG (1984) Central projections of the lateral line nerves in the shovelnose sturgeon. J Comp Neurol 225:129–140

Northcutt RG (1989) The phylogenetic distribution and innervation of craniate mechanoreceptive lateral lines. In: Coombs S, Görner P, Münz H (eds) The mechanosensory lateral line Neurobiology and evolution. Springer, New York, pp 17–78

Plachta D, Hanke W, Bleckmann H (2003) A hydrodynamic topographic map and two hydrodynamic subsystems in a vertebrate brain. J Exp Biol 206:3479–3486

Plachta D, Mogdans J, Bleckmann H (1999) Responses of midbrain lateral line units of the goldfish, *Carassius auratus*, to constant-amplitude and amplitude modulated water wave stimuli. J Comp Physiol A 185:405–417

Pohlmann K, Atema J, Breithaupt T (2004) The importance of the lateral line in nocturnal predation of piscivorous catfish. J Exp Biol 207:2971–2978

Pohlmann K, Grasso FW, Breithaupt T (2001) Tracking wakes: the nocturnal predatory strategy of piscivorous catfish. Proc Nat Acad Sci 98:7371–7374

Pollack GS (1998) Neural processing of acoustic signals. In: Hoy RR, Fay RR, Popper AN (eds) Comparative hearing: insects. Springer, New York, p 341

Prechtl JC, von der Emde G, Wolfart J, Karamürsel S, Akoev GN, Andrianov YN, Bullock TH (1998) Sensory processing in the pallium of a mormyrid fish. J Neurosci 18:7381–7393

Przybilla A, Kunze S, Ruder A, Bleckmann H, Brücker C (2010) Entraining trout: a behavioural and hydrodynamic analysis. J Exp Biol 213:2976–2986

Puzdrowski RL (1989) Peripheral distribution and central projections of the lateral-line nerves in goldfish, *Carassius auratus*. Brain Behav Evol 34:110–131

Rosen MW (1959) Waterflow about a swimming fish. Tech Publ US Naval Test Station, China Lake, California, NOTS TP 2298, pp 1–94

Sand O (1981) The lateral line and sound reception. In: Tavolga WN, Popper AN, Fay RR (eds) Hearing and sound communication in fishes. Springer, New York, pp 459–480

Satou M, Shiraishi A, Matsushima T, Okumoto N (1991) Vibrational communication during spawning behavior in the hime salmon (landlocked red salmon, *Oncorhynchus nerka*). J Comp Physiol A 168:417–428

Song J, Northcutt RG (1991) The primary projections of the lateral-line nerves of the Florida gar, *Lepisosteus platyrhincus*. Brain Behav Evol 37:38–63

Steiner A, Bleckmann H (2012) Responses of fishes to Kármán vortex streets and artificial fish generated wakes. In: 105th annual meeting of the German Zoological Society, Konstanz

Sutterlin AM, Waddy S (1975) Possible role of the posterior lateral line in obstacle entrainment by brook trout (*Salvelinus fontinalis*). J Fish Res Board Canada 32:2441–2446

van Netten SM, Kroese ABA (1989) Dynamic behavior and micromechanical properties of the cupula. In: Coombs S, Görner P, Münz H (eds) The mechanosensory lateral line. Neurobiology and evolution. Springer, New York, pp 247–264

Vogel S (1983) Life in moving fluids. The physical biology of flow. Princeton University Press, Princeton

Voges K, Bleckmann H (2011) Two-dimensional receptive fields of midbrain lateral line units in the goldfish, *Carassius auratus*. J Comp Physiol A 197:827–837

Voigt R, Carton AG, Montgomery JC (2000) Responses of anterior lateral line afferent neurones to water flow. J Exp Biol 203:2495–2502

Walkowiak W, Münz H (1985) The significance of water-surface waves in the communication of fire-bellied toads. Naturwi 72:49–50

Waterman TH, Wiersma CAG (1963) Electrical responses in decapod crustaceans visual systems. J Cell Comp Physiol 61:1–16

Weeg MS, Bass A (2002) Frequency response properties of lateral line superficial neuromasts in a vocal fish, with evidence for acoustic sensitivity. J Neurophysiol 88:1252–1262

Weissert R, Von C Campenhausen (1981) Discrimination between stationary objects by the blind cave fish *Anoptichthys jordani*. J Comp Physiol A 143:375–382

Wieskotten S, Mauck B, Miersch L, Dehnhardt G, Hanke W (2011) Hydrodynamic discrimination of wakes caused by objects of different size or shape in a harbour seal (*Phoca vitulina*). J Exp Biol 214:1922–1930

Will U, Luhede G, Görner P (1985) The area octavo-lateralis in *Xenopus laevis*. I. The primary afferent projections. Cell Tissue Res 239:147–161

Wojtenek W, Mogdans J, Bleckmann H (1998) The responses of midbrain lateral line units of the goldfish *Carassius auratus* to moving objects. Zoology 101:69–82

Wullimann MF, Grothe B (2013) The central nervous organization of the lateral line system. In: Coombs S, Bleckmann H, Fay RR, Popper AN (eds) Springer Handbook of Auditory Research. The Lateral Line System. Springer, New York, pp 195–251

Zittlau KE, Claas B, Münz H (1986) Directional sensitivity of lateral line units in the clawed toad *Xenopus laevis* Daudin. J Comp Physiol A 158:469–477

Part V
Modelling of Flow Sensing and Artificial Flow Sensors

Chapter 16
Hydrodynamic Object Formation: Perception, Neuronal Representation, and Multimodal Integration

J. Leo van Hemmen

Abstract Lateral-line encoding is diffuse, needing at least a large part of the fish body and several detectors to measure the information contained in the velocity or pressure field surrounding a fish or an aquatic frog such as *Xenopus*. This paper presents a careful analysis of the mathematical mechanisms and algorithms underlying neuronal information processing as it is performed by the lateral-line system both in the perception ensuing from neuromasts and in the resulting neuronal representations, the maps. The goal is to explicitly show how the lateral line can simultaneously perceive several objects, e.g., identical ones, which role fish geometry plays in lateral-line detection, and why its direct range is short, about one fish length. A lateral-line 'object' in the outside world has both position and shape and the lateral line can handle both, at the price of having a restricted range. Detection of vortex wakes as hydrodynamic entities exhibiting the consequence of conservation of angular momentum is also analyzed and contrasted with the instantaneous momentum transfer studied as the usual lateral-line stimulus. Finally, it is shown how lateral-line 'objects' may arise neuronally both separately and in the context of a multimodal integration of the lateral-line system and vision, and a concrete theory of map formation in the torus on the basis of neuroanatomy and spike-timing-dependent plasticity (STDP) in conjunction with local excitation and global inhibition is presented. An appendix gives a full and simple mathematical account of surface-wave hydrodynamics, including surface tension.

Keywords Lateral line · Neuronal information processing · Hydrodynamics · Hydrodynamic image · Lateral-line range · Vortex wake · Multimodality · Multimodal integration · Map formation · STDP

J. L. van Hemmen (✉)
Physik Department T35, Technische Universität München, 85747 Garching bei München, Germany
e-mail: lvh@tum.de

H. Bleckmann et al. (eds.), *Flow Sensing in Air and Water*,
DOI: 10.1007/978-3-642-41446-6_16, © Springer-Verlag Berlin Heidelberg 2014

16.1 Introduction

The lateral line is a rather fascinating detection and evaluation system because of at least two reasons. First, the way in which it encodes hydrodynamic signals is diffuse. That is, fish and several aquatic frogs such as *Xenopus* have superficial detectors or superficial neuromasts (surface neuromasts) all over the body that respond to the *velocity* field of the water surrounding the animal. Moreover, fish also have canal neuromasts. Each of the latter is situated between two pores in a duct or "canal" and responds to the canal fluid velocity generated by a pressure difference between the pores. If a hydrodynamic object passes at a distance that is small enough, then neuromasts over a considerable part or even all of the body get excited and respond. This is what one may call *diffuse* encoding since all the neuromasts together encode the information presented to the animal by the object in its neighborhood.

On the other hand, in vision eyes have a decently functioning lens that focuses an outside visual object onto the retina. Discrete aspects remain discrete and this kind of perception is therefore called discrete encoding. In response to a small outside object, only a small amount of detectors gets appreciably excited. Somatosensory perception or haptics (Flanders 2011) is a very explicit example of discrete encoding. The Jeffress scheme of azimuthal sound localization in birds (Schnupp and Carr 2009; Grothe et al. 2010) also operates in a discrete way. The tympani of the ears in birds already do so while catching a precisely and hence discretely defined interaural time difference, but the lateral-line system does not. It is diffuse. As diffuse as active electrolocation (Coombs et al. 2002; Hopkins 2009; von der Emde and Engelmann 2011) used by, e.g., weakly electric fish, which emit electric pulses whose "echo" returns from objects in the water surrounding them to electroreceptors all over the skin. Though based on completely different physical mechanisms, a diffuse encoding such as that of the lateral line or electroreception and a discrete encoding such as vision through the fish eye—think of the fish-eye camera—nevertheless get integrated in the torus semicircularis, or for short torus (Scheich and Ebbesson 1983), and in the optic tectum (Vanegas et al. 1984a, b), and are bound to identify the very same object. That makes understanding multimodal integration a fascinating challenge. Both multimodal integration and the neuronal map formation accompanying it will be analyzed here.

A second intriguing aspect of the lateral-line system is that it operates without direct contact; so to speak, through action at a distance. How far or how near, then, is its range? As is explained below in Sect. 16.3, the answer is one fish length, in a double sense. To see what that means, let us take a predator-prey pair. The former has, say, length L while the latter has length ℓ. For the prey to get localized by for instance a pike, its distance to the predator should be less than approximately L. If the predator then also wants to determine the prey's specific form, the prey fish length ℓ is the determining factor and its distance to the predator, here the pike, should be less than about ℓ.

In a sense, the lateral-line system even presents a third fascinating problem, based on the previous two: How to mathematically model and analyze neuronal *information* processing as it ensues from the lateral-line system and is performed by, for instance, the aquatic frog *Xenopus*, which has superficial neuromasts only, and by fish, which are equipped with both superficial and canal neuromasts. The way in which the latter respond looks far more complicated but, surprisingly, the difference turns out to be marginal. In this paper, the velocity field will be analyzed first by means of studying the neuronal information processing originating from the velocity field as it can be observed in *Xenopus*. In the present context and following a long tradition of continuum mechanics, a *field* is a quantity that depends on the position $\mathbf{x}$ in space, $\mathbb{R}^2$ or $\mathbb{R}^3$ for two or three dimensions. The velocity $\mathbf{v}$ is a vector field that in general depends on both space through the position $\mathbf{x}$ and on the time t so that in full glory it can be written $\mathbf{v} = \mathbf{v}(\mathbf{x}, t)$. The pressure is also a field but a scalar one so that $p = p(\mathbf{x}, t)$.

Sections 16.3 and 16.4 are devoted to the lateral-line system of fish with both canal and superficial neuromasts and analyze whether and how the two types of neuromast are compatible in producing neuronal representations of the outside hydrodynamic world. That gives the right context to mathematically determine the range of the lateral line and explain why it is not much more than a fish length, with hindsight an intuitively reasonable result. It will turn out that in so doing one needs to pay due attention to the underlying geometry.

Until now the focus was on the direct interaction between a hydrodynamic object and the fish lateral line, which is based on momentum transfer from the object to the neuromasts. Interacting with its direct surroundings, a fish has to effectively reconstruct its environment through an *inverse* mapping from the signals coming from the neuromasts on its two-dimensional surface back into the three-dimensional environment. It is nevertheless both advantageous to skim and preferable to skip the mathematics of this reconstruction as it involves a very difficult, "ill-posed," problem (Tikhonov and Arsenin 1977; Isakov 2006). On a deeper level, one now sees one more reason of why the range of the lateral line is to be short: The reconstruction becomes unstable at larger distances. Fish can solve it all very efficiently as, for instance, the blind Mexican cave fish *Astyanax mexicanus* shows off every day. Here too, one steadily sees an exploration of the lateral line's finite range, about one fish length.

At the same time, evolution exhibits a bedazzling richness. A blind Mexican cave fish localizes nearby objects such as stones not because these "wiggle their tail" but since the bow wave it generates (Dijkgraaf 1963) hits the stone, "rebounds," and is perceived by the head lateral-line system. The effective range turns out to be about the radius of curvature R_o of the snout (as seen from above). The bow-wave phenomenon deserves separate study but, since on average it is far less frequently present than the ℓ/L effects discussed above, the issue will not be pursued here. Nor is there need for analyzing the apparent contradiction of the finite range L of the lateral line's direct action and prey localization performed by e.g., the surface-feeding topminnow *Aplocheilus lineatus* (Bleckmann and Schwarz 1982;

Schwarz et al. 2011). *Aplocheilus* localizes prey in the sense of determining both its direction and its distance through surface waves at separations far bigger than the radius of curvature R_o of its head or even its total length L. The mechanism, however, is not the usual one as analyzed below but, as argued by Käse and Bleckmann (1987), it is based on the dispersion of surface waves under the influence of surface tension; compare the Appendix, Eq. (16.68) and Fig. 16.18b.

There is also another stimulus that is very stable and may last in stillwater for several minutes: vortex wakes. They are a consequence not of direct momentum transfer but of conservation of *angular* momentum. Since water is not very viscous some fish can, and do, perform vortex-wake tracking (Pohlmann et al. 2001) and follow vortex wakes along quite a distance. Vortex-wake tracking will be analyzed mathematically as well. Here too it can be seen *how* experiment and mathematical theory agree.

Finally, the focus will be on the question of how the lateral-line system may give rise to a neuronal map and how a lateral-line object can be integrated with, for example, its visual counterpart, an 'object' now being a neuronal representation of its corresponding material object in the outside sensory world.

16.2 *Xenopus* and Surface Waves

The adult clawed frog *Xenopus laevis laevis*, for short *Xenopus*, exemplifies an animal that can perform prey detection exclusively through its lateral-line organs, about 180 altogether; to contrast the relative simplicity of *Xenopus* with fish, see Schmitz et al. (2008). A comfortable aspect simplifying the mathematical analysis is that *Xenopus* has superficial neuromasts (surface neuromasts) only and, though as a nocturnal predator it seems to catch prey on the water surface only, it is meanwhile known (Elepfandt et al. 2006) that on the bottom of the pond it can do so as well but now in three-dimensional space, viz., the pond it lives in. Life quite often simplifies in 2 as compared to 3-dimensions and that is one, but not the only, reason for making *Xenopus'* surface-wave detection such an interesting topic. A key issue of this section is showing that a simple but concrete mathematical model of a map as a neuronal representation of the outside sensory world and based on the biophysics and biology of the animal both allows an explanation of all the experimental data obtained so far on *Xenopus'* prey detection and invites a verification of new ideas it gives rise to, such as how many *identical* wave sources *Xenopus* can discern neuronally; for a mathematically simple but complete exposition of surface waves one may consult the Appendix.

Xenopus is just a typical example. Being a nocturnal predator, the adult animal's lateral-line system has become the central sensory organ for spatial orientation (Tinsley and Kobel 1996). *Xenopus* uses this system for catching prey in water; during night mostly on the water surface. When an insect drops on the water surface, it generates a surface wave that passes along *Xenopus*; cf. Fig. 16.1 and the Appendix. Depending on the waveform, the frog may, or may not, turn towards

Fig. 16.1 The clawed frog *Xenopus laevis laevis*. Its lateral-line organs can be seen as *white* "stitches" on its back and also on its belly, ideal for perceiving surface waves when the eyes are just above but the rest of the body below the water surface. In the mathematical model presented here they are arranged on a circle with a diameter of 4 cm, a convenient but basically irrelevant simplification. Figure courtesy of A. Elepfandt

the wave's origin, its prey. In plain English, the adult *Xenopus* can recognize (Elepfandt et al. 1985; Claas and Münz 1996) a prey-generated waveform as it passes along its lateral-line organs.

This section is devoted to three things, all in the context of a mathematical model. First, showing that *Xenopus can* both determine the direction of a wave source and perform some kind of waveform reconstruction. That is, in the context of a "minimal" model an existence proof is provided. Though as simple as possible (Occam would applaud), it already reproduces the complete experimental repertoire of *Xenopus'* prey catching. Equipped with the insights provided by the minimal model (Sect. 16.2.1), one can then ask, and answer, the question of how the animal realizes such a setup neuronally. Third, *Xenopus* has—so to speak— 180 "ears" through which it can discern two, or more, identical sources. The primate auditory system equipped with only two ears cannot discern two identical sources. Neither can the barn owl (Keller and Takahashi 1996), but *Xenopus* can (Elepfandt 1986). Can, then, the mathematical model of Sect. 16.2.1 discern identical sources too? The work as analyzed below is mainly due to Franosch et al. (2003, 2005a, b).

16.2.1 Minimal Model

A lateral-line organ of *Xenopus* consists of a group of 4–8 linearly arranged, small cupulae with gelatinous flags protruding into the water and being deflected by the local fluid flow (Kalmijn 1988). The group is usually called a stitch, cf. Fig. 1 of

Franosch et al. (2003). The deflection stimulates sensory hair cells at the base of the cupula and in dependence upon the angle between its preferred orientation and the velocity $\mathbf{v}(\mathbf{x}, t)$. In this way a deflection generates spikes in the lateral-line nerves, phase-locked to the stimulus. Simulations (Franosch et al. 2003) have shown similar results to occur when the sensors are assumed to detect the water *pressure*. So the present analysis is applicable to e.g., crocodilian dome pressure receptors (Soares 2002) as well.

16.2.1.1 Transfer Function of Water

Stimuli such as insects struggling on the water surface or a moving stamp in experiments generate surface waves. The deflection $y_i(t)$ of cupula i at time t is proportional to the local velocity (Kalmijn 1988) and so is the neuronal response (Goulet et al. 2012). As shown in the Appendix, water waves can often be analyzed as a linear system; cf. also Acheson (1990) and Billingham and King (2000) so that the *effective* deflection y_i of cupula i is accordingly linear in the stimulus $x^{\mathbf{p}}$ at position $\mathbf{p}$,

$$y_i(t) = \int\limits_0^\infty h_i^{\mathbf{p}}(\tau) x^{\mathbf{p}}(t - \tau)\, \mathrm{d}\tau \equiv (h_i^{\mathbf{p}} * x^{\mathbf{p}})(t) \qquad (16.1)$$

where $h_i^{\mathbf{p}}$ is the so-called impulse response at cupula i while being stimulated by an impulse at position $\mathbf{p}$ on the water surface and $*$ denotes a convolution.

A few words on the existence of the response kernel h_i, are in order. A lateral-line organ i is somewhere on *Xenopus'* skin. As seen from the Appendix, a surface wave emanating from a scrambling insect at position $\mathbf{p}$ can be described formally by a linear operator $\mathscr{L}$ and $x^{\mathbf{p}}(t)$ as a source so as to give $(\partial/\partial t + \mathscr{L})y = x^{\mathbf{p}}(t)$. One now solves the equation, exploiting that $\mathscr{L}$ itself does not depend on the time t (Duhamel's formula), so as to get $(h_i^{\mathbf{p}} * x^{\mathbf{p}})$ as in (16.1) with a continuous response kernel, denoted by h_i. This kernel exists and appears as such in (16.1); evaluating it explicitly is a bit more work. Details as well as a formal derivation of (16.2) below can be found in the Appendix and in the literature (Acheson 1990; Billingham and King 2000; Lamb 1932).

The Fourier transform of the impulse response function $h_i^{\mathbf{p}}$ is the transfer function $H_i^{\mathbf{p}}(\omega) = \int h_i^{\mathbf{p}}(t) \exp(-i\omega t)\, \mathrm{d}t$. A Fourier transform decouples a convolution as in (16.1) into a normal product (Dym and McKean 1972). An approximation of the transfer function between the velocity of a moving stamp with radius, say, $r_0 = 1$ cm exciting the surface of the water and the velocity at a lateral-line organ at distance $r \gg r_0$ is

$$H(\omega) = \sqrt{\frac{r_0}{r}}\, D_{\Delta\varphi} \exp\left[\frac{4\mu k^3}{\omega}(r_0 - r) + ik(r_0 - r)\right] \qquad (16.2)$$

for $\omega > 0$ and $H(-\omega) = H^*(\omega)$ with μ as the water viscosity and $k = 2\pi/\lambda$ as the wavenumber; in c.g.s. units, $\mu_{H_2O} = 0.01$ (poise), in SI units, 10^{-3} (Pa s). The first term on the right in (16.2) describes the $1/r$ reduction of the intensity due to the distance from the source. The second, $D_{\Delta\varphi} = 10^{-2|\Delta\varphi|/\pi}$, is empirical and accounts for the damping or "screening" caused by *Xenopus'* body (Elepfandt and Wiedemer 1986) with $\Delta\varphi$ being the angle between the direction of the lateral-line organ with respect to *Xenopus'* center and the direction of the wave source. The $1/r$ reduction stems from the energy of a surface wave starting somewhere locally and homogeneously distributing itself over a circle with radius r and circumference $2\pi r$ so that the energy content decreases as $1/2\pi r$. The energy being proportional to the square of the amplitude (Acheson 1990; Billingham and King 2000), the result is $1\sqrt{r}$ for the amplitude in (16.2).

The first term in the exponent of (16.2) describes amplitude reduction because of the viscosity μ (Acheson 1990; Billingham and King 2000; Bleckmann 1994) and the second is a phase dependence. The dispersion is given by $\omega^2(k) = (gk + t_S k^3/\rho)$ where g is the gravity acceleration, t_S the water surface tension, ρ the density of water, and $\omega = 2\pi\nu$ with ν as the frequency.

Through its lateral-line system *Xenopus* can determine direction and character of impinging waves. It can also distinguish (Elepfandt et al. 1985) sources of different frequency and probably discern different preys in general. Localization requires a neuronal comparison of inputs from several lateral-line organs. Since each lateral-line organ only encodes the local *superposition* pattern of the waves at the body surface, *Xenopus'* neuronal pattern segmentation ability requires a neuronal comparison of the encoded superposition patterns from several organs and a decomposition of the patterns into their original components. So a natural hypothesis is that *Xenopus* "tries" to determine *what* is going on *where* on the water surface. Let **p** be a position on the water surface and let us assume *Xenopus* determines the temporal waveform of the source at **p.** Furthermore, let $x^{\mathbf{p}}$ be the true time-dependent waveform of the source and $\hat{x}^{\mathbf{p}}$ *Xenopus'* estimate, to be derived in Sect. 16.2.1.2 below.

No known neuroanatomical data suggest or support any specific model yet. A 'minimal' model is therefore our starting point so as to answer the question of how *Xenopus* reconstructs the *waveform*. This is done through a minimum-variance estimator (van der Waerden 1969; Jazayeri and Movshon 2006) that minimizes the expectation value of the least-squares error between the actual stimulus $x^{\mathbf{p}}$ at position **p** and *Xenopus'* estimate $\hat{x}^{\mathbf{p}}$,

$$\|x^{\mathbf{p}} - \hat{x}^{\mathbf{p}}\|^2 = \int_0^{T_1} [x^{\mathbf{p}}(t) - \hat{x}^{\mathbf{p}}(t)]^2 \, \mathrm{d}t. \tag{16.3}$$

Here $T_1 \approx 500$ ms is the frog's minimal response time, including neuronal detection, which is a key topic of the present paper. The expectation value of (16.3) that follows in a minute refers to the omnipresent noise that is to be taken care of. The above expression $\|\ldots\|$ in (16.3) defines what one calls a norm, here a distance between functions in function space (Bachman and Narici 1966). If the distance vanishes, the functions are identical.

16.2.1.2 Neuronal Implementation

The only information available to *Xenopus* for determining the waveform $x^{\mathbf{p}}$ of the wave source are the spikes coming from the nerves of its lateral-line system. The spikes encode a deflection of the lateral-line organs' cupulae. This deflection can be determined only approximately from the spike train because both spike generation and cupula response are afflicted with a large amount of uncertainty that cannot be modeled precisely yet, if ever, and hence needs to be taken into account through "noise" represented by a stochastic process. The stochasticity is modeled conveniently, though not exclusively, by adding independent Gaussian random variables $\sigma_n n_i(t)$ with mean zero and standard deviation σ_n to the deflections $y_i(t)$ of the cupulae. In so doing the robustness of the underlying arguments can also be tested. Moreover, each insect species generates a waveform $x^{\mathbf{p}}$ with a typical mean (Bleckmann 1994, Fig. 5) but also with an intrinsic, stochastic, variation, which is taken to be Gaussian with standard deviation σ_x.

The deflection y_i of cupula i ($1 \leq i \leq 180$) is a linear functional of the wave source x as given by Eq. (16.1) plus noise,

$$y_i = h_i^{\mathbf{p}} * x^{\mathbf{p}} + \sigma_n n_i. \tag{16.4}$$

Knowing the deflections y_i of the cupulae, *Xenopus* has to 'estimate' a waveform $\hat{x}^{\mathbf{p}}$ for a given waveform $x^{\mathbf{p}}$. Putting $\sigma = \sigma_n / \sigma_x$, the first task is averaging (16.3) over the Gaussian noise so as to get $\langle \|x^{\mathbf{p}} - \hat{x}^{\mathbf{p}}\|^2 \rangle$ where $\langle \ldots \rangle$ denotes the stochastic average over the noise. Averaging looks forbidding but it is not because many expressions in nature are self-averaging as a consequence of the strong law of large numbers (Lamperti 1966; van Hemmen 2001). Self-averaging then comes for free once there are many (nearly) independent terms (Lamperti 1966; van Hemmen 2001), as is the case here.

A calculus of variations on $\langle \|x^{\mathbf{p}} - \hat{x}^{\mathbf{p}}\|^2 \rangle$, i.e., a simple directional differentiation in function space (see below), and a simple computation give the solution minimizing the error in (16.3) to be of the form

$$\hat{x}^{\mathbf{p}} = \sum_j s_j^{\mathbf{p}} * y_j \quad \text{with} \quad S_j^{\mathbf{p}}(\omega) = \frac{H_j^{\mathbf{p}*}(\omega)}{\sum_i |H_i^{\mathbf{p}}(\omega)|^2 + \sigma^2}. \tag{16.5}$$

The above $*$ denotes a convolution with respect to time, which becomes an ordinary multiplication after Fourier transformation (Dym and McKean 1972). The functions $S_j^{\mathbf{p}}$ are the Fourier transforms of the *reverse* transfer functions $s_j^{\mathbf{p}}$, which follow from the $S_j^{\mathbf{p}}$. The transfer functions $H_i^{\mathbf{p}}$, as given by (16.2), depend on the, in general, prey position $\mathbf{p}$ the animal, here *Xenopus*, is interested in.

A few words on what is called 'calculus of variations' are maybe in order. There is a gigantic literature but the idea is simple. Instead of minimizing functions on $\mathbb{R}$ or $\mathbb{R}^n$ in n dimensions, the game is now played in an infinite-dimensional normed function (for the experts: Banach, often just Hilbert) space $\mathbb{B}$ with a norm $\|f_1 - f_2\|$ such as the one in (16.3) to measure the 'distance' between two functions f_1 and f_2 in $\mathbb{B}$. Furthermore, instead of a normal function one is given a functional, a "function of functions", $f \mapsto \mathcal{E}(f)$ mapping $\mathbb{B}$ into the real numbers. To find a local minimum or a maximum of a "normal" function $x \mapsto f(x)$ at $x_\circ$ one adds a small perturbation h to $x_\circ$ and requires that at $x_\circ$ the derivative $[f(x_\circ + h) - f(x_\circ)]/h$ as $h \to 0$ vanishes. If f is a function of n variables $(x_1, \ldots, x_n) = \mathbf{x} \in \mathbb{R}^n$, then all directional derivatives $\partial f / \partial x_i$ with $1 \leq i \leq n$ must vanish at $\mathbf{x}_\circ$.

What, then, is a necessary condition that $f_\circ$ minimizes, or maximizes, $\mathcal{E}(f)$ on $\mathbb{B}$ in function space? To this end, a small perturbation $\lambda\phi$ is added to $f_\circ$ and the derivative of the scalar function $\lambda \mapsto \mathcal{E}(f_\circ + \lambda\phi)$ is required vanish at $\lambda = 0$ for *every* $\phi \in \mathbb{B}$, which is what everybody knows from calculus. Being in function space instead of on the real axis with, as above, $x_\circ$ and $h \in \mathbb{R}$ or in $\mathbb{R}^n$ joined by n partial derivatives, the extra condition "for every $\phi \in \mathbb{B}$" now needs to be included as all directions in function space need to get scanned. In terms of a plain equation, one therefore demands

$$\frac{\mathrm{d}}{\mathrm{d}\lambda}\mathcal{E}(f_\circ + \lambda\phi)|_{\lambda=0} = 0, \quad \text{for all } \phi \in \mathbb{B} . \tag{16.6}$$

Since the starting point was (16.3), the result of the differentiation in (16.6) is a vanishing integral of the form $\int \mathrm{d}t \, G(f_\circ, t)\phi(t) = 0$, which must vanish for all directions ϕ in function space. Here $G(f_\circ, t)$ is a real function of t containing $f_\circ$. The integral vanishing for *all* ϕ, one can take $\phi = G$ so that $\int \mathrm{d}t \, G^2(f_\circ, t) = 0$ and hence $G \equiv 0$, which by doing the algebra directly leads to (16.5). For insightful mathematical information on variational calculus, see Clegg (1968) or Gelfand and Fomin (1963).

Equation (16.5) shows that *Xenopus* could estimate the original waveform of the source by simply taking the convolution of deflections y_i of its lateral-line organs with built-in reverse transfer functions s_j^{p}. A question to be resolved later is how one can imagine this "built in". The deflections y_i are represented more or less accurately by the spike trains of the lateral-line nerves. An approximate convolution can be performed efficiently and easily in neuronal hardware, as will be shown below.

To get a decent approximation of the reverse transfer functions s_i^{p} with as few function values s_{ik}^{p} as possible, a decent choice is $s_i^{\mathrm{p}}(t) \approx \sum_k s_{ik}^{\mathrm{p}} \, \delta(t - t_{ik}^{\mathrm{p}})$ where the $s_{ik}^{\mathrm{p}} = s_i^{\mathrm{p}}(t_{ik}^{\mathrm{p}})$ are maxima and minima of s_i^{p} labeled by k; see the inset of Fig. 16.2, which also shows that this discrete approximation makes a lot of sense. Because the system is causal, $s_i^{\mathrm{p}}(t) = 0$ for $t > 0$; see the inset of Fig. 16.2. Hence s_{ik}^{p} with $t_{ik}^{\mathrm{p}} > 0$ do not exist.

It is now time to verify explicitly that neuronal wave-form reconstruction is possible. To this end, the neurons that are directly connected to the lateral-line

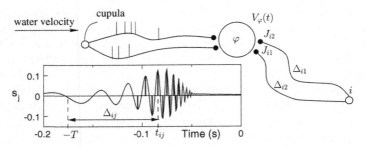

Fig. 16.2 Connections of a model neuron with preferred direction φ to be defined in (Fig. 16.3), to the lateral-line organs i (*open circles*). *Solid lines* denote axons, *small filled circles* synapses with strengths J_{ik}. Axonal delays Δ_{ik} and synaptic strengths are such that the membrane potential $V_\varphi(t)$ of the neuron approximates the original waveform $x(t)$ at the source of the water wave. The Δ_{ik} and J_{ik} are provided by the model. Its localization and source-reconstruction ability need no precise tuning of Δ_{ik} and J_{ik}. The inset shows a typical reverse transfer function s_j provided by Eq. (16.5) and its approximation through Dirac delta functions

nerves (Fig. 16.2) need to get modeled and the convolution appearing in (16.5) asks for explicit computation. For the sake of both convenience, particularly when labeling the lateral-line organs, and ease of imagination it is convenient, but not a severe restriction, to imagine the neuromasts on a circle of diameter 4 cm, of the same order of magnitude as an adult *Xenopus* itself. Every spike in a lateral-line nerve causes a postsynaptic potential ε in the neuron. A practical choice of the postsynaptic potential is the alpha function $\varepsilon(t) = t/\tau \exp(1 - t/\tau)$ for $t \geq 0$ and $\varepsilon(t) = 0$ elsewhere, with $\tau = 10$ ms. The membrane potential $V^{\mathbf{p}}$ of the neuron, modeled as a spike-response neuron (Gerstner and van Hemmen 1994), is

$$V^{\mathbf{p}}(t) = \sum_{i,k,f} J_{ik}^{\mathbf{p}}\varepsilon\left(t - t_i^f - \Delta_{ik}^{\mathbf{p}}\right) + \sum_{i,k,f'} J_{ik}^{\mathbf{p}'}\varepsilon\left(t - t_i^{f'} - \Delta_{ik}^{\mathbf{p}'}\right) \qquad (16.7)$$

where the t_i^f are the firing times of the nerve from lateral-line organ i and $\Delta_{ik}^{\mathbf{p}}$ is the delay of synapse number k with synaptic strength $J_{ik}^{\mathbf{p}}$. There are two lateral-line nerves for each lateral-line organ i, accounting for the two 'opposite' directions of deflection (Görner 1963). One nerve becomes active for deflections $y_i(t) > 0$ and so does the other one for $y_i(t) < 0$, as indicated by primed quantities f', $J_{ik}^{\mathbf{p}'}$, and $\Delta_{ik}^{\mathbf{p}'}$ in (16.7).

A bit of algebra gives that the membrane potential $V^{\mathbf{p}}(t)$ approximately equals the wave-form estimate $\hat{x}^{\mathbf{p}}(t - T)$ in (16.5), if one sets $J_{ik}^{\mathbf{p}} = s_{ik}^{\mathbf{p}}$, $J_{ik}^{\mathbf{p}'} = -s_{ik}^{\mathbf{p}}$ and $\Delta_{ik}^{\mathbf{p}} = T + t_{ik}^{\mathbf{p}}$, demonstrating explicitly that neuronal wave-form reconstruction *is* possible; cf. Fig. 16.3. Delays can be taken less than 100 ms. Figure 16.2 shows a *distribution* of axonal delays. Such a distribution is essential (Kühn and van Hemmen 1995) to a proper learning of spatiotemporal neuronal activity patterns that by their very nature evolve in time and hence to subsequent adequate functioning, here of the ensuing neuronal maps in Fig. 16.3.

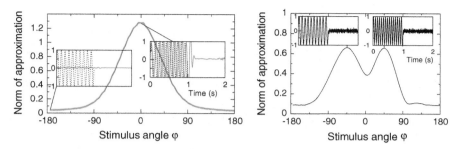

Fig. 16.3 a 'Map' of norms $\|V_\varphi\|$ of membrane potentials of 72 model neurons, each representing a different direction φ (*horizontal axis*). A sinusoidal wave source with frequency of 10 Hz is positioned 10 cm in front of *Xenopus* ($\varphi = 0$). *Xenopus* is to head for the direction φ where $\|V_\varphi\|$ and hence the neuronal firing rate has its maximum, which in this case is indeed $\varphi \approx 0$. As one can see, $\|V_\varphi\|$ has a maximum in the neighborhood of $\varphi = 0$. The membrane potential being maximal, so is a neuron's firing rate, *ceteris paribus*. Moreover, *Xenopus'* "reconstruction" resembles the original waveform $x(t)$ (*dotted lines* in the insets) quite well. **b** 'Map' like that in (**a**) but now for *two simultaneous* wave sources of 10 Hz and 15 Hz, positioned at $\varphi = -45°$ and $45°$ 10 cm away from *Xenopus*. The animal's approximations are shown in the insets. In this way the animal could easily distinguish both position *and* waveform of the sources, as in experiment (Elepfandt 1986)

In the description according to present (minimal) model, *Xenopus* is heading for the direction φ where $\|V_\varphi\|$ has its maximum, which in the case of Fig. 16.3a is $\varphi \approx 0$. A sinusoidal wave source with natural, low frequency is positioned 10 cm in front of *Xenopus* ($\varphi = 0$). After a global delay T for all, each individual neuron of the map in Fig. 16.3 A reconstructs the wave form of the stimulus by means of its membrane potential $V_\varphi(t + T)$ (solid lines in the insets), "assuming" that the actual stimulus comes from direction φ. Once this assumption is correct (at $\varphi = 0$), *Xenopus'* approximation resembles the original waveform $x(t)$ (dotted lines in the insets) quite well. For wrong directions (e.g., $\varphi = 180°$), *Xenopus'* reconstruction is noise with a small amplitude. The present model therefore serves two purposes. First, neurons responding most strongly tell *Xenopus* the direction of the wave source. Second, the membrane potential of these neurons gives *Xenopus* an approximation to the actual wave form and allows the animal to distinguish different kinds of prey. Why, then, should *Xenopus* reconstruct the waveform? It has been shown (Elepfandt et al. 1985) that the animal learns to remember specific frequencies associated with edible or nonedible objects and in view of the wide variety of signals generated by natural prey (Bleckmann 1994, Fig. 5) a natural and more or less minimal assumption is waveform reconstruction. If less would suffice, the present argument proves that this would work too as it can be learned.

Remarkably, the synaptic efficacies $J_{ik}^{\mathbf{p}}$ and axonal delays $\Delta_{ik}^{\mathbf{p}}$ only depend on the maxima and minima $s_{ik}^{\mathbf{p}}$ at times $t_{ik}^{\mathbf{p}}$ of the reverse transfer functions of Eq. (16.5), which in turn depend on the transfer functions of (16.2); cf. Fig. 16.2. As these positions are arranged on a circle, they are characterized by the direction φ. So there is a map of neurons with membrane potentials $V_\varphi(t) := V^{\mathbf{p}(\varphi)}$, each responsible for a specific direction φ. It is assumed that *Xenopus* turns to the angle φ represented by the neuron that has *maximal* average firing rate, i.e., where the norm $\|V_\varphi\|$ as defined by (16.3) is maximal.

With hindsight, the above assumption is reasonable because $\|V_\varphi\|$ indeed has its absolute maximum at the angle φ where the wave comes from, as shown in Fig. 16.3. The reason is that, when the animal is reconstructing the waveforms through (16.5) while picking a wrong direction φ, the wrong transfer functions $H_i^{\mathbf{p}}$ are used and, hence, a reconstruction gives noise only. By the way, a maximum of $\|V_\varphi\|$ also implies a firing-rate maximum of the corresponding neuron coding the direction φ.

A comparison of Figs. 4b and 5b of Claas and Münz (1996) and Fig. 4 of Franosch et al. (2003) shows the agreement between theoretical and experimental distributions of *Xenopus'* turning angles in the lesioned and unlesioned case; lesioning means that some or even all neuromasts have been silenced. Despite partial leasioning (Elepfandt 1982, 1984) the animal can still localize prey. To account for the results of the experiments through the minimal model, it has also been assumed that *Xenopus* exploits no intensity but only phase information for its approximations in that it "uses"

$$H_j^{\mathbf{p}}(\omega) = \sqrt{\frac{r_0}{r_d}} \; \exp\left[-\frac{4\mu k^3}{\omega}(r_d - r_0) - ik\left(r_j^{\mathbf{p}} - r_0\right)\right]$$

in (16.5) with $r_d = 10$ cm instead of the real transfer functions of water in Eq. (16.2); $r_j^{\mathbf{p}}$ is the distance from lateral-line organ j to position $\mathbf{p}$. The above minimal assumptions indeed suffice to give a fair explanation of experimental reality; cf. Fig. 4 of Franosch et al. (2003)

For spike generation in the lateral-line nerves a natural choice is an inhomogeneous Poisson process (van Hemmen 2001) as an approximation to the real input–output characteristics (Strelioff and Honrubia 1978) of the lateral line organ. Nerve i fires in $[t, t + dt]$ with a probability of $[\pm Ry_i(t) + R_s]\, dt$, $R = 300$ Hz, so that realistic spike rates lower than about 150 Hz and spontaneous rates $R_s = 10$ Hz (Görner 1963; Elepfandt and Wiedemer 1986) are the result. The negative sign is for nerves that are excited by negative deflections $y_i(t) > 0$. Alternatively, one could take Poisson neurons (van Hemmen 2001) with firing rate proportional to $\|V_\varphi\|$.

16.2.1.3 Discerning Simultaneous Wave Sources

In accordance with both experiment (Elepfandt 1986) and the present model (Fig. 16.3b), a trained *Xenopus* can easily distinguish two simultaneous but different wave sources. Not only can *Xenopus* determine the *positions* of the two wave sources but it can also distinguish different frequencies, and so can the model. If *Xenopus* is trained to always swim to the wave source with a frequency of, say, 15 Hz with an angle between the two stimuli of, say, 90°, and the two stimuli are then presented with an arbitrary but different angle between them, *Xenopus still* turns to the 15 Hz stimulus without any further training (Elepfandt et al. 2004). The fact that *Xenopus* generalizes appropriately as described supports the model assumption that the actual wave form is somehow approximated and accordingly recognized by the frog.

16.2.1.4 Discussion

As shown in this section, a *large* number of sensory detectors allow an animal equipped with a lateral-line system such as *Xenopus* to perform some kind of waveform reconstruction so as to determine both the prey's direction and its character. A simple neuronal algorithm with realistic firing rates, number of synapses (in the simulations presented here, 100 per neuron), and time constants of postsynaptic potentials suffices to perform localization; see Fig. 16.3a. Further-more, as Fig. 4 of Franosch et al. (2003) illustrates, the model is robust in that it successfully incorporates the effect of lessoning a part of the cupulae, i.e., elim-inating them as described by, for instance, Elepfandt (1982). The present theory also explains *Xenopus'* distinguishing *simultaneous* wave sources, i.e., its per-forming pattern segmentation, as in Fig. 16.3b, and in this way it can both guide and verify future experiments. For the crocodilians' dome pressure receptors the question of their neuronal information processing is still open but a similar potency suggests itself. Finally, the method of the 'minimal model' of Sect. 16.2.1 needs minimal assumptions to allow exploring an animal's response to sensory input efficiently, even though detailed anatomical data are not available yet—as is often the case in practice.

Nevertheless, it is nice to have an existence proof showing that the lateral line can indeed do all this but why not ask: How, then, can an animal with a lateral-line system such as *Xenopus* or fish, the latter possessing canal neuromasts as well, acquire a neuronal setup as described above? That is, can it acquire such a functioning neuronal setup through "learning" (by doing)? And, if so, what are the underlying learning algorithms? That is the topic the next subsection is devoted to.

16.2.2 Learning Algorithms to Get the Neuronal "Hardware"

In nature animals interact with their surroundings and, conversely, their environ-ment influences the way in which their brain and, hence, their handling sensory data develops. In so doing they "learn" and one of the questions that has tantalized learning theory since long is whether or not the neuronal system needs a teacher and, if so, who is teaching what, and how?

The neuronal model presented here reveals how the two simplest tasks of a remote sensory system, namely, to determine *what* happens *where,* may be performed and *learned* by a single neuronal circuit. Computer simulations support the underlying ideas and show that the learning algorithms are feasible, given the constraints of neuronal hardware.

Xenopus' lateral-line system has 180 receptors consisting of 4-8 cupulae that are deflected (Kalmijn 1988) and may be expected to respond (Goulet et al. 2012) proportionally to the local water velocity. Each receptor projects through afferent fibers (with practically identical response characteristics) to the Medulla and from there via the *torus semicircularis* to the optic tectum. The optic tectum is the first station where all modalities, amongst them the visual and the lateral-line system (Zittlau et al. 1986), get together and form maps in register; see also Sect. 16.6.1. Here one finds bimodal neurons that react to visual as well as lateral-line stimuli; see for instance Lowe (1987). Later on, in Sect. 16.6, we will face the question as to whether maps such as a lateral-line one can exist and how such an integration can be realized. The present aim is to see whether and how the simple neuronal setup of Sect. 16.2.1 can be acquired neuronally.

The 180 lateral-line organs on *Xenopus'* skin being deflected, some more, some less, each of them transforms its deflection into spikes and transfers its neuronal activity to the nerve fibers attached to it. To obtain a mathematical description with spikes being at the basis of the neuronal processing, one first needs to quantify the stimulus that generates them. The deflection $y_i(t)$ of the cupulae of lateral-line organ i at time t is proportional to the local water velocity and thus a linear function of the stimulus $x^{\mathbf{p}}$ at position $\mathbf{p}$ on the water surface. The input signal is given by Eq. (16.1),

$$y_i(t) = (h_i^{\mathbf{p}} * x^{\mathbf{p}})(t) = \int\limits_{-\infty}^{\infty} d\tau\, h_i^{\mathbf{p}}(\tau)\, x^{\mathbf{p}}(t - \tau) \tag{16.8}$$

where $h_i^{\mathbf{p}}$ is the impulse response at cupula i while being stimulated by a Dirac delta impulse at position $\mathbf{p}$ on the water surface; as before, the star $*$ denotes convolution. The Fourier transform of the impulse response $h_i^{\mathbf{p}}$ is the transfer function of (16.2), $H_i^{\mathbf{p}}(\omega) = \int h_i^{\mathbf{p}}(t) \exp(-i\omega t)\, dt$. For the sake of simplicity, the directions $\mathbf{p}$ are discretized, here 180 altogether.

Let the prey be at position $\mathbf{p}$. The symbol $\delta_{\mathbf{p},\mathbf{p}'}$ has the value 1 if the positions $\mathbf{p}$ and $\mathbf{p}'$ are identical, else 0. To learn where the prey is as well as what waveform $x^{\mathbf{p}}$ it generates, it is appropriate to assume in the context of a minimal model that *Xenopus* somehow minimizes the expectation value of the error functional

$$E = \sum\nolimits_{\mathbf{p}'} \int dt \left[\hat{x}^{\mathbf{p}'}(t) - \delta_{\mathbf{p},\mathbf{p}'} x^{\mathbf{p}}(t)\right]^2, \tag{16.9}$$

which is a generalization of (16.3). The sum in (16.9) is overall different, in the mathematical description discretized, directions $\mathbf{p}'$ on the water surface. As before in (16.3), $x^{\mathbf{p}}$ is the true time-dependent waveform of the source and $\hat{x}^{\mathbf{p}}$ *Xenopus'* estimate, for the moment to be computed in the style of Eqs. (16.4)–(16.6). The minimization of (16.9) ensures that, first, the reconstruction $\hat{x}^{p'}$ resembles the source x^p as closely as possible if $\mathbf{p}' = \mathbf{p}$ and, second, when reconstructing at the wrong position $\mathbf{p}' \neq \mathbf{p}$, it is as close to zero as possible. The source can therefore be localized by calculating $\hat{x}^{p'}$ for every position and choosing a position $\mathbf{p}$ where, as in (16.3), the norm $\|\hat{x}^{\mathbf{p}}\|$ of the *reconstruction* $\hat{x}^{\mathbf{p}}$ is maximal, in this way signaling where the stimulus is.

The solutions $s_j^{p'}$ of the minimization problem for the error specified by E in (16.9) can be computed by solving (16.12) for $\delta E/\delta s_j^{\mathbf{p}'}(\tau) = 0$ directly, which leads, after Fourier transformation, to the linear system of equations

$$\sum_i \left[\sum_{\mathbf{p}} H_j^{\mathbf{p}*} H_i^{\mathbf{p}} + \sigma^2 \delta_{ij} \right] S_i^{\mathbf{p}'} = \sum_{\mathbf{p}} \delta_{\mathbf{p},\mathbf{p}'} H_j^{\mathbf{p}*}. \tag{16.10}$$

It is good to know the explicit form of the solution as given by (16.10) but how can the animal attain it in terms of synaptic efficacies? Suppose a learning neuronal system uses the simplest optimization procedure, a gradient or "steepest" descent. Though it sounds trivial, its main advantage is that its simplicity and one can use gradient descent to even derive (Pfister et al. 2006) spike-timing-dependent plasticity (STDP).

For the moment, learning is starting somewhere with a set of $s_j^{\mathbf{p}'}(t_\circ)$ at time $t_\circ$ and adapting the reverse impulse responses $s_j^{\mathbf{p}}$ in the direction of the negative functional derivative of the error functional E of (16.9) at $s_j^{\mathbf{p}'}(t')$ for $t' \geq (t_\circ)$, viz.,

$$\Delta s_j^{\mathbf{p}'}(\tau) = \eta \delta E/\delta s_j^{\mathbf{p}'}(\tau) \tag{16.11}$$

where η is a (small) learning parameter and the δ on the right instead of the usual ∂ indicates a functional derivative; cf. (16.6). That is, taking a functional derivative boils down to starting with E as given by (16.10) at $\left\{ s_i^{\mathbf{p}'}(t') \right\}$ where $1 \leq i \leq N$ and N is the total number of neuromasts, putting a step "forward" in the direction $\lambda s_j^{\mathbf{p}'}(t')$, and taking the partial derivative of E with respect to λ at $\lambda = 0$. Using the expression for $\hat{x}^{\mathbf{p}}$ from (16.5) one finds

$$\frac{\delta E}{\delta s_j^{\mathbf{p}'}(\tau)} = -2\eta \int dt \left[\sum_i \left(s_i^{\mathbf{p}'} * y_i \right)(t) - \delta_{\mathbf{p},\mathbf{p}'} x^{\mathbf{p}}(t) \right] y_j(t - \tau). \tag{16.12}$$

The gradient descent of (16.12) is then repeated until a minimum has been reached. Here the procedure is bound to stop. Being at a minimum the solution then has to satisfy $\delta E/\delta s_j^{\mathbf{p}'}(\tau) = 0$, of course for all $1 \leq j \leq N$.

What the animal needs is a neuronal reconstruction $\hat{x}^{\mathbf{p}} = \sum_j s_j^{\mathbf{p}} * y_j$ of the signal $x^{\mathbf{p}}$ at position $\mathbf{p}$ and depending (linearly) on the neuromast input $y_i = h_i^{\mathbf{p}} * x^{\mathbf{p}}$ as given by (16.4). So the art consists in determining $s_j^{\mathbf{p}}$, which is what Eq. (16.12) makes explicit. The present algorithm that is based on Eq. (16.12) does converge but convergence to the minimum specified by (16.10) is very slow, signaling that "something" is wrong. Furthermore, the reconstruction quality is strongly influenced by noise, which is no good either. Hence it seems reasonable to widen—so to speak—the uptake and replace the "sharp" Kronecker-δ expression $\delta_{\mathbf{p},\mathbf{p}'}$ in (16.9) and thus in Eqs. (16.12) and (16.10) by a larger 'window of reference' $F(\mathbf{p},\mathbf{p}')$, e.g., by the Gaussian

$$F(\mathbf{p},\mathbf{p}') = \exp\left\{-[\varphi(\mathbf{p}) - \varphi(\mathbf{p}')]^2 / 2\sigma_\varphi^2\right\}, \qquad (16.13)$$

depending on the angular difference $\varphi(\mathbf{p}) - \varphi(\mathbf{p}')$ between the actual position $\mathbf{p}$ of a prey and the reconstruction position $\mathbf{p}'$.

The underlying idea is the following and, in fact, quite biological. The norm $\|\hat{x}^{\mathbf{p}'}\|$ is a smooth function of $\mathbf{p}'$ so that at neighboring positions the reconstructions are bound to "look" similar. Hence one cannot expect the error E of (16.9) to vanish exactly for $p' \neq p$. For discrete directions, the Kronecker delta is therefore replaced by a finite window $F(\mathbf{p},\mathbf{p}')$ while a new, "wider," functional

$$E_W = \sum_{\mathbf{p}'} \int \left[\hat{x}^{\mathbf{p}'}(t) - F(\mathbf{p},\mathbf{p}')x^{\mathbf{p}}(t)\right]^2 dt \qquad (16.14)$$

needs to get minimized. Figure 16.4a shows the result of a numerical simulation in conjunction with a window of reference à la (16.13). Apparently looking at the world through a "wider" observation window such as the one of (16.13) pays off in that also the temporal resolution improves as compared to, for instance, the one of Fig. 16.3a. In other words, learning improves performance. How, then, can *Xenopus* or any being with a lateral line implement its lateral-line learning?

16.2.2.1 Neuronal Implementation of Learning

A neuronal implementation of the model developed so far has to solve two problems. First, how is the wave source reconstruction $\hat{x}^{\mathbf{p}}$ computed according to (16.5), *given* the functions s_i? Second, how are the latter obtained through a learning algorithm?

A neuron whose membrane potential approximates the convolution of (16.5) is depicted in Fig. 16.5. Its spike train will be denoted by $y_{ik}^{\mathbf{p}}$. This is the sequence of spikes that arrive from lateral-line nerve fiber i at synapse k of the neuron that is responsible for a reconstruction of the wave source at position $\mathbf{p}$. The spike train is the same as the one coming from the lateral-line nerve, only delayed by a time $\Delta_{ik}^{\mathbf{p}}$. If f_i denotes times when spikes are generated in lateral-line nerve i, then

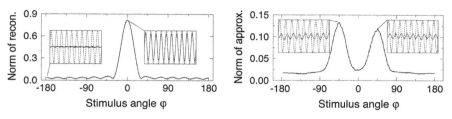

Fig. 16.4 **a** Model performance after minimizing the error E_W of Eq. (16.14) iteratively with the window of reference $(\sigma_\varphi = 14°)$ of Eq. (16.13) for 200,000 times. The figure shows a "map" of norms of reconstructions $\|\hat{x}^{\mathbf{p}}\|$ (*thick solid line*) at different angles $\varphi(\mathbf{p})$ around *Xenopus*. As expected, the norm and accordingly the neuronal response is maximal where the actual test stimulus is, viz., at 0°, at a distance of 10 cm right in front of *Xenopus*. Moreover, at 0° the model reconstructs (*solid line* in the inset on the *right*) the given test stimulus (sinusoidal, *dashed line* in the inset) quite well. At angles far off (−180° in the inset on the *left*), the reconstruction is basically noisy with low amplitude. The Gaussian white noise added to the deflections y_i of the lateral-line organs—cf. Eq. (16.4)—has a standard deviation of $\sigma_n = 0.01$. The temporal resolution is better than the ones shown in Fig. 16.3. **b** A map of 180 neurons that have learned to reconstruct the source at 180 positions arranged on a circle with radius 10 cm around *Xenopus*. After 50,000 iterative learning steps, i.e., presentations of Gaussian white-noise stimuli during 2 s, performed according to Eq. (16.19), the map is able to accurately localize both of two *simultaneous* wave sources at $\varphi = −45°$ (17 Hz) and 45° (18 Hz). In addition, the neurons indicating the respective angles *reconstruct* the wave source (*solid lines* in the insets). The delays $\varDelta_{ik}^{\mathbf{p}}$ of Eq. (16.15), with $1 \leq k \leq 100$, are randomly distributed in an interval of 500 ms. In this way, the animal could easily distinguish position and waveform of the sources. As compared to Fig. 16.3b where the two stimuli were quite similar, the segregation of the two sources has become far better (the local minimum between the two maxima at $\varphi = \pm45°$ is much deeper) but the waveform reconstruction is less good, though still acceptable

$$y_{ik}^{\mathbf{p}}(t) = \sum\nolimits_{f_i} \delta(t - f_i - \varDelta_{ik}^{\mathbf{p}}) \qquad (16.15)$$

where δ now indicates a Dirac delta function describing an action potential. Equation (16.15) is a sum of delta functions, which is nothing but a spike train. Imagine that each spike arriving at a synapse generates a membrane potential of the form $\varepsilon(t)$, an alpha function. Denoting synaptic strengths by $J_{ik}^{\mathbf{p}}$, one can compute the membrane potential $V^{\mathbf{p}}$ of a neuron, whose behavior is modeled as a spike-response neuron (Gerstner and van Hemmen 1994; van Hemmen 2001),

$$V^{\mathbf{p}}(t) = \sum_{ik} \int_{-\infty}^{\infty} d\tau J_{ik}^{\mathbf{p}} \varepsilon(t - \tau) y_{ik}^{\mathbf{p}}(\tau). \qquad (16.16)$$

For simulations, the alpha function

$$\varepsilon(t) = \left(t/\tau_s^2\right) \exp(-t/\tau_s) \qquad (16.17)$$

for $t \geq 0$ and $\varepsilon(t) = 0$ for $t < 0$ is a natural choice, with a synaptic time constant $\tau_s = 10\,\text{ms}$; because of τ_s^{-2} in the prefactor its integral equals 1. The membrane

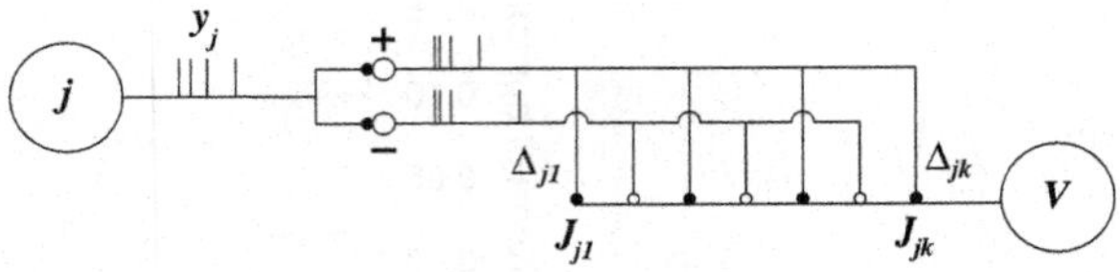

Fig. 16.5 Circuit diagram of a neuron with membrane potential V that is connected to lateral-line nerve fiber j by excitatory (*full circles*) and inhibitory (*open circles*) synapses with synaptic strengths J_{jk} and delays Δ_{jk}. The neurons indicated with "+" and "−" signs accommodate the observation (Dale's principle) that a single fiber can only have synapses of one type, excitatory or inhibitory

potential $V^{\mathbf{p}}(t)$ of the neuron at time t should closely resemble the deflection $x^{\mathbf{p}}(t - T)$ of the source at a time $t - T$ in the *past* so that a *causal* system can be constructed with $\Delta^{\mathbf{p}}_{ik} \geq 0$. Accordingly, the neuronal learning process should minimize the error functional

$$E_N = \sum_{\mathbf{p}'} \int dt \left[V^{\mathbf{p}'}(t) - F(\mathbf{p}, \mathbf{p}') x^{\mathbf{p}}(t - T) \right]^2. \tag{16.18}$$

Upon substituting $V^{\mathbf{p}}$ from (16.16) into (16.18) and keeping an eye on (16.16) and (16.11), a straight forward minimization of (16.18) á la (16.6), now with respect to $J^{\mathbf{p}'}_{ik}$, leads to the learning equation

$$\Delta J^{\mathbf{p}'}_{ik} = -2\eta \int \int_{-\infty}^{\infty} dt \left[V^{\mathbf{p}'}(t) - F(\mathbf{p}, \mathbf{p}') x^{\mathbf{p}}(t - T) \right] y^{\mathbf{p}'}_{ik}(\tau) \varepsilon(t - \tau). \tag{16.19}$$

The "teacher's" feedback $F(\mathbf{p}, \mathbf{p}') x^{\mathbf{p}}(t - T)$ is provided through synapses with strengths $F(\mathbf{p}, \mathbf{p}')$ and—assumption—projecting from the visual system through the optic tectum to the lateral-line "map". Equation (16.19) is similar to the more general STDP learning rule (Gerstner et al. 1996; van Hemmen 2001)

$$\Delta J = \eta \int \int_{-\infty}^{\infty} dt \, d\tau \, y_{\text{in}}(t) W(t - \tau) y_{\text{out}}(\tau) \tag{16.20}$$

where y_{in} is the input spike train of (16.15), y_{out} the output spike train, and W the *learning window*. Upon identifying y_{in} with $y^{\mathbf{p}'}_{ik}$ and setting $W(t) = -2\varepsilon(-t)$, Eqs. (16.19) and (16.20) match. The only proviso is that the original stimulus $x^{\mathbf{p}}$ is given as additional inhibitory input to the neuron under consideration, delayed by T and weighted by the window of reference so that its output spike train y_{out} approximates the function $V^{\mathbf{p}'}(t) - F(\mathbf{p}, \mathbf{p}') x^{\mathbf{p}}(t - T)$. As it is not known yet for sure which learning rule real neurons in the lateral-line system implement, the more exact Eq. (16.19) has been used for the simulations whose results are shown in Fig. 16.4.

16.2.2.2 Response to More Than Two Identical Stimuli

Barn owls cannot discern two *identical* sound sources (Keller and Takahashi 1996) but, as was noted earlier, *Xenopus* can (Elepfandt 1986). Biologically, this behavior seems quite natural since a barn owl has only two ears, whereas *Xenopus* has 180 lateral-line organs distributed all over the body surface. For two identical sources, Figs. 16.3 and 16.4 suggest that the model that has been developed here is able to resolve the question of whether *Xenopus* can discern even more than two identical sources. Two precautions have to be taken care of. First, "identical" means for hair cells that the stimuli appear simultaneously. Second, if two stimuli are too near, no organ is ever able to segregate them. That is, the stimuli have to be separated by a minimal angle, the size of which is an open question as we will soon see. To resolve the problem, one may therefore put n sources on a circle of a radius d that is large enough (say, a few decimeters) and separated from each other by an angle of $360°/n$.

Proceeding (Franosch et al. 2007) in the style and hence with a map as discussed before in this section, we find as indicated in Fig. 16.6 that indeed for $n \geq 3$ segregation due to the map is still possible. Furthermore, for 7–8 the suggestion is that stimulus segregation may become critical. Whether the animal in terms of its behavior can discern that many sources is what one cannot decide yet on the basis of the model and the anatomical and neurophysiological data available at the moment.

16.2.2.3 *Xenopus'* Experimental Performance

Action potentials in the lateral-line nerves have been modeled by an inhomogeneous Poisson process (van Hemmen 2001) approximating the real input–output characteristics (Strelioff and Honrubia 1978) of *Xenopus'* lateral-line organs. Nerve i generates an action potential in the time interval $[t, t + dt)$ with a probability of $[\pm R y_i(t) + R_s] \, dt$, $R = 300$ Hz, giving realistic spike rates lower than about 150 Hz and spontaneous rates $R_s = 10$ Hz (Elepfandt and Wiedemer 1986). The negative sign is for nerves that are excited by negative deflections $y_i(t) < 0$. Action potentials generated by Gaussian random noise produce the feedback $F(\mathbf{p}, \mathbf{p}')x^\mathbf{p}(t - T)$ in Eq. (16.19). The postsynaptic potential ε for the feedback is the same as in (16.17).

In experiments (Elepfandt 1986; Claas and Münz 1996), the frog turns in the direction of a wave caused by a moving plunger. In Fig. 16.3b, the model frog can even distinguish two *simultaneous* stimuli, just as in experiment (Elepfandt 1986). In the model of Fig. 16.4b, *Xenopus* has learned to localize the prey with high accuracy after 50,000 learning steps, about one day of practice. It could also be shown (Lingenheil 2005) that the map as learned for arbitrary directions instead of $\varphi = 0°$ gives equally good performance. Not only does the adult animal equipped with its superficial neuromasts localize its prey on the surface of a pond or discern different or even identical preys but the model does so too. Being two-dimensional

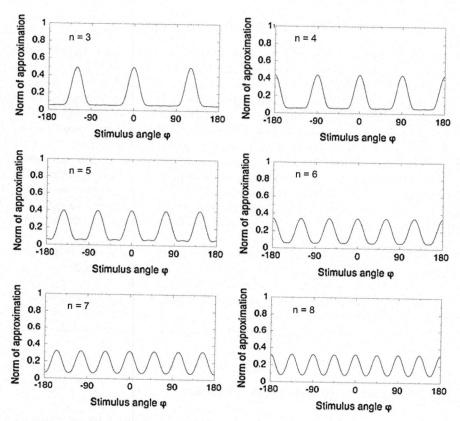

Fig. 16.6 Model performance as in Fig. 16.4 but now for $3 \leq n \leq 8$ equidistant *identical* sources on a circle with radius $d = 20$ cm and with *Xenopus* at the center. The map is clearly (to the beholder) able to segregate the stimuli as different objects but evidently the performance becomes worse as n increases. On the basis of present-day neurobiological data, the model cannot decide from which n onwards *Xenopus* stops discerning different identical stimuli

it has the great advantage of allowing a detailed mathematical modeling. What, then, can be said about map formation as neuronal-object representation of the outside sensory world?

16.2.2.4 Discussion

The model presented in this section has provided a neurobiological and simultaneously a mathematical mechanism of how the lateral-line system of the clawed frog *Xenopus* can learn to localize prey. Though "minimal" but therefore simple, the model is based on realistic properties of a neuronal system. Moreover, it can also learn an internal representation as generated by the wave-source signal, which would allow the frog to discern different kinds of prey. The model, based on a minimization principle functioning as in (16.6), is so general that it can learn any linear relationship between a reference input and a sensory input. It solves the problem of determining *what* is going on *where* through a map of neurons indicating by their spatial activity where the source is and by their temporal activity pattern what kind of stimulus it is, independently of its origin.

Not only is the model discussed here minimal and thus fairly transparent. Far more importantly, it is also mechanistic in that it is based on concrete mathematics and hence algorithms that ask for, and allow, verification. In a sense, it is remarkable that a minimal model already explains one of the astounding properties of *Xenopus'* (and any) lateral-line system: discerning $n \geq 2$ identical sources. In the final Sect. 16.7, schooling with typically four identical neighbors surrounding a fish laterally and one in front of it, will be another focus.

Moreover, as shown in Fig. 16.7, a lateral-line map has been found at least in *Xenopus'* optic tectum (Zittlau et al. 1986, Fig. 7). Glancing the right part of the figure one directly notices the empty space at the center. *Xenopus* either spends its time at the water surface waiting for prey, typically an insect that drops into the water and starts scrambling on the surface, in this way generating surface waves. To map the *direction,* all *Xenopus* needs is a one-dimensional arrangement, not only ideally but also in reality positioned on a circle. Because animals have a left–right symmetry in their brain's representation, there is a break in the middle, the dashed line. As for the rest, it is all a continuous map from a circular representation onto a corresponding "circle" in the optic tectum. That is exactly what we see in Fig. 16.7, as does *Xenopus* at the water surface.

It is known (Elepfandt et al. 2006) that *Xenopus* can also localize small prey fish while spending its day time at the bottom of the pond it lives in. It then does so in three-dimensional space, as does any fish. As one will see in more detail in the next section, this kind of prey can be perceived by the lateral-line system as well, provided the prey is in its range, viz., that of the frog length L. Interestingly, an adult *Xenopus* with $L = 8$ cm stops swimming if it has not reached yet the prey it was heading for, which was observed to happen at a distance of about 10 cm. It is therefore natural to hypothesize that the empty space at the center of Fig. 16.7, is reserved for a completion of the two-dimensional direction map corresponding to a half-sphere, indicated symbolically by the dashed arc in Fig. 16.7a.

All in all, the mathematical model presented in this section may well be applicable to other animals, such as crocodilians that use about 2,000 dome pressure receptors on their snout to localize prey (Soares 2002); in so doing they use a similar hunting position as *Xenopus* in Fig. 16.7a. As the present volume shows, fish use both canal and superficial neuromasts to determine the pressure or velocity field surrounding them; see also Bleckmann (1994). The canal neuromasts respond to the pressure *difference* $p(\mathbf{x} + \varDelta\mathbf{x}) - p(\mathbf{x}) = [\nabla p(\mathbf{x})] \cdot \varDelta\mathbf{x} + \cdots$ between two adjacent pores separated by $\varDelta\mathbf{x}$; the center dot $\cdot$ denotes a scalar product and ... is the rest of a Taylor series. Once $p(\mathbf{x}) \equiv C^{\text{st}}$, the gradient vanishes and so does the pressure difference. So canal neuromasts respond effectively to a discretized

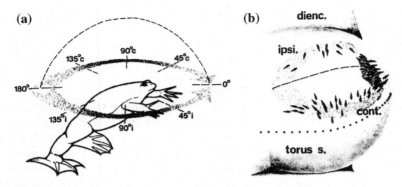

Fig. 16.7 Map of best directions of tectal neurons in *Xenopus*. **a** The directions around the frog in typical nocturnal hunting position with its eyes just above the water surface; *i* means ipsilateral and *c* contralateral. **b** Directional map in the optic tectum, effectively a one-dimensional representation corresponding to the rotational angle in A, a circle enclosing an empty area without response. The surface of the optic tectum and the *torus semicircularis* (torus s.) has been drawn as seen from above. *Solid arrows* indicate best directions of 60 units that were recorded in the positions where the *arrows* start. The *dashed line* is the meridian, which divides the ipsi-(ipsi.) from the contralateral (cont.) part. The *dotted line* at the *bottom* indicates the border between optic tectum and torus. The *dark line* at the *top* borders the diencephalon. The present figure is due to Zittlau et al. (1986, Fig. 7)

gradient that is made out of the pressure field. For any analysis of the lateral line's pressure field, the corresponding discretization is to be constantly borne in mind.

Many animals use independent sensory systems such as vision, audition, the vestibular system, and tactile senses that have to work together (Calvert et al. 2004; Stein 2012) so as to form some integrated neuronal representation or "picture" of the outer world in the animal's brain. For instance, the barn owl uses visual feedback to match its auditory prey localization to its visual system (Gutfreund et al. 2002; Gutfreund and King 2012). This multimodal, also called multisensory, integration belongs to what one then calls neuronal *object formation*. The very same object in the outside world is perceived by many senses exploiting totally different physical processes, which are then transformed into spikes, a kind of digital coding, out of which the optic tectum or the torus has to re-create a *single* object as neuronal representation. This integrative aspect will be treated in the one but final Sect. 16.6. Hydrodynamic object perception as performed by the lateral line of fish in three dimensions will be analyzed first, proving that its natural range is "one fish length". Then it will be time to see how fish do it in practice, either directly through momentum transfer, or indirectly by vortex-wake tracking due to conservation of angular momentum.

16.3 Range of the Lateral-Line System

Aquatic animals such as fish or *Xenopus* use their lateral line as a passive detection system to localize predator, prey, obstacles, or conspecifics. The lateral line enables even the Mexican tetra or blind cave fish to navigate efficiently through its environment and discriminate different structures and obstacles (von Campenhausen et al. 1981). Fish and aquatic amphibians such as *Xenopus* analyze the hydrodynamic structure of the velocity or pressure field, or both, so as to localize objects and determine their size and presumably shape. Can fish, then, really do so and also determine both size and shape? And, if so, what is the range of their lateral-line system?

Focusing on concepts, here an explanation is given of the underlying ideas to the mathematical proofs that fish can and, in so doing, ascertain mathematically the range of the lateral line as a system exploiting momentum transfer: Just one fish length. Behavioral experiments already indicated that the lateral-line system of fish can only be used to detect objects that are "near" the fish. More in particular, several authors, either through qualitative arguments (Harris and van Bergeijk 1962; Denton and Gray 1983), or in combination with the auditory system (Kalmijn 1988; Coombs 1994), or via indirect evidence (Coombs et al. 1996; Coombs and Conley 1997; Ćurčić-Blake and van Netten 2006) but all restricted to dipolar stimuli, had already suggested that the range of the lateral line is about 1–2 fish lengths. More recent theoretical studies (Sichert et al. 2009; Sichert and van Hemmen 2010; Urban et al. 2013) have then provided a consistent mathematical formalism with two different proofs that about one fish length is indeed the range, extending the mathematics to general instead of dipolar stimuli and showing the important distinction between predator and prey fish length in determining the shape of a hydrodynamic object. The arguments below are due to Sichert et al. (2009) and Urban et al. (2013).

Most experimental studies on the lateral-line system use a vibrating or translating sphere as stimulus. Because of the special symmetry of the sphere, the resulting velocity field is exactly that of a dipole—no matter whether vibrating or translating; cf. Guyon et al. (2001, §6.2.3.4) and Lamb (1932). Of course, not all objects in nature are spheres. Only for vibrating bodies could one identify (Kalmijn 1988) the influence of their shape on the flow field, showing a reasonable dominance of the dipole. This section presents a short discussion of how much, or little, information is available in the velocity field, whether fish can extract it, and to what extent distance plays a role, viz., through both the length L of the predator and ℓ of the prey.

Any hydrodynamic description and certainly that of fish starts with the Navier–Stokes equation (Lamb 1932; Acheson 1990; Billingham and King 2000),

$$\rho \left(\frac{\partial}{\partial t} + \mathbf{v} \cdot \nabla \right) \mathbf{v} = -\nabla p + \rho \mathbf{g} + \mu \Delta \mathbf{v} \tag{16.21}$$

where $\mathbf{v}$ is the water velocity field, p the pressure, ρ the water density, $\mathbf{g}$ the acceleration due to gravity, and μ the viscosity while Δ is the Laplace operator. Water being incompressible (Lamb 1932; Acheson 1990; Billingham and King 2000), $\mathrm{div}(\mathbf{v}) = 0$. Dropping the final term $\mu\Delta\mathbf{v}$, which will be done in a minute anyway, the Euler equation is obtained, of which the physics is straightforward. The velocity *field* $\mathbf{v}(\mathbf{x}, t)$ depends not only on the time t but also on the position, viz. $\mathbf{x}$. Newton's second law tells us $m\mathrm{d}\mathbf{v}/\mathrm{d}t = \mathbf{F}$ with $\mathbf{F}$ being the force. In a fluid, the mass m is replaced by the mass density ρ. The right-hand side of Eq. (16.21) is then the force, $-\nabla p + \rho\mathbf{g}$ due to pressure difference and gravity. The left-hand side is simply m, now ρ, times $\mathrm{d}\mathbf{v}/\mathrm{d}t$. Moving with the particles in the fluid one obtains the bracketed term on the left of (16.21) and one is done.

Viscosity, here for an incompressible fluid with $\mathrm{div}\,\mathbf{v} = \nabla \cdot \mathbf{v} = 0$, has been taken care of mathematically by Navier and Stokes (1822/1845) through $\mu\Delta\mathbf{v}$. This is an experimental fact (Acheson 1990; Billingham and King 2000) and there is no proof. What also changes as compared to Euler is the boundary condition: Instead of having $\mathbf{v} \cdot \mathbf{n}_S = 0$ everywhere at the surface S of the region outside the fish but inside the container with $\mathbf{n}_S$ as normal vector one now simply uses the *no-slip* boundary condition: $\mathbf{v}|_S = 0$ at the boundary. In fluids with relatively small viscosity μ, such as water with $\mu_{\mathrm{H_2O}} = 0.01$ in c.g.s. units, viscosity only has an effect in a small *boundary layer,* which is at most 2–3 mm thick under the conditions considered here.

Regarding the fluid two further, reasonable, assumptions are needed. First, viscosity plays no dominant role. For laminar flow in water and times that last relatively short, say less than a few minutes, this is quite reasonable. As a consequence $\mu = 0$ and the Navier–Stokes equation (16.21) reduces to Euler. Second, the velocity field $\mathbf{v}$ is rotation-free or irrotational with $\nabla \times \mathbf{v} = \mathrm{curl}\,\mathbf{v} = 0$. Consequently there exists a potential Φ so that (Acheson 1990; Billingham and King 2000) $\mathbf{v} = -\nabla\Phi$.

The problem of determining the range of the lateral line can now be analyzed from two completely different but complementary points of view. First, that of a prey that needs to avoid the predator. Second, that of the predator, which for the sake of convenience will also carry the origin but now at the position of a neuromast on its scales so that the shape of a prey is given by $\mathbf{r}$. The key question is: What is the strength of the velocity field generated by the prey as stimulus in dependence upon the distance $r = \|\mathbf{r}\|$? If the size of the predator as detecting animal is denoted by L, then only two length scales are relevant, viz., a predator's L and a prey's ℓ.

1. *Signal emitted by the prey as stimulus* A velocity field $\mathbf{v}(\mathbf{r}) = -\nabla\Phi(\mathbf{r})$ represents an adequate and natural stimulus to the lateral-line system and can be described by a multipole expansion of the velocity potential Φ (Lamb 1932) because the relevant fluid dynamics—outside the boundary layer—is well described by the Euler equation (Goulet et al. 2008). Using the real spherical

harmonics Y_{lm}^R and the multipole moments $\mathbf{q} = (\dots, q_{lm}, \dots)$, the velocity potential Φ can now be expanded in so-called multipoles,

$$\Phi(\mathbf{r}) = \sum_{l=1}^{\infty} \sum_{m=-l}^{l} q_{lm} \Phi_{lm} = \sum_{l=1}^{\infty} \sum_{m=-l}^{l} q_{lm} \frac{1}{2l+1} \frac{Y_{lm}^R(\theta, \varphi)}{r^{l+1}}. \tag{16.22}$$

Since a fish is usually far longer than wide the math of the multipole expansion gives that the q_{lm}'s scale as ℓ^{l+1} so that, whatever $-l \leq m \leq l$, q_{lm}/ℓ^{l+1} is of order one, $\mathcal{O}(1)$.

Introducing the characteristic length ℓ, the body-length of a stimulus such as prey, into Eq. (16.22) one can write the velocity field $\mathbf{v}(\mathbf{r})$ in the form

$$\mathbf{v}(\mathbf{r}) = \sum_{l,m} \hat{q}_{lm} (\ell/r)^{l+1} f_{lm}(\theta, \varphi) \tag{16.23}$$

where the $f_{lm}(\theta, \varphi)$ are now functions depending *only* on the angular coordinates θ and φ and directly follow from Eq. (16.22). Through the substitution $q_{lm} = \hat{q}_{lm} \ell^{l+1}$ "dimensionless" multipole moments $\hat{q}_{lm}$ are obtained, that are independent of the stimulus size. One sees explicitly from $\mathbf{v}(\mathbf{r}) = \sum_{l,m} (\ell/r)^{l+1} \hat{q}_{lm} f_{lm}(\theta, \varphi)$ how the influence of each multipole varies in dependence upon the size ℓ of the stimulus and the distance r to the detecting animal with length L, such as a predator. In view of Eq. (16.23), for $\ell/L \ll 1$ only the dipole term $l = 1$ survives whereas for $r \geq 1$ higher multipole moments come in and can provide information about the specific stimulus shape since different shapes are represented by different sets of multipole moments. On the other hand (Sichert et al. 2009), beyond $r \approx L$ the lateral-line information to localize a prey or whatever object is practically gone and this holds the more so since omnipresent noise needs to be handled as well (Sichert et al. 2009).

If, then, a fish can localize prey once the distance to it is less than approximately its own fish length L, how do they do so? Or, phrased differently, where is the information to determine the distance? Section 16.2.2.1 has presented detailed analytics to obtain the direction of a hydrodynamic object neuronally and Sect. 16.4 will show how distance information can be extracted from the lateral-line excitation pattern on the fish body. Direction and distance ($\leq L$) localize the object together. The shape of the stimulus is an important factor only if its size ℓ and the distance r are at least of the same order of magnitude, or shorter. It is for instance obvious that plankton, which is ten to hundred times smaller than the detecting animal, will not transmit any shape information through the flow field—though in this case shape information is not needed either. However, as stated before, in schooling, mate finding, or predator–prey behavior object information may well make a difference.

2. *Three-Dimensional prey to be reconstructed from Two-dimensional lateral-line input* What actually happens is that a fish needs to reconstruct the stimulus position and shape from the input provided by the velocity-field changes induced by the stimulus, say prey, on its body surface, where the neuromasts

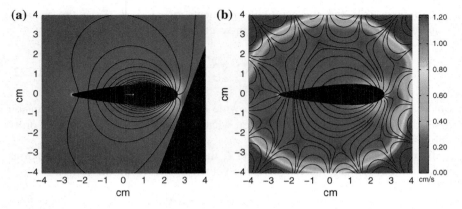

Fig. 16.8 a The *left figure* shows a situation where a swimming fish, which generates a flow field, meets a slanting wall on the *right*. **b** Reconstruction through a regularization as sketched in Tikhonov and Arsenin 1977 and showing (1) a decent reconstruction quality for the wall and (2) the natural range of stability of the reconstruction, a circle with a radius of about one "fish" length. The color code for both plots is on the *right*. The *solid lines* are stream lines as seen from the fish. Picture courtesy of Urban et al. (2014)

are. In other words, the inverse problem that is to be solved reads: How to reconstruct a three-dimensional prey from two-dimensional lateral-line input? As a reminder, a space is two- or three-dimensional, if it is spanned by two or three (independent) vectors. The fish surface is called a two-dimensional manifold as we can get it by "suitably" bending a two-dimensional plane.

In mathematics, the present kind of reconstruction of three-dimensional space out of data on a two-dimensional manifold is a notoriously difficult, so-called *ill-posed*, problem (Tikhonov and Arsenin 1977; Isakov 2006). It is characterized by high numerical instabilities that severely restrict the range of validity of the reconstruction procedure. What appears in mathematics as an ill-posed problem is what fish have to solve through their lateral-line system. Hence the lateral-line range is bound to be restricted by… the length L of the reconstructing surface; cf. Fig. 16.8. Beyond L instability of reconstruction takes over and thus reconstruction is of no use anymore.

Once the spatial restrictions of the lateral-line system are known, one can ask how fish exploit the lateral-line geometry in direct momentum transfer and how they take advantage of conservation of angular momentum in wakes that may survive in stillwater for a surprisingly long time, viz., minutes, since the viscosity of water is low. The next section is devoted the former question and vortex-wake tracking is the main topic of Sect. 16.5.

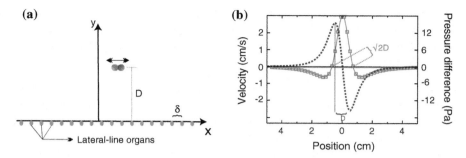

Fig. 16.9 **a** Two-dimensional minimal model where the lateral-line organs are arranged in a linear array on the *x*-axis. Note the distance between two consecutive organs is δ; for a canal lateral line these are the pores. An oscillating sphere is located at position $(0, D)$. It oscillates either parallel (‖, as in the figure) or orthogonally ($\perp$) to the *x*-axis, in both cases generating a dipolar velocity field. **b** Velocity v_x near the skin of the fish in dependence upon the *x*-position in (**a**) and the dipolar stimulus orientation, viz., parallel [‖, *gray dashed line*, due to Eq. (16.25)] or perpendicular [$\perp$; *solid line*, due to Eq. (16.26)] as compared to the *x*-axis; The stimulus is an oscillating sphere, a dipole, at distance $D = 1$ cm from the skin. The oscillating sphere generates either a triphasic or a biphasic response. The distance between the zeroes is $\sqrt{2}D$ for the triphasic response while the distance between maximum and minimum is D for the biphasic and $\sqrt{6}D/2$ for the max/min in the triphasic response. The *squares* represent the pressure difference. The velocity field and pressure difference field are *proportional* to each other; see also Eq. (16.31)

16.4 Geometry of Lateral-Line Perception

As in Sect. 16.2.1, a minimal model is the starting point but the response of both canal and superficial neuromasts in goldfish can now be treated from a mathematical as well as from an experimental point of view. In so doing, the focus is on the geometry of velocity and pressure field surrounding a fish since the lateral line takes advantage of a diffuse coding so that it is relatively safe to assume that it needs the total, diffuse, input delivered at all neuromasts for further information processing. For the moment, the setup is two-dimensional (Fig. 16.9a) but the reasoning can, and will, be extended to three dimensions; see also van Netten (2006). The arguments and figures below are due to Goulet et al. (2008), whom the reader can also consult for further information. For exact solutions of the velocity field exploiting the technique of conformal mapping—typically two-dimensional—while providing a 'hydro-dynamic toolbox' the reader may also consult Sichert and van Hemmen (2010). The present focus is on the ensuing neuronal information processing based on a detailed analysis of the biological physics of the underlying hydrodynamics.

Dipolar stimuli generated by a, in general, small oscillating sphere are by far the simplest and hence the most popular in experiment. It has to be constantly borne in mind, though, that neuronal reality can, and does, handle more complicated stimuli, as has been illustrated by the previous section. Canal neuromasts have been treated by Ćurčić-Blake and van Netten (2006, Eqs. (1)–(3)), who also show the simple sine/cosine linear dependence of the stimulus input on the dipole direction. Following Goulet et al. (2008), the approach of this section aims at handling both superficial and canal neuromasts and showing that the neuronal system has no reason to discern the two.

In first approximation, a superficial neuromast responds to the velocity field whereas a canal neuromast reacts to the pressure difference between the two pores surrounding it, i.e., acceleration; see also Coombs and van Netten (2006). It is important to realize that the Navier–Stokes equation (16.21) governs the fluid dynamics *in* the canal whereas the pressure field on the skin of the fish may be taken to result from the Euler approximation because the perpendicular pressure is almost constant in the boundary layer near the skin (Acheson 1990; Schlichting and Gertsen 2003). Outside the boundary layer the Euler description holds to high precision as long as the viscosity is low and the flow is laminar (Acheson 1990; Billingham and King 2000). The pressure difference between two adjacent pores is the relevant external force driving the fluid through the canal, where the fluid velocity is proportional to the pressure difference between two pores (Denton and Gray 1982). Physically speaking, water being incompressible pressure differences directly propagate through the (thin) boundary layer. In contrast, a superficial neuromast's response is proportional to the water velocity Goulet et al. (2012) near the skin of the fish so that one needs to verify to what extent boundary-layer effects play a role; see for example, Goulet et al. (2008, Fig. 4).

As in the previous section, it is assumed that the fluid is irrotational so that $\nabla \times \mathbf{v} = 0$ and the dipole flow field can be described by a velocity potential ϕ with $\mathbf{v} = -\nabla\phi$. Because water is incompressible, $\nabla \cdot \mathbf{v} = 0$. With $\mathbf{v} = -\nabla\phi$ one directly obtains that ϕ is a so-called harmonic function satisfying $\Delta\phi = 0$ with Δ as the Laplace operator. For a sphere oscillating with frequency v at position (D_x, D_y) *parallel* to the x-axis and in the plane of the lateral-line organ (Fig. 16.10a) with $\mu(t) := 2\pi\omega s a^3 \sin(\omega t)$ and $\omega = 2\pi v$ the two-dimensional potential ϕ is given by (Lamb 1932, §92),

$$\phi_{\|}(x,y,t) = \frac{\mu(t)}{4\pi}\left\{\frac{(x-D_x)}{\left[(x-D_x)^2+(y-D_y)^2\right]^{3/2}} + \frac{(x-D_x)}{\left[(x-D_x)^2+(y+D_y)^2\right]^{3/2}}\right\}$$

(16.24)

which satisfies the boundary condition $\mathbf{v} \cdot \mathbf{n}|_S = 0$ of the Euler equation on the surface S of the sphere. The resulting water velocity $v_x = \partial\phi_{\|}/\partial x$ in x-direction is

$$v_{x,\|}(x, y = 0, t) = \frac{\mu(t)}{2\pi}\frac{(2x^2 - D^2)}{(x^2 + D^2)^{5/2}}.$$

(16.25)

The distance between the two zeroes directly around the maximum is $\sqrt{2}D$; cf. Fig. 16.9b.

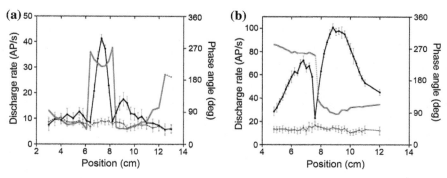

Fig. 16.10 Physiological data, provided by Jacob Engelmann and recorded as a function of sphere position for a (**a**, presumably canal neuromast) triphasic (parallel oscillations), and (**b**, superficial neuromast) biphasic (perpendicular oscillations) receptive field. In both figures the mean (over $n = 10$ trials) discharge rate (*solid line*) and the ongoing discharge rate (*thin dashed line, bottom*) has been plotted (y-axis on the *left*) versus the sphere's position. *Error bars* show standard deviations. Furthermore, the mean phase angle has been plotted as a solid *gray line*; cf. y-axis on the *right* of both (**a**) and (**b**). Phase values with non significant phase locking ($Z < 4.6$) are shown as *black dotted lines*. In the example shown in (**a**) the sphere vibrated with an amplitude of 80 μm at a distance of 15 mm to the neuromasts (trout), whereas in (**b**) the distance was 26.5 mm and the amplitude 200 μm (goldfish). Triphasic receptive fields such as (**a**) have two consecutive phase jumps of approximately 180°. The discharge rate has a central peak that is flanked by two lower side peaks. The distance $\Delta_\parallel$ between the zeroes was measured as the distance between the two phase jumps. The discharge rate of biphasic receptive fields such as (**b**) has two consecutive peaks of almost equal amplitude separated by a phase jump of 180°. The agreement with Fig. 16.9b is more than just accidental

In the case of a sphere oscillating *perpendicularly* to the skin of the fish, i.e., in the direction of the y-axis (Fig. 16.9a) and denoted by $\perp$, the result is

$$\phi_\perp(x, y, t) = \frac{-\mu(t)}{4\pi} \left\{ \frac{y - D_y}{\left[(x - D_x)^2 + (y - D_y)^2\right]^{3/2}} - \frac{y + D_y}{\left[(x - D_x)^2 + (y + D_y)^2\right]^{3/2}} \right\}$$

with water velocity $v_x = -\partial\phi_\perp/\partial x$ in x-direction given by

$$v_{x,\perp}(x, y = 0, t) = \frac{3\mu(t)Dx}{2\pi(x^2 + D^2)^{5/2}} \tag{16.26}$$

If the axis of vibration in the *xy*-plane makes an arbitrary angle α with the x-axis, then one directly gets (Ćurčić-Blake and van Netten 2006) $\phi(x, y, t) = \phi_\parallel(x, y, t) \cos\alpha + \phi_\perp(x, y, t) \sin\alpha$.

Figure 16.9b shows the water velocity near the skin of the fish as computed from the above Eqs. (16.25) and (16.26) for $v_{x,\parallel}$ and $v_{x,\perp}$. In the case of a dipole oscillation parallel to the fish surface, the water velocity in dependence upon the position at the skin is an even function whereas for perpendicular oscillation it is odd. To use the symmetry of the water velocity in both cases, it is normal to put $D_x = 0$, $D = D_y$, $y = 0$ at the skin and let x vary between, say, -5 and 5 cm. Accordingly, D henceforth denotes the distance between fish and submerged object, moving—cf. Goulet et al. (2008), [Eqs. (16.21)–(16.28)]—or not. For v_x as the relevant biophysical quantity stimulating superficial neuromasts, Eqs. (16.25) and (16.26) show that in principle fish could use the minima, i.e., zeroes, or maxima/minima of the velocity and hence firing-rate maxima to determine the distance D to an outside object. Whether it does so and, if so, how remains to be discovered.

Equipped with so much insight regarding dipolar stimuli, which as shown in the previous section dominate the lateral-line input world of fish, it is time to turn to experimental confirmation and verify whether superficial and canal neuromasts provide neuronal information processing with different, maybe even conflicting, information. For a different analysis focusing on canal neuromasts, one may consult (Ćurčić-Blake and van Netten 2006).

A thorough mathematical understanding of how lateral-line detectors respond to the velocity field (superficial neuromasts) and the pressure field (canal neuromasts) is essential to further progress of both experimental and theoretical neuroscience as they can now interact with each other in a positive, quantitative, way. As a further illustration of why it is now time to concentrate on neuronal information processing since the neuromast description has reached a satisfying quantitative description, a preview of a few canal/superficial neuromast responses as shown in Fig. 16.11 is timely, underlining also that canal and superficial neuromasts effectively give the same response to an oscillating dipole as stimulus.

16.4.1 Superficial and Canal Neuromasts

Canal neuromasts respond to the pressure *difference* between two adjacent pores. A focus of interest therefore needs to be the pressure field. A bit of surprise is then the analytical result that the pressure difference distribution along the canal lateral line has a form *identical* to the velocity field. For distance determination, this means that both surface and canal neuromasts could function by the same mechanism.

To calculate the pressure in a *non*viscous fluid at the pores of the canal lateral line, one can use the Bernoulli equation (Billingham and King 2000)

$$\left[p(x, y, t) + \frac{1}{2}\rho \mathbf{v}^2(x, y, t) + \rho \frac{\partial \phi_\parallel}{\partial t}\right]_{y=D} = p_0(t) \tag{16.27}$$

where $\mathbf{v}^2$ is the square of the water-flow velocity at the body and $p_0(t)$ is a constant depending on the time t only. Since the pressure p occurs in the Navier–Stokes/ Euler equation (16.21) only in ∇p one can gauge $p_0(t)$ "away," put $p_0(t) \equiv 0$ or, if desired, define it to be the hydrostatic pressure so that $p(x, y, t) - p_0$ is the

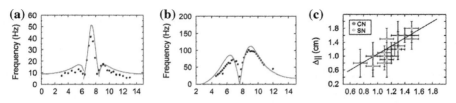

Fig. 16.11 **a** and **b** Experimentally determined firing rates (*dots*) at the afferent nerve and theoretical predictions (*solid lines*) of Eqs. (16.38) and (16.39). **a** Triphasic field of a canal neuromast for an oscillating sphere at distance $D = 1.5$ cm from the skin of the fish, with instantaneous firing rate $\mathscr{I} = 8$ and the free parameter $A = 170$ to fit the amplitude. **b** Biphasic field of a superficial neuromast for a sphere at distance $D = 2.6$ cm from the skin of the fish, with $\mathscr{I} = 13$ and $A = 6{,}000$. **c** Distance $\Delta_{\parallel}$ between the zeroes of the velocity field and hence minimal firing rate as a function of sphere distance D for afferents with triphasic receptive fields—see (a) or Fig. 16.10a—belonging to 15 canal neuromasts and 10 superficial neuromasts, The 25 samples stem from a population of 17 goldfish and 8 trout while the error bars represent standard deviations in x-direction. The *solid line* $y = x$ is just a guide to the eye. The agreement between predictions derived from canal and superficial neuromasts is striking. Experimental-data courtesy of Jacob Engelmann

deviation. As the velocity $v_y = \partial \phi_{\parallel}/\partial y$ perpendicular to the skin is zero at the skin because of the Euler boundary condition, one directly finds at all pores $\mathbf{v}^2 = v_x^2$. Using Eq. (16.26) for a sphere at distance D from the skin oscillating parallel to the x-axis one gets

$$\mathbf{v}^2(x, y = 0, t) = v_x^2 = (\partial \phi_{\parallel}/\partial x)^2 = \frac{a^6 \omega^2 s^2 (2x^2 - D^2)^2}{(x^2 + D^2)^5} \sin^2(\omega t). \qquad (16.28)$$

Again, the notion of a *thin* boundary layer is important since outside this layer the Euler equation can be used in conjunction with the Bernoulli equation (16.27) to obtain the pressure. On the skin the relation $\partial_t \phi_{\parallel} \gg \mathbf{v}^2$ holds.

The time derivative ∂_t of the velocity potential for an oscillating sphere is

$$\frac{\partial \phi_{\parallel}(x, y, t)}{\partial t} = \frac{1}{2\pi} \frac{d\mu(t)}{dt} \frac{x}{(x^2 + D^2)^{3/2}} \qquad (16.29)$$

where $d\mu(t)/dt = 2\pi \omega^2 a^3 s \cos(\omega t)$. The water velocity within the canal and thus the deflection of a canal neuromast is proportional to the pressure difference Δp between two adjacent pores at position $(x, y = 0)$ on the skin

$$\Delta p(x, y = 0, t, \delta) = \frac{\partial \phi_{\parallel}(x + \delta, 0, t)}{\partial t} - \frac{\partial \phi_{\parallel}(x, 0, t)}{\partial t} \qquad (16.30)$$

where δ the distance between two adjacent pores; cf. Fig. 16.9b. As δ is small, one obtains

$$\Delta p(x, y = 0, t, \delta) \approx \frac{\partial}{\partial t} \frac{\partial \phi_{\parallel}(x, y = 0, t)}{\partial x} \cdot \delta = \omega v_{x-\text{max}}(x, y = 0) \cos(\omega t) \cdot \delta.$$

$$(16.31)$$

In general, using Eq. (16.24) one gets

$$\Delta p(x, y = 0, t, \delta) = -\omega^2 a^3 s \rho \left\{ \frac{\delta + x}{\left[(\delta + x)^2 + D^2\right]^{3/2}} - \frac{x}{(x^2 + D^2)^{3/2}} \right\} \cos(\omega t).$$

$$(16.32)$$

Figure 16.9b shows a plot of the maximum amplitude of $\Delta p(x, y = 0, t, \delta)$ as a function of x for $\cos(\omega t) = -1$. It reveals that the pressure-difference distribution has a form *identical* to that of the velocity field, a remarkable result.

As a side remark, the distance between the zeroes is proportional to the distance D between sphere and lateral line. An explicit mathematical expression for the distance between the zeroes can be found by equating (16.32) to zero and then keeping only the terms linear in δ since $\delta \ll x$ and $\delta^2 \ll \delta$. This gives

$$-2D^6 + 6D^2 x^4 + 4x^6 = 0. \qquad (16.33)$$

The above equation has two real-valued zeroes $x = \pm D/\sqrt{2}$. Hence $\Delta_{\parallel} = \sqrt{2}D$ which corresponds to what has been shown before for the velocity field governing the response of superficial neuromasts. As in Eqs. (16.25) and (16.26), the distance between the zeroes has been derived in the context of the Euler equation and only depends on the distance between oscillating dipole and fish, *and nothing else*.

Furthermore, though the straight line in Fig. 16.9a looks fairly artificial for, say, the trunk lateral line of many fish, it can be shown that the same argument as used until now holds for a curved surface and hence for (nearly) any fish; see Goulet et al. (2008, Fig. 6).

Finally, the response of the pressure-difference distribution having a form *identical* to that of the velocity field, at least for a dipole field, one may wonder why nature has created superficial *and* canal neuromasts. By their very construction, canal neuromasts are rather insensitive to both turbulence along the fish body and to fish eigenmotion. The advantage is increased by the boundary layer transmitting pressure variations to the pores in a practically unchanged manner. These two properties give canal neuromasts distinctive detection advantage to both predator and prey in e.g., turbulent water of creeks. For a short review of the nature and presence of superficial neuromasts, one may consult Song et al. (2011).

16.4.2 Three Dimensions

Two dimensions do not restrict the above arguments as they all hold in three dimensions as well. Life only becomes a bit more involved. A few hints may suffice. As the previous subsection has outlined, there are good reasons for analyzing a problem in two dimensions. First and foremost, a dimension reduction provides both an understanding of the potential of the lateral line such as simultaneously discerning several objects and a lot of analytical insight (Sichert and van Hemmen 2010) while there is no, or hardly any, dimensional reduction of validity. Most of real life, however, happens in three dimensions and, for example, there is evidence (Janssen et al. 1990) that fish can use their lateral line not only to detect a source in a horizontal plane but also to determine the source's *elevation* such as its z-component in Fig. 16.9a. Anatomical data suggest that both canal and superficial neuromasts are organized orthogonally to each other (Schwartz and Hasler 1966; Schmitz et al. 2008). In the following it is shown how fish could compute the *3-dimensional* position, i.e., location and distance, of a dipole from lateral-line data.

For a dipole at position $(0, D_y, D_z)$ in three dimensions one finds (Lamb 1932)

$$\phi_\parallel(x,y,z,t) = -\frac{\mu(t)}{4\pi} \left\{ x\left[x^2 + (D_y - y)^2 + (D_z - z)^2\right]^{-3/2} + x\left[x^2 + (D_y + y)^2 + (D_z - z)^2\right]^{-3/2} \right\}.$$
(16.34)

Simplifying the anatomy of the lateral line, a natural assumption is taking two lines of receptors arranged perpendicularly to each other along the x and z axis (with $y = 0$) so as to get

$$v_x = \frac{\mu(t)}{2\pi} \left\{ 2x^2 - \left[(D_y)^2 + (D_z - z)^2\right] \right\} \left[x^2 + D_y^2 + (D_z - z)^2\right]^{-5/2}.$$
(16.35)

Since our main interest is at present in the velocity along the line of receptors on the x-axis, one may put $z = 0$ and calculate the distance between the zeroes to be

$$\Delta_{3D} = \sqrt{2}\left(D_y^2 + D_z^2\right)^{1/2}$$
(16.36)

For the line of receptors along the z-axis one then finds

$$v_z = \frac{\mu(t)}{2\pi} \frac{3x(D_z - z)}{\left[x^2 + D_y^2 + (D_z - z)^2\right]^{5/2}}.$$
(16.37)

If $x \neq 0$ (i.e., the line of detectors is not at $x = D_x = 0$), there is only one point with zero velocity on the z-axis, viz., at $z = D_z$.

A fish may then determine the position of a dipole in three dimensions as follows. The x-position D_x of the dipole is between the two zeroes of v_x.

The z-position D_z is at the zero of v_z. The y-position can be determined from Eq. (16.36). For canal neuromasts one can use Eq. (16.31) and show that the pressure difference distribution along the canal lateral line has an identical form to that of the velocity field. It may be well to realize, though, that the present argument does not imply at all that fish proceed this way. What is meant is that the *information* to localize a hydrodynamic object is available.

16.4.3 Outlook

It is of course essential to have a thorough mathematical understanding of how lateral-line detectors respond to the velocity field (superficial neuromasts) and the pressure field (canal neuromasts) before one can turn to analyzing the ensuing neuronal information processing. As a further illustration of how well the neuro-mast description has reached a satisfying, quantitative level, a quick look at a few canal/superficial neuromast responses as shown in Fig. 16.11 may suffice.

To determine the extent of agreement between theoretical predictions and measured responses, let us compare a "firing rate" function of the form of a triphasic field as in Fig. 16.10,

$$F(x) = \left| \mathscr{I} + A \, \frac{[2(x - x_0)^2 - D^2]}{[(x - x_0)^2 + D^2]^{5/2}} \, \Theta \right|, \tag{16.38}$$

with actual receptive fields. Here $\mathscr{I}$ is the experimentally determined instanta-neous firing rate, A denotes a scaling parameter and x_0 is the position of the sphere. The variable Θ is 1 when the neuronal response is in phase with the vibrating sphere and -1 when there is a $180°$ phase difference; cf. Figs. 16.9 and 16.10.

A few remarks concerning the nature of the fit in Fig. 16.10 are in order. Similar results, though for canal neuromasts only, have been obtained by Ćurčić-Blake and van Netten (2006) who have used a wavelet fit for their measurements of extracellular receptor potentials or for short ERPs, which in a crude way are summed potentials of the hair-cell synapses, showing the average activity in neuromasts. The technique is closely related to that of the microphonic potential (Jielof et al. 1952), which was brought to perfection by Kuiper (1956). The fits of Ćurčić-Blake and van Netten (2006) are empirical parameter fits based on their Eq. (4). For a discussion of the wavelet-transform fit as used by Ćurčić-Blake and van Netten (2006), the reader is referred to the literature (Goulet et al. 2008, p. 14). The fit is nice but the biophysical basis of using the wavelet notion as such is still unclear. In addition and as illustrated by Sect. 16.3, the lateral line may well handle stimuli far more general than dipolar ones.

Taken together, Ćurčić-Blake and van Netten (2006) have presented an empirical hypothesis as to how canal neuromasts may extract distance cues from flow fields originating from dipoles but the first experimental proof that a specific algorithm may be suited to extract this information from actual spike trains has been provided by Goulet et al. (2008). These works all show that both canal neuromasts (Coombs et al. 1996; Curcic-Blake and van Netten 2006; Goulet et al. 2008) and surface neuromasts (Goulet et al. 2008, 2012) can provide the information needed to extract source distances.

In passing two aspects are worth mentioning. First, Fig. 16.10, which has been obtained by recordings, also confirms nicely the more global ERP results. Second, Coombs et al. (1996) have performed analogous recording experiments on the trunk canal but with a slightly different fit in mind. They also find phase reversals as depicted in Fig. 16.10b but refrain from a clear-cut interpretation as offered here. A vanishing velocity is namely hard to measure precisely whereas maxima/minima with their phase difference of 180° or equivalently π are far easier to determine. There being no explicit "distance map", distance determination is still an open problem begging for a neurophysiological solution. Concrete algorithms are finally in position but an experimental verification of what really happens is now needed more than ever

In the case of an oscillating sphere moving perpendicularly to the skin of the fish (biphasic field as in Figs. 16.9 and 16.10), the firing rate is

$$F(x) = \left| \mathscr{I} + A\, \frac{D(x - x_0)}{[(x - x_0)^2 + D^2]^{5/2}}\, \Theta \right|. \tag{16.39}$$

As shown in Fig. 16.11a, b, the agreement between the modeled firing rate F and two arbitrarily chosen neuronal receptive fields is satisfying.

An at least puzzling coincidence offered by Fig. 16.11c is the nice linear relationship between the distance $\Delta_{\parallel}$ of the zeroes of the velocity field on the fish surface and the distance D from an oscillating sphere to afferents with tri- or biphasic receptive fields, as shown, e.g., in Fig. 16.10a. Goulet et al. (2008, Eq. (34)) present the full-blown three-dimensional derivation and expression for a curved fish but the essence of their message is already contained in the discussion following (16.33). Within the error bars, there is no experimental distinction between superficial and canal neuromasts, as shown theoretically in Sect. 16.4.1.

16.5 Vortex-Wake Tracking

When a fish swims, it has to give itself momentum so as to move forward through the water surrounding it. In so doing it uses its tail, which then produces a string of vortices (Blickhan et al. 1992; Bleckmann 1994; Hanke et al. 2000; Drucker and Lauder 2002) that often merge into a vortex ring (Dickinson 2003).

There is a remarkable difference between the action induced by the lateral line as a consequence of direct momentum transfer, characterized by a finite range (cf. Sect. 16.3), and vortex-wake tracking, which exists preferentially in stillwater and is a consequence of conservation of angular momentum in fluids of low viscosity such as water. Momentum transfer is direct and practically instantaneous

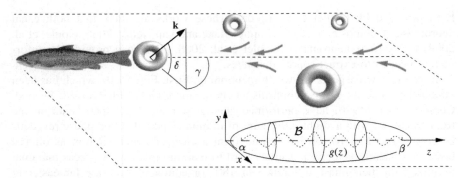

Fig. 16.12 A fish's wake consists of vortex rings originating from the tail beats with a velocity component (*thick arrows*) that compensates the fish's displacement; cf. Blickhan et al. (1992, Fig. 1). The vector **k** describes the orientation of a vortex. A second, for computational convenience (Geer 1975) rotationally symmetric, fish with body $\mathscr{B}$ is tracking the wake structure. As Geer 1975 could prove, a continuous distribution of hydrodynamic poles $g(z)$ located on the axis of the fish body between α and β—lower picture—ensures that the flow through its surface $\partial\mathscr{B}$ is zero, in this way satisfying the Euler boundary condition. Picture courtesy of Franosch et al. 2009

whereas interaction through angular momentum is indirect and long-lasting (Hanke et al. 2000; Hanke and Bleckmann 2004). Moreover, Figs. 16.11a, b clearly show a dependence of the neuromast firing-rate response upon the position of the stimulus with respect to the animal, suggesting a kind of somatosensory neuronal representation or, for short, mapping. No such response is to be expected from vortex wakes; see Fig. 16.12. While tracking a vortex wake, which is what predators may do [see, e.g., Pohlmann et al. (2001) for catfish], the vortex-wake pattern, though specific, simply moves along their body. Since fluid motion in a canal is driven by the pressure difference between neighboring pores, the canal lateral-line system essentially monitors discretized pressure gradients alongside the fish body through the fluid velocity in the canals (Kalmijn 1988; Bleckmann 1994).

As has been shown in Sect. 16.4.1, in neuronal processing of hydrodynamic objects canal and superficial neuromasts exhibit many parallels. How, then, can vortex-wake tracking be described? To answer this question, one now needs to analyze pressure difference distributions $\partial p/\partial x$ along the (trunk) lateral line due to a vortex ring (Dickinson 2003) as a function of position along the body relative to the body length. During wake tracking vortices with different orientations γ move, say, in z direction along the fish. Any such vortex generates an excitation pattern that often looks like a vortex ring whose normal, a unit vector, has two orientation angles, γ and δ. In numerical simulations, a rotationally invariant "cigar" fish is quite a decent approximation—with the extra advantage that, to a large extent, it can be handled analytically (Geer 1975)—so that $\delta = 0$ is a harmless assumption. Then $\gamma = 0$ means that the vortex ring orientation, i.e., its rotation axis, is parallel to the fish axis of symmetry.

Ćurčić-Blake and van Netten (2006, Eqs. (1c) and (2a)) have shown that the stimulus patterns of $\gamma = 0$ and $\gamma = \pi/2$ are orthogonal to one another and that any excitation pattern is a linear combination of the two, allowing a fish to uniquely identify the vortex' orientation by projecting the stimulus in question onto pre-determined excitation patterns. The point is that this is a linear operation and can be performed by a neuronal network with neurons operating—to decent approximation—in their linear range. Fish can thus determine the direction they must follow to track their target.

Mathematically, the problem is doable but fairly intricate so that for details the reader is referred to Franosch et al. (2009). These authors could take advantage of brilliant work of Geer (1975) and Hassan (1993). The latter, whose preparative work (Hassan 1985, 1992a, b) is equally impressive, and Kalmijn (1988) were the first to provide a complete mathematical description of the hydrodynamics underlying the lateral-line system's functioning.

Let us consider a fluid without viscosity ($\mu = 0$) and hence work with the Euler equation. With (the fair approximation of) zero viscosity conservation of energy holds that for horizontal motion in fluids reappears in the form of the Bernoulli equation (16.27) with only two terms, viz., $p = \rho \partial \phi / \partial t + \rho \mathbf{v}^2 / 2$, with ϕ as the potential and p as the deviation from hydrostatic pressure, which can be taken constant here.

The result of the rather involved computation, whose details have been presented elsewhere Franosch et al. (2009), is that in the case of a self-propagating vortex and a fixed fish the pressure (deviation) is given by

$$p = \rho \tilde{U} \left(\cos \gamma \frac{\partial \phi}{\partial z} + \sin \gamma \frac{\partial \phi}{\partial x} \right) + \frac{1}{2} \rho \mathbf{v}^2. \tag{16.40}$$

Experimentally, the core structure of a vortex ring is hardly altered by the physical presence of the fish body from a distance of at least a few (≥ 3) cm onwards (Chagnaud et al. 2006). Thus, for larger distances, circular vortices are the appropriate stimulus to examine. As a real vortex ring propagates, its core radius increases by viscous diffusion, but only so slowly that it retains its structure for over a minute (Hanke et al. 2000; Hanke and Bleckmann 2004). In nature, various vorticities and vortex radii have been observed. The propagation speed $\tilde{U}$ of a vortex ring results from the vortex-ring radius R and the vortex-core radius R'—together R and R' specify a torus in three dimensions—and the vortex strength Γ as defined by Lamb (1932, §163). In computing the stimulus one can nearly always neglect viscosity and take $\mu = 0$ because a small μ only alters the flow field in a so-called boundary layer near the fish's body surface while the pressure is constant in any cross section of the boundary layer that is orthogonal to the flow and to the surface (Lamb 1932, §371a).

Furthermore, vortices at different distances from the fish generate excitation patterns that only differ in amplitude. The same holds true for changes in Γ, R, and U. As shown in Fig. 16.13a for an experiment with *fixed* fish, a vortex ring with $\gamma = 0$ or π and $\delta = 0$ generates excitation patterns that are either four-phasic or

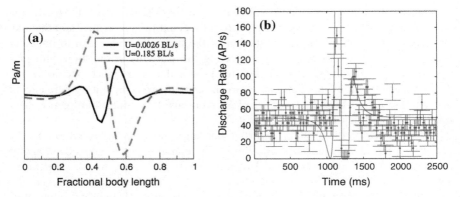

Fig. 16.13 **a** Theoretical prediction (Franosch et al. 2009) of the pressure along a *fixed* fish, when a vortex ring with $\gamma = 0$ or π and $\delta = 0$ passes and generates excitation patterns that are either four-phasic or two-phasic depending on the vortex propagation speed $\tilde{U}$. **b** Neuronal response ensuing from a canal neuromast, experimental results. The dots with error bars represent time-dependent spike rates of a canal neuromast measured when a vortex with $\gamma = 0$ or π and $\delta = 0$ passed a fixed fish, averaged over ten runs. As the vortex has approximately constant velocity, the recorded time-dependent data match the nervous excitation pattern along the lateral line at one moment of time. Canal neuromasts respond to the pressure difference $\Delta p \approx (\partial p/\partial x)\Delta x$ between two adjacent pores shown in (**a**). The *solid line* is a fit to the theory; because of negative hair-cell polarity, it is in the present case a mirror image of the *black solid line* in (**a**). Both pictures are due to Franosch et al. (2009) and they are one out of many pairs of theoretical prediction and experimental verification

two-phasic depending on the vortex propagation speed $\tilde{U}$. Accordingly, in this case there is no one-to-one relation between γ and stimulus but it is easy to see why. As illustrated by Fig. 16.13, a faithful hydrodynamic description of hydrodynamic vortices leads to a faithful prediction of the neuronal response.

16.6 Lateral Line, Map Formation, and Multimodal Integration

A vortex wake provides a fish with a global stimulus to its lateral-line system and wake tracking is the ability to follow such a globally stimulating vortex trace in space and time; see, for instance, Pohlmann et al. (2001). Momentum transfer is direct and, as seen in Sect. 16.3, its lateral-line input is diffuse and the range is short. Exactly because of its short range a lateral-line object can also be seen once a clear sight is available. Then the eye with its lens is (nearly always) the most accurate modality. Hence, it is quite safe to assume that vision, when it is available, is a key factor in integrating a lateral-line map with those stemming from the other modalities through feedback.

A 'map' is defined to be a neuronal representation of the outside sensory world (van Hemmen 2006). As indicated in Sect. 16.2, lateral-line maps with a rather intricate performance may well exist but, in view of their diffuse encoding strategy, can they exist, what should they then look like, and how can they arise? To phrase it more colloquially, is vision (with its "fish-eye camera") playing the role of a teacher telling the lateral-line system how to generate a map, or can the latter first produce the map itself, which is then integrated into the visual map in the tectum opticum, or is there a mix, or …? To answer these natural but rather intricate questions, it may well pay off to make a small detour and analyze a simpler case first, which seems to be understood completely since long, viz., that of the barn owl's integration of auditory and visual maps, fascinating work of Knudsen et al. (1987), Gutfreund et al. (2002) and Gutfreund and King (2012).

16.6.1 Neuronal Integration of Different Modalities

Each animal is unique but one can learn quite a bit from the way in which evolution has equipped modalities in different animals with similar, or different, neuronal algorithms to handle information provided by the senses. How are, for instance, the auditory and visual system of nonmammals integrated in the optic tectum, where for the first time the inputs to *all* modalities come together? For the barn owl, this is known in detail and hence can be taken as valuable orientation for understanding what may happen in, e.g., fish. Three weeks after hatching the young barn owl's head is adult but its sound localization is not. After two more weeks it is. This stage belongs to the juvenile sensitive period (Knudsen et al. 1987; Knudsen 2002). During integration of audition and vision the visual system "ought" to dominate, and so it does (Gutfreund et al. 2002; Gutfreund and King 2012) because it is far more precise (Bürck et al. 2010). The big advantage of understanding neuronal integration of visual and azimuthal sound localization maps in the barn owl is that both encodings are *discrete:* The visual one due to the lens generating a sharp image on the retina through focusing while the auditory one starts in the cochlea with frequency decomposition in conjunction with high temporal precision (if the frequency is low enough, say, <1 kHz) and ends up in a discrete encoding through the Jeffress scheme (Schnupp and Carr 2009) realized in the intertwining delay lines of the laminar nucleus, a few nuclei before the optic tectum (Knudsen 2002).

To perform azimuthal sound localization, a barn owl needs to compare the interaural time difference (ITD) and hence the phases of the sound arriving at left and right ear. Hence, it needs a precise phase locking since otherwise it cannot "compare" the phases as they arrive at left and right ear, and it does so up to about 9 kHz. Right in front of the animal the ITD vanishes (= 0) and it in- or decreases by about 2 µs per degree. The left and right cochlea perform a frequency decomposition and after the cochlear (and magnocellular) nucleus the signals reach the laminar nucleus. This is the first nucleus where signals from left and right ear come together. It is here that a neuronal representation or for short a map is formed (Schnupp and Carr 2009). The inferior colliculus (IC) is the next stage and through the central nucleus of the inferior colliculus (ICC) and the external nucleus of the inferior colliculus (ICX), where all frequency signals are again combined, the action potentials reach the optic tectum (OT); see in particular Fig. 7 of Knudsen (2002).

Different modalities use *different* physical techniques to perceive their surroundings. For audition, it is sound and for normal vision it is a certain part of the electromagnetic spectrum. A natural question is then: How should several modalities combine the information emitted by the very same (external) object into the very same 'object' as neuronal representation in the optic tectum? To get these different representations fit into a single neuronal object, nature has provided the optic tectum (or superior colliculus in mammals) with a feedback to the stage before integration, for audition the ICX. How to imagine what kind of map evolves at the ICX so as to prepare the input to the optic tectum, where all modalities get integrated? The optic tectum now has the role of teacher to guide the evolution of the (as yet) auditory map in the ICX under the influence of the auditory input from the preceding ICC.

Phrasing it all slightly differently, how can one imagine the action of a teacher, particularly, a neuronal one? A first suggestion is that it be positive in the sense of excitatory and local, the *LE* model. So to speak, a model of the "ideal" teacher. As the relevant learning mechanism, one can profitably take STDP (Gerstner et al. 1996; van Hemmen 2001) and let it run. What comes out is quite surprising (Friedel and van Hemmen 2008): It works but not that well. If, on the other hand and following experimental results on the barn owl (Knudsen et al. 1987; Gutfreund et al. 2002; Gutfreund and Knudsen 2007; Gutfreund and King 2012), the interaction is taken locally excitatory but also globally inhibitory so that a net inhibition remains except there where things "look okay," viz., where excitatory teacher input from the optic tectum acts, then it all works far better and more robustly (Friedel and van Hemmen 2008). Clearly, a combination of local excitation and global inhibition (LEGION) except there where the excitation is, which one may also call 'local disinhibition,' seems to be the optimal combination in nature, as is indeed also seen in the neuroanatomy of fish (Vanegas et al. 1984a, b).

Local disinhibition is what Gutfreund et al. (2002) and Gutfreund and King (2012) call "gating" since the other ICX neurons outside the OT (teacher) input get inhibited. To quote these authors, "visual activity is useful only when it occurs simultaneously with auditory activity and both activities represent a common source. Thus, only visual activation that is associated with a salient auditory stimulus should be permitted through the gate," a philosophy we may well agree with.

In Fig. 16.14a the top layer (OT) is defined to be the "teacher," with the external direction (the angle φ) as horizontal axis. For the resulting activity one can take a rate coding. When under normal conditions the young barn owl hears the fox, a signal comes from the ICC (bottom) as input, then there is also a corresponding signal from where it sees the fox, which dominates the OT (top), the

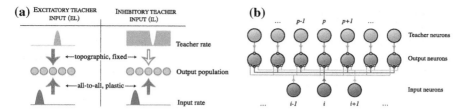

Fig. 16.14 **a** Firing-rate representation of the basic idea behind input and teacher population in multimodal integration. Both the input and the teacher population are organized in a map. *Left* In the *LE* model, the output neurons receive excitatory input from both the input and the teacher population. *Right* In the *LEGION* model, the output neurons are all silenced by the inhibitory teacher input, except for a small number of neurons receiving no inhibition, which are now driven by the input population, as in the *LE* model on the left. **b** Network topology of the layers as modeled. There are three layers: (1) The input layer (*bottom*) corresponds to ICC and is labeled by i. (2) The teacher layer (*top*) corresponds to OT and is labeled by p; here the neurons fire at a rate that depends on the input position y. (3) The output layer corresponds to ICX and is driven by spikes from both the input and the teacher layer. The synaptic connections from input to output layer are subject to *plastic* weight changes and the connections from the teacher layer to the output layer remain fixed. In the model, every output neuron receives spikes from one corresponding teacher neuron and from all input neurons. The above figures have been taken from Friedel and van Hemmen (2008)

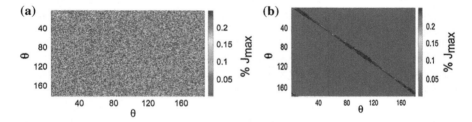

Fig. 16.15 **a** Synaptic weight distribution between Medulla and torus semicircularis before learning, a random distribution between 0 and an upper bound, J_{max}; cf. Fig. 16.14b. **b** Map of synaptic connectivity between torus and Medulla cells *after* learning according the local disinhibitory STDP as explained in Fig. 16.14, called for short *LEGION*. If a map is ideal, the labeling of the horizontal axis representing directional input corresponds to that of the vertical one, its topographic neuronal representation. Accordingly, an ideal map is a 'diagonal' one and indeed a well-defined lateral-line map of the outside world appears, here a directional one; cf. Fig. 16.9a. Figure courtesy of Goulet (2010, Fig. 6.6)

teacher. It is a natural, and correct, guess that the visual and auditory object are there where the firing rates are maximal. Therefore, the ICX is called the "output" layer, the OT the teacher, and the ICC the input layer (Fig. 16.15).

For an external stimulus at a certain location, the firing rate in Fig. 16.14a peaks strongly at a well-defined location in the input and the teacher map. The input and teacher map can, however, be misaligned., e.g., since the young barn owl wears binocular prisms displacing the visual field to the right (Knudsen and Knudsen 1989).

The output (ICC) population receives spikes from the input (ICX) and from the teacher (OT) layer. In the *LE* model, the output neurons receive excitatory input from both the input and the teacher population. The output neurons corresponding to the teacher neurons with the highest rate will fire most and the connections from the active input neurons to this set of output neurons will therefore be strengthened by STDP.

In the *LEGION* model, the output neurons are all silenced by the inhibitory teacher input, except for a small number of neurons receiving no inhibition. These neurons are driven by the input population and—as for the *LE* model—the connections between the active input neurons and the active output neurons are strengthened as a consequence of STDP. Both models are able to form a high-quality map based on the input and the teacher signal. The resulting map is automatically aligned with that in the teacher population.

To explain how map formation operates mechanistically and finish the argument, it is best to imagine three layers of neurons; cf. Fig. 16.14b. The input layer (corresponding to ICC) and the teacher layer (corresponding to OT) fire at a rate that depends on the input position y. The output layer (corresponding to ICX) is driven by spikes from the input and teacher layer. The synaptic connections from input to output layer are subject to plastic weight changes according to STDP and the connections from the teacher layer to the output layer remain fixed. Every output neuron receives spikes from one corresponding teacher neuron—could also be more in a finite surroundings—and from *all* input neurons. One could replace 'all' by a wide input connectivity but for modeling and simulation 'all' is far more convenient to handle.

Friedel and van Hemmen (2008) have shown that the *LEGION* model with its local disinhibition performs far better and is far more robust than the *LE* model with merely local excitation. Therefore, sticking to the former and enquiring how local disinhibition can contribute to map formation in the lateral-line system is reasonable. Admittedly it looks as if a *combination* of local excitation and global inhibition as teacher signal for the ICX in the barn owl and analogous centers in, say, fish were the best of all worlds to get an optimal preparation of the maps of separate modalities such as vision, the lateral line, and the vestibular system, before their integration in the optic tectum. As shown below for fish, experimental evidence (Vanegas et al. 1984a, b) hints in the direction of local excitation plus global inhibition (LEGION) so as to gate the "right" excitation while keeping, sloppily formulated, the excitatory garbage (also called "noise") from elsewhere under control.

16.6.2 Lateral-Line Objects and Map Formation

Whether multimodal integration in fish happens in the *torus semicircularis* as suggested by Scheich and Ebbeson (1983) or in the optic tectum as exposed in detail in the monumental volumes of Calvert et al. (2004) and Stein (2012), the argument of local disinhibition remains equally valid. In the neuroanatomy of fish one faces a feed forward sequence starting with the mechanoreceptors of superficial and canal neuromasts—for *Xenopus* without the latter—and leading via the *medial octavolateral nucleus* (MON) or for short the Medulla to the *torus semicircularis* and from there to the optic tectum. Given that diffuse encoding exists, one may therefore wonder how the different diffusely encoding modalities manage to get their maps sorted.

Xenopus In *Xenopus* the lateral-line encoding is diffuse but, as shown in Sect. 16.2.2.3 and particularly in Fig. 16.7b, clear directional maps can be found in the optic tectum. Furthermore, Sect. 16.2.2.1 has provided explicit evidence that the clawed frog not only has a map such as the one shown in Fig. 16.7b but also that it can exhibit a rather subtly discerning performance as in Fig. 16.4b and, equally important, that it can be understood mathematically how such a map may function. How, then, does it arise? Here too a tentative suggestion is vision as a helpful teacher. *Xenopus* eyes are covered by wide-angle lenses with an all-round view, which—it is generally believed—is optimally adapted to vision in air above the water surface. This would be totally consistent with the above Fig. 16.7b, where central vision is missing.

On the bottom of the pond the rest of the world is above the animal. Young *Xenopus* preferentially hunts in clear water and during sunset, optimal conditions for learning a lateral-line map with vision as teacher (Udin 2007), both at the surface and underwater.

Since *Xenopus* acquires about 70 % of the oxygen through its skin it can spend quite a long time underwater while watching its surroundings at the bottom of its favorite pond.

In an Appendix to Chung et al. (1975) Land and Stirling establish that the eyes of both larval and adult *Xenopus* are somewhat farsighted (hypermetropic) in water, but behave practically identically in both air and water, so that *Xenopus* can use its eyes for teaching the lateral line until its adulthood.

Fish There is some analogy between the lateral-line and somatosensory system in that both exhibit local maxima in pressure or velocity—see, for instance, Fig. 16.9b—signaling the presence of an object but... In somatosensory perception the object boundaries are sharp whereas in the lateral-line one they are diffuse, so to speak, "smeared out". Figure 16.9b underlines explicitly why global encoding may be essential. To determine a distance, the value of a maximal/minimal velocity itself is irrelevant. It is all relative and the maximum/minimum by itself indicates at best two of the three coordinates of the hydrodynamic object on the fish body, the haptic aspect. Exactly this is also the difference between *Xenopus* with just 180 lateral-line organs and fish, which have a few thousand all over the body; in a true sense, somatotopic. What is left is the distance between the object and the scales of the fish; it is measured along the normal vector. To compute the distance to an outside object neuronally, a fish needs more global data sampled from many neuromasts, such as the distance between maxima and minima of the velocity field or zeroes; see, for instance, Fig. 16.9b.

16.6.3 Does a Lateral-Line Map Exist and, if so, How Does It Look?

There are two closely related questions: Does a lateral-line map exist and, if so, how does it look and where can it be found? Until now a true *somatotopic* organization in the Medulla could not be shown but this is not too surprising either since one also has direction and excitation strength that need to get encoded as well. There is evidence that toral cells receive input from flow-insensitive Medulla neurons that may receive input from canal-neuromast afferents and map the position of a moving object (in the experiment a sphere) into a kind of somatotopic map (Plachta et al. 2003; Engelmann and Bleckmann 2004; Meyer et al. 2012). The data for flow-sensitive (probably receiving input from superficial neuromasts) units presented by Plachta et al. (2003) are in favor of some somatotopic encoding of water motion. In short, there is evidence for a map showing a roughly somatotopic arrangement, which is reasonable since the input is diffuse as it stems from the velocity/pressure field surrounding the whole animal and transmitted through the lateral line system on the animal's body surface.

Far more importantly, as in the barn owl and the clawed frog, a map is to develop on a genetically determined substrate. It is important to realize this since primary maps, though steered by sensory input, develop independently of the other sensory modalities and under the guidance of spike-timing-dependent plasticity or for short STDP (Gerstner et al. 1996; Bi and Poo 2001; van Hemmen 2006). The surprising result (Krippner 2012) is that a crude, genetically determined, somatotopic organization of the synaptic input already suffices to generate a lateral-line map. This map formation then functions by itself on the basis of diffuse neuromast input and a visual teacher is not needed, in agreement with first results of Plachta et al. (2003) and Engelmann and Bleckmann (2004). Then comes integration of the different modalities in the optic tectum or, in fish, maybe already in the torus.

Continuing their way through the different modalities, sensory input arrives in the optic tectum, where (Sect. 16.6.1) local disinhibition may well be the driving force to obtaining a coherent neuronal representation of sensory input from the lateral-line, visual, auditory, and (for electric fish) electric sense (Bastian 1982; Scheich and Ebbesson 1983). Electric fish (von der Emde and Engelmann 2011) show in many of their optic tectum (OT) neurons a response to direction and speed of water motion, which just indicates that velocity is a vector and that superficial neuromasts respond to the velocity field.

There is evidence that in the torus some cells sensitive to a mechano-sensory input will also respond to auditory stimuli. The projections between torus and optic tectum match topographically (Vanegas et al. 1984a, b), which means that the lateral-line map in the torus and the multimodal map in the optic tectum should have an organization compatible with each other and be organized in a retinotopic way, which agrees nicely with the fish-eye- camera character of the eyes observing a lateral-line object in their direct neighborhood. Here it is important to realize that fish have two practically independent eyes (with negligible visual overlap) that function literally as "fish-eye" cameras. The optic-tectum map is similar to the image projected by the lens onto the retina (Heiligenberg and Rose 1987).

Interestingly, the feedback from the tectum opticum to the torus semicircularis is believed to be at least in part inhibitory. Vanegas et al. (1984a, b) describe it as a signal exciting the few neurons at the exact position of the stimulus and inhibiting the other neurons around it. That is, local excitation and global inhibition (LEGION). Hence a decent amount of evidence appears in favor of a powerful algorithm to arrive at multimodal integration and, hence, lateral-line map formation in fish.

16.7 Conclusion

It is time to summarize the insights we have obtained from the point of view of theoretical neurobiology. Schooling, or shoaling, fish may provide quite suitable a test. It is known since long (Pitcher et al. 1976; Faucher et al. 2010) that the lateral line is essential for schooling. The fish density is to be high to allow collective shoaling behavior and thus the average distance is to be small, say, a fish length, the average amount of neighbors a fish needs to observe is five, viz., four neighbors around and one in front of it, they are all of the same size and character, and vision is dispensable.

Section 16.2 shows in detail that a lateral-line system can indeed discern several *identical* objects by using direct momentum transfer. Puzzling as it may look, being able to fetch one out of several identical objects by means of mechanosensory perception, there is now a clean explanation; see for instance Fig. 16.4b. In addition to the way in which lateral-line information is gained, the precise upper bound of how many objects can be distinguished depends on "higher" neuronal areas and is beyond this paper. Both experimentally (Elepfandt 2012) and theoretically, the theoretical method exploiting techniques of Sect. 16.2.1.3, it has been found that this minimal number is at least five.

Exploiting the notion of multipole expansion, Sect. 16.3 proves why for momentum transfer the range of the lateral line is short, one fish length. One needs to distinguish predator with length L and prey with size ℓ. For L's localizing an object the latter should be within the range L away from its lateral line. For discerning the *shape* of a lateral-line object ℓ as well, the latter has to be at most one fish length (its own, viz., ℓ) away from the predator of length L.

The multipole expansion starts with the source and develops the signal as a series expansion (16.23) on the surface of the fish body, i.e., the lateral line. Conversely, one may pose the question as to how one can reconstruct a three-dimensional object on the basis of two-dimensional data, which is what the lateral line provides. The ensuing mathematical problem is ill-posed (Tikhonov and Arsenin 1977; Isakov 2006) and consequently the reconstruction is numerically unstable so that the range of the reconstruction algorithm is quite finite,... about one fish length L; cf. Fig. 16.8.

What is the precise role of geometry in the lateral line's processing? Section 16.4 provides information on both canal and superficial neuromasts, paying due attention to a dipolar stimulus. It shows explicitly how hydrodynamic excitation of neuromasts can be computed. Furthermore, it is proven that within one fish length L there is indeed enough information to determine not only a lateral-line object's direction but also the distance to the observing animal. How, then, the animal encodes the distance and whether there is a map is still unknown. Finally, if the hydrodynamic object's length ℓ is less than the distance to the observing lateral line, then the latter can determine not only ℓ but also the shape that belongs to it.

Returning to schooling, one may wonder how a fish localizes its conspecific just in front of it. Is that done directly or indirectly through the vortex trail the latter generates? Though the question seems to be open, there are now enough quantitative tools to analyze how the lateral line can handle both direct and instantaneous momentum transfer (Sects. 16.2, 16.3) and indirect, long-living vortex-wake stimuli (Sect. 16.5). In addition, the response of the neuromasts has meanwhile been shown (Goulet et al. 2012) to be proportional to the velocity field, an aspect greatly simplifying the mathematical analysis.

Developing explicit mathematical models and specifying algorithms of how the lateral-line system works is valuable for at least two reasons. First, and well known from the history of science (Dijksterhuis 1969), an intensive *interaction* of experimental analysis and theoretical verification while probing new ideas through mathematical modeling expedites scientific progress considerably. Second, and equally important, technological implementation (Pandya et al. 2006; Yang et al. 2006, 2010, 2011; van Hemmen 2010) is possible only if a full-blown mathematical theory of natural phenomena and the algorithms that join them exists.

Once one has understood the lateral-line sensory system through the response of its neuromasts to the velocity and pressure field surrounding a fish or any aquatic animal such as *Xenopus,* an obvious question is whether and, if so, how the outside hydrodynamic world is represented neuronally through one or several maps on the basis of the plethora of sensory, and often diffuse, input. Section 16.6 provides ample information that in torus and optic tectum a conglomerate of consistent maps centered around lateral line and vision exists and that its integration may well be driven by local excitation and global inhibition (LEGION). How they look and whether directional maps such as the one suggested by Fig. 16.7, in conjunction with another one for distance, are fascinating questions that seem to near an answer. This paper provides ample evidence that we now face concrete mechanisms and concrete ideas that need to be tested, complemented, and implemented, a wonderful challenge to both experimental and theoretical neuroscience and biological physics.

Fig. 16.16 The coordinate system as it is used here. A pond in equilibrium has the (x, z)-plane as its surface. The *bottom* of the pond is at $y = -h$. The wave $\{y = \eta\,(x, t)\}$ is 'plane' in the sense that it is infinitely extended along the z-axis

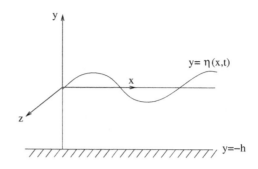

Appendix: Water Waves and Surface Tension

The present Appendix is devoted to a mathematically succinct and simple description of water waves. It is complete but assumes wave amplitudes to be small so that the theory will effectively turn out to be linear. Furthermore, most of the time we will consider a 'plane' wave that propagates in one, say the x, direction and is translationally invariant along the z axis, orthogonal to it; cf. Fig. 16.16. For *Xenopus,* insects dropping on the water surface generate (practically) circular waves. If the distance is large enough, say ≥ 10 cm, a plane wave is a decent approximation. Its mathematics is much simpler than a description with cylindrical coordinates, which are natural to a circular wave.

The appendix consists of three parts: Basics, gravity waves, and capillary-gravity waves, the final part taking into account surface tension. For details, the reader can profitably consult some highly readable books such as Acheson (1990) and Billingham and King (2000). To make the arguments below reasonably self-contained, a few proofs have been added.

A.1 Basics

Conservation of mass implies the continuity equation

$$\partial_t \rho + \mathrm{div}(\rho\mathbf{v}) = 0 \tag{16.41}$$

where $\partial_t = \partial/\partial t$ denotes a partial differentiation with respect to time and $\mathbf{v}$ is the local velocity of the fluid. Water in a pond can safely be taken to be incompressible so that the density ρ is constant and we are left with $\mathrm{div}(\mathbf{v}) = 0$.

Newton's law also holds for a fluid. Focusing on a small blob of fluid in a pressure field p and a gravitational field $\mathbf{g}$, Euler found as early as 1752

$$\rho D_t \mathbf{v} = -\nabla p + \rho\mathbf{g} + \mu\varDelta\mathbf{v}. \tag{16.42}$$

Here

$$D_t \mathbf{v} = \partial_t \mathbf{v} + (\mathbf{v} \cdot \nabla)\mathbf{v} \qquad (16.43)$$

is the rate of change of $\mathbf{v}$ 'following the fluid,' as it behooves a Newtonian particle with mass m (here a blob of fluid with constant density ρ) obeying the law $\mathbf{F} = m d\mathbf{v}/dt$. The $\cdot$ in $(\mathbf{v} \cdot \nabla)$ denotes a scalar product and will henceforth be dropped when appearing between two vectors. The final term $\mu \Delta \mathbf{v}$ is not due to Euler but to Navier (1822) and Stokes (1845), describes viscosity, and will be neglected here, i.e., $\mu = 0$.

Finally, we assume our fluid to be vortex-free, which definitely holds in a pond during night, if you wait long enough and take into account that a small $\mu > 0$ then counts. Consequently, rot $(\mathbf{v}) = 0$ and we can find a function φ such that $\mathbf{v} = \nabla\varphi$. Thus, we arrive at

$$\mathbf{v} = \nabla\varphi \quad \text{and} \quad \text{div}\,\mathbf{v} = 0 \Rightarrow \Delta\varphi = 0 \qquad (16.44)$$

where $\Delta = \partial_x^2 + \partial_y^2$ is the Laplace operator. That is to say, φ is a harmonic function that can be obtained once we have specified the boundary conditions. To do so, we need the Bernoulli equation, dating back to 1738.

Exploiting $\text{rot}(\mathbf{v}) = 0$ together with the vector identity $0 = \mathbf{v} \times \text{rot}\,\mathbf{v} = \nabla(\mathbf{v}\mathbf{v}/2) - (\mathbf{v}\nabla)\mathbf{v}$, where $\times$ is a vector product, we can write the second term on the right in Eq. (16.43)

$$(\mathbf{v} \cdot \nabla)\mathbf{v} = \nabla\left(\frac{1}{2}\mathbf{v}\mathbf{v}\right). \qquad (16.45)$$

The gravitation field can be written $\mathbf{g} = -g\nabla\gamma$ for some function γ and $g = 10 \text{ ms}^{-2}$ being a decent approximation of the acceleration value. For instance, in the coordinate system of Fig. 16.16 we have $\gamma(\mathbf{x}) = y$ for $\mathbf{x} = (x, y, z)$. Collecting terms we then get for the Euler equation (16.42) in conjunction with (16.43)–(16.45)

$$\partial_t(\nabla\varphi) + \nabla\left(\frac{1}{2}\mathbf{v}\mathbf{v}\right) = -\rho^{-1}\nabla p - g\nabla\gamma \qquad (16.46)$$

so that

$$\nabla\left(\partial_t\varphi + \frac{1}{2}\mathbf{v}\mathbf{v} + \rho^{-1}p + g\gamma\right) = 0. \qquad (16.47)$$

The expression between the brackets in (16.47) is hence a constant $C(t)$ that can only depend on the time t. By performing the transformation $\varphi \mapsto \varphi + \int_0^t ds\,C(s)$, which does not change the velocity $\mathbf{v} = \nabla\varphi$, we can even assume $C = 0$ and are left with the Bernoulli equation

$$\partial_t\varphi + \left(\frac{1}{2}\mathbf{v}\mathbf{v} + g\gamma\right) + \rho^{-1}p = 0. \qquad (16.48)$$

We can interpret the two terms in the middle as the sum of kinetic and potential energy. Note, though, that $\mathbf{v} = \nabla\varphi$ so that φ determines the pressure p, and conversely.

A.2 Gravity Waves

To find the appropriate boundary conditions, we realize that a pond has a bottom and a free water surface with water in between. For the bottom we take $\{y = -h\}$ and turn to it later. The free surface is taken to be a 'plane' water wave described by the function $y = \eta(x, t)$. Neglecting for the moment the water's surface tension we have the atmospheric pressure $p = p_{\text{atm}}$ at the surface and putting $p \mapsto p - p_{\text{atm}}$, which is allowed since only ∇p is occurring in our equations, we can take $p = 0$ at the surface. Because of the Bernoulli equation (16.48) we get

$$\partial_t \varphi + \frac{1}{2}\mathbf{v}\mathbf{v} + g\gamma = 0|_{y=\eta(x,t)}, \tag{16.49}$$

which is known as the *dynamic boundary condition* at a free surface. The term $|_{y=\eta(x,t)}$ reminds us of the fact that we are just there.

We now turn to the so-called *kinematic boundary condition*, which follows from the observation that water at the surface remains at the surface. In other words, following the fluid $D_t[y - \eta(x, t)] = 0$. Taking advantage of (16.43), with $\mathbf{v}$ replaced by the function $f(\mathbf{x}) = y - \eta(x, t)$, and doing the partial derivatives we are left with

$$\partial_y \varphi = \partial_t \eta + \partial_x \varphi \partial_x \eta|_{y=\eta(x,t)}. \tag{16.50}$$

Once again this holds at the water surface, where $y = \eta(x, t)$. Finally, we consider a pond with a flat bottom at $\{y = -h\}$. For a watertight bottom the normal velocity is bound to be zero so that

$$v_y = \partial_y \varphi = 0|_{y=-h}; \tag{16.51}$$

see Fig. 16.16.

So far everything is exact but not really accessible to an analytic treatment. We therefore make the simplification that the wave amplitude at the (free) water surface is much smaller than the wave length. For insects *Xenopus* is hunting for, this is a sensible approximation. We then develop everything in sight and drop all terms of order two and higher. At the water surface we have

$$\varphi(x, y; t)|_{y=\eta(x,t)} = \varphi(x, 0; t) + (\partial_y \varphi)(x, 0; t)\eta$$
$$\mapsto \varphi(x, 0; t) \tag{16.52}$$

since we can drop the term $(\partial_y \varphi)(x, 0; t) \eta$ as being small of second order. Using (16.49)–(16.52) we have to solve $\Delta \varphi = 0$ with the following boundary conditions at the surface $\{y = \eta(x, t)\}$ and the bottom $\{y = -h\}$,

$$\gamma = \eta \ \ \& \ \ (16.49) \ \ \& \ \ (16.52) \Rightarrow \partial_t \varphi|_{y=0} + g\eta = 0, \tag{16.53}$$

$$(16.50) \ \ \& \ \ (16.52) \Rightarrow \partial_y \varphi|_{y=0} = \partial_t \eta, \tag{16.54}$$

$$(16.51) \Rightarrow \partial_y \varphi|_{y=-h} = 0. \tag{16.55}$$

The boundary conditions are on the right and the equations they come from on the left. In (16.53) we have used that $\gamma(\mathbf{x}) = y$ is to be taken at the surface $y = \eta$ so that $\gamma = \eta$. Once we know φ, we also know η as a consequence of (16.53).

To describe a plane wave propagating in one direction, say the x axis, we make the ansatz $\varphi(\mathbf{x}; t) = F(x - ct)Y(y)$ and substitute this in the Laplace equation (16.44) so as to find $F''Y + FY'' = 0$ where F'' is the second derivative of F and Y'' that of Y. In other words, we obtain

$$\frac{F''}{F} = -\frac{Y''}{Y} = -k^2 \tag{16.56}$$

for some constant $-k^2$ since F''/F only depends on $x - ct$ and Y''/Y on y and both have nothing to do with each other. The differential equations following from (16.56) are trivial. Their solutions F and Y contain free constants that can be used to satisfy the boundary conditions (16.53)–(16.55) via φ. The upshot is a progressive gravity wave

$$\eta(x, t) = a \cos k(x - ct), \tag{16.57}$$

with amplitude a, belonging to the potential

$$\varphi(\mathbf{x}, t) = \frac{ag}{kc} \cdot \frac{\cosh k(y + h)}{\cosh(kh)} \cdot \sin k(x - ct) \tag{16.58}$$

and propagating at the speed $c(k)$ given by

$$c^2(k) = \frac{g}{k} \tanh(kh), \tag{16.59}$$

a dispersion relation depending on the wave number $k = 2\pi/\lambda$ with λ being the wavelength, an interpretation directly following from (16.57). It is easy to verify that (16.58) satisfies $\Delta \varphi = 0$ plus boundary conditions. Proving uniqueness is a bit harder and finishes the job. The problem being linear, one can match any situation by a linear superposition of solutions (16.58) for as many k as one may wish.

Rotational invariance Until now we had plane waves with translational invariance in the z direction so that for example Fig. 16.16 provided full information. Suppose now that our pond has a circular shape with a sprawling insect positioned in the middle, which is also the origin, and assume rotational invariance; hence cylindrical coordinates $\{r, \theta, y\}$ are natural. Equation (16.44) then reads

$$\Delta \varphi = 0 \Rightarrow \frac{\partial^2 \varphi}{\partial r^2} + \frac{1}{r}\frac{\partial \varphi}{\partial r} + \frac{1}{r^2}\frac{\partial^2 \varphi}{\partial \theta^2} + \frac{\partial^2 \varphi}{\partial y^2} = 0. \tag{16.60}$$

Because of rotational invariance $\partial_\theta \varphi = 0$ and we therefore make the ansatz $\varphi(r, \theta, y) = R(r)Y(y)\psi(t)$ where, as before, y is the vertical coordinate; cf. Fig. 16.16. In a finite pond with prescribed boundary conditions proving uniqueness of a solution is not too hard (Acheson 1990; Billingham and King 2000) so that the only thing left is verifying that a specific ansatz satisfies all equations in sight.

As in (16.56) we directly get for $\varphi = RY\psi$

$$(\partial_r^2 + r^{-1}\partial_r + k^2)R = 0, \tag{16.61}$$

which is Bessel's differential equation (Watson 1922). Its solutions are Bessel functions of order zero. Only $J_0(kr)$ is regular at $r = 0$ whereas the other one (of Hankel-function type) is singular and hence not allowed since an insect at the origin cannot be sent to infinity.

A natural pond being large we will not analyze the boundary condition for R at the border, say, at $r := r_b \gg 5\,\text{m}$, but just notice that $v_r|_{r=r_b} = \partial_r \varphi|_{r=r_b} = 0$. Because of damping due to viscosity ($\mu > 0$) it is simpler, and also more physical, to assume an outgoing wave in a large pond and neglect reflection. For an experiment with absorbing boundaries (Elepfandt et al. 2000) the same holds true.

As for the dynamic boundary condition (16.49), it is reasonable to model an insect sprawling at the water surface at $r = 0$ by a delta function $f(t)\delta(r)$ with the amplitude f being function of the time t and replace the pressure $(p_{\text{atm}} - p) = 0$ in the right-hand side of (16.49) by $f(t)\delta(r)$. Instead of (16.53), we then arrive at $\partial_t \varphi|_{y=0} + g\eta = \rho^{-1}f(t)\delta(r)$ where ρ is the fluid's density. The rest is clear, though maybe a bit clumsy, sailing.

A.3 Capillary-Gravity Waves

Until now gravity was the only force driving a fluid back to equilibrium. The waves originating from this process are called gravity waves. As the wavelength becomes smaller, the *surface tension* t_S starts playing an important role (Isenberg 1992). It is a consequence of intermolecular attractive forces not compensating each other at the water surface. The surface tension t_S has the dimension of force divided by length, i.e., Newton/meter or a convenient equivalent. For a 'plane' wave, which is effectively determined by a function $y = y(x)$ as shown in Fig. 16.16, t_S is determined by the local curvature κ. If it is large enough, capillary-gravity waves arise. So what is curvature?

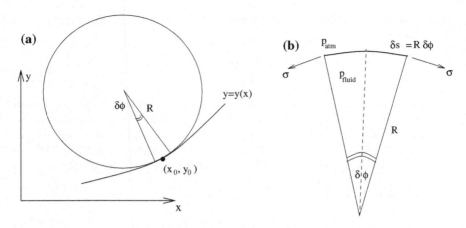

Fig. 16.17 a The osculating circle with radius R approximates a smooth curve such as $y = y(x)$ up to second order at the point (x_0, y_0), i.e., locally. The curve's curvature at this point is given by $\kappa = R^{-1}$. A *straight line* would have $\kappa = 0$ everywhere. **b** Mechanical equilibrium requires $p_{\text{fluid}}\, \delta_s = p_{\text{atm}}\, \delta_s + 2t_S \sin(\delta/2)$ along the vertical axis. We automatically have equilibrium along the horizontal axis in this situation. Translational invariance reigns in the direction orthogonal to the plane of the picture, where we then take one unit of length. Multiplied by t_S this gives a force

A curve such as $y = \eta(x)$, with arc length s, can be approximated locally by a circle with radius R given by $R = ds/d\phi$; cf. Fig. 16.17a. It is called the circle of curvature or also, more aptly, the 'osculating' circle. In terms of the original function $y = \eta(x)$ one finds (Thomas 1972)

$$\kappa = R^{-1} = \eta''[1 + (\eta')^2]^{-3/2} \approx \eta'' \tag{16.62}$$

where the sign of κ tells us whether the curve is above or below its tangent. The example $\eta(x) = ax^2/2$ at, say, $x = 0$ trivially illustrates this. The approximation $\kappa \approx \eta''$ holds in the present context of small-amplitude waves.

As Fig. 16.17b shows, the pressure $p = p_{\text{fluid}} - p_{\text{atm}}$ for a piece of cylinder with radius R and curve length $\delta s = R\delta\phi$ spanned by an angle $\delta\phi$ is given by

$$(p = p_{\text{fluid}} - p_{\text{atm}})\delta s = 2t_S \sin(\delta\phi/2) \approx t_S\delta\phi \tag{16.63}$$

where $\delta s = R\delta\phi$ so that in the limit $\delta\phi \to 0$ we locally find $p = t_S/R = -t_S\eta'' > 0$, as should be the case for the water wave in Figs. 16.17a and 16.16 with normal vector pointing downwards.

For a general two-dimensional surface (Isenberg 1992), the Laplace-Young (1806) equation $p = p_{\text{fluid}} - p_{\text{atm}} = t_S(1/R_1 + 1/R_2)$ holds, where $H \equiv (1/R_1 + 1/R_2)$ is the mean curvature, the sum of the two principal curvatures $1/R_\alpha$ taken along the two orthogonal principal directions. There exist convenient differential-geometric expressions for H (Isenberg 1992). In the situation of Fig. 16.17b one of the principal curvatures vanishes, the other one equals $1/R$.

The only boundary condition containing the pressure p is the dynamic one stemming from the Bernoulli equation (16.48). Properly extended it reads in the linear approximation $\partial_t \varphi + g\eta + p/\rho = 0$. That is, since now $p = -t_S \partial_x^2 \eta$,

$$\partial_t \varphi + g\eta - \frac{t_S}{\rho} \partial_x^2 \eta = 0 \ . \tag{16.64}$$

The rest remains as is. We differentiate (16.64) with respect to t, use (16.54), insert $\partial_x^2 \varphi = -\partial_y^2 \varphi$ since $\Delta \varphi = 0$, and find a partial differential equation for φ only,

$$\partial_t^2 \varphi + g\partial_y \varphi + \frac{t_S}{\rho} \partial_y^3 \varphi = 0|_{y=0} \ , \tag{16.65}$$

a combination of the dynamic and kinetic boundary condition. Being interested in a progressive wave, we make the same ansatz $\varphi(\mathbf{x}; t) = F(x - ct)Y(y)$ as before, immediately cash the bottom boundary condition (16.55), and end up with

$$\begin{aligned} \varphi(\mathbf{x}; t) =& [A \cos k(x - ct) + B \sin k(x - ct)] \\ & \times \cosh k(y + h). \end{aligned} \tag{16.66}$$

Finally, we substitute this into (16.65) so as to find the dispersion relation for a capillary-gravity wave with wavelength $\lambda = 2\pi/k$,

$$c^2(\lambda) = \frac{g\lambda}{2\pi} \left[1 + \left(\frac{t_S}{g\rho} \right) \frac{4\pi^2}{\lambda^2} \right] \tanh\left(\frac{2\pi h}{\lambda} \right) \ . \tag{16.67}$$

For small λ the phase speed $c(\lambda)$ diverges as $1/\sqrt{\lambda}$ whereas for large λ, as long as $2\pi h/\lambda \gg 1$, it diverges as $\sqrt{\lambda}$. So there is a *minimum* in between; more precisely, at $\lambda_{\min} = 2\pi \ell_c$ where $\ell_c = (t_S/g\rho)^{1/2}$. For water, $t_S = 0.07$ Nm^{-1} and $\ell_c = 2.6$ mm so that $\lambda_{\min}^{H_2O} = 1.6$ cm while in a deep pond $c_{\min}^{H_2O} = 23$ cm/s. For the clawed frog *Xenopus*, insects are a favorite prey generating waves in exactly this range; cf. Fig. 16.18a.

In experiment it is often handier to know the phase speed c as a function of the frequency v instead of the wavelength λ. A closer look at Fig. 16.18a suffices to convince us that $\lambda = 10$ cm is so large that we can safely assume to be in the gravity regime. Since for frogs such as *Xenopus* the mean depth of the pond they live in is 40–70 cm we take h infinite in (16.67) and replace[1] $\tanh(2\pi h/\lambda)$ by 1.

We now have to solve c for the frequency v. For the moment, it is advantageous to work with the wave number $k = 2\pi/\lambda$. By definition $c = v\lambda$ while (16.67) gives us $c = c(\lambda)$. We can therefore solve v for λ so as to get an equation of the form $v(\lambda) = \lambda^{-1} c(\lambda)$, which can be written

---

[1] Great care is needed in the case of shallow baths with a depth h of only a few cm. On the other hand, the experimental setup of, e.g., Elepfandt et al. (2000) is a nearly optimal.

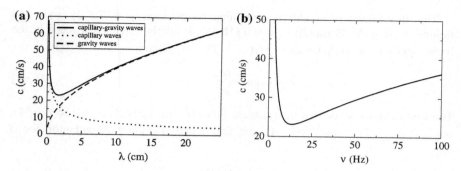

Fig. 16.18 **a** Phase speed c given by (16.67) has been plotted as a function of the wavelength λ in a deep pond filled with water. For small wavelengths (ripples) capillarity is dominating, for larger wavelengths gravity governs the phase speed. Accordingly there is a minimum $c_{min}^{H_2O} = 23$ cm/s at $\lambda_{min}^{H_2O} = 1.7$ cm. It is here where for instance *Xenopus*'s favorite insects are active when they are sprawling at the water surface. **b** Phase speed c plotted as a function of the frequency v for waves in a deep pond filled with water. As in **a** the minimum is due to surface tension

$$v(k) = \frac{1}{2\pi} \left\{ gk \left[1 + \left(\frac{t_S}{g\rho} \right) k^2 \right] \right\}^{1/2} \tag{16.68}$$

Here $v(k)$ is a monotonically increasing function of k. For $k \to 0$ it behaves like $k^{1/2}$ whereas $v(k) \propto k^{3/2}$ for $k \to \infty$. So for all v (≥ 0) we find a *unique* positive solution that is the single real root $k(v)$ (the other two are complex conjugate) of the cubic polynomial $ak^3 + bk - v^2 = 0$ with positive coefficients a and b. In fact, rewriting our polynomial in the form $k^3 + 3pk - 2q = 0$ we get (Mostowski and Stark 1964) $k(v) = \alpha - p/\alpha$ with $\alpha := \left[\sqrt{p^3 + q^2} + q \right]^{1/3}$ and $2q = v^2/a$.

Through $c = v\lambda = 2\pi v/k(v)$ we finally obtain the desired dependence of the phase speed c upon the frequency v. The result has been plotted in Fig. 16.18b.

Acknowledgments The author sincerely thanks his collaborators over the years on many lateral-line issues: His colleagues, Professors Horst Bleckmann, Jacob Engelmann, and Andreas Elepfandt as well as his former graduate students who have been involved in several projects discussed here; particularly, Drs. Moritz Franosch, Paul Friedel, Julie Goulet, Andy Sichert, and Andreas Vollmayr. Financial support from the BMBF through BCCN—Munich is gratefully acknowledged.

References

Acheson DJ (1990) Elementary fluid dynamics. Oxford University Press, Oxford
Bachman G, Narici L (1966) Functional analysis. Academic, New York; (2000) Dover, Mineola
Bastian J (1982) Vision and electroreception: integration of sensory information in the optic tectum of the weakly electric fish *Apteronotus albifrons*. J Comp Physiol A 147:287–297

Bi G-Q, Poo M-M (2001) Synaptic modification by correlated activity: Hebbõs postulate revisited. Annu Rev Neurosci 24:139–166

Bleckmann H (1994) Reception of hydrodynamic stimuli in aquatic and semiaquatic animals. Fischer, Stuttgart

Bleckmann H, Schwarz E (1982) The functional significance of frequency modulation within a wave train for prey localization in the surface-feeding fish *Aplocheilus lineatus* (Cyprinodontidae). J Comp Physiol A 145:331–339

Billingham J, King AC (2000) Wave motion. Cambridge University Press, Cambridge

Blickhan R, Krick C, Zehren D, Nachtigall W, Breithaupt T (1992) Generation of a vortex chain in the wake of a subundulatory swimmer. Naturw 79:220–221

Bürck M, Friedel P, Sichert AB, Vossen C, van Hemmen JL (2010) Optimality in mono- and multi-sensory map formation. Biol Cybern 103:1–20

Calvert G, Spence C, Stein BE (eds) (2004) The handbook of multisensory processes. MIT Press, Cambridge

von Campenhausen C, Riess I, Weissert R (1981) Detection of stationary objects by the blind cave fish *Anoptichthys jordani* (Characidae). J Comp Physiol A 143:369–374

Chagnaud BP, Bleckmann H, Engelmann J (2006) Neural responses of goldfish lateral line afferents to vortex motions. J Exp Biol 209:327–342

Chung SH, Stirling RV, Gazei RM (1975) The structural and functional development of the retina in larval *Xenopus*. J Embryol Exp Morph (now: Development) 33(4):915–940. The Appendix *Optics of* Xenopus *eyes during development* by M. Land and R. V. Stirling appears on pp 934–940. The author thanks Susan Udin (SUNY at Buffalo) for drawing his attention to this wonderful piece of work

Claas B, Münz H (1996) Analysis of surface wave direction by the lateral line system of *Xenopus*: source localization before and after inactivation of different parts of the lateral line. J Comp Physiol A 178:253–268

Clegg JG (1968) Calculus of variations. Oliver and Boyd, Edinburgh

Coombs S (1994) Nearfield detection of dipole sources by the goldfish (*Carassius auratus*) and the mottled sculpin (*Cottus bairdi*). J Exp Biol 190:109–129

Coombs S, Conley RA (1997) Dipole source localization by mottled sculpin I: approach strategies. J Comp Physiol A 180:387–399

Coombs S, van Netten S (2006) The hydrodynamics and structural mechanics of the lateral line system. Chapter 4 in: fish physiology. In: Shadwick RE, Lauder GV (eds) Fish biomechanics, vol 23. Academic

Coombs S, Hastings M, Finneran J (1996) Modeling and measuring lateral line excitation patterns to changing dipole source locations. J Comp Physiol A 178:359–371

Coombs S, New JG, Nelson M (2002) Information-processing demands in electrosensory and mechanosensory lateral line systems. J Physiol (Paris) 96:341–354

Ćurčić-Blake B, van Netten SM (2006) Source location encoding in the fish lateral line canal. J Exp Biol 209:1548–1559

Denton E, Gray JAB (1982) The rigidity of fish and patterns of lateral line stimulation. Nature 297:679–681

Denton E, Gray JAB (1983) Mechanical factors in the excitation of clupeid lateral lines. Proc R Soc Lond B 218:1–26

Dickinson M (2003) How to walk on water. Nature 424:621–622

Dijksterhuis EJ (1969) The mechanization of the world picture. Oxford University Press, Oxford

Dijkgraaf S (1963) The functioning and significance of the lateral-line organs. Biol Rev 38:51–105

Drucker EG, Lauder GV (2002) Experimental hydrodynamics of fish locomotion: Functional insights from wake visualization. Integr Comp Biol 42(2):243–257

Dym H, McKean HP (1972) Fourier series and integrals. Academic, New York

Elepfandt A (1982) Accuracy of taxis response to water waves in the clawed toad *(Xenopus laevis* Daudin) with intact or with lesioned lateral line system. J Comp Physiol A 148:535–545

Elepfandt A (1984) The role of ventral lateral line organs in water wave localization in the clawed toad (*Xenopus laevis*). J Comp Physiol A 154:773–780

Elepfandt A (1986) Detection of individual waves in an interference pattern by the clawed frog *Xenopus laevis* Daudin. Neurosci Lett 26:S380

Elepfandt A (2012) private communication

Elepfandt A, Wiedemer L (1986) Lateral-line responses to water surface waves in the clawed frog, *Xenopus laevis*. J Comp Physiol A 160:667–682

Elepfandt A, Seiler B, Aicher B (1985) Water wave frequency discrimination in the clawed frog, *Xenopus laevis*. J Comp Physiol A 157:255–261

Elepfandt A, Kroese ABA, van Netten SM, private communications. The latter two have performed their experiments independently of Elepfandt; their lateral-line object was a passing zebrafish

Elepfandt A, Lebrecht A, Schroedter K, Brudermanns B (2004) Discrimination of two water waves presented simultaneously in the clawed frog, *Xenopus laevis laevis*. In: Abstract 7th congress of the international society for neuroethology; Schroedter K, Staatsexamensarbeit, Humboldt Universität zu Berlin, (2002) under the direction of A. Elepfandt

Elepfandt A, Eistetter I, Fleig A, Günther E, Hainich M, Hepperle S, Traub B (2000) Hearing threshold and frequency discrimination in the purely aquatic frog *Xenopus laevis* (Pipidae): measurement by means of conditioning. J Exp Biol 203:3621–3629; see in particular Fig. 1

von der Emde G, Engelmann J (2011) Active electrolocation. In: Farrell AP (ed) Encyclopedia of fish physiology: from genome to environment, vol 1. Elsevier, Amsterdam, pp 375–386

Engelmann J, Bleckmann H (2004) Coding of lateral line stimuli in the goldfish midbrain in still and running water. Zoology 107(2):135–151

Faucher K, Parmentier E, Becco C, Vandewalle N, Vandewalle P (2010) Fish lateral system is required for accurate control of shoaling behaviour. Animal Behav 79:679–687

Flanders M (2011) What is the biological basis of sensorimotor integration? Biol Cybern 104:1–8

Franosch J-MP, Sobotka MC, Elepfandt E, van Hemmen JL (2003) Minimal model of prey localization through the lateral-line system. Phys Rev Lett 91:158101

Franosch J-MP, Lingenheil M, van Hemmen JL (2005a) How a frog learns what is where in the dark. Phys Rev Lett 95:078106

Franosch J-MP, Sichert AB, Sobotka MC, Elepfandt A, van Hemmen JL (2005b) Model of amphibian prey localization through the lateral-line system. Physik Department T35, Technische Universität München, internal report

Franosch J-MP, Hagedorn HJA, Goulet J, Engelmann J, van Hemmen JL (2009) Wake tracking and the detection of vortex rings by the canal lateral line of fish. Phys Rev Lett 103:078102

Friedel P, van Hemmen JL (2008) Inhibition, not excitation, is the key to multimodal sensory integration. Biol Cybern 98:597–618

Geer J (1975) Uniform asymptotic solutions for potential flow around a slender body of revolution. J Fluid Mech 67:817–827

Gelfand IM, Fomin SV (1963) Calculus of variations. Prentice-Hall, Englewood Cliffs

Gerstner W, van Hemmen JL (1994) Coding and information processing in neural networks. In: Do-many E, van Hemmen JL, Schulten K (eds) Models of neural networks II. Springer, New York, pp 39–47

Gerstner W, Kempter R, van Hemmen JL, Wagner H (1996) A neuronal learning rule for sub-millisecond temporal coding. Nature 383:76–78

Görner P (1963) Untersuchungen zur Morphologie und Elektrophysiologie des Seitenlinienorgans vom Krallenfrosch (*Xenopus laevis* Daudin). Z vergl Physiol 47:316–338

Goulet J (2010) Information processing in the lateral-line system of fish. Doctoral Dissertation, Physik Department T35, Technische Universität München. http://mediatum2.ub.tum.de/node?id=959089

Goulet J, Engelmann J, Chagnaud BP, Franosch J-MP, Suttner MD, van Hemmen JL (2008) Object localization through the lateral line system of fish: theory and experiment. J Comp Physiol A 194:1–17

Goulet J, van Hemmen JL, Jung SN, Chagnaud BP, Scholze B, Engelmann J (2012) Temporal precision and reliability in the velocity regime of a hair-cell sensory system: the mechanosensory lateral-line of goldfish, *Carassius auratus*. J Neurophysiol 107:2581–2593

Grothe B, Pecka M, McAlpine D (2010) Mechanisms of sound localization in mammals. Physiol Rev 90:983–1012

Gutfreund Y, Knudsen EI (2007) Visual instruction of the auditory space map in the midbrain. In: Calvert et al. (2004) Chapter 38

Gutfreund Y, Zheng W, Knudsen EI (2002) Gated visual input to the central auditory system. Science 297:1556–1559

Gutfreund Y, King A (2012) What is the role of vision in the development of the auditory space map? Chapter 32. In: Stein BE (ed) The new handbook of multisensory processes. Cambridge, MIT Press

Guyon E, Hulin J-P, Petit L, Mitescu CD (2001) Physical hydrodynamics. Oxford University Press, Oxford

Hanke W, Brücker C, Bleckmann H (2000) The ageing of the low-frequency water disturbances caused by swimming goldfish and its possible relevance to prey detection. J Exp Biol 203:1193–1200

Hanke W, Bleckmann H (2004) The hydrodynamic trails of *Lepomis gibbosus* (Centrarchidae), *Colomesus psittacus* (Tetraodontidae) and *Thysochromis ansorgii* (Cichlidae) investigated with scanning particle image velocimetry. J Exp Biol 207:1585–1596

Harris GG, van Bergeijk WA (1962) Evidence that the lateral-line organ responds to near-field displacements of sound sources in water. J Acoust Soc Am 34(12):1834–1841

Hassan El-S (1985) Mathematical analysis of the stimulus for the lateral line organ. Biol Cybern 52:23–36

Hassan EI-S (1992a) Mathematical description of the stimuli to the lateral line system of fish derived from a three-dimensional flow field analysis: I The cases of moving in open water and of gliding towards a plane surface. Biol Cybern 66:443–452

Hassan EI-S (1992b) Mathematical description of the stimuli to the lateral line system of fish derived from a three-dimensional flow field analysis: II The case of gliding alongside or above a plane surface. Biol Cybern 66:453–461

Hassan El-S (1993) Mathematical description of the stimuli to the lateral line system of fish, derived from a three-dimensional flow field analysis: III The case of an oscillating sphere near the fish. Biol Cybern 69:525–538

van Hemmen JL (2001) Theory of synaptic plasticity In: Moss F, Gielen S (eds) Handbook of biophysics, vol 4; see in particular §2 and Appendices A and B. Elsevier, Amsterdam, pp 771–823

van Hemmen JL (2006) What is a neuronal map, how does it arise, and what is it good for? In: van Hemmen JL, Sejnowski TJ (eds) 23 Problems in systems neuroscience. Oxford University Press, New York, pp 83–102

van Hemmen JL (2010) Lateral-line detection of underwater objects: from goldfish to submarines. Bull Am Phys Soc 55(2):V10.00008. See also TUM Faszination Forschung 7(10):70–75

Heiligenberg W, Rose GJ (1987) The optic tectum of the gymnotiform electric fish, *Eigenmannia*: labeling of physiologically identified cells. Neuroscience 22:331–340

Hopkins CD (2009) Electrical perception and communication. In: Squire LR (ed) Encyclopedia of neuroscience, vol 3. Academic Press, Oxford, pp 813–831

Isakov V (2006) Inverse problems for partial differential equations, 2nd edn. Springer, New York

Isenberg C (1992) The science of soap films and soap bubbles. Dover, Mineola

Janssen J, Coombs S, Pride S (1990) Feeding and orientation of mottled sculpin, *Cottus bairdi*, to water jets. Environ Biol Fishes 29:43–50

Jazayeri M, Movshon JA (2006) Optimal representation of sensory information by neural populations. Nat Neurosci 9:690–696

Jielof R, Spoor A, de Vries Hl (1952) The microphonic activity of the lateral line. J Physiol 116:137–157

Käse RH, Bleckmann H (1987) Prey localization by surface wave ray-tracing: fish track bugs like oceanographers track storms. Experientia 43:290–293

Kalmijn AJ (1988) Hydrodynamic and acoustic field detection. In: Atema J, Fay RR, Popper AN, Tavolga WN (eds) Sensory biology of aquatic animals. Springer, New York, pp 83–130

Keller CH, Takahaski TT (1996) Binaural cross-correlation predicts the responses of neurons in the owl's auditory space map under conditions simulating summing localization. J Neurosci 16(13):4300–4309

Kuiper JW (1956) The microphonic effect of the lateral line organ. Ph. D. Thesis, Natuurkundig Laboratorium, University of Groningen, The Netherlands

Knudsen EI (2002) Instructed learning in the auditory localization pathway of the barn owl. Nature 417:322–328

Knudsen EI, Knudsen PF (1989) Vision calibrates sound localization in developing barn owls. J Neurosci 9(9):3306–3313

Knudsen EI, du Lac S, Esterly SD (1987) Computational maps in the brain. Annu Rev Neurosci 10:41–65

Krippner M (2012) Multimodales Lernen im blinden Mexikanischen Hohlenfisch. Diploma thesis, Physik Department T35, Technische Universität München

Kühn R, van Hemmen JL (1995) Temporal association. In: Domany E, van Hemmen JL, Schulten K (eds) Models of neural networks, 2nd edn. Springer, Berlin, pp 213–280 (particularly, §7.4)

Lamb H (1932) Hydrodynamics, 6th edn. Cambridge University Press, Cambridge. See in particular Sects. 92, 226 ff., 246, 331, and Chaps. 5 and 7

Lamperti J (1966) Probability. Benjamin, New York; 2nd edn. (1996) Wiley, New York

Lingenheil M (2005) Theorie der Beuteortung beim Krallenfrosch. Diploma thesis, Physik Department T35, Technische Universität München

Lowe DA (1987) Single-unit study of lateral line cells in the optic tectum of *Xenopus laevis*: evidence for bimodal lateral line/optic units. J Comp Neurol 257:396–404

Meyer G, Klein A, Mogdans J, Bleckmann H (2012) Toral lateral line units of goldfish, *Carassius auratus*, are sensitive to the position and vibration direction of a vibrating sphere. J Comp Neurol 198:639–653

Mostowski A, Stark M (1964) Introduction to higher algebra. Pergamon, Oxford. Particularly, Chap. 7, §4. This book is a mine of useful information and clear exposition

van Netten SM (2006) Hydrodynamic detection by cupulae in a lateral-line canal: functional relations between physics and physiology. Biol Cybern 94:67–85

Pandya S, Yang Y, Jones DL, Engel J, Liu C (2006) Multisensor processing algorithms for underwater dipole localization and tracking using MEMS artificial lateral-line sensors. EURASIP J Appl Signal Proc 076593

Pfister J-P, Toyoizumi T, Barber D, Gerstner W (2006) Optimal spike-timing-dependent plasticity for precise action potential firing in supervised learning. Neural Comput 18:1318–1348

Pitcher TJ, Patridge TL, Wardle CS (1976) A blind fish can school. Science 194:963–965

Plachta DTT, Hanke W, Bleckmann H (2003) A hydrodynamic topographic map in the midbrain of goldfish *Carassius auratus*. J Exp Biol 206:3479–3486

Pohlmann K, Grasso FW, Breithaupt T (2001) Tracking wakes: the nocturnal predatory strategy of piscivorous catfish. Proc Natl Acad Sci USA 98:7371–7374

Press WH, Teukolsky SA, Vetterling WT, Flanery BP (1995) Numerical recipes in C, 2nd edn. Cambridge University Press, Cambridge; see in particular p. 34 for an explanation of the pseudo-inverse

Scheich H, Ebbesson SOE (1983) Multimodal torus in the weakly electric fish Eigenmannia. Springer, Berlin

Schlichting H, Gertsen K (2003) Boundary layer theory. Springer, Berlin

Schmitz A, Bleckmann H, Mogdans J (2008) Organization of the superficial neuromast system in goldfish, *Carassius auratus*. J Morphol 269:751–761

Schnupp JWH, Carr CE (2009) On hearing with more than one ear: lessons from evolution. Nat Neurosci 12(6):692–697

Schwartz E, Hasler AD (1966) Superficial lateral line sense organs of the mudminnow. Z vergl Physiol 53:317–327

Schwarz JS, Reichenbach T, Hudspeth AJ (2011) A hydrodynamic sensory antenna used by killifish for nocturnal hunting. J Exp Biol 214:1857–1866

Sichert AB, van Hemmen JL (2010) How stimulus shape affects lateral-line perception: analytical approach to analyzing natural stimulus characteristics. Biol Cybern 102:177–180

Sichert AB, Bamler R, van Hemmen JL (2009) Hydrodynamic object recognition: When multipoles count. Phys Rev Lett 102:058104

Soares D (2002) An ancient sensory organ in crocodilians. Nature 417:241–242

Song J, Fan C, Wang X, Zhang X (2011) A phylogenetic survey of morphological patterns of superficial neuromasts in teleost fish. Brain Behav Evol 78:190

Stein BE (ed) (2012) The new handbook of multisensory processing. MIT Press, Cambridge

Strelioff D, Honrubia V (1978) Neural transduction in *Xenopus laevis* lateral-line system. J Neurophysiol 41:432–444

Thomas GB (1972) Calculus and analytic geometry, 3rd edn. Addison-Wesley, Reading (§12.6)

Tikhonov AN, Arsenin VY (1977) Solutions of ill-posed problems. Winston, Washington

Tinsley RC, Kobel HR (eds) (1996) The biology of *Xenopus*. Oxford University Press, Oxford

Udin SB (2007) The instructive role of binocular vision in the *Xenopus* tectum. Biol Cybern 97:493–503

Urban S, Vollmayr AN, van Hemmen JL (2014) Hydrodynamic imaging on a 1-dimensional manifold and its inversion in 2-dimensional potential flow. TUM preprint

Vanegas H, Ebbeson SOE, Laufer M (1984a) Morphological aspects of the teleostean optic tectum. In: Vanegas H (ed) Comparative neurology of the optic tectum. Plenum, New York, pp 93–120

Vanegas H, Williams B, Essayac E (1984b) Electrophysiological and behavioral aspects of the teleostean optic tectum. In: Vanegas H (ed) Comparative neurology of the optic tectum. Plenum, New York, pp 121–162

van der Waerden BL (1969) Mathematical statistics. Springer, Berlin. See Jazayeri and Movshon (2006) for a comprehensive explanation of how references like van der Waerden's classic may serve theoretical neuroscience

Watson GN (1922) A treatise on the theory of Bessel functions (Chaps. 2, 3). Cambridge University Press, Cambridge

Yang Y, Chen J, Engel J, Pandya S, Chen N, Tucker C, Coombs S, Jones DL, Liu C (2006) Distant touch hydrodynamic imaging with an artificial lateral line. Proc Natl Acad Sci USA 103:18891–18895

Yang Y, Nguyen N, Chen N, Lockwood M, Tucker C, Hu H, Bleckmann H, Liu C, Jones DL (2010) Artificial lateral line with biomimetic neuromasts to emulate fish sensing. Bioinsp Biomim 5:016001

Yang Y, Klein A, Bleckmann H, Liu C (2011) Artificial lateral line canal for hydrodynamic detection. Appl Phys Lett 99:023701

Zittlau KE, Claas B, Münz H (1986) Directional sensitivity of lateral line units in the clawed toad *Xenopus laevis* Daudin. J Comp Physiol A 158:469–477

Chapter 17
Crickets as Bio-Inspiration for MEMS-Based Flow-Sensing

Gijs J. M. Krijnen, Harmen Droogendijk, Ahmad M. K. Dagamseh,
Ram K. Jaganatharaja and Jérôme Casas

Abstract MEMS offers exciting possibilities for the fabrication of bio-inspired mechanosensors. Over the last few years, we have been working on cricket-inspired hair-sensor arrays for spatio-temporal flow-field observations (i.e. flow camera) and source localisation. Whereas making flow-sensors as energy efficient as cricket hair-sensors appears to be a real challenge we have managed to fabricate capacitively interrogated sensors with sub-millimeter per second flow sensing thresholds, to use them in lateral line experiments, address them individually while in arrays, track transient flows, and use non-linear effects to achieve parametric filtering and amplification. In this research, insect biologists and engineers have been working in close collaboration, generating a bidirectional flow of information and knowledge, beneficial to both, for example, where the engineering has greatly benefitted from the insights derived from biology and biophysical models, the biologists have taken advantage of MEMS structures allowing for experiments that are hard to do on living material.

G. J. M. Krijnen (✉) · H. Droogendijk
MESA Research Institute for Nanotechnology, University of Twente,
Enschede, The Netherlands
e-mail: gijs.krijnen@utwente.nl

H. Droogendijk
e-mail: h.droogendijk@utwente.nl

A. M. K. Dagamseh
Electronics Engineering Department, Hijjawi Faculty for Engineering Technology,
Yarmouk University, Irbid, Jordan
e-mail: a.m.k.dagamseh@yu.edu.jo

R. K. Jaganatharaja
ASML, Veldhoven, The Netherlands
e-mail: ram.kottumakulal@asml.com

J. Casas
Institut de recherche sur la biologie de l'insecte, Université de Tours, Tours,
France
e-mail: jerome.casas@univ-tours.fr

H. Bleckmann et al. (eds.), *Flow Sensing in Air and Water*,
DOI: 10.1007/978-3-642-41446-6_17, © Springer-Verlag Berlin Heidelberg 2014

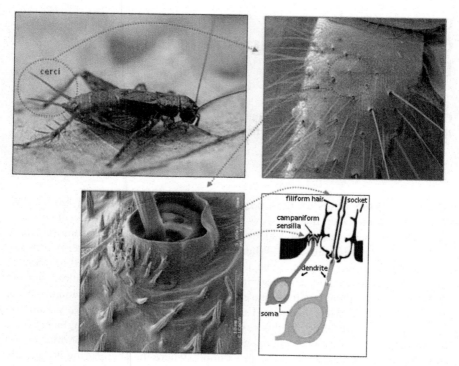

Fig. 17.1 Illustration of the hair-sensory system of crickets

17.1 Introduction

The filiform hairs of many insects, spiders and other invertebrates are among the most delicate and sensitive flow sensing structures: they measure displacement of less than a hydrogen diameter (sensitivity ca. 10×10^{-10} m = 1 Å) and react to flow speeds down to 30 μm/s. If one considers the energy needed to trigger a cell reaction, one finds that they react with a thousandths of the energy contained in a photon, so that they surpass photoreceptors in energy sensitivity. In fact, these mechanoreceptors work at the thermal noise level (Shimozawa et al. 2003). These hairs pick up air motion, implying that they measure both direction and speed of airflow, in contrast to pressure receivers, i.e. ears. Since several decades, the biomechanics of the filiform hairs has been studied with care by several groups worldwide, based on the analogy with a single degree of freedom inverted pendulum (see e.g. the review of hair biomechanics in Humphrey and Barth 2008).

Among insects, mainly cockroaches and crickets have been studied, because their airflow sensitive hairs are put on two antenna-like appendages, the cerci (cercus in singular). Insect hairs usually have a high aspect ratio, with a length of a few hundreds of microns up to 2 mm, and with a diameter of less than a dozen microns (Fig. 17.1). Their longitudinal shape is conical which has been proposed to have an important influence on the ratio of drag forces to moment of inertia and is supposed to be subjected to evolutionary pressure (Shimozawa et al. 2003). Hairs and sockets are ellipsoidal in cross-section, which leads to a preferential direction of movement. The base of the hairs is complex, and its mechanics poorly understood. In crickets, only a single sensory cell is below the hair shaft (Fig. 17.1).

Single hairs, or groups of hairs, are not placed at random on the body (Miller et al. 2011; Heys et al. 2012). Exact shape of a hair, its position on the sensory organ and its relative position within a group of hairs have been driven by natural selection. This aspect of mechanosensory research is, however, badly neglected. As for the positions along the cercus, the presence of a potential acoustic fovea (i.e. a location with particularly high acuity) at the base of the cercus has been hinted at (Dangles et al. 2008), not only due to the highest hair density in this region, but also because it corresponds to the region with the largest flow velocities, due to the cercus being the largest there. Putting hairs radially around the cercus enables crickets also to pick up transversal flows. In such an arrangement, the received peak flow velocities are larger than if the hairs would be placed on a flat surface; in the latter case, hairs are submitted to longitudinal flow with lower peak velocities (Dangles et al. 2008). In summary, where sensors are placed relative to body geometry matters a lot.

Filiform hairs are innervated by a single mechanosensory receptor neuron. These neurons get mechanically stressed by the hair movement and transduce the mechanical signals received by the hair into neuroelectrical signals, i.e. action potentials. The axons of the receptor neurons are bundled in the cercal nerve and their terminals form excitatory connections onto neurons within the terminal abdominal ganglion (TAG). Information from all hairs, as well as from other sensors, converges and is processed by interneurons. The convergence of information at this stage is enormous: the about 1,500 afferent neurons of hairs are connected to only some 20 interneurons (Jacobs et al. 2008). The fact that invertebrates possess few large, singly identifiable neurons enabling comprehensive mapping and repeated recordings of activities in identified neurons is a unique asset which explains the interest in this mechanosensory system. Descending information from the central brain and higher order ganglia also reaches the TAG. Once processed, the combined information moves up towards higher neuronal centres, in particular the ganglia in which the hind-leg movements are triggered. This local feedback loop enables the animal to process vital information and act accordingly very quickly. Thus, as is observed often among invertebrates, what can be processed locally will be done so. This type of distributed processing explains why biomimetism has so much to gain from this group of animals.

After processing, the perceived sensory information must be converted into an appropriate behavioral action. Flow sensing is known to be of importance in predator and prey perception, mate selection and most likely other context, such as perceiving its own speed and movement. Predator avoidance is obviously a major selection force, where speed is of paramount importance. Jumping or running away is the behavior which is elicited in response to appropriate stimuli.

The cricket possesses in the TAG an internal map of the direction of the stimuli from the outside world and the geometric computation of the direction of incoming flow by the cercus is one of the nicest case studies of spatial representation in the central nervous system (Jacobs et al. 2008). Computation of the speed of an approaching predator is also carried out by the TAG, a task that has been only recently studied using appropriate stimuli (Dangles et al. 2006). Where to jump is a different question, and how and where in the brain this decision is made, is presently unknown. Presumably, directing stimuli and other conditions intervene in this process.

Natural selection acts along the full chain of information transfer, from acquisition and processing, up to actuation. This is important to restate in a biomimetic context, as the extreme sensitivity on the biomechanical side of the hair shaft, which has been almost the exclusive focus of the engineer attention, could be otherwise lost into an inefficient sequence of information transfer. However, as of today, we have very little information about the constraints acting on the different parts of the chain, and hence no idea about their optimisation levels.

17.2 Hair-Sensors

Although hair-sensors are abundant in nature and are found on animals of virtually all scale, they seem to be especially important on smaller animals, e.g. small invertebrates such as insects and arachnids. Important in this respect is that the, e.g. predator–prey, interactions between the animals take place at distances that are small relative to the wavelength of the sounds emitted, i.e. they take place in near-field conditions.

17.2.1 Flow as Information Source

Whereas for many (larger) animals flow may not provide a very information-rich modality to probe the environment, nature seems to have numerous examples of species that exactly do this. To put this in perspective, it may be helpful to look at the fields produced by a harmonically moving sphere (dipole), which, in a somewhat simplistic view, resembles natural sources such as wing-beats of flying insects or tail movements of fish.[1]

Obviously the fields produced by a dipole entail both pressure and flow fields. Depending on the medium in which the dipole resides, e.g. water or air, and more specifically on the mediums compressibility, it may seem that pressure and flow

---

[1] Although emission patterns may be far more complex and need description by a multipole expansion, higher order poles tend to drop off faster with distance.

Fig. 17.2 Theoretical dipole-field as function of normalised distance

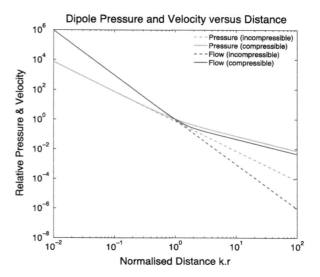

could play comparably important roles. However, pressure is a scalar field only, whereas flow-fields inherently carry directional information (being vector-fields) helping a flow-sensory system to more easily determine the direction of a source than a pressure based one, especially at low frequencies. Further, although pressure and flow may have a fixed ratio at relatively long distances[2] near the source flow-fields are 'comparatively stronger'. This is shown in Fig. 17.2, where on-axis pressure and velocity, normalised to their respective values at $k \cdot r = 1$ (with k the wavenumber and r the distance), are plotted as a function of normalised distance. Both the situations of compressible (solid lines) and incompressible (dashed lines) media, as calculated from the equations in Lamb (1910), are shown.[3] The ratio between pressure and flow-velocity[4] equals $j\rho\omega r/2$ (ρ being the density of the medium and ω the angular frequency) for $k \cdot r \leq 1$ indicating that pressure becomes comparatively small at shorter distance and lower frequencies. As an example for an interaction of a flying wasp with a wing beat of about 150 Hz with, say, a caterpillar (Tautz and Markl 1978) the condition $k \cdot r = 1$ corresponds to $r \approx 0.34$ m a relative large distance as measured in body lengths of the animals. In other words: at the scale of the interaction distances of e.g. crickets and their predators, and at the frequencies of concern, it is easier to monitor the near-field environment from spatio-temporal flow-profiles than from pressure signals.

---

[2] For compressible media one could think of at least a few wavelengths away from the dipole.

[3] Clearly, for $k \cdot r \leq 1$, i.e. distances small compared to the wavelength, there are only minor differences between the expressions for compressible and incompressible media, a fact that can be readily exploited when modelling relative complex aerodynamic predator–prey interactions at short distances Kant and Humphrey (2009).

[4] The ratio of pressure and flow-velocity would be the acoustic impedance for compressible media.

Fig. 17.3 Schematic of the
hair-sensor mechanical
system (after Shimozawa
et al. 1998)

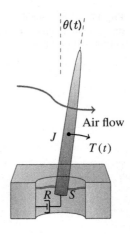

17.2.2 Hair-Sensor Mechanics

Both natural as well as artificial hair-sensors, can be understood mechanically as a
so-called "inverted pendulum" (Shimozawa et al. 1998). That is, a second-order
rotational-mechanical system with moment of inertia J, [kgm^2], rotational stiffness
S, [Nm/rad], and rotational damping R, [Nms/rad], (see Fig. 17.3). Airflow generates a torque on the hair-shaft, primarily by viscous drag since at the velocities
and geometries normally encountered pressure drag is small. Note that under most
conditions artificial hairs can be assumed infinitely stiff and that the rotation angles
are rather small (of the order of 1–10 mrad amplitude per m/s flow-velocity
amplitude).

17.2.3 Hair-Sensor Physics

The physics of flow-sensing by hair-sensors has been unravelled independently by
Humphrey et al. (1993) and Shimozawa et al. (1998) with initial contributions by
Tautz and Markl (1978) and Gnatzy and Tautz (1980). In general, these flows have
been assumed to be small which, in combination with the small hair diameters,
yields rather small Reynolds (Re) numbers. Moreover, the frequencies are limited
to a few 100 Hz causing the Strouhal number ($St = \omega d/2V_r$) to be small as well.
For a hair-diameter of 25–50 μm,[5] an air-oscillation frequency of 250 Hz and a
flow-velocity amplitude of 10 mm/s, Re varies between 0.008 and 0.016 and St
between 1.96 and 3.92 (for the flow around the hairs).

---

[5] These are diameters encountered in MEMS-based hairs with a length of up to 1 mm; natural
hairs tend to be much smaller in diameter.

Here we reiterate the modelling by Humphrey et al. as given in Humphrey and Barth (2008). The analysis starts with the abstraction of a hair-sensor being a second-order mechanical system, with the rotational angle θ the single degree of freedom:

$$J\frac{d^2\theta(t)}{dt^2} + R\frac{d\theta(t)}{dt^2} + S_0\theta(t) = T(t) \tag{17.1}$$

where $T(t)$ is the driving drag torque due to the airflow. In general, the rotation of the hair will lack behind the flow and the drag-torque needs to be calculated using the difference in airflow and hair velocity, which is dependent on the distance to the substrate z.

The rather small Re and the large hair-length to hair-diameter ratio allows using the Stokes expressions for the drag-torque exerted by the airflow on the hairs Stokes (1851) and to integrate it over the length of the hairs. This results in:

$$J\frac{d^2\theta(t)}{dt^2} + R\frac{d\theta(t)}{dt^2} + S_0\theta(t) = 4\pi\mu G \int_0^L \left[v(z,t) - z\frac{d\theta}{dt}\right]zdz$$
$$+ \left(\frac{\pi\rho D^2}{4} - \frac{\pi^2}{g\omega}\right) \int_0^L \left[\frac{dv(z,t)}{dt} - z\frac{d^2\theta}{d^2}\right]dz \tag{17.2}$$

where $v(t)$ is in the direction $\parallel$ with the substrate ($\perp$ to the hair), μ is the dynamic viscosity, D the diameter of the hair, $g = \ln(s) + \gamma$, with γ Euler's constant (0.5772...) and $S = (D/4)\sqrt{\omega/v}$, and $G = -g/(g^2 + \pi^2/16)$. Since the terms on the right-hand side with time derivatives of θ do not depend on $v(z, t)$ they can be transfered to the left-hand side where they appear as additional terms to the moment of inertia (J_ρ, J_μ) and rotational damping (R_μ) (Humphrey and Barth 2008).

$$\left(J + J_\rho + J_\mu\right)\frac{d^2\theta(t)}{dt^2} + (R + R_\mu)\frac{d\theta(t)}{dt} + S_0\theta(t) = 4\pi\mu G \int_0^L v(z,t)zdz$$
$$+ \left(\frac{\pi\rho D^2}{4} - \frac{\pi^2}{g\omega}\right) \int_0^L \frac{dv(z,t)}{dt}zdz \tag{17.3}$$

For cricket hairs, the values of these additive terms can be significant and should not be neglected (Humphrey and Barth 2008). For MEMS based bio-inspired hair-sensors these contributions are relatively small for operation in air. However, for sensors working in water, both for bio-inspired as well as natural sensors, these terms may be of comparable magnitude as J and R.

Fig. 17.4 Mechanical
response versus frequency for
a MEMS-based hair-sensor
(unpublished data)

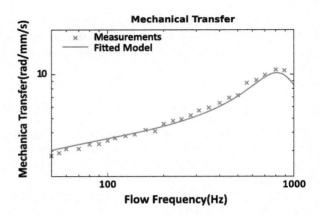

Once the flow velocity $v(z, t)$ is known the driving torque $T(t)$ can be calculated. However, only relative simple flows allow for straightforward modelling, i.e., the boundary layer effects of transient flows are intricate and need to be determined by numerical schemes like Finite Element Modelling (FEM). Our artificial hair-sensors are mounted on flat substrates allowing the use of the Stokes expressions for harmonic flows along the hairs (Humphrey et al. 1993). We have shown previously that the Stokes expressions can be usefully employed for our artificial hair-sensors (Dijkstra et al. 2005). Based on the no-slip condition, these expressions predict a viscous flow over an infinite substrate to be harmonic in time with 45° phase advance near the substrate interface and a boundary layer thickness (δ_b) proportional to the inverse of $\beta = \sqrt{\omega/2v}$ where v is the kinematic viscosity (1.79×10^{-5} m^2/s for air at room-temperature) and ω is the radial frequency. As an example, at 100 Hz the boundary layer is roughly 0.5 mm thick.

$$v(z, t) = V_0 \sin(\omega t) - V_0 e^{-\beta z} \sin(\omega t - \beta z) \qquad (17.4)$$

Substituting 17.4 in 17.3 provides all the information needed to calculate the response of the hair-sensors to steady state harmonic driving. Comparison between calculation and model is shown in Fig. 17.4. Obviously, the agreement is rather good.

17.2.4 Hair-Sensor Design

Despite the beauty and sensitivity of natural hairs, it is not possible[6] to copy them one-to-one to artificial versions. The basic principle of capturing flow by drag-forces exerted on a hair can be used. However, lacking artificial neurons, the

---

[6] Note that despite the fact that 'copying' is not possible it is not desirable either since the artificial hair-sensors have to operate under much different conditions and with different purposes as well.

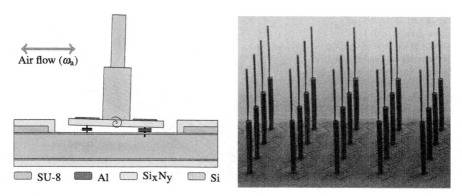

Fig. 17.5 *Left* schematic of the artificial hair-sensors using differential capacitive read-out. *Right* Scanning Electron Micrograph (SEM) of a realised hair-sensor array

transduction mechanism needs to be entirely different. In our work, we have decided to use differential capacitive read-out: we fabricate hairs on top of a membrane, suspended by torsional beams, allowing the hair plus membrane to rotate when exposed to drag-torque (see Fig. 17.5). On the membrane we have an aluminium electrode, roughly 100 μm long by 100 μm wide, on each of the two sides. In combination with the underlying higly conductive silicon substrate, from which the membranes are only separated by a gap of 600–1,000 nm, these electrodes form capacitors which can be read-out using AC signals.[7] On membrane rotation one capacitance will increase, on the side where the gap reduces, and on the side where the gap increases the capicatance decreases. These changes are read-out differentially to reduce parasitic effects.

17.2.4.1 Hair-Sensor Design Optimisation

From a biological standpoint, one may want to understand the interrelationship between the geometric and physical parameters of the hair-sensor system *as they are*. However, from an engineering viewpoint, things look slightly different since (a) not all detail and interplay of all the involved parameters of the insects hair-sensor system are well-known (i.e. plain mimicking of the cricket cercal system is no option) and (b) MEMS fabrication technology offers a latitude of size possibilities and material choices that only partly overlaps the natural system. Therefore, the values of various design parameters need to be determined from additional analyses.

---

[7] The membranes are connected to the outside world by aluminium electrodes running over the torsion beams.

Fig. 17.6 Drag-torque versus hair-length for harmonic airflow of various frequencies as predicted by Stokes' equations. Initial dependence is cubic for $L < \delta_b$) and turns over into quadratic for $L > \delta_b$

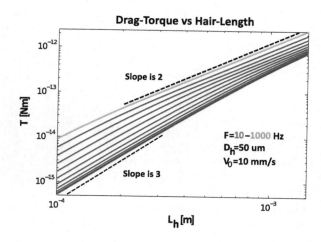

Hair Length and Boundary Layer

The length of the hairs (L) plays a dominant role in the overall performance of the hair-sensors. Obviously, when exposed to a uniform flow, the total drag torque on cylindrical hairs would increase proportional with the hair-length squared. However, due to the boundary layer, in which the flow-velocity increases with distance to the substrate, the drag-torque first increases with the third power of hair-length, i.e. $O(L^3)$, up to about the boundary layer thickness δ_b. Then when the hairlength is above δ_b the drag-torque follows a quadratic dependence (see Fig. 17.6). But at the same time the hair inertial moment (J) is of order $O(L^3)$. Hence, the hair-length needs to be chosen judiciously to balance drag-torque versus inertial moment.

Hair Diameter and Viscous Drag

When increasing the diameter of the hairs (D) the resulting drag-force will increase as well. Using numerical evaluation of the Stokes expressions for drag-force shows that the dependence on diameter is weak in the range of interesting hair-diameters, of the order of $O(D^{1/3})$. At the same time, the hair moment of inertia J is of order $O(D^2)$, negatively affecting the resonance frequency (i.e. bandwidth). Therefore, it is beneficial to have thin hairs. Technologically it turns out to be rather difficult to make hairs with aspect ratios of more than about 10–20. We have tackled this problem by segmenting our artificial hairs with a lower part diameter of 50 μm and a top part diameter of 25 μm, reducing the hair inertial moment by about 65 % (Jaganatharaja et al. 2009), see Fig. 17.7.

Fig.17. 7 Extruded SOI-
based hair-sensor structure
(Dagamseh et al. 2010)

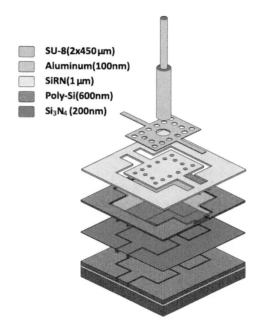

SU-8(2x450 µm)
Aluminum(100nm)
SiRN(1 µm)
Poly-Si(600nm)
Si₃N₄ (200nm)

Torsional Stiffness

Obviously, when looking for the largest rotation angle for any given drag-torque one may want to choose the lowest possible torsional stiffness (S). But for given hair inertial moment, a reduction of S also leads to a reduction of the resonance frequency which is given by $\omega_0 = \sqrt{S/J}$.

Damping

Damping of the hair-sensors comes in multiple forms. For the crickets, the hair-sockets provide some torsional damping (R) by visco-elastic material properties (see McConney et al. 2007 for such material properties in spider hair mechano sensors) whereas for the artificial hair-sensors torsional damping is caused by both material as well as squeeze film damping due to the small gap between the silicon-nitride plates and the substrate. On top of these damping contributions, the hairs themselves incur damping by viscous forces when the hairs move relative to the surrounding air. In the case of crickets, the total damping seems to be appropriately controlled (Bathellier et al. 2012) by the organism yielding hairs that are approximately critically damped. It is hypothesized that mechanical impedance matching helps the sensors to obtain maximum energy from the surroundings (Shimozawa et al. 1998). On the other hand, on critically damping a second-order system one also maximizes its agility to respond to (transient) flows. Nevertheless, the evolutionary pressures driving the appropriate damping for cricket hair-sensors have not yet been identified. In the artificial hair-sensors, except for adding specific holes to the membranes to tailor the squeeze film damping, not much can be done to optimise the damping without far-reaching consequences for the fabrication technology.

Torsional Spring Material

The mechanical sensitivity of our hair-sensors is currently about two orders of magnitude less than those of crickets, primarily due to a much larger rotational stiffness: 1.5×10^{-11} Nm/rad for crickets versus 4.85×10^{-9} Nm/rad for our sensors. But reducing the torsional stiffness comes with two difficulties. In order to conserve bandwidth, the moment of inertia of the hairs needs to be further reduced. The second complication is that the suspension beams provide torsional (S) as well as vertical stiffness (K). Both decrease with increasing length l but S decreases with $O(l^{-1})$ whereas K decreases with $O(l^{-3})$. The result is that a large rotational compliance combined with a large vertical stiffness can only be obtained using compliant materials, i.e. with low Youngs modulus and appropriate beam-cross-sections. A nice reference to natural hairs, where the materials in the hair-sockets are rather soft compared to the stiff silicon-nitride beams used in the artificial sensors, despite the fact that our torsional suspension and the cricket hair-sockets have little in common.

Figure of Merit

Optimisation of our hair-sensors has been driven by a Figure of Merit (FoM) (Krijnen et al. 2007), being the product of mechanical responsivity[8] and bandwidth: $\text{FoM} = \sqrt{L/\rho SD^{4/3}}$. This has emphasized what could be learned directly from observation of cricket hair-sensors, i.e. that hairs should be long and thin, and mounted on very compliant suspensions. However, with respect to damping the optimum damping factors still need to be identified. Compared to crickets, for a 1 mm long hair, the FoM of our hair-sensors is about a factor of 70 smaller due to the larger rotational stiffness and thicker hairs.

Capacitive Read-Out

The angular rotations induced by harmonic flows normally encountered are rather small: on the order of 1 mrad/mm/s. Therefore, the capacitive read-out needs to be judiciously implemented. Our hair-sensors are based on a differential read-out,

---

[8] For the hair-sensor system, one can define the mechanical responsivity as the angular rotation of the hair per m/s of flow-velocity.

Fig. 17.8 Single-hair
threshold at 250 Hz
(Dagamseh et al. 2010)

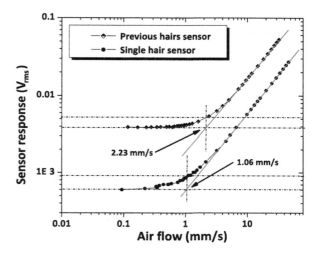

using a 1 MHz interrogation signal, a charge amplifier and a multiplier to retrieve
the base-band information. Since parasitics due to bond-pads and wires are relatively large the fractional capacitance changes, which ultimately determine the
sensitivity of the sensor, need to be optimised. Because the sensor's membrane
area close to the rotational axis does not generate much capacitance change the
membrane should primarily be long. Also, the smaller the effective gap between
the capacitor electrodes, the larger the effect. Eventually the fractional capacitance
change is given by $\partial C/\partial\alpha \times 1/C = l/d$, i.e. one should aim for a long membrane
and a small (effective) gap. Early generations of our sensors were affected with
stress-induced upward curvature of the membranes, negatively influencing the
capacitive sensitivity. In later generations, aluminium is used as electrode material
since it has a high electrical conductivity (and therefore the layer can be thin), has
a low Young's modulus (thus will cause relative little bending when under stress)
and can be deposited at low temperatures (reducing residual thermal stress).

The latest generation of our artificial hair-sensors is based on silicon-on-
insulator (SOI) technology (see Fig. 17.7), which helps to reduce parasitic
capacitances. The performance of this type of sensors is shown in Figs. 17.8 and
17.9 where results are displayed for a single hair-sensor. The threshold flow-
amplitude value is at about 1.00–1.25 mm/s for frequencies between 100 and
400 Hz which is currently determined by electronics noise (thermal–mechanical
noise is predicted to be more than two orders of magnitude smaller). There is also a
clear directivity pattern, closely matching a theoretically ideal figure of eight,
Fig.17.9 (Dagamseh et al. 2010).

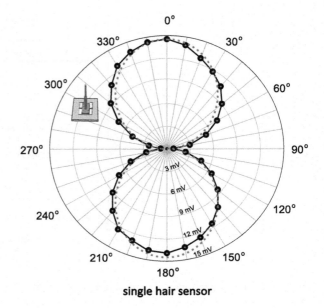

Fig. 17.9 Single hair-sensor directivity (Dagamseh et al. 2010)

single hair sensor

17.2.4.2 Optimisation of Hair Geometry

Artificial hairs on top of the sensor membranes are the mechanical interfaces that cause the flow information to result in membrane rotations inducing capacitance changes, which are eventually transformed into equivalent electrical signals. Making long hairs in micro-fabrication technology is by no means an easy challenge since both the absolute length as well as the aspect ratio of these structures are non-standard for MEMS.

Importance of Hair Shape

For the effective operation of our artificial hair sensors, the shape of the hair plays a central role. The hair geometry serves two basic purposes: (1) it determines the amount of flow-induced drag-torque acting upon the hair and (2) it contributes to the mass moment of inertia, which determines the mechanics of the sensory system. Finding the optimum balance between the drag-torque reception and the hair moment of inertia has been the primary motivation for such optimisation.

Taking a closer look at the shape of cercal filiform hairs themselves, could guide towards the first steps of hair shape optimisation. The hairs on the cerci are found to appear in a wide range of lengths from 30 to 1,500 μm with diameters occurring from 1 to 9 μm (Humphrey et al. 1993). Initially, the structural effects of the cercal hairs were analysed by assuming a cylindrical (Humphrey et al. 1993) or linearly tapered conical shape (Shimozawa and Kanou 1984). But upon accurate

Fig. 17.10 Initial attempts to make hairs were based on Deep Reactive Ion etching in silicon with subsequent conformal nitride overgrowth and silicon removal

measurements, the hair shape was found to follow extremely elongated paraboloids, i.e. the diameter of the hair increases with the square-root of the distance from the tip (Shimozawa et al. 1998). Other electron microscope studies showed the hairs to be hollow tubes, with the diameter of the inner hollow cavity being approximately one-third of the outer diameter (Gnatzy 1978). The elongated-paraboloid shape of the filiform hairs apparently strikes a fine balance between the drag-torque reception capability and its moment of inertia. The goal is, thus, to fabricate artificial hairs which closely resemble the natural filiform hairs.

Artificial Hairs: Past and Present

Artificial hairs were initially fabricated by etching structures on bare silicon wafers and covering them with silicon-nitride using low-pressure chemical vapour deposition (LP-CVD). Upon selectively etching the silicon substrate, silicon-nitride hairs were fabricated (Fig. 17.10) (van Baar et al. 2003). These hairs were very complex to be integrated into a functional sensor. Subsequent generations of our sensor arrays comprised an artificial hair made of SU-8, a negative-tone, epoxy-based photoresist, fabricated by means of standard photo-lithography. First, a single layer of SU-8 photoresist was used resulting in hairs of about 450 μm length (Dijkstra et al. 2005) (Fig. 17.11 left). In later generations of our hair sensors, two subsequent layers of SU-8 were spun and hairs of length of up to 900 μm were photo-patterned by top-side exposure (Fig. 17.11 right). The result was that hairs were long and had a cylindrical shape with a uniform diameter of about 50 μm.

For the current generation of artificial hairs (as discussed before and shown in Fig. 17.7), a new geometry was chosen in order to reduce the hair moment of inertia. The idea is to fabricate artificial hairs in two parts (i.e. two layers of SU-8 photoresist), where the hair diameter of the upper part is only half of that of the bottom part. Such a hair geometry effectively reduces the hair moment of inertia by about 65 % while reducing the drag torque by only about 20 %.

Fig. 17.11 First (*left*) and second (*right*) generation artificial hair-sensors

One of the difficulties in using multi-spun SU-8 layers for artificial hairs is that alignment of structures on the subsequent layers becomes critical. Further, the standard top-side exposure of SU-8 lithography limits us to to only two or three hair length variations on one device. Therefore, to fabricate artificial hairs with a more nature like shape, a new and less-complex fabrication technology is sought in order to realise hairs of varying hair-lengths in a wider range, all within the same device and made at the same time in the same process.

Nature-Like Hairs: Future?

Bottom-side exposure of SU-8 layers is a well-known technique, commonly used as molds in the fabrication of micro-needle arrays for drug-delivery applications (Huang and Fu 2005; Kim et al. 2004; Yu et al. 2005). For our requirements, we used the above-mentioned bottom-side exposure for fabrication of hairs with gradually tapering tips aimed to resemble the shape of actual filiform hairs of cerci. A simple, proof-of-concept process-flow was developed to fabricate SU-8 hairs by a bottom-side exposure method. For the fabrication, a patterned aluminium layer with circular openings on top of a standard glass substrate is used. Two layers of SU-8 are spun to a thickness of 900–1,000 μm, after which the glass substrate is flipped and the SU-8 exposed through the circular openings. Owing to the illumination dynamics inside the SU-8 layer, upon the development nature-like SU-8 hairs, resembling their natural counter-parts closer than previous versions of our artificial hairs (Fig. 17.12) are created. Further, the variations in the diameter of the circular openings of the patterned aluminium layer allow us to achieve a wide range of hair length variations, all in a single photolithographical step (Fig. 17.13).

Fabricated nature-like hair samples were analysed to find the optimal exposure time and the effect of different design parameters of the aluminium pattern on the hair geometry. The next challenge is to develop a new process scheme to integrate the nature-like SU-8 hairs into the existing sensor fabrication flow. The process

Fig. 17.12 Scanning Electron Microscope image of nature-like hairs with variations in hair-length determined by the diameter of the open circles used for exposure

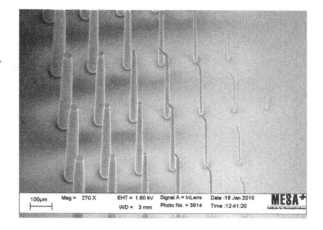

Fig. 17.13 The length of the nature-like hairs depends on the diameter of the opening through which illumination takes place, larger diameters causing longer and wider hairs. The illumination process of nature-like hairs is slightly dependent on illumination time (indicated in seconds in the legend) but still rather robust

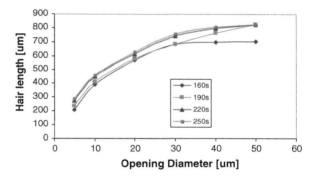

flow for wafer-through etch-holes on the silicon wafers for back-side exposure and the applicability of aluminium as both capacitor electrode and hair mask should be tested and optimised.

17.3 Viscous Coupling

Arthropods are very often quite hairy, and the high density of flow sensing hairs implies that these hairs may interact with each other through viscous effects, i.e. when the distance between hairs is too small, one hair may reduce the drag-torque on its nearby neighbour. This fluid-dynamic interaction between hairs, called viscous coupling, has been studied only recently and was found to be highly dependent on the geometrical arrangement of hairs, of their respective lengths and preferential planes of movement, as well as on the frequency of the input signal (Casas et al. 2010). Hairs often interact over long distances, up to 50 times their radius, and usually negatively. Short hairs in particular 'suffer' substantially from

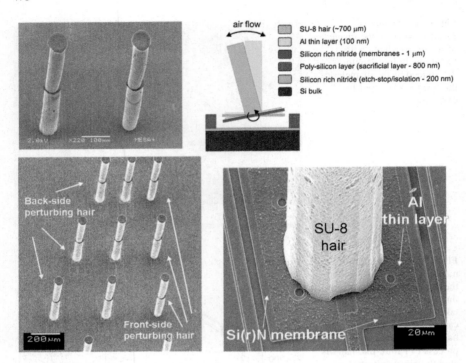

Fig. 17.14 SEM images showing different portions of the chip for characterisation of viscosity-mediated coupling. *Top-left* free-moving reference hair-sensor placed near a fixed perturbing hair of same length. *Top-right* schematic of a hair-sensor. *Bottom-left* hair-sensors are arranged at various inter-hair distances. *Bottom-right* close-up of the Si(r)N membrane and hair base

the presence of nearby longer hairs. Positive interactions, where the flow velocity at one hair is increased by the presence of other nearby hairs, have, however, been observed in biological cases and reproduced numerically (Lewin and Hallam 2010). The biological implications of these interactions have only very recently been addressed, and hint towards a coding of incoming signals which relies strongly on the specific sequence of hairs being triggered (Mulder-Rosi et al. 2010). In other words, the signature of the incoming signal may be mapped into a given sequence of recruited hairs, which in turn produces a typical sequence of action potentials.

On the physical level, it is rather hard to determine viscous coupling effects on real animals due to the pseudo randomness of hair-position, hair-length and hair-orientation. Here the MEMS capabilities to form regular structures with well-defined inter-hair distances present a way to tackle the problem. We have made various structures to systematically investigate viscous coupling effects. Both the flow-profiles (Casas et al. 2010) as well as the hair-rotations in the presence of perturbing hairs (Jaganatharaja et al. 2011) have been studied. Figure 17.14 shows the layout of hair-sensor devices that were particularly made for viscous coupling

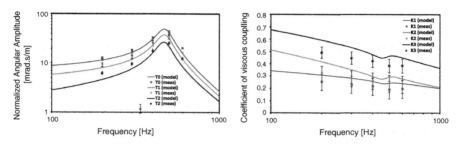

Fig. 17.15 *Left* Influence of inter-hair viscous coupling on hair-rotation amplitude. *Right* frequency dependence of the viscous coupling constant for 2 versus 0 perturbing hairs (*black*), 2 versus 1 perturbing hair (*red*) and 1 versus 0 perturbing hairs (*blue*). The normalised distance (*D/S*) between the hairs is ≈ 2.1. Lines are predictions based on a modified model introduced in Bathellier et al. (2005) in the limit of arrested hairs, dots are measurements with uncertainty intervals

experiments: the membranes were rather short and only covered with aluminium to get a good reflection in order to enable laser-vibrometer measurements. The results of the measurements are shown in Fig. 17.15. The left graphs shows a frequency response of a hair-sensor with two perturbing hairs (black), one perturbing hair (red) and no perturbing hairs (blue). The hair sensors spacing was about 2.1 times the hair diameter. Figure 17.15, right, shows the coupling constant, as introduced in Bathellier et al. (2005), for the various situations with one or two perturbing hairs. Clearly, with one or two perturbing hairs the hair-rotations are smaller than without perturbing hairs. Other experiments (not shown here) have confirmed the overall tendency reported that viscous coupling increases with decreasing frequency an decreasing inter-hair distance.

17.4 Array-Sensing

Whereas a single hair-sensor is an interesting object by itself, a collection of such sensors, forming geometric arrays, enables an entire different class of measurements: the determination of spatio-temporal flow-fields. The array compares to the sensor as a camera chip to a single photodiode; with the array it becomes feasible to observe the environment and the movement of objects in it. Therefore, we often use the name *flow-camera* when referring to an array of flow-sensors.

17.4.1 Interfacing Array Sensors

The SOI-based technology not only serves to reduce parasitic capacitance but also allows for crossing electrodes since both the silicon device-layer of the SOI wafer as well as the top aluminium layer, mutually separated by silicon-nitride, can serve

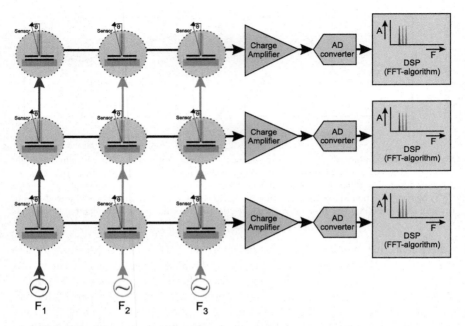

Fig. 17.16 Frequency division multiplexing reduces the number of electrical connections while retaining the original sensor SNR (Dagamseh et al. 2011)

as electrical connection. The technology allows for frequency division multiplexed (FDM) interfacing to individual sensors in a rectangular array, reducing the number of required electrode connections from $3(n \times m)$ to $2n + m$ for a $n \times m$ array of hair-sensors (Dagamseh et al. 2012). Further, this scheme retains the signal-to-noise ratio (SNR) of the hair-sensors at the level they have when each single hair-sensor is individually connected (see Fig. 17.16).

The FDM technique allows for real-time read-out of many sensors in parallel. Therefore, it enables the observation of spatio-temporal flow-patterns in which the details carry information of the source of the field, i.e. this type of flow-sensor ar-ray in principle allows for the observation of moving objects in the near-field en-vironment, thus acting as a flow-camera. Figure 17.17, shows an array of sensors, each individually interfaced by FDM.

17.4.2 Transient Airflow Measurements Using Artificial Hair Sensors

Airflow patterns observed hair-sensors carry highly valuable information about the sources of these flows. The successful extraction of the characteristics of these spatio-temporal airflow patterns will give us insight in their features and information

Fig. 17.17 Microphotograph of an 8 × 8 array of individually FDM addressable hair-sensors, ordered in pairs with orthogonal directivity (Dagamseh et al. 2011)

contained in them. In nature, there are numerous examples representing transient airflow stimuli such as spider motion (Dangles et al. 2006) and (passing) humming flies (Barth et al. 1995).

In most investigations on our artificial hair-sensors, measurements were conducted using sinusoidal airflows (Dagamseh et al. 2010). Obviously, using transient signals spatio-temporal flow-structures become richer and array-measurements will allow to capture important flow events. Here we describe measurements of spatio-temporal airflow fields generated by a transient airflow by means of our artificial hair-sensor arrays. The measurements show the hair-sensors ability to determine the flow field with sufficient temporal and spatial resolution.

We measured responses of our biomimetic hair sensors to airflow transients using a sphere with 3 mm radius attached to a piston system to represent the motion of a spider at a distance (D) from the substrate. A single-chip array consisting of single hair sensors is used for flow-detection. Motion direction of the sphere was parallel to the x-axis. i.e. the line of the sensors. Figure 17.18 shows a photograph of the measurement setup.

The results show that our hair sensor is able to capture the essential features of the transient airflow field generated by the moving sphere. Interestingly, the hair-sensor response shows strong similarities with the field shape generated by a dipole source. Figures 17.19 and 17.20 show an example of a theoretical and a measured hair-sensor response as caused by a passing sphere. The distance to the sphere is encoded in the characteristic points[9] of the flow-field (Dagamseh et al. 2010). Hence, it can be derived from the sensor output. In the transient response, the time difference between the characteristic points can be translated into position using the piston speed, and subsequently into an estimated distance (D_{est}) between sphere and hair sensor. Figure 17.21 shows D_{est} versus D using the transient hair response.

To exclude the effects of the hair-mechanics, a deconvolution was performed to recover the flow-velocity. The results show that the deconvolved sensor data

---

[9] The characteristic points are the points in the flow profile where the parallel component of the flow either changes direction or where it is maximum. The points are easily recognisable in Fig. 17.19.

Fig. 17.18 Setup used for transient measurements

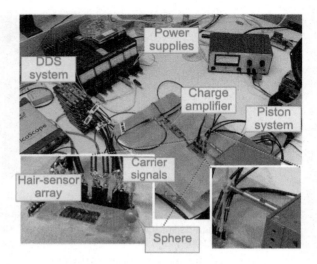

Fig. 17.19 Theoretical transient dipole flow-field parallel to the direction of orientation of the flow-sensor

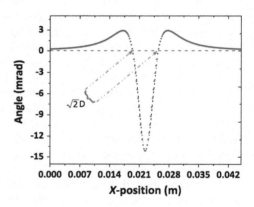

Fig. 17.20 Example of measured hair-sensor response (*solid*) when exposed to a transient flow

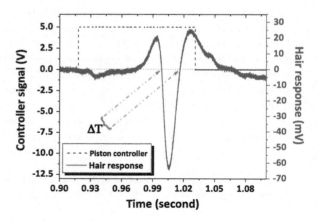

Fig. 17.21 D_{est} versus D using transient hair-response measurements, before (*solid-squares*) and after (*solid-circles*) deconvolving the hair-sensor response. The best linear-line fit for both measurements are compared with ideal linear-line (*dotted*). D represents the height of the sphere centre relative to the substrate. The *error bars* represent the uncertainty in determining the zero-crossing points of the measured dipole profile

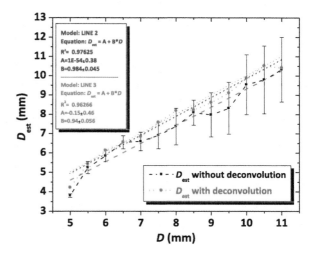

nearly matches the raw sensor data with a slight increase in the distance between the characteristic points. Hence, we assume that the hair sensor is following the temporal course of the flow profile rather well, a consequence of the nearly critical damped system with best frequency in the range of 250–300 Hz. The deconvolved sensor data (Fig. 17.21) indicate that the linear-line fit of D_{est} more closely matches the physical D while for the raw sensor data the D_{est} seems to closer resemble the distance to the centre of the hair shaft. This is due to the effect of the mechanics and the hair-shaft of the sensor. Since we know of no way to correct for the integrated drag-force on the length of the hair, we consider the torque as a reasonable representation of the flow field at between 1/2 and 2/3 the hair length (i.e. 600–700 μm above the substrate).

Arrays of hair sensors offer us spatial information, specifically if they are measured simultaneously. Here we integrated frequency division multiplexing (FDM) to simultaneously measure the transient response of multiple hairs, i.e. spatio-temporal airflow pattern measurements. Figure 17.22 shows the response of four single-hair sensors in one row, when they are exposed to a transient airflow produced by a moving sphere. Using the signal profiles as detected by the entire array would allow us to determine an increasing number of source properties. By virtue of the piston velocity, the delay represents the separation distance in between two hairs divided by the sphere velocity. Thus, the sphere velocity can be determined independently of the distance to the sphere.

Using the signal profiles as detected by an entire array would allow us to determine a number of source properties. By virtue of the piston velocity, the delay represents the separation distance in between two hairs divided by the sphere velocity. Thus, the sphere velocity can be determined independently of the distance to the sphere. As a first trial, we determined the delays between the signals from four hairs in one row. As an example, Fig. 17.22 displays the normalised responses for four hair-sensors in line where the sphere was moved along.

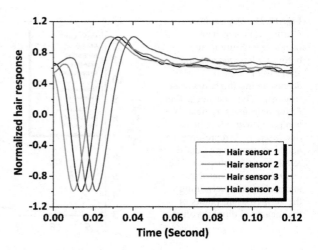

Fig. 17.22 Normalised output of four simultaneously measured sensors when exposed to a sphere passing by at certain distance (Dagamseh et al. 2012)

The measurements show about 4 ms time delay between each two subsequent hair-sensor responses. From the sensor-responses and the distance between the sensors, a speed of 512 mm/s at a distance of 5–7 mm was inferred (Franosch et al. 2005; Dagamseh et al. 2012). This demonstrates the possibility to perform spatio-temporal flow pattern measurements using a single-chip hair sensor array with FDM and to, subsequently, use the features of these flow profiles to determine source parameters (i.e. size, speed and position).

Measurements like these, in principle, allow to extract the following information: (a) The projection of the velocity of the passing sphere in the direction parallel to the row of the sensor array can be determined using the distance between the sensors and the time of flight; (b) Once the velocity is known, the distance of the sphere trajectory perpendicular to the row of sensors can be determined from the characteristic points of the dipole-induced signal (Franosch et al. 2005; Dagamseh et al. 2012); (c) With the distance to the sphere and its velocity known, the amplitude of the signal can be used to determine the size of the sphere; (d) Additional sensors allow to track the motion of the sphere in other directions as well. We do not know whether crickets use their hair-sensors in a similar way as we use our artificial sensors. Nonetheless, it is highly instructive to see what information a multitude of sensors in principle can uncover.

17.5 Beyond Bio-Inspiration: Parametric Effects

Apart from using the capacitively interrogated hair-sensors strictly for sensing, one may achieve parametric effects by application of additional DC or AC bias-voltages to the electrodes (Fig. 17.23). These voltages will produce electrostatic forces, which in a balanced situation (i.e. no tilt of the hair) do not change the rotational angle, but in a tilted situation will produce the largest forces on the side

Fig. 17.23 Electrodes can also be exploited for electrostatic actuation (Droogendijk et al. 2012a)

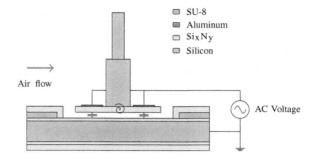

with the smallest gap. So, the electrostatic torque tends to add to the flow-induced torque and therefore the applied voltages will serve as an electronic means to adaptively change the spring-stiffness of the hair-sensor system, i.e. electrostatic spring softening (ESS).

17.5.1 DC-Biasing

A DC-bias voltage can be used to change the system's torsional stiffness. Experimentally the mechanical transfer is determined for flow frequencies from 100 to 1,000 Hz with and without the application of a DC-bias voltage. During this measurement, a DC-bias voltage U_{dc} of 2.5 V is used, giving an increase in sensitivity of about 80 % for frequencies within the sensor's bandwidth.[10] Also lowering of the resonance frequency ω_r, is observed (about 20 %). Overall, measurements are in good agreement with modelling and it is shown clearly that DC-biasing leads to a larger sensitivity below the sensor's resonance frequency (Fig. 17.24) (Droogendijk et al. 2012a). With respect to the FoM, it may be remarked that the responsivity is proportional with $1/S$ and the bandwidth with $\sqrt{S/I}$ so that the FoM, being the product of both, increases with $\sqrt{S}$. Hence, for the measurements of Fig. 17.24, the FoM increases by a factor of 1.44 due to the U_{dc} bias current of 2.5 V.

17.5.2 Parametric Amplification

To improve the performance of these sensors even further and implement adaptive filtering, we make use of non-resonant parametric amplification (PA). Parametric amplification is a mechanism based on modulation of one or more system parameters, in order to control the system's behavior. This leads to complex

---

[10] It was experimentally shown that also AC biasing, using frequencies significantly higher than the resonance frequency of the hair-sensor, can be used to obtain 'virtual DC-biasing' (Droogendijk et al. 2012a).

Fig. 17.24 Improvement of
the mechanical responsivity
and reduction of the
resonance frequency on DC-
bias induced ESS
(Droogendijk et al. 2012a)

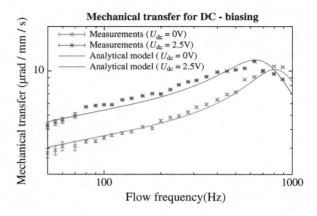

Fig. 17.25 Measured gain of
about 20 dB for the flow
frequency component at
150 Hz determined by FFT.
The AC-bias voltage is fixed
at $f_p = 150$ Hz with an
amplitude of 5 V

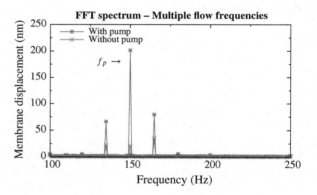

interactions between the modulating signals in which amplitude, frequency and
phase play important roles (Rugar and Grütter 1991). In this work, we obtain the
conditions for PA by changing the DC-bias voltage to an AC-bias voltage (also
called pump signal), which is another way of exploiting ESS.

Parametric amplification can give selective gain or attenuation, depending on
the pump frequency f_p and pump phase ϕ_p. Equal frequencies for flow and pump
($f_p = f_a$) give coherency in torque and spring softening, for which the pump phase
determines whether the system will show relative amplification or attenuation.
Therefore, it is possible to realise a very sharp band pass/stop filter, depending on
the pump settings.

Setting the frequency of the AC-bias voltage to 150 Hz, its amplitude to 5 V
and the pump phase to the value producing maximum gain, and supplying an
oscillating airflow consisting of three frequency components (135, 150 and
165 Hz), filtering and selective gain of the flow signal are demonstrated. See
Fig. 17.25. The presence of bias-signal through the action of non-resonant PA,
increases the frequency-matched signal by 220 dB, whereas the other two components are only amplified by 8–9 dB, resulting in selective gain of the flow signal
(Droogendijk et al. 2011).

Fig. 17.26 Improvement of the quality of the measured RMS-voltage values at low frequency signals using EMAM. In case of EMAM, a clear linear relationship between flow and output voltage is observed above the system's noise level (>5 mm/s)

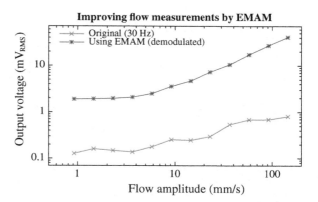

17.5.3 Electro-Mechanical Amplitude Modulation

We also implemented ESS by setting the AC-bias voltage frequency considerably higher than the frequency of the airflow. As a result, the system's spring-stiffness is electromechanically modulated, which results in Electro Mechanical Amplitude Modulation (EMAM). Experimentally, generating a harmonic flow at 30 Hz and setting the AC-bias voltage frequency to 300 Hz the flow is modulated and the flow information is upconverted to higher frequencies (Droogendijk et al. 2012b).

The incoming airflow signal is recovered by demodulation (using synchronous detection) of the measured rotational angle. Without EMAM, a noisy relationship between the flow amplitude and the resulting output voltage is observed. Also, large, undesired, fluctuations are observed (Fig. 17.26). However, with EMAM, a clear linear relationship) is observed for flow velocity amplitudes above 5 mm/s, showing that the measurement quality of low frequency flows too can be improved by ESS.

17.6 Summary and Conclusions

Crickets possess a sensitive, distributed hair-sensor system with near to thermal–mechanical noise-threshold sensitivities, which is an interesting example for sensory-system engineering. Engineers and biologists working together on this system have been able to make artificial hair-sensor systems and quantify the effects of viscosity mediated coupling. Interfacing arrays of sensors by means of FDM has delivered systems with simultaneous read-out of many sensors, which can be used as flow cameras. The capacitive structure for read-out doubles as a means for actuation of the hair-sensors and allows such exciting things as parametric amplification and filtering, adaptive-reversible sensor-modifications and electromechanical amplitude modulation (frequency shifting of signals). Future work will encompass studies on the use of stochastic resonance and application of our technology to other bio-inspired sensing modalities.

Despite all advancements in artificial hair-sensor systems the biological example is still far more complex, evolved and capable. The full three-dimensional shape of the cricket cerci, the large number of innervated hairs, the robust generation of neural signals and subsequent intricate processing in the TAG are still unattainable in current technology. And even if this were technologically possible, still much of the cricket flow-sensing system is unknown holding both challenges and promises for the future

Acknowledgements The authors would like to thank STW/NWO for funding this research in the framework of the Vici project BioEARS and the EU for funding the Cicada and Cilia projects. Contributions from T. Lammerink and R. Wiegerink have been invaluable. T. Steinmann, E. Berenschot, M. de Boer, R. Sanders and H. van Wolferen have given technical support without which this work would not have existed. Numerous students have contributed to this research, for which they are gratefully acknowledged.

References

van Baar JJ, Dijkstra M, Wiegerink RJ, Lammerink TSJ, Krijnen GJM (2003) Fabrication of arrays of artificial hairs for complex flow pattern recognition,. In: Proceedings of IEEE sensors 2003, Toronto, Canada, pp 332–336

Barth FG, Wastl JAHU, Halbritter J, Brittinger W (1995) Dynamics of arthropod filiform hairs. iii. flow patterns related to air movement detection in a spider (*Cupiennius salei*). Phil Trans R Soc B 347:397–412

Bathellier B, Barth F, Albert J, Humphrey J (2005) Viscosity-mediated motion coupling between pairs of trichobothria on the leg of the spider *Cupiennius salei*. J Comp Physiol A 191(733):746

Bathellier B, Steinmann T, Barth FG, Casas J (2012) Air motion sensing hairs of arthropods detect high frequencies at near-maximal mechanical efficiency. J R Soc Interface 9(71):1131–1143

Casas J, Steinmann T, Krijnen G (2010) Why do insects have such a high density of flow-sensing hairs? Insights from the hydromechanics of biomimetic mems sensors. J R Soc Interface 7(51):1487–1495

Dagamseh A, Bruinink C, Droogendijk H, Wiegerink R, Lammerink T, Krijnen G (2010) Engineering of biomimetic hair-flow sensor arrays dedicated to high-resolution flow field measurements. In: Proceedings of IEEE sensors 2010, Waikoloa, Hawaii, USA, pp 2251–2254

Dagamseh A, Lammerink T, Sanders R, Wiegerink R, Krijnen G (2011) Towards highresolution flow cameras made of artificial hair flow-sensors for flow pattern recognition. In: Proceedings of MEMS 2011, Cancun, Mexico, pp 648–651

Dagamseh AMK, Wiegerink RJ, Lammerink TSJ, Krijnen GJM (2012) Towards a high-resolution flow camera using artificial hair sensor arrays for flow pattern observations. Bioinsp Biomim 7(046):009

Dangles O, Ory N, Steinmann T, Christides JP, Casas J (2006) Spider's attack versus cricket's escape: velocity modes determine success. Anim Behav 72:603–610

Dangles O, Steinmann T, Pierre D, Vannier F, Casas J (2008) Relative contributions of organ shape and receptor arrangement to the design of crickets cercal system. J Comp Physiol A 194:653–663

Dijkstra MA, van Baar JJJ, Wiegerink RJ, Lammerink TSJ, de Boer JH, Krijnen GJM (2005) Artificial sensory hairs based on the flow sensitive receptor hairs of crickets. J Micromech Microeng 15:S132–S138

Droogendijk H, Bruinink CM, Sanders RGP, Krijnen GJM (2011) Non-resonant parametric amplification in biomimetic hair flow sensors: selective gain and tunable filtering. Appl Phys Lett 99(213):503

Droogendijk H, Bruinink CM, Sanders RGP, Dagamseh AMK, J R, Krijnen GJM (2012a) Improving the performance of biomimetic hair-flow sensors by electrostatic spring softening. J Micromech Microeng 22(6):065,026

Droogendijk H, Bruinink CM, Sanders RGP, Krijnen GJM (2012b) Application of electro mechanical stiffness modulation in biomimetic hair flow sensors. In: Proc. MEMS 2012, Paris, France, pp 531–534

Franosch J, Sichert A, Suttner M, Hemmen JV (2005) Estimating position and velocity of a submerged moving object by the clawed frog xenopus and by fish—a cybernetic approach. Biol Cybern. 93:231–238

Gnatzy W (1978) Development of the filiform hairs on the cerci of *Gryllus bimaculatus* deg. (saltatoria, gryllidae). Cell Tiss Res 187:1–24

Gnatzy W, Tautz J (1980) Ultrastructure and mechanical properties of an insect mechanorecep-tor: stimulus-transmitting structures and sensory apparatus of the cercal filiform hairs of cryllus. Cell Tissue Res 213(3):441–463 (URL:http://www.ncbi.nlm.nih.gov/pubmed/7448849)

Heys JJ, Rajaraman PK, Gedeon T, Miller JP (2012) A model of filiform hair distribution on the cricket cercus. PLoS ONE 7(10):e46588

Huang HC, Fu CC (2005) Out-of-plane polymer hollow micro needle array integrated on a micro fluidic chip. In: IEEE sensors 2005, Irvine, USA, pp 484–487

Humphrey J, Devarakonda R, Iglesias I, Barth F (1993) Dynamics of arthropod filiform hairs. i. mathematical modeling of the hair and air motions. Phil Trans: Bio Sci 340:423–444

Humphrey JA, Barth FG (2008) Medium flow-sensing hairs: biomechanics and models. Adv Insect Physiol 34:1–80

Jacobs GA, Miller JP, Aldworth Z (2008) Computational mechanisms of mechanosensory processing in the cricket. J Exp Biol 211:1819–1828

Jaganatharaja R, Bruinink C, Hagedoorn B, Kolster M, Lammerink T, Wiegerink R, Krijnen G (2009) Highly-sensitive, biomimetic hair sensor arrays for sensing low-frequency air flows. In: Proceedings of transducers 2009, Denver, USA, pp 1541–1544

Jaganatharaja R, Droogendijk H, Vats S, Hagedoorn B, Bruinink C, Krijnen G (2011) Unraveling the viscosity-mediated coupling effect in biomimetic hair sensor arrays. In: Proceedings of MEMS 2011, Cancun, Mexico, pp 652–655

Kant R, Humphrey JAC (2009) Response of cricket and spider motion-sensing hairs to air flow pulsations. J R Soc Interface 6(6):1047–1064

Kim K, Park DS, Lu HM, Che W, Kim K, Lee JB, Ahn C (2004) A tapered hollow metallic microneedle array using backside exposure of su-8. J Micromech Microeng 14:597–603

Krijnen G, Lammerink T, Wiegerink R, Casas J (2007) Cricket inspired flow-sensor arrays. In: Proceedings of IEEE sensors 2007, Atlanta, USA, pp 539–546

Lamb H (1910) The dynamical theory of sound. Edward Arnold, London

Lewin GC, Hallam J (2010) A computational fluid dynamics model of viscous coupling of hairs. J Comp Physiol A: Neuroethology Sens Neural Behav Physiol 196(6):385–395. DOI: 10.1007/s00359-010-0524-6 (URL: http://www.ncbi.nlm.nih.gov/pubmed/20383713)

McConney M, Schaber C, Julian M, Barth F, Tsukruk V (2007) Visco-elastic nanoscale properties of cuticle contribute to the high-pass properties of spider vibration receptor (*Cupiennius salei*). J R Soc Interface 4:1135–1143

Miller JP, Krueger S, Heys JJ, Gedeon T (2011) Quantitative characterization of the filiform mechanosensory hair array on the cricket cercus. PLoS ONE 6(11):e27873

Mulder-Rosi J, Cummins G, Miller J (2010) The cricket cercal system implements delay-line processing. J Neurophys 103(4):1823–1832

Rugar D, Grtitter P (1991) Mechanical parametric amplification and thermomechanical noise squeezing. Phys Rev Lett 67:699–702

Shimozawa T, Kanou M (1984) The aerodynamics and sensory physiology of range fractionation in the cercal filiform sensilla of the cricket *Gryllus bimaculatus*. J Comp Physiol A 155:495–505

Shimozawa T, Kumagai T, Baba Y (1998) The shape of wind-receptor hairs of cricket and cockroach. J Comp Physiol A 183:171–186

Shimozawa T, Murakami J, Kumagai T (2003) Cricket wind receptors: thermal noise for the highest sensitivity known. In: Barth FG, Humphrey JAC, Secomb TW (eds) Sensors and sensing in biology and engineering. Springer, Berlin, pp 145–159

Stokes GG (1851) On the effect of the internal friction of fluids on the motion of pendulums. Trans Cambr Phil Soc 9:1–141

Tautz J, Markl H (1978) Caterpillars detect flying wasps by hairs sensitive to airborne vibration. Behav Ecol Sociobiol 4:101–110

Yu H, Shibata K, Li B, Zhang X (2005) Fabrication of a hollow metallic microneedle array using scanning laser direct writing. In: Proceedings of miniaturized systems for chemistry and life sciences, Boston, USA, p 187

Chapter 18
Complex Flow Detection by Fast Processing of Sensory Hair Arrays

Christoph Brücker and Ulrich Rist

Abstract Hair-like sensor structures for detecting near-wall fluid motion can be found in nature as superficial mechanosensors in many organisms. Typically, they appear in characteristic array arrangements. It is suggested that multisensor detection plays a role in the signal filtering. The signal output is apparently influenced by the arrangement of the hairs, which may allow to distinguish complex patterns in their spatiotemporal signature. This is probably important in nature where the multisignal sensor output needs to be selective enough to trigger specific tasks such as flow control or escape behavior under the influence of turbulent background noise. To explore these effects, we investigated the response of arrays of artificial hair sensors to turbulent boundary layer flows with stochastic appearance of coherent flow structures near the wall. Imaging of these structures and online processing using FPDA-technology enables to use the spatiotemporal signature of the signals for flow control in engineering applications. In addition, the arrays are used to test hypotheses of signal processing and masking in turbulent flows in natural sensory systems.

18.1 Introduction

Advances in the manufacture of micromechanical structures allow today to produce new surfaces including arrays of micromechanical sensors, such as the flexible "micro-pillars" reported in Brücker et al. (2005, 2007). These surfaces can not only detect individually the local near-wall flow but in principle those arrays may be able to detect the time-varying topology of the flow structure close to the wall. One

C. Brücker (✉)
Institut für Mechanik und Fluiddynamik, TU Bergakademie, Freiberg, Germany
e-mail: Christoph.Bruecker@imfd.tu-freiberg.de

U. Rist
Institut für Aerodynamik und Gasdynamik, Universität Stuttgart, Stuttgart, Germany

H. Bleckmann et al. (eds.), *Flow Sensing in Air and Water*,
DOI: 10.1007/978-3-642-41446-6_18, © Springer-Verlag Berlin Heidelberg 2014

Fig. 18.1 Typical flow pattern in the near-wall region of a transitional boundary layer flow over a small roughness step (backward facing step, $h = 15y^+$) with local separation showing the complexity of spatiotemporal flow signature. The image is taken from a high-speed recording of an array of small flexible micropillar sensors for wall-shear stress imaging on a flat plate

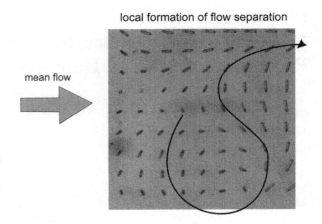

local formation of flow separation

mean flow

major advantage of the micropillar sensory structures compared to others is that they detect not only fluctuations in the mean flow direction but they provide the full information of the local 2D velocity vector tangential to a surface in strength and direction. Therefore, these sensory structures can detect also fluctuations in spanwise direction. This could be of great importance for flow control applications in turbulent boundary layers or to detect flow separation. A characteristic picture of a footprint of flow separation behind a small step in a boundary layer flow is given in Fig. 18.1 which shows the complexity of the pattern.

The flow over such a small step is interesting to study with respect to signal masking. Local flow is influenced by the spatiotemporal signatures imposed by the coherent vortex structures from the transitional boundary layer flow overflowing the step. In addition, flow structures are generated at the step itself. In boundary layer flows with adverse pressure gradient, formation of cells of backflow in this region may be a precursor for the generation of larger flow separation bubbles. It is obvious from Fig. 18.1 that the spanwise component of the hair bending is of the same order as the streamwise component. Therefore, formation of cells can be more easily detected by the twist of the near-wall velocity vectors rather than from their streamwise bending alone. For practical application, it is therefore of interest to detect such flow patterns early in the process of the generation cycle of flow separation to undertake preventive action by active flow control tools such as blowing, suction, etc.

The importance of hair-like sensory structures for flow control was found in bat flight. It was shown (Zook 2005) that removing those hairy filaments from the bat wings changed drastically the patterns and speed of flight. Recently, it was proven that the role of those hairs is related to the detection of local reverse flow conditions (Sterbing-D' Angelo et al. 2008). It has been claimed that sensing incipient stall conditions would allow the bird to modify the flying modality to prevent it. Research on the subject including its eventual application in aeronautics is actually undertaken by various groups (Brücker et al. 2007; Dickinson et al. 2012). It is argued that a fast and reliable decision on incipient flow separation must be based on more than single sensor information. In addition, filtering and selectivity to specific wavelength must be involved.

For a single isolated hair in a laminar flow, the interaction with the flow is already well understood and the signal is well defined. However, for groups of hairs, the interactions with the flow and with each other may lead to a different response, where little research has been done so far. Sensory hair structures occur in nature mostly in characteristic arrangements. That rises the question of what makes a grid of sensor hair better than a single sensor to process the signals and what type of multi-sensor processing is used to reduce noise or decision uncertainty (Casas et al. 2010). On a tandem pair of sensory hairs it was recently shown that the comparative and correlative processing of the signals can suppress the highly nonlinear response of a single hair completely (Chagnaud et al. 2008). In addition, it could be demonstrated, that the actual signal which is transferred is the convective velocity of near-by passing vortex structures rather than the local flow velocity. Recently, Mulder-Rosi et al. (2010) have demonstrated that an intelligent arrangement of the sensor structures with respect to the flow direction can act as a band-pass filter effect. This illustrates the immense potential in the use of multiple sensors for the signal enhancement and decision finding.

Besides the effects on boundary layer separation and stall, the action of hairy coatings on turbulent flow control is also considered by Brücker (2011). Flexible hairs of appropriate dimensions are shown to achieve strong reconfiguration in streamwise direction with their long cylindrical trailing bodies placed in the buffer layer, and lead to a reduction of skin friction via energy transfers and increased coherence in streamwise and spanwise directions.

18.2 Sensors and Experiments

The artificial sensory hairs we used were flexible micropillars made from an elastomeric material. They were fixed at the base on a thin membrane. The membrane can be positioned on a flat plate or over an airfoil to study the flow in situ on the wing. The flow studies were carried out in a low-turbulence oil flow channel which was designed for high-resolution optical measurements of turbulent boundary layer flows, Fig. 18.2. A thin flat plate (total length of 3 m) with a sharp trailing edge was placed in the middle of the test section. A tripping wire (diameter 1.5 mm) was fixed on the plate 0.15 m downstream of the trailing edge to trigger a transition scenario based on the evolution of Tollmien–Schlichting waves (TS).

We focus our studies on a region located 1.8 m downstream of the trailing edge where the boundary layer is in an early turbulent state. The main flow velocity is 5 m/s, oil kinematic viscosity is 12×10^{-6} m²/s at 22 °C. Temperature is controlled with a cooling system. The Reynolds number was Re $= 7.5 \times 10^5$ and the viscous length scale was about 70 μm. An additional source of disturbances is induced by means of a backward facing step in the wall located at x=1.8m, see Fig. 18.3.

Fig. 18.2 Flow channel for flow studies of a transitional boundary layer flow along a flat plate

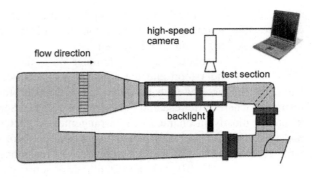

Fig. 18.3 Complex signal detection of near-wall flow is tested using arrays of artificial sensory hairs in form of flexible micropillars mounted on the flat plate downstream of a small backward facing step

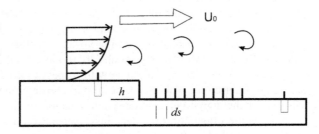

The step height is $h^+ = 15$ wall units. The microhair sensor foil consisted of 60×60 flexible micro-pillars (diameter D $\approx$ 35 µm, height L $\approx$ 450 µm) with a regular spacing of 500 µm in streamwise and 500 µm in spanwise direction. These hairs are used as sensory micropillars to measure the wall shear stress distribution (see Brücker 2008; Brücker et al. 2005). The microhairs have a uniform circular cross-section representing homogeneous circular beams with a planar base and tip. They are made of polydimethylsiloxane (PDMS). A digital high-speed camera (Photron APX-RS, recording rate 3,000 fps) equipped with a long-range microscope ($M = 1$) is used to record the hair motion from top. A typical snapshot picture of the sensory hairs in motion from top and from side view is given in Fig. 18.4. Additional experiments were carried out in a turbulent wall jet flow (Skupsch et al. 2012).

18.3 Detection and Processing

Instead of using mechanoelectric, piezoelectric, or magnetoelectric transducer methods, we use the conversion of the bending signal into a sensory information by imaging the bending of the structures with a high-speed camera, which usually allows to record multiple sensors of the order of $O(10^2)$ simultaneously. As demonstrated with the tandem-sensor analysis by Chagnaud et al. (2008), the signal of near-wall flow structures and their convective velocity is not necessarily

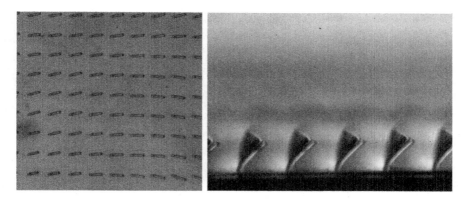

Fig. 18.4 Exemplary pictures of the sensory hairs in motion, taken by a high-speed camera. *Left top* view, *right side* view (mean flow is from *left* to *right*)

provided by the bending of the individual sensor hair. Instead, the convection velocity is obtained by the cross-correlation of natural disturbance patterns passing the sensory structures at different times. Therefore, even a strong nonlinearity in the response behavior of the individual sensory hair does not hamper the actual information detection. This important finding opens a new interpretation of multisensor processing where not the individual sensor signal but the spatio-temporal structure is the major information to be detected. Therefore, it is not necessary to design the individual sensor for optimum response or linear behavior, when characteristic spatiotemporal flow patterns are of interest. In light of this conclusion, we have chosen two different artificial hair sensor configurations, one in which the sensory hairs are soft and bend easily with the flow and the other one with rather stiff hairs. The latter show a linear response over a large bandwidth and act as wall-shear stress sensors. The former are comparable to elastic microtufts with a nonlinear response due to strong reconfiguration of the structures. The detection of characteristic flow patterns is different for both and requires different strategies of image processing.

18.3.1 Sensory Hairs with High Flexibility (Larger Bending)

The image shown in Figs. 18.1 and 18.5 is easy to be interpreted in light of characteristic flow pattern of near-wall flow. It obeys the style of a nice flow visualization picture with a well-defined disturbance pattern in contrast to the unidirectional mean flow shown in Fig. 18.4. This disturbance pattern has a characteristic streamwise and spanwise scale which is of order of the size of the observation window. The observer is intuitively able to detect a "S"-type structure of swirl flow pattern near the wall.

Fig. 18.5 Fast vectorization
using standard image
processing on segmented
regions of interest ROI's,
with background subtraction,
binarization, closing, and
vectorization

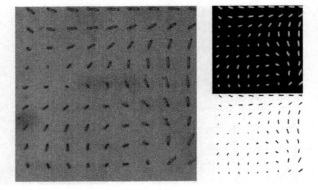

Because of the high quality of image capture and lighting conditions, the hairs
are easy to see at high contrast to the background. This offers realizing a simple
image binarization method which can be run online. In addition, since the sensory
hairs are arranged in form of a regular 2D grid, processing is only done in a
number of squared ROI's (ROI = region of interest) centered at the base of the
hairs. The result of binarization and downsampling is shown in Fig. 18.5 in the
right column on top. Thereafter, a closing procedure is applied and finally, vectorization is achieved by processing of orientation and length of the bright stripes.
Since all procedures are standard in image processing and can be implemented
nowadays in current hardware technology using field programmable gate arrays
(FPGA), one can achieve the spatiotemporal signal detection online while
recording. The scale of the swirling flow fluctuations is much larger than the step
height; it is, therefore, argued that it stems from coherent flow structures running
with the mean flow over the step.

18.3.2 Sensory Hairs with High Stiffness (Smaller Bending)

When the flow forces become relatively small in comparison to the elastic
restoring forces, the hairs bend less than the order of their own diameter. Imaging
of the small bending amplitudes of the sensory hairs under load is then difficult to
be detected optically for a larger array of sensory hairs and leads to noisy results
because of limited spatial resolution. On the other hand, vectorization is reduced to
the simple algorithm of peak fitting at the tip of the micropillars. In order to
achieve high spatial resolution and fast vectorization of the sensory array, we
developed a segmented imaging principle using microlenses (Skupsch et al. 2012).
Therein, each sensory hair is magnified with a microlens in the array and all
segments are imaged together on the sensor chip. This ensures a high local
magnification at each sensory hair, see Fig. 18.6.

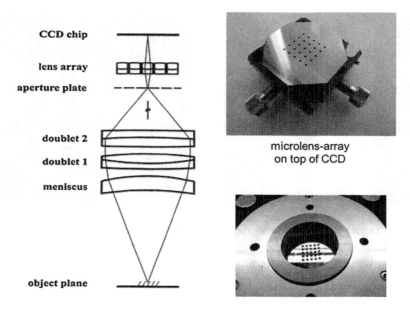

CCD chip

lens array

aperture plate

doublet 2

doublet 1

meniscus

object plane

microlens-array
on top of CCD

Fig. 18.6 Principle of segmented imaging using a microlens array. The arrangement of the lenses is adapted to the arrangement of the micropillars. Manufacturing of the optics is made in-house (Skupsch et al. 2012)

The resulting images of the sensory hairs are displayed in Fig. 18.7 at rest and under typical flow conditions for a turbulent wall jet flow. Image processing includes the procedures of binarization, ROI segmentation, centroid detection, and finally a peak fitting at the centroid position in the original image. Again, programming on FPGA enables online processing of the bending information at all sensory hairs.

18.3.3 Spatiotemporal Signal Analysis Using POD

The analysis of events with characteristic spatiotemporal signature of the near-wall flow structures is done with the proper orthogonal decomposition (POD, see e.g., the review of Berkooz et al. 1993). The flow is composed as the weighted sum of fundamental modes, where the weighting factors vary in time. The fundamental modes represent flow structures with high energy of fluctuation. They are determined by solving an eigenvalue problem in the POD.

The eigenvalues correspond to the fraction of kinetic fluctuation energy in the mode. Figure 18.8 shows the first three modes in a turbulent boundary layer flow over a micropillar array as in Fig. 18.7. The modes contain 73.7, 17.2, and 4.2 %

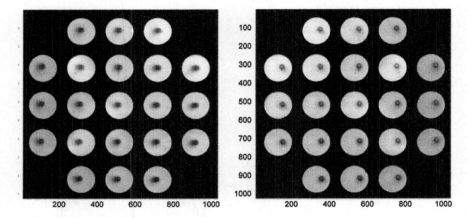

Fig. 18.7 Images of artificial sensory hairs using segmented imaging optics. Picture at rest and under full flow conditions (Skupsch et al. 2012)

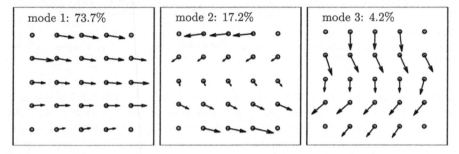

Fig. 18.8 Exemplary results of first three POD modes of turbulent wall-bounded flow over a micropillar array (Skupsch et al. 2012)

of the total fluctuating kinetic energy, respectively. The magnitude of deflection is coded in the arrow length and direction in the angle as a typical vector plot. Again, the "S"-type structure is seen in mode 3, which contains 5 % of the total energy.

Events with complex flow structures can be seen in the time-series when the amplitudes of the second and third modes become similar to the primary mode. Figure 18.9 shows the time-series of the POD coefficients for the results of Fig. 18.8. It is seen that such events appear during the period of 1,300–1,400 or 1,800–1,900 time units. Similar data can be obtained from numerical flow simulation where POD has been used to detect characteristic events in a transitional boundary layer flow with and without flow control (Gunes and Rist 2004).

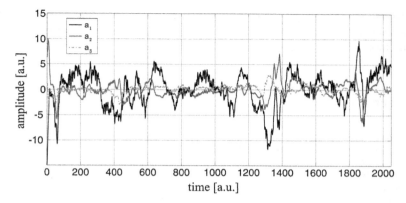

Fig. 18.9 Exemplary results of the time-series of amplitude of the first three POD modes in a turbulent wall-bounded flow (Skupsch et al. 2012)

18.4 Conclusions

Near-wall flow sensory systems with superficial neuromasts or wind hairs in nature are able to detect characteristic spatiotemporal flow signature patterns and distinguish those from background noise, for example in turbulent environments. First hints on the function of signal transformation of two or more sensory hairs using delay line or correlation-based information for complex pattern recognition has been shown for tandem pairs of sensory hairs in laboratory experiments (Mulder-Rosi et al. 2010; Chagnaud et al. 2008). Their analysis is limited to unidirectional flows, no further effect of 2D or 3D flows around the sensory hairs is investigated so far. Herein, we used artificial sensory hairs which respond to the main flow direction as well to lateral or spanwise flow disturbances. This gives a much richer picture of the flow pattern moving across the sensory array. In principle, such arrays are able to detect the complex spatio-temporal flow pattern along a 3D surface of a body in flow. However, there is still a large gap in physical understanding of the nature of the pattern-based signal acquisition and processing. A better understanding of these processes opens considerable potential for the efficient application of such sensory arrays in technical applications, e.g., for flow control.

Imaging of arrays of sensory hairs has been proven as a valuable tool for fast and reliable flow pattern detection in laboratory experiments and has the potential to be implemented in real-scale flow situation such as the flow over an airfoil using transparent windows and subcutaneous sensor imaging. With the help of fast online processing using high-speed FPGA, pattern recognition can be done online for complex event detection, see also Woods et al. (2010). In this way, a control circuit for flow control can be implemented. The most promising approach for signal separation is the use of POD and comparison of strength of different modes which can be trained a priori on characteristic events, where flow control is necessary. Further improvement of the sensory arrays with respect of hair arrangement, orientation, and mutual interaction may ease the processing step furthermore. This is subject of future work in our group.

References

Berkooz G, Holmes P, Lumley JL (1993) The proper orthogonal decomposition in the analysis of turbulent flows. Ann Rev Fluid Mech 25:539–575

Brücker C (2008) Signature of varicose wave packets in the viscous sublayer. Phys Fluids 20:061701

Brücker C (2011) Interaction of flexible surface hairs with near-wall turbulence. J Phys: Condens Matter 23:184120. doi:10.1088/0953-8984/23/18/184120

Brücker C, Spatz J, Schröder W (2005) Feasability study of wall shear stress imaging using microstructured surfaces with flexible micropillars. Exp Fluids 39:464–474

Brücker C, Bauer D, Chaves H (2007) Dynamic response of micro-pillar sensors measuring fluctuating wall-shear-stress. Exp Fluids 42:737–749

Casas J, Steinmann T, Krjinen G (2010) Why do insects have such a high density of flow-sensing hairs? Insights from the hydromechanics of biomimetic MEMS sensors. J Royel Soc Interface 7:1487–1495

Chagnaud BP, Brücker C, Hofmann MH, Bleckmann H (2008) Measuring flow velocity and flow direction by spatial and temporal analysis of flow fluctuations. J Neurosci 28:1–15

Dickinson B, Singler J, Batten B (2012) Mathematical modeling and simulation of biologically inspired hair receptor arrays in laminar unsteady flow separation. J Fluids Struct 29:1–17. doi:10.1016/j.jfluidstructs.2011.12.010

Gunes H, Rist U (2004) Proper orthogonal decomposition reconstruction of a transitional boundary layer with and without control. Phys Fluids 16:2763–2784

Mulder-Rosi J, Graham I, Cummins GI, Miller JP (2010) The cricket cercal system implements delay-line processing. J Neurophysiol 103:1823–1832. doi:10.1152/jn.00875.2009

Skupsch C, Klotz T, Chaves H, Brücker C (2012) Channelling optics for high quality imaging of sensory hair. Rev Sci Instrum 83:045001. doi:10.1063/1.3697997

Sterbing-D' Angelo S, Chadha M, Moss C (2008) Representation of the wing membrane in somatosensory cortex of the bat. Eptesicus fuscus. In: Neuroscience Abstracts 370.4

Woods L, Teubner J, Alonso G (2010) Complex event detection at wire speed with FPGAs. In: Proceedings of 36th international conference very large data bases, vol 3(1), Singapore

Zook J (2005) The neuroethology of touch in bats: cutaneous receptors of the wing. In: Neuroscience Abstracts 78.21

Chapter 19
Stress-Driven Artificial Hair Cell for Flow Sensing

Francesco Rizzi, Antonio Qualtieri, Lily D. Chambers,
Gianmichele Epifani, William M. Megill and M. De Vittorio

Abstract Bio-inspiration from natural structures and systems can be used to design innovative engineered solutions. Here natural sensor architectures inspire the design of micro-electronic-mechanical systems (MEMS) for flow sensing. In this chapter, we introduce an innovative approach to artificial flow sensing based on mimicking stereocilia and their mechanical properties. This method exploits the intrinsic differences in material properties of multilayered thin films such as thermal expansion properties, crystalline lattice order and interatomic distances. If a cantilever beam is multilayered, these properties create a stress gradient along the cantilever cross section, allowing an upwards bending, defined as 'stress-driven geometry'. When inserted in a superficial fluid stream, the cantilever beam is deformed by the flow and acts as a fluid flow velocity sensor. It is shown that a Parylene post-processing conformal coating not only waterproofs the device, but also sets the flexural stiffness of the beam, thus tuning the dynamic range for flow measurements optimisation.

Keywords Artificial hair cells · Biomimetics · Bioengineering · Mechano-sensors · Cilium · Piezoresistivity · Parylene conformal coating encapsulation · Stress-driven cantilevers · Strain gauge · Micro-fabrication · Silicon · Silicon Nitride · Multi-layered cantilever beam

F. Rizzi (✉) · A. Qualtieri · M. De Vittorio
Center for Biomolecular Nanotechnologies@UniLe, Istituto Italiano di Tecnologia,
Via Barsanti sn 73010 Arnesano, Lecce, Italy
e-mail: francesco.rizzi@iit.it

L. D. Chambers · W. M. Megill
Ocean Technologies Laboratory, Mechanical Engineering Department,
University of Bath, Bath BA2 7AY, UK

G. Epifani · M. De Vittorio
National Nanotechnology Laboratory, Istituto Nanoscienze-CNR,
73100 Arnesano, Lecce, Italy

M. De Vittorio
Dip. di Ingegneria dell'Innovazione, Università del Salento, 73100 Arnesano, Lecce, Italy

H. Bleckmann et al. (eds.), *Flow Sensing in Air and Water*,
DOI: 10.1007/978-3-642-41446-6_19, © Springer-Verlag Berlin Heidelberg 2014

19.1 Introduction

Mechanoreceptors enable vertebrates to sense movement in the fluid that surrounds their bodies. These receptors are often in the form of hair cells, on or near the surface of the skin. The fluid is often the air or water that surrounds the animal, but can also be the fluid in the internal cavities such as the inner ear. The mechanical signal created by the relative motion of the body and fluid is converted to a neuronal one through the action of the hair cells.

Hair cell mechanoreceptors consist of a cell body and a set of hair-like structures which extend into the fluid. The cells respond to mechanical deflections of these hair-like appendages by producing neuro-electrical pulses, which are transmitted to the animal's nervous system. As in all sensors, the sensitivity of these cells has to be tuned to the magnitude of the incoming signal of interest. Recent studies of fish flow sensors have shown that this sensitivity is controlled by morphology and material properties of the natural mechanoreceptors, which ultimately determine their flexural stiffness (McHenry and van Netten 2007). This has also been shown for hair cells in the human inner ear, where the stiffness of the hair bundle can be adjusted in order to reset sensitivity. As a consequence, the hair cell responds to novel stimuli even if the hair bundle is deflected, i.e. sensitivity is shifted in an adaptation process to prevent saturation (Howard and Hudspeth 1987; Hudspeth 2008).

The evolutionary pressure to adapt to survive in a wide variety of environments has led to a wide variation in sensor morphology and sensitivity. Roboticists designing their machines to operate in similar environments have correspondingly similar sensor requirements, so it makes good sense to look to biological systems for inspiration (Bleckmann et al. 2004; Fratzl and Barth 2009). In this chapter, we describe a new type of biomimetic artificial hair cell (AHC) and a manufacturing process which allows the mechanical properties of the sensor to be tuned to the design requirements.

19.1.1 Fish Flow Sensors

Fish possess a unique organ called the lateral line, which consists of an array of flow and pressure gradient sensors distributed over the surface of their bodies. Each sensor unit, or neuromast, contains bundles of hair cells embedded within a cupula (Fig. 19.1a). The cupula is a gelatinous matrix and aids in the transfer of the mechanical stimuli to the hair cell membranes (Kuiper 1956; Flock and Wersall 1962; Coombs 2001). The hair cells contain a kinocilium and a group of stereovilli. Individual hair cells within a neuromast have an axis of best sensitivity which points from the smallest stereovillum to the kinocilium. The lateral line system can take advantage of this directionality by orientating the hair cells axis of best sensitivity parallel to the oncoming flow signal (Schmitz et al. 2008), which for bulk water flow means parallel to the rostro-caudal axis of the fish.

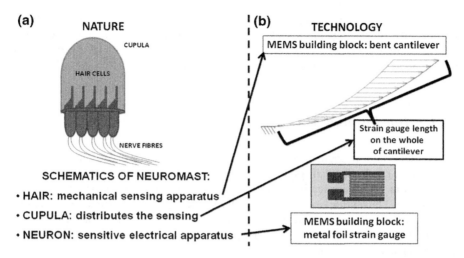

Fig. 19.1 A comparison between the neuromast and the artificial flow sensor. **a** The schematics of a natural hair cell where its single sensing systems are shown. **b** Stress-driven artificial neuromast with its building blocks mimicking the neuromast features indicated in (**a**)

The mechanoelectrical transduction by hair cells is done via ion channels (Hudspeth 1989). Tiplinks tether the stereovilli and the kinocilium together. It takes energy to overcome the tension in these tips and open the ion channels. In the hair cell, this energy takes the form of mechanical work in the displacement of the hair bundle by fluid hydrodynamics. The hair cell membranes contain sodium and potassium ion channels which control the electrochemical response of the nerve cell. The dynamic non-linearity of the cupula can be attributed to the opening and closing of the ion channels of the hair cell by the action of the tip links (van Netten and Khanna 1994).

The transduction of the mechanical work into a nerve pulse also depends on the structure of the neuromast. Using Atomic Force Microscopy (AFM), the flexural stiffness of a neuromast was correlated with the cupula material properties and the number of hair cells embedded within it (McHenry and van Netten 2007). In research by McHenry et al. (2008) the kinocilium and cupula height were modelled as a two-part bending beam, to represent their respective contributions to the overall flexural stiffness of the surface neuromast. The proximal beam represented the behavior of the kinocilium, while the more flexible distal tip represented that of the cupula. This successful model related the theoretical role of neuromast stiffness and height for sensitivity to a range of stimulus (McHenry et al. 2008).

The ability of this sensory structure to adapt to the needs of the animal within a particular environment is highlighted by the research of Teyke (1990), which by investigates the morphological difference in the lateral line of the sighted (*Astyanax mexicanus*) over the blind cave fish (*Astyanax hubbis*). The lateral line neuromasts of the blind cavefish are double the length, leading to the hypothesis that evolutionary adaptation of the sensory system increased the animals' flow sensitivity, to accommodate for the loss of vision. In behavioral studies the length of the neuromast on larvae of the willow shiner *Gnathopogan elongatus caerulescens* is linked to their success in capturing prey. Prey capture increased with increased regrowth of the cupula (Mukai et al. 1994).

19.1.2 Bio-Inspired Micro-Electro Mechanical Systems

Micro-electronic-mechanical systems (MEMS) have attracted particular attention due to their small physical footprint, responsivity, dynamic range and bandwidth. Advances in micro-fabrication have allowed the development of several approaches to flow sensing devices based on different technological principles: pressure distribution (Kalvesten et al. 1996), torque transfer (Svedin et al. 1998) and thermal transfer (Yang et al. 2006). Recently, bio-inspired engineering has made strong advances in AHC sensing devices designed to exploit micro-electro mechanical system (MEMS) technology. MEMS engineers are following biology's lead to design new miniaturised sensors and actuators which can, like natural sense organs, similarly cope with very large changes in the environment. Cilia-inspired MEMS technology has been under dynamic development for the past decade; Zhou and Liu (2008) covered this comprehensively in a review where the different devices are categorised as micro-actuators as well as micro-sensors.

All of the bio-inspired artificial mechanoreceptors in the past decade (Fan et al. 2002; Dijkstra et al. 2005; Krijnen et al. 2006; Xue et al. 2007; Yang et al. 2010; Qualtieri et al. 2011) are based on a strain gauge measurement principle. When a flow impinges on the sensor, a mechanical element is bent, inducing a deformation on a strain gauge placed at the maximum deformed position. The most extensively investigated structure is based on a vertical cilium: an SU8 vertical pillar is placed perpendicular to the flow, on the free end of an in-plane fixed-free cantilever (Fan et al. 2002; Yang et al. 2010), in the middle of a four-beam microstructure (Xue et al. 2007), or on a Si_xN_y suspended membrane (Dijkstra et al. 2005; Krijnen et al. 2006). In the latter case, the sensor relies on capacitive detection of the membrane torque while in the former examples, the mechanical bending moment is transferred from the cilium to the horizontal cantilever beam, which bends in response, and the signal is measured using a piezoresistance placed at the location of the greatest strain. In both types of sensors the fluidic moment is transferred with high efficiency to the sensing mechanism; high responsivities are obtained, making this approach very attractive. However, because the cilium in these sensors is rigid and the strain concentration at its base is very high, these kinds of sensors are liable to fracture upon mechanical overloading of the vertical cilium.

In this chapter a new bio-inspired approach will be introduced. It is a stress-driven flow cantilever, which also reaches perpendicularly into a flow stream, but which does not make use of the rigid vertical pillar, instead consisting of a continuous bent beam which flattens or curls to generate a signal related to the fluid flow speed, in a manner not unlike that of the kinocilium of a hair cell. Moreover, the mechanical properties of this cantilever can be altered after manufacture so as to tune the responsivity of the flow sensor to the dynamic range required. This post-processing method can be exploited to optimise the stress-driven flow sensor performance.

19.2 Stress-Driven Out-of-Plane Cantilever Technology

Here we describe the design of a stress-driven flow sensor which is based on an out-of-plane bent cantilever projecting into a flow stream. First, the design and technological principle will be introduced. Secondly, a Parylene-coated Silicon Nitride/Silicon cantilever beam, suitable for underwater operations, will be described and characterised in water.

19.2.1 The Stress-Driven Approach to Flow Sensing

The sensor consists of an out-of-plane bent cantilever, obtained by releasing a long narrow beam from the substrate by upward stress. The stress-driven cantilever maintains a curved equilibrium position at a fixed height with respect to the substrate. If the cantilever is deflected along its length, it bends in response, either rolling up (increasing its curvature), or flattening out (decreasing it). Notably, the stress-driven cantilever distributes the load applied by the fluid along the whole length of the beam instead of concentrating the torque near the fixed end (*hinge*). A stress-driven cantilever in a fluid flow streaming from left to right at a fixed velocity is shown in Fig. 19.1b. The vector lengths describe the load profile along the beam. This approach shows a high robustness to deformation and a fast response in real time to external stimuli.

A micro strain-gauge sensor, affixed to the cantilever, senses the deformation caused by an applied force. The positive or negative slope of the resistance change allows the sensor to distinguish the direction of the applied force. In a neuromast, the distal tip of the cupula extends beyond the graded height of the stereovilli and kinocilium both to reach through the boundary layer, and to distribute the signal over more sensing elements and thereby increase the sensitivity. In our artificial configuration, strain is distributed along the whole length of the cantilever, rather than concentrated at the *hinge*. For this reason we extended the strain gauge over the full length of the cantilever. This geometry provides the sensor with a working principle which is close to the natural hair cell apparatus (Fig. 19.1b). The increased strain gauge length enhances its resistance and, therefore, increases thermo-electrical noise background, coming from the following sources: Johnson noise, shot noise and $1/f$ noise (Liu 2012). Optimizing the Signal-to-Noise Ratio in piezoresistive cantilevers is a complex procedure, related to the sensor design and processing (Yu et al. 2002). On the other hand, in a motion-based MEMS sensor (like a flow sensor), vibration of micro structures is a fundamental source of noise in addition to electrical ones. In our experimental set-up, the largest noise generation comes from micro-turbulences in the flow.

The MEMS architecture mimics the single hair cell in its mechano-electrical characteristics and in its shape. As a consequence of this bio-inspired engineering approach to sensor design, the following characteristics have been obtained:

- the sensor performs a direct measurement of fluid flow velocity through the strain gauge deformation;
- it performs bidirectional flow measurements thanks to two bending orientations (flatten out or curl up);
- it easily adapts to flow variations due to its deformability in all points along the cantilever beam;
- the "distributed load" design assures robustness of the beam up to 0.50 m s^{-1}.

The technological principle behind this design consists of embedding and overlaying thin films of different elastic and thermal properties, taking advantage of the fact that differences in thin film material properties create internal mechanical stresses. For example, the mismatch of atomic sizes between different layered materials or lattice-mismatch developed in the thermal cycling of the chip during material deposition and micro-fabrication are the main causes of intrinsic stress (Hu 1991). This generates deformation among MEMS structural elements and it is usually considered detrimental. In contrast, in a stress-driven design approach, this phenomenon is exploited by the MEMS engineers as an intrinsic release mechanism for the out-of-plane bending of the cantilever. As shown in Fig. 19.2, once the cantilever beam length is set, the stress must be high enough to bend the cantilever to a tip height that can reach through the flow boundary layer over the sensor surface (McHenry and van Netten 2007).

First attempts at this approach have used single material layer structural beams such as chemical vapor deposition (CVD) silicon nitride (Wang et al. 2007), polycrystalline silicon (Kao et al. 2007) and silicon dioxide (Zhang et al. 2010), that by virtue of their growth processes, exhibit an intrinsic stress difference along their thickness. The intrinsic stress that can be generated is essentially inherent to the growth technology and is difficult to control: the tip height reached by the cantilever beam can be only designed by changing the cantilever length. This characteristic is critical, especially when the application calls for very high tip height (hundreds of micrometers). Moreover, the above examples can be unstable, are not waterproof and have limited dynamic ranges, unsuited for high fluxes.

To deal with these challenges, a new design for stress-driven flow sensors has been recently developed, based on multilayered cantilevers (Qualtieri et al. 2011, 2012). The upwards bending is obtained by carefully engineering the material stress difference among each multilayer component which results from the difference among internal stresses in CVD-based growth or from the crystalline lattice difference among each component layer of the cantilever beam in epitaxial growth. All the beam components are grown on a sacrificial layer. Then, once the U-shaped region defining the cantilever beam is etched inside the multilayered grown material

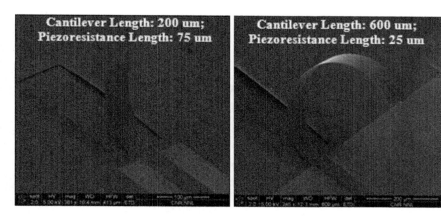

Fig. 19.2 The flow sensor tip height is dependent on geometry. The picture shows two examples with different cantilever beam length

and the sacrificial layer underneath is etched away, the unbalanced stresses relax, causing the cantilever beam to bend upward in a curved shape. This tensile bending moment can be specifically designed: once the multilayer materials are set, different growth and micro-fabrication parameters can be selected to design suitable characteristics into the sensor, such as cantilever curvature and height. The degree of bending of cantilevers can be controlled by the lithographic patterning dimensions (beam length, width and layer thicknesses) of the unreleased cantilever beam, by the sacrificial layer thickness and by growth parameters which include substrate temperature, deposition pressure, flows of gases involved in the deposition process and radio frequency-electric field power bias applied to the substrate.

19.2.2 Parylene-Coated Stress-Driven SiN/Si Cantilever-Based Flow Sensor

The eventual goal is to mass-produce these sensors so that large arrays can be developed. Therefore, it was important to select materials and processes that were compatible with current commercial technologies. To facilitate this, the micro-fabrication of a device using silicon-on-insulator (SOI) substrates was chosen. This is a mature technology, suitable for the transfer of the sensor fabrication to an industrial mass production platform.

The SOI substrate fabrication processing involves: a bulk support wafer (handle) silicon layer (thickness: 400 μm) coated on one side with a thermal SiO_2 insulating layer (thickness: 2 μm). Another silicon wafer was bonded on the SiO_2 coated side of the substrate and machined to a thickness of 2 μm (thickness: 2 μm, device layer). The whole SOI substrate was coated with a 300 nm thick stoichiometric silicon nitride layer by low pressure chemical vapour deposition (LPCVD) on both

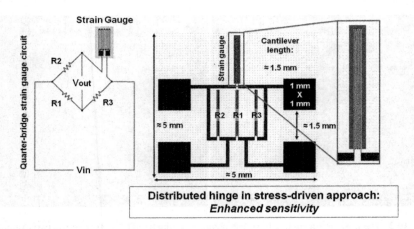

Fig. 19.3 Wheatstone bridge-equipped strain gauge. *Left* "quarter-bridge" circuitry design; *right* integration of the strain gauge on the flow sensor design. The enlargement highlights the strain gauge extension along the whole cantilever beam, exploiting the distributed *hinge* design

sides. This coating behaves as an electrical insulating layer and due to the thermal cycling connected with its deposition has an intrinsic 800 MPa residual stress. This silicon nitride layer behaves as the stressor with respect to the silicon device layer, inducing a stress difference throughout the thickness of the multilayered cantilever. This intrinsic stress, measured by the production company (Si-Mat Silicon Materials, Germany), in combination with a beam length of 1,500 μm and a beam width of 100 μm, is enough to allow the tip of the released cantilever beam to reach a height of ≈ 1,200 μm. This height extends the tip beyond the flow boundary layer and into the free stream over the sensor surface. The multilayered cantilever beam consists of two layers: the 2 μm thick silicon structural layer (device layer) and the 300 nm thick silicon nitride (insulating and stressor Layer).

To integrate the sensor with external circuitry, a Wheatstone-bridge circuit was directly incorporated into the sensor. Figure 19.3 shows the circuit diagram and its realization in the single sensor.

There are four identical resistances with one along the cantilever beam acting as the deformable strain gauge, while the other three remain flat on the substrate. This is the typical design of a "quarter-bridge" strain gauge circuit and measures the direct proportionality between the voltage signal read-out and the resistance variation with the strain. The resistors consisted of eight periods of a continuous grid of narrow wires (4 μm diameter). The strain gauge was distributed along the cantilever beam, with seven periods and a pitch of 12 μm, and was designed to take advantage of the changing curvature design, and assure an enhanced sensitivity with the flow. Figure 19.3 shows the dimensions of the sensor circuit: the single sensor dimensions are 5 × 5 mm. In order to obtain a high sensitivity a nichrome 80/20 metal alloy (80 % nickel and 20 % chrome, thickness 50 nm) with a piezoresistivity gauge factor up to 2.63 was chosen as the strain gauge material (Higson 1964).

Fig. 19.4 Processing steps for sensor fabrication. SiN/Si-based cantilever beam processing. From *top to bottom*: **a** metal evaporation for piezoresistance and contact pads, **b** dry-etching for cantilever definition, **c** back side wet-etching, **d** release of cantilever and, finally, **e** conformal coating for water-proof operations

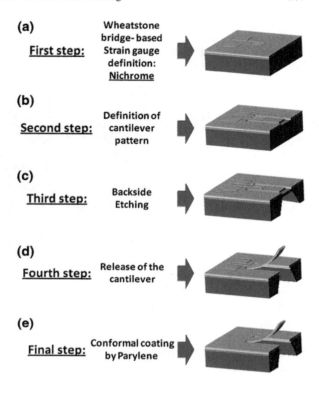

(a)
First step:
Wheatstone bridge-based Strain gauge definition: Nichrome

(b)
Second step:
Definition of cantilever pattern

(c)
Third step:
Backside Etching

(d)
Fourth step:
Release of the cantilever

(e)
Final step:
Conformal coating by Parylene

The microfabrication process proceeds as follows (Fig. 19.4). In the first step (Fig. 19.4a), the Wheatstone bridge strain gauges and the four gold contact pads are laid down by metal evaporation on the substrate top side. Then a dry etching of the silicon nitride layers (both on the top and bottom sides of the substrate) is performed by a $SiCl_4$-based plasma by inductively coupled plasma (ICP) dry etching using a photoresist as a mask: a $SiCl_4/N_2/Ar$ (20/25/7 sccm) gas mixture at a working pressure of 1 mTorr was used. On the substrate top side (Fig. 19.4b) dry etching is performed to define the cantilever geometry (100 μm wide with a length of 1,500 μm), stopping at the silicon device layer. The same recipe is used on the substrate bottom side to open apertures in the silicon nitride layer under the cantilever positions. Next (Fig. 19.4c), a wet etching by potassium hydroxide (KOH) 28 % solution at 80 °C is performed on the back side to open an empty space, 400 μm thick, underneath the cantilever beams inside the handle layer. In this step, the 2 μm thick thermal oxide insulating layer works as a wet etching barrier. After this step, the oxide layer is removed by hydrofluoric acid (HF) wet etching. Now (Fig. 19.4d), because of the residual stress which was designed in, the complete KOH wet etching of the exposed silicon topmost device layer around the cantilever allows the Si_3N_4/Si cantilever beam to bend out of the plane. As a final step (Fig. 19.4e), a waterproof Parylene conformal coating by Room

Fig. 19.5 SEM of a
Parylene-coated SiN/Si
cantilever (tilted @ 30°) with
height up to 1,200 μm [the
value reported should be
corrected by a factor
$(\sin 30°)^{-1}$]. The strain gauge
along the cantilever beam is
visible (Modified from
Qualtieri et al. 2012—
Reproduced by permission of
Elsevier VB)

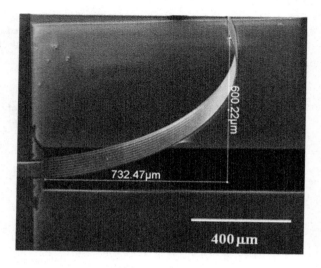

Temperature Chemical Vapour Deposition is added which creates a 2 μm thick
layer, covering all sides of the cantilever beam and all of the circuitry, allowing for
its underwater operation.

In Fig. 19.5, a SEM image of the cantilever shows the maximum height reached
for a cantilever beam 1,500 μm long. The upwards bending generated by the
residual stress inside the cantilever beam extends the cantilever to a height of
1,200 μm. The wet etching release step thins the silicon structural layer on the tip,
promoting further upwards bending motion of the cantilever tip. As a consequence,
the curvature is not constant along the length and therefore difficult to measure.
However, as fabrication occurs in a single silicon wafer processing batch, their
shapes are repeatable. The Parylene conformal coating used here is virtually
unstressed (Harder et al. 2002), and it does not affect the bending and the curved
shape of the cantilever. In contrast it does affect the stiffness of the cantilever, as
will be discussed in Sect. 19.3.

The cantilevers have been tested underwater to explore the sensor responsivity
and validate the Parylene-based waterproofed packaging. The set-up consisted of a
closed loop system in which a continuous water flow of up to 0.5 m s^{-1} was
maintained. A 5 V signal was applied to the Wheatstone-bridge circuitry as the
power supply, while a multimeter monitored the sensor read-out voltage signal. As
a reference, a commercial water-flow meter (Bürkert fluid control systems, type
8035T combined with Type S030 inline fitting) was installed to measure the flow
velocity during the experiment.

Figure 19.6 shows the electrical behavior of the sensor in a continuous water
flow. The dynamic range of this sensor was between 0.05 and 0.35 m s^{-1}. The
dependence of the responsivity on bulk flow velocity, v, shows a quadratic relationship, due to the drag force applied on the beam by the water flow (Denny
1995). The equation of the line of best fit, a parabola with the intercept fixed in the

Fig. 19.6 Electrical characteristic behavior of the 2 μm thick Parylene-coated sensor under continuous water flow conditions. A quadratic best fit is calculated, and the signal saturation region is highlighted (Modified from Qualtieri et al. 2012— Reproduced by permission of Elsevier VB)

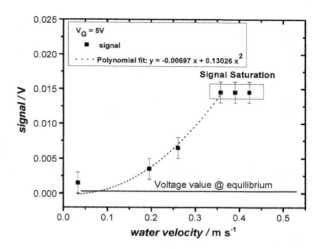

axis origin, is approximately $(0.13\ v^2 – 0.007\ v)$ V m^{-1} s and a 0.07 V m^{-1} s responsivity on the linear part of the calibration curve. The fitting resulted in an adjusted R^2 of 0.978. The measured signals were affected by noise ranging around 20 % of the signal values, due to micro-turbulence. For water velocities higher than 0.35 m s^{-1}, the sensor showed saturation in the output signal. In fact the cantilever was bent downwards by the flow and no longer deformable. The sensor returned to its equilibrium position when the flow was switched off and no hysteresis behavior was observed (Qualtieri et al. 2012).

The coating process is key to establishing the flexural stiffness of the beam and, consequently, the mechanical behavior of the sensor for a specific dynamic range. In Sect. 19.3 it will be shown that adjustment of the Parylene conformal coating thickness and the use of other materials can be effective for tuning the flexural thickness and setting the non-linear behavior to a specific dynamic range.

19.3 Responsivity Tuning of a Biomimetic Artificial Hair Cell by Parylene Conformal Coating Encapsulation

In Sect. 19.2, a new biomimetic AHC design was introduced, based on stress-driven MEMS device fabrication. Here it is shown that post-fabrication processing is effective for tuning the flexural stiffness of this miniaturised MEMS device. The process is based on a Parylene CVD conformal coating, usually designed for electrical insulation and for underwater operation. Parylene conformal coating thickness was shown to be an efficient method to control the mechanical and sensory properties of the device, tuning its responsivity from a sub-linear to a super-linear behavior (Rizzi et al. 2013). By controlling the thickness of the coating and hence tuning the flexural stiffness the sensor can be modified to behave linearly, or can be provided with non-linear sensitivity for specific applications. First, a brief introduction to stiffness tuning of the structural element of a MEMS as a mechanism to set dynamic range and responsivity of a sensor will be described. Then, encapsulation of a stress-driven flow sensor will be detailed and the signal/flow velocity characterization was analyzed. These results will be explained on the basis of flexural stiffness tuning of the out-of-plane cantilever beam.

19.3.1 Flexural Stiffness Tuning as an Adaptation Mechanism for a Cantilever Based Biomimetic Sensing System

There is considerable interest in understanding the design of natural structural elements which allows them to be tuned to a wide range of loading conditions while still retaining their sensing capability. Engineering does have some examples of similar designs, at the macroscopic end of the scale, where, for example, beams with controllable flexural stiffness have been designed to adapt the response of a macro-structure to external excitations, preventing structural collapses (Murray and Gandhi 2010). At the microscopic end of the scale, solutions have proven more challenging, and biomimetics is providing one successful strategy for designing highly sensitive sensors, adapted to a variety of environmental conditions. Both previously described SU8 pillar-based flow sensor approaches have been combined with other technologies to modify their mechanical sensitivities. McConney et al. (2009) used a drop-casting method to encase the pillar in a hydrogel polymer that mimicked the mechanical behavior of the cupula of the fish lateral line. Their stiffening procedure enhanced the performance of flow detection by about two orders of magnitude (McConney et al. 2009). Although structurally very similar to the biological design, because of its single element capping process, it is difficult to envisage how this biomimetic design could be mass-produced. Droogendijk et al. (2012) reported on an electrostatic spring mechanical softening design, used to adaptively change the mechanical transfer function of a system based on a Si_xN_y capacitive membrane flow sensor with the SU8 pillar cantilevered from its center. They were able to actuate their flow sensors using airflow or electrostatics. Application of bias voltages to the sensor's capacitive structures was used to improve the detection limit and sensor's responsivity to airflows (Droogendijk et al. 2012). This successful approach moves away from trying to mimic the natural mechanism, based on flexural stiffness changes in the cupula (McHenry and van Netten 2007).

In the stress-driven sensor design described here, changes in the flexural stiffness of the multi-layered beam can be introduced by choosing the right combinations of materials or by coating the element in various ways. The available parameter space makes it possible to tune the device to a wide range of environmental conditions.

The curved stress-driven sensor is not a simple cantilever loaded perpendicularly at its end, so the equations of simple beam theory are inadequate to describe its behavior. However, González and Llorca (2005) showed that its deflection resistance is nonetheless a similar function of the geometry (curvature, length, and moment of inertia) and Young's modulus. According to Murray and Gandhi (2010), the reduction of the coating material shear modulus, and hence the Young's modulus, decreases the overall flexural stiffness. Under deformation, when the coating layer is stiff the behavior is predominantly dominated by normal strain, while for a soft coating the behavior results in a shear strain and cross-sectional area deformation. Different deformation regimes are able to tune the sensor behavior to the working dynamic range through a transition from super-linear to a sub-linear dependence, allowing engineers to design for a specific operational environment and optimise performances. In the following section, two different materials with a similar coating technique will be investigated to understand how this mechanism works on a strain-gauge distributed on an upwards bent cantilever beam.

19.3.2 Responsivity Dependence from the Flexural Stiffness of Calibration Curves of the Stress-Driven Flow Sensor

In order to investigate the dependence of beam flexural stiffness on the material properties of the coated composite, two different materials were selected. Even though they belong to different material families both can be deposited by a CVD, in order to ensure integrated circuit compatible (IC-compatible) technological processing (Qualtieri et al. 2012): Parylene, type C (samples A and B), a soft polymer material, virtually unstressed (Harder et al. 2002), and Si_xN_y (sample C), a very hard ceramic material, deposited as a covering layer on the top side of the substrate, after the Wheatstone Bridge definition but before the cantilever beam fabrication. Parylene was deposited, by Room-Temperature CVD (RT-CVD) conformal coating, on both sides of the cantilever. Two thicknesses of Parylene were investigated: 0.5 μm (sample A, providing a 1 μm total addition to the starting structural layer's thickness) or 2 μm thick (sample B providing a 4 μm total addition to the starting structural layer's thickness). This technique resulted in the complete encapsulation of the sensor. The Parylene thin-film had a typical Young's modulus of $E_{\text{Parylene}} = 2.8$ GPa and a Poisson ratio $v = 0.4$, resulting in a shear stress modulus approximately equal to $G = 1$ GPa (Harder et al. 2002). In contrast, the Si_xN_y thin-film was a 300 nm thick layer, deposited by plasma enhanced chemical vapour deposition (PECVD); it insulated the cantilever top side including the micro strain-gauge. PECVD growth of a Si_xN_y thin-film, conducted at lower processing temperature, presented a lower intrinsic stress (approximately 300 MPa) (Hu 1991). The film showed a typical Young's modulus of $E_{SixNy} = 250$ GPa and a Poisson ratio $v = 0.23$ resulting in a shear stress modulus approximately equal to

$G = 101.6$ GPa (Callister 1997; Sharpe 2003). In both deposition techniques, the coating layers had a much lower intrinsic stress than LPCVD Si_3N_4, negligibly affecting the stress-gradient through the cantilever thickness, and resulting in no further bending of the beam. Table 19.1 summarises these data.

The flow sensor characterization was done in the same continuous water flow setup described in Sect. 19.2.2. From the calibration curves analysis, it was evident that the material properties determine the sensing responsivity during fluid–structure interaction. The electrical behavior of the three sensors (denoted as sample A, B and C) during the continuous flow in water is reported in Fig. 19.7. In this figure, the relative sensor signal (V) is plotted as a function of the water flow velocity (m s^{-1}).The background signal measured at equilibrium (negligible flow) was subtracted from each curve. The output signal from sample A (Parylene 1 μm) extended up to approximately 0.03 V, while sample B (Parylene 4 μm) output signal reached 0.015 V and sample C (Si_xN_y) output signal reached 0.16 V. The input dynamic range, extending from negligible flow up to approximately 0.35 m s^{-1}, was comparable for the three different sensors. However, the different output signal ranges of the different sensors and the specific sensor responsivities depended on the different coating materials and related beam flexural stiffnesses. The thinner Parylene-coated cantilever (sample A) showed a sub-linear behavior as function of the water flow with a linear responsivity of ≈ 0.2 V m^{-1} s at low flow velocities (up to 0.20 m s^{-1}), the thicker Parylene and Si_xN_y-based coated sensors (samples B and C) showed a quadratic super-linear signal/flow velocity characteristic with a responsivity of ≈ 0.07 and ≈ 0.9 V m^{-1} s, respectively, at higher flow velocities (on the linear part of the curve between 0.25 and 0.35 m s^{-1}). For flow values higher than 0.35 m s^{-1}, when the cantilevers were flattened by high flow speed and no longer deformable, samples B and C showed saturation while sample A exhibited oscillations (not shown) around 0.03 V, suggesting that the cantilever beam was vibrating. Optical and micro-particle image velocimetry (PIV) measurements should be performed in future to measure this vibration and compare it to the vortex shedding frequency. All the sensors returned to their equilibrium positions when the flow was switched off.

The different behavior can be explained by analyzing the flexural stiffness of each single bent cantilever. All coated cantilevers reached a similar tip height, around approximately 1,200 μm from the substrate and exhibited a comparable curvature radius: the degree of bending of the cantilevers was related to the geometrical dimensions and to the intrinsic residual stress inside the structural layers. In fact, in Parylene-coated sensors the bending curvature is fixed after release of the bare cantilever. Parylene, as showed in Table 19.1, is virtually unstressed and does not affect the cantilever curvature after coating. In the case of low-stressed PECVD Si_xN_y (300 MPa, Table 19.1) coated cantilever, its intrinsic stress and thickness are not enough to change considerably the bending curvature with respect to the uncoated device. This was confirmed using finite element method (FEM) analysis of cantilever bending simulations. The saturation level was common to all sensors as it is strongly dependent on the geometrical features of the beam. In fact, the electrical saturation region can be defined as the velocity range

Table 19.1 Summary of the material characteristics and beam mechanical properties for different coating layers

Material properties	Type	Coating deposition technique	Thickness (μm)	E, young modulus (GPa)	Poisson ratio	G, shear modulus (GPa)	Layer residual tensile stress (MPa)	Total beam thickness (μm)	Effective beam flexural stiffness (10^{-11} N m^2)
Parylene (Sample A)	Soft polymer	RT-CVD conformal coating	1 (0.5 per side)	2.8	0.4	1	≈0	3.3	1.73
Parylene (Sample B)	Soft polymer	RT-CVD conformal coating	4 (2 per side)	2.8	0.4	1	≈0	6.3	2.12
Si_xN_y (Sample C)	Hard ceramic	PECVD	0.3 (on top side)	250	0.23	101.6	300	2.6	2.55

Fig. 19.7 Electrical behavior of the three coated sensors in a continuous water flow. The *dashed lines* are intended as a guide for the eye. The different coating material characteristics give a different curve shape, tuning from a sub-linear to a super-linear trend. A common signal saturation region is shown (Modified from Rizzi et al. 2013—Reproduced by permission of The Royal Society of Chemistry)

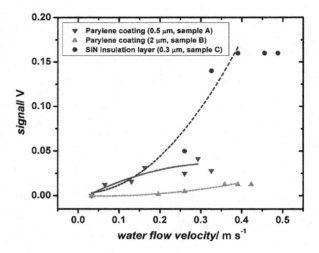

at which the fluid, distributing its drag force on the whole beam surface, is intense enough to overcome the elastic restoring force in the beam, deflecting it to flat, such that no further bending occurs as velocity increases further. As FEM analysis confirmed, all cantilevers reach the same tip height, therefore projecting the same frontal area to the fluid flow: for each velocity value the drag force acting on each sensor is consequently the same and bends the cantilevers in a similar manner. In the following section, the different sensors characteristics and how they relate will be described. As Murray and Gandhi (2010) highlighted, different coating material shear modulus and thicknesses, set the effective flexural stiffness of the beam and its different deformation regimes.

To analyze the flexural stiffness related to the coating material and thickness, FEM simulations have been exploited by applying a downward directed point force at the free end of the simulated cantilever, F_{point}, and calculating the force value which is able to fully flatten the beam. As stated earlier, the mechanics of a bent cantilever undergoing large strains cannot be fully represented by simple beam theory (González and Llorca 2005), but for comparison between materials, the simplified theory is sufficient for calculating an effective flexural stiffness, which is obtained using the following equation

$$EI_{\text{beam}} = (1/3) \cdot \left(F_{\text{point}}/\delta_{\text{tip}}\right) \cdot L^3 \tag{19.1}$$

where L is the cantilever length and δ_{tip} the bent cantilever tip height (Park et al. 2010). Figure 19.8 shows, the total thickness of the cantilever, accounting for the structural layer and the insulating layers, and the relative effective flexural stiffness. The arrow indicates the direction of increasing effective flexural stiffness. The effective flexural stiffness value depends on the Young's modulus of the beam layer material and on the beam's second moment of area. The RT-CVD conformal coating, controlling the deposition of Parylene layers at different thicknesses, allowed us to tune the final flexural stiffness. The 0.5 μm thick coating on each

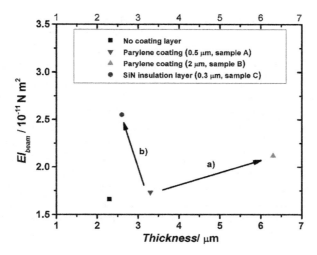

Fig. 19.8 Effective flexural stiffness values related to the three different coating thicknesses. *Arrows* highlight two different ways of contributing to the flexural stiffness increase: **a** flexural stiffness dependence by increasing coating thickness for the same Parylene coating material; **b** flexural stiffness dependence by increasing shear modulus at similar coating thicknesses (Modified from Rizzi et al. 2013—Reproduced by permission of The Royal Society of Chemistry)

side of the beam (sample A) resulted in an effective flexural stiffness of approximately 1.73×10^{-11} N m^2, for a final beam thickness of 3.3 μm, very close to the value of the uncoated beam (1.66×10^{-11} N m^2, 2.3 μm total thickness). The thicker layer (2 μm on each beam side, sample B) leads to an effective flexural stiffness value of 2.12×10^{-11} N m^2 for a final beam thickness of 6.3 μm. On the other hand, the thin layer (300 nm on top beam side, sample C) of a very hard Si_xN_y layer attributes to the whole beam an effective flexural stiffness of approximately 2.55×10^{-11} N m^2 for a final beam thickness of 2.6 μm. These values are reported on in Table 19.1.

The electrical response of the sensors, depicted in Fig. 19.7, can be explained by analyzing the effective flexural stiffness values. Cross-sectional area deformation in the strain gauge, due to shear strain and normal strain in response to the fluid drag force, combine to generate the linear part in the characteristic signal curve. Different coating materials with similar coating thicknesses (samples A and C, arrow b) in Fig. 19.8 are shown to modify the behavior to be directly dependent on the shear modulus G enhancement: the shear modulus increases the effective flexural stiffness allowing the cross-sectional area deformation of the coating layer only at high flow velocities (Murray and Gandhi 2010). Therefore, the non-linear behavior can be explained by the flexural stiffness-dependent fluid–structure interaction: the thin soft Parylene-coated beam (0.5 μm on each beam side) shows the sub-linear compliant behavior, with a cross-sectional area deformation starting at low velocity and low tangential fluid drag forces, and diminishing at higher velocities. The response saturation occurs when the maximum shear deformation is reached and the cantilever is flat. In contrast, the stiffer Si_xN_y coated beam (0.3 μm) limits the strain detected by the micro strain-gauge to that due to the fluid drag force. This results in a super-linear trend, where cross-sectional area deformation in the coating layer was prevented till stronger tangential fluid drag forces were acting on the beam i.e. higher flow velocities. Comparing the two curves obtained from the same Parylene-coated devices (samples A and B) shows that the electrical behavior change is only dependent on the thickness (Eq. 19.1 and arrow a) in Fig. 19.8). Thicker polymeric coating materials become stiffer (Kim 2009), as they near their polymeric critical thickness (where a transition from ductile to brittle mechanical characteristics occurs for a polymeric thin film); for Parylene this value is 1 μm (Kim 2009). Therefore, a reduction in the cross section area deformation effect occurs and limits the strain suffered by the micro strain-gauge in the same way as in the case of the Si_xN_y coated beam. Comparable effective flexural stiffness beams (samples B and C), irrespective of materials and/or thickness, result in similar trends.

19.4 Conclusion

A bio-inspired design for an AHC for flow sensing, based on a stress-driven technology, has been developed. The microfabrication process for a Si_3N_4/Si stress-driven cantilever-based flow sensor has been shown. Due to the stress-driven out-of-plane bending, the fabricated cantilever showed a behavior close to the natural flow sensor, being sensitive to the directionality (Qualtieri et al. 2012) and to low values of applied force. In case of a reversed flow, the signal changes will be negative with respect to the resting signal value, allowing detection of an opposite flow orientation. On the other hand, it showed a good mechanical resistance at high flow values and an easy recovery to the resting position at negligible flow.

The insulation post-processing step for underwater operations has been described. A Parylene conformal coating fabrication step made it possible to change the mechanical properties of this artificial flow sensor, through the beam flexural stiffness control, and to set its dynamic range behavior. In fact, while the dynamic range can be set by appropriately designing the cantilever geometry, the flexural stiffness of the stress-driven AHC makes it possible to tune the flow sensor responsivity within the cantilever's dynamic range. The increase of the beam flexural stiffness tunes the responsivity from sub-linear to super-linear behavior: making a biological parallel, flexural stiffness increase sets the AHC behavior. This bio-inspired mechanism is analogous to the hair cell mechanosensors in nature and suggests new material-based strategies for artificial mechanoreceptors design.

Acknowledgements This work was carried out under the robotic FIsh LOcomotion and SEnsing (FILOSE) project, supported by the European Union, seventh framework programme (FP7-ICT-2007-3).

References

Bleckmann H, Schmitz H, von der Emde G (2004) Nature as a model for technical sensors. J Comp Physiol A 190:971–981. doi:10.1007/s00359-004-0563-y

Callister WD (1997) Materials science and engineering, an introduction, 4th edn. Wiley, New York

Coombs S (2001) Smart skins: information processing by lateral line flow sensors. Auton Robot 11:255–261. doi:10.1023/A:1012491007495

Denny MW (1995) Air and water: the biology and physics of life's media, Reprint edition edn. Princeton University Press, Princeton

Dijkstra M, Van Baar JJ, Wiegerink RJ, Lammerink TSJ, De Boer JH, Krijnen GJM (2005) Artificial sensory hairs based on the flow sensitive receptor hairs of crickets. J Micromech Microeng 15:S132–S138. doi:10.1088/0960-1317/15/7/019

Droogendijk H, Bruinink CM, Sanders RGP, Dagamseh AMK, Wiegerink RJ, Krijnen GJM (2012) Improving the performance of biomimetic hair-flow sensors by electrostatic spring softening. J Micromech Microeng 22:065026–065034. doi:10.1088/0960-1317/22/6/065026

Fan Z, Chen J, Zou J, Bullen D, Liu C, Delcomyn F (2002) Design and fabrication of artificial lateral line flow sensors. J Micromech Microeng 12:655–661. doi:10.1088/0960-1317/12/5/322

Flock Å, Wersäll J (1962) A study of the orientation of the sensory hairs of the receptors cells in the lateral line organ of fish, with special reference to the function of the receptors. J Cell Biol 15:19–27. doi:10.1083/jcb.15.1.19

Fratzl P, Barth FG (2009) Biomaterial systems for mechanosensing and actuation. Nature 462:442–448. doi:10.1038/nature08603

González C, Llorca J (2005) Stiffness of a curved beam subjected to axial load and large displacements. Int J Solids Struct 42:1537–1545. doi:10.1016/j.ijsolstr.2004.08.018

Harder TA, Yao TJ, He Q, Shih CY, Tai YC (2002) Residual stress in thin-film parylene-C. In: 15th ieee international conference on micro electro mechanical systems. Proceedings: IEEE micro electro mechanical systems workshop. IEEE, Piscataway, NJ, p 435. doi:10.1109/MEMSYS.2002.984296

Higson GR (1964) Recent advances in strain gauges. J Sci Instrum 41:405–414. doi:10.1088/0950-7671/41/7/301

Howard J, Hudspeth AJ (1987) Mechanical relaxation of the hair bundle mediates adaptation in mechanoelectrical transduction by the bullfrog's saccular hair cell. Proc Natl Acad Sci USA 84:3064–3068

Hu SM (1991) Stress-related problems in silicon technology. J Appl Phys 70:R53–R80. doi:10.1063/1.349282

Hudspeth AJ (1989) How the ear's works work. Nature 341:397–404. doi:10.1038/341397a0

Hudspeth AJ (2008) Making an effort to listen: mechanical amplification in the ear. Neuron 59:530–545. doi:10.1016/j.neuron.2008.07.012

Kalvesten E, Vieider C, Lofdahl L, Stemme G (1996) An integrated pressure-flow sensor for correlation measurements in turbulent gas flows. Sensor Actuat A: Phys 52:51–58. doi:10.1016/0924-4247(96)80125-7

Kao I, Kumar A, Binder J (2007) Smart MEMS flow sensor: theoretical analysis and experimental characterization. IEEE Sens J 7:713–722. doi:10.1109/JSEN.2007.894910

Kim N (2009) Fabrication and characterization of thin-film encapsulation for organic electronics. Dissertation, Georgia Institute of Technology

Krijnen GJM, Dijkstra M, van Baar JJ, Shankar SS, Kuipers WJ, de Boer RJH, Altpeter D, Lammerink TSJ, Wiegerink R (2006) MEMS based hair flow-sensors as model systems for acoustic perception studies. Nanotechnology 17:S84–S89. doi:10.1088/0957-4484/17/4/013

Kuiper JW (1956) The microphonic effect of the lateral line organ. Dissertation, University of Groningen

Liu C (2012) Foundations of MEMS, 2nd edn. Prentice Hall, Upper Saddle River

McConney ME, Chen N, Lu D, Hu HA, Coombs S, Liu C, Tsukruk VV (2009) Biologically inspired design of hydrogel-capped hair sensors for enhanced underwater flow detection. Soft Matter 5:292–295. doi:10.1039/b808839j

McHenry MJ, van Netten SM (2007) The flexural stiffness of superficial neuromasts in the zebrafish (Danio rerio) lateral line. J Exp Biol 210:4244–4253. doi:10.1242/jeb.009290

McHenry MJ, Strother JA, van Netten SM (2008) Mechanical filtering by the boundary layer and fluid–structure interaction in the superficial neuromast of the fish lateral line system. J Comp Physiol A 194:795–810. doi:10.1007/s00359-008-0350-2

Mukai Y, Yoshikawa H, Kobayashi H (1994) The relationship between the length of the cupulae of free neuromasts and feeding ability in larvae of the willow shiner gnathopogon elongates caerulescens (Teleostei, Cyprinidae). J Exp Biol 197:399–403

Murray G, Gandhi F (2010) Multi-layered controllable stiffness beams for morphing: energy, actuation force, and material strain considerations. Smart Mater Struct 19:45002–45012. doi:10.1088/0964-1726/19/4/045002

Park SJ, Doll JC, Pruitt BL (2010) Piezoresistive cantilever performance—Part I: analytical model for sensitivity. J Microelectromech S 19:137–148. doi:10.1109/JMEMS.2009.2036581

Qualtieri A, Rizzi F, Todaro MT, Passaseo A, Cingolani R, De Vittorio M (2011) Stress-driven AlN cantilever-based flow sensor for fish lateral line system. Microelectron Eng 88:2376–2378. doi:10.1016/j.mee.2011.02.091

Qualtieri A, Rizzi F, Epifani G, Ernits A, Kruusmaa M, De Vittorio M (2012) Parylene-coated bioinspired artificial hair cell for liquid flow sensing. Microelectron Eng 98:516–519. doi:10.1016/j.mee.2012.07.072

Rizzi F, Qualtieri A, Chamber LD, Megill WM, De Vittorio M (2013) Parylene conformal coating encapsulation as a method for advanced tuning of mechanical properties of an artificial hair cell. Soft Matter 9:2584–2588. doi:10.1039/c2sm27566j

Sharpe WN (2003) Murray lecture tensile testing at the micrometer scale: opportunities in experimental mechanics. Exp Mech 43:228–237. doi:10.1007/BF02410521

Schmitz A, Bleckmann H, Mogdans J (2008) Organisation of the superficial neuromast system in goldfish, *Carassius auratus*. J Morphol 269(6):751–761. doi:10.1002/jmor.10621

Svedin N, Kalvesten E, Stemme E, Stemme G (1998) A new silicon gas-flow sensor based on lift force. J Microelectromech S 7:303–308. doi:10.1109/84.709647

Teyke T (1990) Morphological differences in neuromasts of the blind cave fish *Astyanax hubbsi* and the sighted river fish *Astyanax mexicanus*. Brain Behav Evolut 35:23–30. doi:10.1159/000115853

van Netten SM, Khanna SM (1994) Stiffness changes of the cupula associated with the mechanics of hair cells in the fish lateral line. Proc Natl Acad Sci USA 91:1549–1553. doi:10.1073/pnas.91.4.1549

Wang YH, Lee CY, Chiang CM (2007) A MEMS-based Air flow sensor with a free-standing microcantilever structure. Sensors 7:2389–2401. doi:10.3390/s7102389

Xue C, Chen S, Zhang W, Zhang B, Zhang G, Qiao H (2007) Design, fabrication, and preliminary characterization of a novel MEMS bionic vector hydrophone. Microelectr J 38:1021–1026. doi:10.1016/j.mejo.2007.09.008

Yang Y, Chen J, Engel J, Pandya S, Chen N, Tucker C, Coombs S, Jones DL, Liu C (2006) Distant touch hydrodynamic imaging with an artificial lateral line. Proc Natl Acad Sci USA 103:18891–18895. doi:10.1073/pnas.0609274103

Yang Y, Nguyen N, Chen N, Lockwood M, Tucker C, Hu H, Bleckmann H, Liu C, Jones DL (2010) Artificial lateral line with biomimetic neuromasts to emulate fish sensing. Bioinspir Biomim 5:16001–16009. doi:10.1088/1748-3182/5/1/016001

Yu X, Thaysen J, Hansen O, Boisen A (2002) Optimization of sensitivity and noise in piezoresistive cantilevers. J Appl Phys 92:6296–6301. doi:10.1063/1.1493660

Zhang Q, Ruan W, Wang H, Zhou Y, Wang Z, Liu L (2010) A self-bended piezoresistive microcantilever flow sensor for low flow rate measurement. Sensor Actuat A-Phys 158:273–279. doi:10.1016/j.sna.2010.02.002

Zhou ZG, Liu ZW (2008) Biomimetic cilia based on MEMS technology. J Bionic Eng 5:358–365. doi:10.1016/S1672-6529(08)60181-X

Chapter 20
Snookie: An Autonomous Underwater Vehicle with Artificial Lateral-Line System

Andreas N. Vollmayr, Stefan Sosnowski, Sebastian Urban, Sandra Hirche and J. Leo van Hemmen

Abstract In this work we present *Snookie*, an autonomous underwater vehicle with an artificial lateral-line system. Integration of the artificial lateral-line system with other sensory modalities is to enable the robot to perform behaviors as observed in fish, such as obstacle detection and geometrical-shape reconstruction by means of hydrodynamic images. The present chapter consists of three sections devoted to design of the robot, its lateral-line system, and processing of the ensuing flow-sensory data. The artificial lateral-line system of *Snookie* is presented in detail, together with a simple version of a flow reconstruction algorithm applicable to both the artificial lateral-line system and, e.g. the blind Mexican cave fish. More in particular, the first section deals with the development of the autonomous under-water vehicle *Snookie*, which provides the functionality and is tailored to the requirements of the artificial lateral-line system. The second section is devoted to the implementation of the artificial lateral-line system that consists of an array of hot thermistor anemometers to be integrated in the nozzle. In the final section, the information processing ensuing from the flow sensors and leading to conclusions about the environment is presented. The measurement of the tangential velocities at the artificial lateral-line system together with the no-penetration condition provides

A. N. Vollmayr (✉) · S. Urban · J. L. van Hemmen
Physik Department, Lehrstuhl für Theoretische Biophysik, T35, TU München,
James-Franck-Straße 85748 Garching, Germany
e-mail: andreas.vollmayr@tum.de

S. Urban
e-mail: sebastian.urban@tum.de

J. L. van Hemmen
e-mail: lvh@tum.de

S. Sosnowski · S. Hirche
Lehrstuhl für Informationstechnische Regelung, Fakultät für Elektro- und
Informationstechnik, TU München, Barer Straße 21 80333 München, Germany
e-mail: sosnowski@tum.de

S. Hirche
e-mail: hirche@tum.de

H. Bleckmann et al. (eds.), *Flow Sensing in Air and Water*,
DOI: 10.1007/978-3-642-41446-6_20, © Springer-Verlag Berlin Heidelberg 2014

the robot with Cauchy boundary conditions, so that the hydrodynamic mapping of potential flow onto the lateral line can be inverted. Through this inversion information is accessible from the flow around the artificial lateral line about objects in the neighbourhood, which alter the flow field.

Keywords Artificial lateral-line system · Autonomous underwater robot · Blind Mexican cave fish · Flow-field reconstruction · Flow sensing · Highly manoeuvrable AUV · Hot thermistor velocimetry · Hydrodynamic image · Submarine dynamics

Acronyms

2D	2-dimensional
3D	3-dimensional
6D	6-dimensional
ALL	Artificial lateral-line system
AUV	Autonomous underwater vehicle
BEM	Boundary-element method
BFS	Body-fixed system
CAD	Computer-aided design
FOR	Frame of reference
PD	Proportional-derivative
PVDF	Polyvinylidene fluoride fibers
PWM	Pulse-width modulation
SLAM	Simultaneous localisation and mapping
TFR	Thikonov-regularised flow-field reconstruction

20.1 Introduction

Even if completely blind fish are able to locate obstacles and avoid them under poor visual conditions (Dijkgraaf 1933, 1963). Studies on the blind cave form of *Astyanax mexicanus* and the closely related *Astyanax jordani* (previously known as *Anoptichthys jordani*) show that these fish are able to detect and also discriminate objects, if gliding past or towards them at close distance. The objects are perceived by means of the lateral-line organ, which is distributed along the body of the fish and responds to the movement of the water relative to the fish's skin (Van Trump and McHenry 2008; Mogdans and Bleckmann 2012). The presence of objects leads to an alteration of the flow field around the fish, which creates a 'hydrodynamic image' (Hassan 1989) of the surroundings on the fish's body.

On the basis of behavioral experiments, some of the tasks the lateral system is involved in and some of the features of stimuli that are reconstructed by the lateral-line system have been identified. For most objects moving towards the lateral-line system at some distances the flow field may be approximated by that of a dipole—a moving sphere of equal volume (Howe 2006, p. 24 ff)—since higher multipoles decrease much more rapidly with increasing distance to the moving object. It has been shown that goldfish and mottled sculpin are able to determine the position of the dipole (Coombs 1994). Mottled sculpins respond to the presentation of an oscillating sphere as lowest-order representation of the flow field of prey with hunting behavior and a strike towards the dipole source (Coombs and Conley 1997a; Conley and Coombs 1998; Coombs et al. 2001; Coombs and Patton 2009). Experiments carried out on goldfish (Vogel and Bleckmann 2001) suggest that fish are in principle also able to distinguish the direction of motion, speed, shape and size of solid objects. As a natural example schooling can be done solely by perception of the flow fields of neighbouring fish (Partridge and Pitcher 1980).

The blind cave form of *Astyanax mexicanus* can detect, avoid and also discriminate objects, when gliding past or towards them at close distance (Teyke 1985; Hassan 1986; Windsor et al. 2008). For the blind Mexican cave fish, on the basis of behavioral experiments (von Campenhausen et al. 1981; Weissert and von Campenhausen 1981; Hassan 1986), there is no doubt about its elaborate capabilities sensing its environment by means of the lateral-line system. Although it is not quite clear what the capabilities of a lateral-line system really are and what tasks it can be used for, the example of the blind Mexican cave fish shows that it is obviously possible to make vital decisions based solely on information mediated by the surrounding fluid motion.

Flow sensing is usually split up in either the reception of a vortex structure or the reception of irrotational flow. Major objects moving in water produce a wake provided that they are not streamlined. In a large range of Reynolds numbers, the wake releases vortex structures, which are called von Kármán vortex street. Especially, the strokes of the tail fin of fish leave behind prominent flow structures. These vortex structures mark the trace of swimming fish for quite a while (Hanke et al. 2000; Hanke 2004). Because of the low viscosity and the high mass of water, vortices are quite stable and may remain up to several minutes. Catfish have been shown to sense the vortex street with the lateral-line system during prey capture (Pohlmann et al. 2001, 2004). Wake tracking can in principle also be done by other flow sensory systems such as the whiskers of harbour seals (Dehnhardt et al. 2001; Wieskotten et al. 2011). While the mapping of a vortex on the fish's lateral-line system is understood quite well (Franosch et al. 2009), it is a non-trivial task to determine the properties of the vortex-producing object (Akanyeti et al. 2011).

The flow field of these vortex structures—seen from the perspective of extracting information—is completely different from the flow field, e.g. the blind Mexican cave fish produces (Handelsman and Keller 1967; Geer 1975; Hassan 1985, 1992a, b, 1993) to sense its environment. The flow field in front of and besides the blind Mexican cave fish may be treated irrotational as long it is moving through nearly undisturbed water. The vorticity produced by the action of viscosity at the surface of the fish is convected to the rear with the incident flow. This is usually expressed by a high Reynolds number in front of the fish resulting in a inviscid and irrotational region of flow around the snout well described by a velocity potential Φ for the incompressible Euler equations (Panton 2005; Oertel and Mayes 2004). Of course, the frequencies and the velocities involved in the problem guarantee incompressibility at every time.

Any object in the near surrounding disturbs the flow field on the surface of the fish compared to open water, the *hydrodynamic image*. The properties of a hydrodynamic image of a moving body mapped through an incompressible inviscid irrotational fluid are discussed in Sichert et al. (2009) by performing a multipole expansion of the flow field of varying shapes. The flow field is measured by a transparent artificial lateral line, meaning that the presence of the artificial lateral line does not disturb the flow field of the moving body. Then from the estimated multipole moments, basic information about the shape of the moving body is extracted, e.g. the volume is represented by the dipole moment. The conclusions are that—given a realistic resolution of the lateral-line sensors—the upper bounds for the range of localisation and shape reconstruction are roughly the size of the lateral-line system and the size of the moving object. The hydrodynamic image therefore only provides information about the environment in a very close range.

The present project of *Snookie* aims at the integration of an artificial lateral-line system (ALL) to a technical system, more specifically an underwater robot. Transferring the mentioned capabilities of the lateral-line organ to a robotic system would be beneficial in a number of ways. It complements existing established sensor technology. For instance, sonar sensors have a minimum distance at which to measure, with a blind zone within that distance, whereas camera- or laser-based systems are dependent on visual conditions. Its function is passive in the sense that it uses information that is present anyway due to the physics of bodies moving in water. The motion control of a group of several platforms equipped with flow sensory systems is possible without sensor interference or the need for data exchange. Also reflections, like with sonar systems in narrow spaces, do not interfere with the measurement. Moreover, the mapping of the hydrodynamic properties of the environment is enabled. As will turn out in the course of this work, reconstruction of the environment is doable with little computational effort. An explicit mathematical description of the underlying algorithms is presented.

From the constraints of the hydrodynamic image severe requirements follow for the implementation of an ALL on a moving robot. The lateral-line sensors must be capable of detecting small, slowly varying (Bleckmann et al. 1991) changes in the comparably strong flow field around the moving robot. The information processing must be very fast to enable the robot to react on detected changes of the immediate environment and the robot must be highly manoeuvrable in order to change the state of motion appropriately within this narrow range.

20.2 Related Work

Research on ALLs, the processing of the sensory data and the transfer to technical systems is not a completely new idea, but faces some inherent challenges, which are still to be solved. The biomimetic process covered in this work can be separated into three different stages: the development of an ALL (Sect. 20.4) the design of a robot for it (Sect. 20.3) and the process of acquiring information about the environment for the robot (Sect. 20.5). Each stage has been subject to previous or parallel research on the topic, which will be related to the proposed approaches in the following. The first stage is to build a sensory system that can mimic the function of the lateral-line system.

The basic functionality and morphology of the lateral-line system are well known (Coombs and van Netten 2005; Bleckmann 2008; Bleckmann and Zelick 2009). However, the exact transfer from the hydrodynamic stimulus to the excitation of the sensor (Coombs et al. 1996; Curcić-Blake and van Netten 2005; van Netten 2006; Goulet et al. 2008; McHenry et al. 2008), the resulting neuronal signals (Coombs and Conley 1997b; Engelmann et al. 2002; Chagnaud et al. 2006), and their processing is still under investigation (Kröther et al. 2002; Plachta 2003; Engelmann and Bleckmann 2004; Bleckmann 2008; Künzel et al. 2011; Meyer et al. 2012).

This means that so far attempts to rebuild the lateral-line system can only lead to an approximation or abstraction of the biological counterpart. For air, building biomimetic flow sensors is significantly simpler due to the properties of the medium, especially the viscosity and conductivity. Research in biomimetic flow sensing is driven by the upcoming interest in insect-like microflight. A review on different technologies in this sector is given by Motamed and Yan (2005) highlighting sensor design and experiments. The focus is on the determination of forces acting on the micro-robot as a feedback for control. One step further in terms of object/stimulus localisation are projects utilising arrays of biomimetic hair cells (*cilia*) as sensors. Work by Izadi et al. (2010) and Dagamseh et al. (2013) show the localisation of a dipole source—a vibrating sphere—in air by measuring the deflection of artificial hair sensors. The deflection of the hair induces a capacitive change in the hair base of the sensor, which can be related to the flow velocity. Other artificial cilia are based on the piezoelectric effect, for example, with polyvinylidene fluoride fibers (PVDF) (Li et al. 2010). The sensors are either used as surface neuromasts (Liu 2007; Hsieh et al. 2011; Qualtieri et al. 2011), or integrated in a canal (Yang et al. 2011; Klein and Bleckmann 2011; Klein et al. 2011, 2013). Both approaches can in principle be used for dipole localisation (Nguyen et al. 2011; Yang et al. 2011). An extension of the cilia approach is encapsulating them with a hydrogel cupula (Peleshanko et al. 2007). While biomimetic cilia might come close to the biological source of inspiration, the robustness, manufacturing complexity and signal-to-noise ratio are still challenges that prevent the application in an autonomous underwater vehicle (AUV).

A different approach for underwater sensing is to use thermal transport as a means for detecting the flow velocity. Hot-wire anemometers have been used for measuring flow velocities in gases and fluids (Middlebrook and Piret 1950) for a long time, but advances in the miniaturisation make them applicable to ALLs. First trial runs were done by Coombs et al. (1989), as a means of 'measuring water motions used in stimulating the mechanosensory lateral-line system of a teleost fish'. Micromachined arrays of hot-wire elements show the ability of localising dipole sources as good as biomimetic cilia (Chen et al. 2006; Yang et al. 2006; Pandya et al. 2007; Yang et al. 2010).

Only recently, there have been some works on the integration of ALLS or comparable sensors on underwater robots. A general overview on the state of underwater robotics is given by Kinsey et al. (2006) and Nicholson and Healey (2008). Hsieh et al. (2011) describe the implementation of PVDF sensors on a robotic fish, in which the robot is supposed to sense pressure deviations due to the presence of a wall. The modelling of the wall presence is done with an image charge method that is similar to the method proposed in Sosnowski et al. (2010). Fernandez et al. (2007, 2009, 2011) place a pressure sensor array along a submarine dummy. Using principal component analysis, two different cross section shapes (round and square) can be classified if the object is moved along the dummy.

Classification of a sensor reading proofs the usability of a sensory concept. For the usage of an ALL in an a priori unknown environment a minimal number of assumption about the environment is desirable to process the sensory data. A much more general method is required, to extract shape and location of unknown objects.

Studies focused on the Mexican cave fish usually only consider the forward problem, modelling the stimulus that occurs from the hydrodynamic interference with objects on the fish's body (Hassan 1985, 1992a, b, 1993; Windsor et al. 2010a, b). To utilise data gathered from the sensors on the robot and to obtain information about the environment the inverse problem has to be solved. Attempts to reconstruct the environment from the hydrodynamic image so far are limited to special cases with strong assumptions or prior knowledge about the environment.

The previously mentioned estimation of multipole moments of a 3-dimensional (3D) moving body requires exactly one body moving though an unbounded inviscid incompressible fluid initially at rest. The flow velocities are measured by a transparent—or more accurately, virtual—flow sensory system, which may not disturb the flow field of the moving body by its presence. Position and multipole moments up to order three are estimated simultaneously by means of a maximum-likelihood estimator given the flow velocities measured by the virtual lateral-line system. The generalisation to further incorporate an estimate of the velocity of the moving object is straightforward. The moving object may move in an arbitrary fashion.

A similar analysis has been carried out by Boufianais et al. (2010) for the 2-dimensional (2D) pressure field of the stationary potential of one body. The pressure is successively approximated by substitution of the leading terms of a Laurent series for the complex velocity potential in the stationary Bernoulli equation. This allows to discuss the dependence of pressure on distance and shape of the moving object. As an advantage, the method provides an estimate of the position and orientation of the object independent of its shape.

Quite often, the shape of one object—usually a sphere—moving relative to the physical present flow sensory system is assumed to be known. The position of the dipole can then be extracted easily from its hydrodynamic image (Coombs and van Netten 2005; Curcić-Blake and van Netten 2006; Franosch et al. 2005; Pandya et al. 2006; Goulet et al. 2008). In Sect. 20.5 a more general method is presented that can deal with arbitrary solid stationary boundaries. The method is applicable to quasi-2D incompressible, inviscid and irrotational flow around the lateral-line system.

20.3 The Autonomous Underwater Vehicle *Snookie*

The submarine *Snookie* is an AUV specifically designed as a test bed for the ALL. In this section, the concept of the robot is presented and a brief overview of its structure is given.

Special care has to be taken of the dynamics due the limited range of view of the lateral-line system. The robot must be capable of precise motion in the close vicinity of other objects at a distance of typically the diameter of the snout. Therefore, high manoeuvring capabilities are crucial. It must also be able to react on the sudden appearance of other objects in the range of view. A careful design of the hull, the fins and the thruster arrangement, driven by an accurate physical model of its dynamics is required, which also serves the tuning of the controllers of motion.

One of the most important design considerations concerns the shape of the robot. Not only must the components of the robot fit inside, but also the outline has direct influence on the functionality of the ALL. The expected sensing quality is dependent on the hydrodynamic properties of the shape of the robot, which determine the properties of the flow field of the surrounding fluid. A good compromise among ease of realisation, a simple mathematical treatise and the quality of the hydrodynamic image is a cylindrical shape with hemispheres as caps on both ends.

The sensors of the ALL are intended to be placed in a cross in allocated mountings in the front sphere. They extend 2–3 mm above the surface to avoid boundary layer effect introduced by the surface of the hull. The spherical shape of the sensory system allows to perform approximative analytic calculations; see Sect. 20.5.

20.3.1 The Robot

The robot consists of a cylindrical watertight main compartment, in which all of the electronics is encapsulated, two half-spheres at the end of the cylinder and six thrusters. It has a total length of $L = 74$ cm and a diameter of $2R = 25$ cm. The overall mass is 32.234 kg including the flooded bow and stern, which can be fine tuned to match the water displacement of the robot for neutral buoyancy (Fig. 20.1).

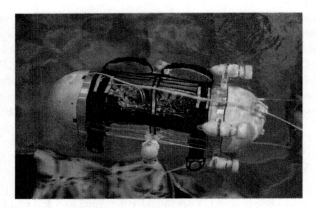

Fig. 20.1 *Snookie*, an autonomous underwater vehicle with an artificial lateral-line system

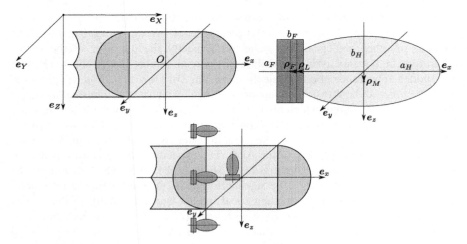

Fig. 20.2 *Left*: Frame of reference $\{e_X, e_Y, e_Z\}$ and body-fixed system $\{e_x, e_y, e_z\}$ with origin O at the geometric centre of the hull. *Right*: Geometry of the fluid-mechanically active parts of the robot with equivalent fin and equivalent hull without the thrusters. *Bottom*: Arrangement of the 6 thrusters in the body-fixed system of the robot

To achieve high manoeuvrability, a helicopter-like multi-propeller propulsion system is adapted from the AMOUR V robot (Vasilescu et al. 2010). The basic layout incorporates four thrusters arranged in a symmetric cross in the stern pointing in forward direction; see Fig. 20.2. This allows direct control over the forward/backward movement along the robot's longitudinal axis, the pitch angle and the yaw angle. All four motors work in combination for acceleration/deceleration. Additionally, two vertically mounted thrusters control depth and the roll angle.

A low-level control unit, based on an autopilot board by Ascending Technologies, is the central hub for the embedded systems and controls the 6-dimensional (6D) motion underwater. It consists of two 60 MHz ARM7 RISC processors: one of them is freely programmable. The other one combines three micro-electro-mechanical systems (MEMS) gyroscopes, a three-axis acceleration sensor, a three-axis magnetometer, and a pressure sensor to an inertia–force-measurement unit and preprocesses the data from these sensors.

The command unit of the high-level control can utilise this angular and translational data over a direct on-board link. The high-level control is done on a standard personal computer in a small form factor integrated in the robot. It provides the Robot Operating System (ROS) infrastructure to decide on the desired speed and direction, the processing of the sensor data, object avoidance and recognition, data logging and interfacing to command and control.

A land-based station can be used to monitor the status of the robot and to give new commands. Direct control of the movement of the robot is also possible via either a wiiMote, Joystick or keyboard. The robot can operate tethered via a Cat5 Ethernet cable for a high bandwidth communication. Alternatively, for untethered operation the link between command and control and the robot can be established via an acoustic modem by Tritech.

The ALL is described in detail in Sect. 20.4. The flow sensors are arranged in two equidistant array arranged to a cross on the frontal hemisphere with 17 sensors in total.

20.3.2 Model of the Dynamics of Snookie

As mentioned in the introduction to this section, for a proper design and control of the robot capable of manoeuvring on the basis of flow sensing, a careful description and analysis of the dynamics is crucial. Estimating the forces acting on a body in a fluid is a non-trivial problem. The traditional approach to describe the dynamics of bodies moving in fluids is an approximative analytic one. Such an approach delivers correct estimates of orders of magnitude and reasonable bounds of the relevant forces.

To avoid both, additional contributions due to the wave drag and complications in the calculation of the flow field due to a non-linear boundary condition at the forcefree surface, the robot is assumed to dive sufficiently deep. This condition is met in good approximation at a depth larger than five times the diameter of the moving object (Brennen 1982, Sect. 3.8) below an undisturbed water surface. The forces acting on the body are empirically split up into contributions of viscous drag, pressure drag (also called form drag), lift and increased inertia expressed by *added masses* as a consequence of the acceleration of displaced fluid. The viscous and the pressure drag contributions may be considered as corrections due to viscosity of the stationary motion of a body in an ideal fluid, which otherwise would not experience any forces. The lift contribution stems from lifting surfaces with sharp trailing edges such as the fins. This decomposition is actually an attempt of a low order series approximation in velocity and acceleration of the forces acting on a body moving in an infinite viscous fluid initially at rest. In general, the drag as well as fluid inertia depend on the current and previous velocity and acceleration (Obasaju et al. 1988). Additional effects like Basset forces and drag and lift forces, arising from the shedding of vortices, which in principle can be accounted for by a semi-analytic model, are ignored throughout this work. The velocities and time constants of the motion of the robot, of course, guarantee incompressibility at any time. In the following subsections the force components are described and computed for *Snookie*.

All quantities measured and all action performed by the robot are with respect to the coordinate system of the robot, the *body-fixed system* (BFS). However, for self localisation, path planning and navigation, the robot also requires its own velocity and position, and the position and velocity of surrounding objects in a global frame, the *frame of reference* (FOR). It provides an inertial system with the undisturbed fluid at rest. In the following section both frames, the relevant kinematic and dynamic quantities, and their transformations are briefly introduced.

20.3.2.1 Coordinate Frames of Reference

A suitable coordinate system for the description of the environment of *Snookie*, the FOR $\{e_X, e_Y, e_Z\}$, is fixed in space. It is an orthonormal inertial system. Stationary objects like the walls of a basin, which enter the fluid-mechanics as boundary conditions, are fixed in the FOR and therefore independent of time. The coordinate system is defined in a way, such that the directions e_X and e_Y are in the plane of the undisturbed water surface and e_Z is in positive direction pointing downwards into the fluid.

The second coordinate system is the *BFS*, see Fig. 20.2. It is prescribed by the robot, a rigid body, carrying out arbitrary motion relative to the laboratory system. The system is orthonormal. It is defined by the basis vectors $\{e_x, e_y, e_z\}$. The orientation of the basis vector e_x of the BFS shall coincide with the longitudinal axis of *Snookie* pointing to the bow, the e_y direction points to the starboard side, and e_z is given by their cross product. The natural choice for the position of the origin O of the BFS, expressed in coordinates of the FOR by the vector O, is the *centre of mass* of a *rigid body*, since no coupling between rotational and translations degrees of freedom in the equations of free motion (20.4) occurs in this system in the sense that resultant forces only affect the translational momentum and resultant torques only the angular momentum. For now, the real position of the centre of mass of the robot is unknown. A good choice for O is in the centre of volume of the hull of the robot. All quantities measured by devices on-board of *Snookie* are provided in the BFS. The state of a rigid body is fully described by the location and orientation of the BFS relative to the FOR. In order to clearly distinguish between a quantity expressed in the FOR and the same quantity in the BFS, capital symbols are used for coordinate vectors and matrices in the FOR $\{e_X, e_Y, e_Z\}$ and lower symbols for the BFS $\{e_x, e_y, e_z\}$.

The BFS may be rotated against the laboratory system, which is described by a modified set $\boldsymbol{\Phi} = (\Phi, \Theta, \Psi)$ of implicit Euler angles (Tait-Bryan angles). If the orientation of the body and the body-fixed basis vectors was initially parallel to the ones of the FOR, the following procedure describes the rotation of the body at a given instance of time. Rotate around e_Z about the yaw angle Ψ onto $\{e'_X, e'_Y, e_Z\}$ with $-\pi < \Psi \leq \pi$. Next, perform a rotation around e'_Y about the pitch angle Θ onto $\{e_X, e'_Y, e'_Z\}$ with $-\frac{\pi}{2} \leq \Theta \leq \frac{\pi}{2}$ and finally rotate around e_X about the roll angle Φ onto $\{e_x, e_y, e_z\}$ with $-\pi < \Phi < \pi$. Then, the rotation R of a vector from the FOR to the BFS is given by

$$
R = \begin{pmatrix} 1 & 0 & 0 \\ 0 & \cos\Phi & \sin\Phi \\ 0 & -\sin\Phi & \cos\Phi \end{pmatrix} \begin{pmatrix} \cos\Theta & 0 & -\sin\Theta \\ 0 & 1 & 0 \\ \sin\Theta & 0 & \cos\Theta \end{pmatrix} \begin{pmatrix} \cos\Psi & \sin\Psi & 0 \\ -\sin\Psi & \cos\Psi & 0 \\ 0 & 0 & 1 \end{pmatrix}
$$

(20.1)

and its inverse by $R^{-1} = R^T$. The definition of the Euler angles depends on the order in which the transformation is carried out and finite rotations are therefore not commutative. The angular velocities are computed from the Euler angles (Lewandowski 2003) by

$$
\Omega = \begin{pmatrix} 1 & 0 & -\sin\Theta \\ 1 & \cos\Phi & \sin\Phi\cos\Theta \\ 0 & -\sin\Phi & \cos\Phi\cos\Theta \end{pmatrix} \begin{pmatrix} \dot{\Phi} \\ \dot{\Theta} \\ \dot{\Psi} \end{pmatrix}.
$$

(20.2)

The total time derivative of a vector-valued quantity in an accelerated system expressed in BFS coordinates is given by

$$
\frac{\mathfrak{D}}{\mathfrak{D}t} = \frac{d}{dt} + \omega\times
$$

(20.3)

with $\omega = R\Omega R^T$. The additional term $\omega\times$ stems from the time derivative of the basis vectors of the accelerated BFS.

20.3.2.2 Rigid-Body Motion

As will be seen in Sect. 20.3.2.4, the fluid-mechanical forces on the body due to acceleration in an ideal fluid can be formulated in the framework of rigid body dynamics. Before the forces on the robot exerted by the fluid are treated in depth in the following sections, the inertial forces of the rigid body due to its body mass in the BFS shall be briefly introduced.

The velocity V of the origin O of the BFS expressed in coordinates of the BFS is denoted by $\vec{v}$, the angular velocity Ω of the BFS about O by ω, the acceleration $\dot{V}$ of O by a, the angular acceleration by $\dot{\Omega}$ and α, the moments of inertia I of the rigid body computed about its centre of mass by ι, and the position of the centre of body mass relative to O by ρ_M. The mapping between the two systems is given by $O, \Phi = (\Phi, \Theta, \Psi)^T$, (20.1) and (20.2).

Instead of formulating the inertial force f^I and inertial torque t^I as a sum or integral over the inertial forces acting on each particle, especially in case of changing masses, it is easier to take the total time derivative of the momentum

$m = m(v + \omega \times \rho_M)$ and the angular momentum $l = \iota\omega + m\rho_M \times (\omega \times \rho_M)$, where m is the body mass and l is the moment of inertia computed about the centre of mass ρ_M relative to O, which results in

$$\begin{pmatrix} f^l \\ t^l \end{pmatrix} = \begin{pmatrix} m\mathbb{1} & -m\rho_M\times \\ m\rho_M\times & \iota - m\rho_M\times \rho_M\times \end{pmatrix} \begin{pmatrix} a \\ \alpha \end{pmatrix} + \begin{pmatrix} \omega \times mv \\ \omega \times (\iota - m\rho_M\times \rho_M\times)\omega \end{pmatrix},$$

$$(20.4)$$

whereby the totally anti-symmetric matrix representation

$$c\times = \begin{pmatrix} 0 & -c_3 & c_2 \\ c_3 & 0 & -c_1 \\ -c_2 & c_1 & 0 \end{pmatrix} \tag{20.5}$$

of a cross product $c\times$ is used. Any other location of the origin of the BFS than the centre of mass couples linear and angular motion. Equation (20.4) incorporates Steiner's theorem (Meirovitch 2004)—also called parallel axis theorem—through the transformation of the moments of inertia $\iota - m\rho_M \times \rho_M\times$. With shifted centre of mass an external force resultant not only changes the velocity of the BFS but also induces a change in angular velocity. The same holds true for an external torque resultant.

The equations of motion (20.4) describing the change of momentum and angular momentum may be further unified to a single equation in a very compact notation, which will be extended in the following section to incorporate fluid-mechanical forces acting on the robot. By definition of the 6×6 mass matrix

$$\lambda = \begin{pmatrix} m\mathbb{1} & -m\rho_M\times \\ m\rho_M\times & \iota - m\rho_M\times \rho_M\times \end{pmatrix}, \tag{20.6}$$

the anti-symmetric 6×6 matrix

$$\varpi\times = \begin{pmatrix} \omega\times & 0 \\ 0 & \omega\times \end{pmatrix}, \tag{20.7}$$

where $\omega\times$ denotes the anti-symmetric 3×3 matrix representation (20.5) of the cross product, the generalised 6D velocity vector $u = (v, \omega)^T$, and the generalised 6D force vector $f = (f^l, t^l)^T$, the equations of rigid body motion without external forces then yield

$$f^l = \frac{\mathcal{D}\lambda u}{\mathcal{D}t} = \frac{d\lambda u}{dt} + \varpi \times \lambda u = 0. \tag{20.8}$$

20.3.2.3 Motion in an Ideal Fluid

Inertia forces cannot be neglected in comparison to drag forces. However, not only the robot itself but also the fluid displaced by the robot needs to be accelerated. The contribution of the displaced fluid to inertia in general depends on the flow field around the robot, which in turn depends on the initial state of the surrounding fluid, the shape of the moving body, the surrounding boundary conditions and in general on the Reynolds number. Some of the difficulties can be rendered inactive by a proper choice of the settings. *Snookie* is supposed to move through an unbounded fluid that is initially at rest. For the almost symmetric flow field around an ellipsoid without distortion by the thrusters, a good and straightforward mathematical description of the forces acting on the accelerated robot is available (Lewandowski 2003).

For the case of a rigid body moving with translational velocity V and rotational velocity $\boldsymbol{\Omega}$ through an inviscid, incompressible and unbounded fluid $\mathcal{D}$ initially at rest, the velocity $U = \nabla\Phi$ of the fluid is fully described by the velocity potential Φ fulfilling (Lamb 1945)

$$\Delta\Phi = 0 \tag{20.9}$$

on $\mathcal{D}$ and the no-penetration condition

$$\left.\frac{\partial\Phi}{\partial N}\right|_{S} = (V + \boldsymbol{\Omega} \times \mathbf{P}) \cdot N \tag{20.10}$$

at any point P on the surface S of the rigid body and the corresponding unit surface normal N inward $\mathcal{D}$. The velocity potential is supposed to vanish at infinity. Following the naming conventions from Sect. 20.3.2.1, capital symbols denote quantities with respect to the FOR. Therefore, P and N are functions of time. The time course of Φ is solely given by the right-hand side of (20.10). The potential Φ is a linear function of the velocity of the surface of the moving body.

20.3.2.4 Added Masses

The linearity of (20.9) and the linearity of Φ in the translational and rotational velocities of the body (20.10) allows to separate the potential into

$$\Phi = V_X\varphi_1 + V_Y\varphi_2 + V_Z\varphi_3 + \Omega_X\varphi_4 + \Omega_Y\varphi_5 + \Omega_Z\varphi_6. \tag{20.11}$$

After the separation, given the setting described in Sect. 20.3.2.3, the harmonic functions φ_i do not depend on the velocity of the moving body any more, just on its shape, since

$$\left.\frac{\partial\varphi_i}{\partial N}\right|_{S} = N_i \quad \text{for } i = 1, 2, 3 \quad \text{and} \quad \left.\frac{\partial\varphi_i}{\partial N}\right|_{S} = (\mathbf{P} \times N)_{i-3} \quad \text{for } i = 4, 5, 6, \tag{20.12}$$

and the choice of the coordinate system they are computed in. For a simpler notation the translational and rotational velocities are gathered in the 6D velocity vector $\mathcal{U} = (V, \boldsymbol{\Omega})$ as in Sect. 20.3.2.2, and the potential is given by

$$\Phi = \sum_{i=1}^{6} \mathcal{U}_i\varphi_i. \tag{20.13}$$

The kinetic energy stored in the flow field $\nabla \Phi$ around the moving body

$$T = \frac{1}{2} \varrho \int_{\mathcal{D}} |\nabla \Phi|^2 dV$$

can be rewritten using Green's second identity as $T = \frac{1}{2} \lambda_{ij} \mathcal{U}_i \mathcal{U}_j$ with the coefficients

$$\lambda_{ij} = -\varrho \oint_S \frac{\partial \varphi_i}{\partial N} \varphi_j dS. \tag{20.14}$$

and the density ϱ of water. The effects of the ideal incompressible irrotational fluid on the motion of the body are fully accounted for by additional inertia λ_{ij} (Kirchhoff 1870; Korotkin 2010), the so-called *added-masses*.

20.3.2.5 Added Mass Matrix Under Coordinate Transformations

For the sake of simplicity, the added mass matrix λ is computed with respect to the point of maximum symmetry. It is usually necessary to adapt λ to a given geometry, e.g. shift the axis of rotation, since other constraints such as the choice of the origin of the BFS force the body to rotate around a different point than the one assumed for the computation of the added masses.

The kinetic energy of the flow field around the body must be invariant under a change of the coordinate system the added masses are computed in. All the summands are quadratic forms of the velocities. The idea of the derivation of the transformation formulas for the added masses is to express the kinetic energy of the flow field in a new coordinate system and collect all term of a certain product $\mathcal{U}_i \mathcal{U}_j$. The coefficients are the added masses in the new system.

In the most general case the new system $\{e_x, e_y, e_y\}$ is shifted by the vector $\boldsymbol{\xi}$ moves with velocity $\boldsymbol{V}$, and rotates about its origin with angular velocity $\boldsymbol{\Omega}$ relative to the old system $\{e_X, e_Y, e_Z\}$. Due to the analogy to the transformation between a FOR and a BFS, the same notation as in Sect. 20.3.2.2 is used. The added masses shall be known in the system $\{e_X, e_Y, e_Z\}$, and are supposed to be determined in $\{e_x, e_y, e_y\}$. The velocity expressed in coordinates of the new system is given by $\boldsymbol{u} = R(\boldsymbol{V} + \boldsymbol{\Omega} \times \boldsymbol{\xi})$ with rotation matrix R as defined by (20.1). Again, 6D vectors are used to express the velocity of the new system in coordinates of the old $\mathcal{U} = (\boldsymbol{V}, \boldsymbol{\Omega})^T$ and new $\mathfrak{u} = (\boldsymbol{v}, \boldsymbol{\omega})^T$ system. Written component-wise the transformation of the velocity and angular velocity is given by

$$\mathcal{U}_{i=1\dots3} = \sum_{m=1}^{3} \mathfrak{u}_m R_{mi} - \varepsilon_{ijk} \Omega_j \xi_k \quad \text{and} \quad \mathcal{U}_{3+i} = \sum_{m=1}^{3} \mathfrak{u}_m R_{mi} \quad \text{for } i = 1, 2, 3$$

with the Levi–Civita symbol ε_{ijk}, and the transformation matrix $R_{mn} = e'_m \cdot e_n$, where $m \in \{x, y, z\}$ and $n \in \{X, Y, Z\}$.

The transformation of the added masses turns out to be quite simple, if the coordinate transformation consists of just a shift about ξ, i.e. a shift of the axis of rotation by $-\xi$, and if the added mass matrix to be transformed has diagonal shape. This means that the original mass matrix was computed about the centre of the fluid-mechanical forces of an ideal fluid acting on the body and the axes of the BFS coincide with the principal axes of the moment of inertia sub-matrix of the added mass tensor. The transformation directive becomes $\lambda'_{mn} = \lambda_{mn}$ for $m = n = 1, 2, 3$; $\lambda'_{15} = -\lambda_{11}\xi_3$, $\lambda'_{24} = \lambda_{22}\xi_3$, $\lambda'_{34} = -\lambda_{33}\xi_2$, $\lambda'_{35} = \lambda_{33}\xi_1$, $\lambda'_{16} = \lambda_{11}\xi_2$ and $\lambda'_{26} = \lambda_{22}\xi_1$ for $m = 1, 2, 3$ and $n = 4, 5, 6$ or $m = 4, 5, 6$ and $n = 1, 2, 3$; $\lambda'_{44} = \lambda_{44} + \lambda_{22}\xi_3^2 + \lambda_{33}\xi_2^2$, $\lambda'_{45} = -\lambda_{33}\xi_1\xi_2$, $\lambda'_{46} = -\lambda_{22}\xi_1\xi_3$, $\lambda'_{55} = \lambda_{55} + \lambda_{33}\xi_1^2 + \lambda_{11}\xi_3^2$, $\lambda'_{56} = -\lambda_{11}\xi_2\xi_3$, and $\lambda'_{66} = \lambda_{66} + \lambda_{11}\xi_2^2 + \lambda_{22}\xi_1^2$ for $m = 4, 5, 6$ and $n = 4, 5, 6$. Any entry not covered by the symmetry of λ and not listed vanishes. A careful comparison of the transformed added masses with the inertia of a rigid body (20.4) shows, that the added masses behave like body masses in every way. These properties of the added masses are a consequence of the fact, that the potential of the flow field around the moving rigid body is a linear function of the translational and rotational velocity of the moving body.

In Sect. 20.3.2.9, the *total inertia* of *Snookie* will be composed from the body mass and inertia and the added masses of independent geometrical primitives, resembling the fluid-mechanically active parts of the robot (Fig. 20.2), using these transformation directives for added masses.

20.3.2.6 Forces on the Hull

As discussed in Sect. 20.3.2.4, the favourable system to compute the added masses is a body-fixed FOR using all available symmetries. For the purpose of computational simplicity the trunk of *Snookie* shall be approximated by a prolate spheroid with long axis $2a_H$ and short axis $2b_H$, whose axis of revolution lies along the x-axis, with a being the semi-length along the axis and b the radius in the equatorial plane at $x = 0$. Then, the added masses with respect to the geometric centre of the ellipsoid are given by (Newman 1977, p. 144)

$$m_{H1} := \lambda_{11} = \frac{4}{3}\pi\varrho a_H b_H^2 \frac{\alpha_H}{2 - \alpha_H}, \quad m_{H2} := \lambda_{22} = \lambda_{33} = \frac{4}{3}\pi\varrho a_H b_H^2 \frac{\beta_H}{2 - \beta_H},$$

$$\iota_H := \lambda_{55} = \frac{4}{3}\pi\varrho a_H b_H^2 \left(a_H^2 + b_H^2\right) \frac{e^4(\beta_H - \alpha_H)}{(2 - e^2)[2e^2 - (2 - e^2)(\beta_H - \alpha_H)]},$$

$$\lambda_{66} = \lambda_{55}, \quad \lambda_{44} = 0, \quad \text{and} \quad \lambda_{ij} = 0 \quad \text{for} \quad i \neq j, \qquad (20.15a)$$

where

$$
\alpha_H = \frac{2(1 - e^2)}{e^3} \left[\frac{1}{2} \ln\left(\frac{1 + e}{1 - e}\right) - e \right], \quad \beta_H = \frac{1}{e^2} - \left(\frac{1 - e^2}{2e^3}\right) \ln\left(\frac{1 + e}{1 - e}\right),
$$

$$
e^2 = 1 - \left(\frac{b_H}{a_H}\right)^2. \tag{20.15b}
$$

The parameters a_H and b_H are chosen so that, first, the volume $V_E = 4/3 a_H b_H^2$ of the prolate spheroid is equal to the volume of the trunk of *Snookie* consisting of the water tight cylinder and the two semi-spheres at the bow and the stern. And second, the surface of the ellipsoid

$$
S_E = 2\pi b_H^2 + \frac{2\pi a_H b_H}{\sqrt{1 - \frac{b_H^2}{a_H^2}}} \sin^{-1}\left(\sqrt{1 - \frac{b_H^2}{a_H^2}}\right)
$$

has to be equal to the surface $S_H = 4\pi R^2 + 2\pi R(L - 2R)$ of *Snookie*. Numerical solution of these two conditions yields $a_H = 41.484$ cm and $b_H = 13.620$ cm. The added masses are $m_{H1} = 5.922$ kg and $m_{H2} = 23.573$ kg, the added moment of inertia is $\iota_H = 1.822$ kg m^2.

The viscosity induced drag on the surface of the moving body is hard to determine analytically and usually described by empirical drag coefficients (Panton 2005). The main drag on *Snookie* stems from the separation of the boundary layer around the hull and the breaking of the symmetry of the flow field, which finally leads to a wake with reduced pressure at the stern. For *Snookie* a different drag force is expected for forward and sideward motion

$$
F_{\mathrm{fd}} = -\frac{1}{2} \varrho C_{\mathrm{fd}} A_H |u_x| u_x, \quad \mathbf{F}_{\mathrm{sd}} = -\frac{1}{2} \varrho C_{\mathrm{sd}} B_H \begin{pmatrix} u_y \\ u_z \end{pmatrix} \sqrt{u_y^2 + u_z^2}, \tag{20.16}
$$

where C_{fd} and C_{sd} are the forward and sideward drag coefficients and $A_H = \pi R^2$ and $B_H = \pi R^2 + 2R(L - 2R)$ the respective cross sections with $L = 74$ cm being the overall length and $R = 12.5$ cm the radius of the robot. The total pressure drag is approximated by a linear composition (Newman 1977, p. 13 ff.) of forward and sideward drag. It acts on the geometric centre of the hull, the origin O of the BFS. The drag coefficient is a non-trivial function of the Reynolds number. It varies, e.g. for a circular cylinder between $C \approx 20$ at $Re = 1$ and $C \approx 1.1$–1.3 at $Re = 1,000$. In absence of measured data and better assumptions, forward and sideward drag coefficients are set to $C_{\mathrm{fd}} = C_{\mathrm{sd}} = 0.3$. For the Reynolds numbers considered here, this value is a safe estimate of the lower bound of the pressure drag coefficient of a sphere (Oertel and Mayes 2004, p. 165 ff). Overestimation of the pressure drag would lead to underestimated thruster forces required to stop the robot when an obstacle appears. The drag on the robot due to rotation is neglected, since in general high angular velocities are not intended to occur. The drag force on the hull $\mathbf{d}_H = (F_{\mathrm{fd}}, \mathbf{F}_{\mathrm{sd}})^{\mathsf{T}}$ acts opposite to the direction of motion with force resultant attacking at the geometric centre of the hull and does not produce any torque. Thus, the resultant drag force vector of the hull is given by $\mathfrak{f}_H = (d_H, 0)^T$.

20.3.2.7 Forces on the Fins

The motion of an elongated blunt body like *Snookie* in the direction of its main axes, even if it were perfectly symmetric with respect to the main axis and the fluid were perfectly at rest, is unstable. Any small disturbance in pitch or yaw causes torque about the centre of mass, called Munk's moment (Lewandowski 2003, p. 39), throwing the body out of the desired trajectory. The yaw and pitch instability of forward motion is balanced by a vertical and a horizontal fin at the stern of *Snookie*. The fins consist of thin plates of length 20.2 cm and height $b_F = 2R = 25$ cm equal to the diameter of the robot, with a cut-out for the spherical stern. This results in an effective surface of 260 cm^2 per fin with a mean effective length $a_F = 10.4$ cm of each fin. The geometric centre of the rectangular equivalent fin is located $\rho_F = (\rho_F, 0, 0), \rho_F = -39.3$ cm behind the geometric centre of the hull. The vertical and horizontal fin are arranged to form a symmetric cross-like shape.

The incident flow to the fins is taken to be homogeneous at large distance, and it is assumed that it is not affected by the presence of the hull or the thrusters. The relative velocity between fin and undisturbed fluid is approximated by the velocity $u_F = -v - \omega \times \rho_F$ of the geometric centre of the fin. Analytic expressions are available for the lift on a 2D cross section of a plate of zero thickness and cord length l in an homogeneous free stream and angle of attack α (Breslin and Andersen 2008, p. 66 ff). The circulation around a cross section of the fin according to the 2D theory is given by $\Gamma_F = \pi a_F u_F \sin \alpha$, which results in the total lift $F_F = \pi \varrho A_F u^2 \sin \alpha$ on the fin with fin area $A_F = a_F b_F$. In 3D the lift on a plate of finite length h is overestimated depending on the ratio h/l. Due to the absence of analytic expressions for the full 3D case, the 2D expression is widely used for hydrofoils. The section of the plate also experiences a torque $T_L = 1/4 \pi \varrho A_F^2 \sin \alpha$ per unit length about the centre of pressure forces at the so-called quarter-cord point, located halfway between the leading edge and the centre line of the fin. The quarter-cord point of the fins is located at $\rho_Q = (-36.7, 0, 0)^T$ cm behind the geometrical centre O of the hull on the long axis of the robot. The occurring torques on the fin are balanced by the mounting and do not affect the dynamics.

The z (y) component of the incident flow due to the presence of the horizontal (vertical) fin does not significantly contribute to the lift produced by the vertical (horizontal) fin. The incident flow u_F is therefore decomposed into the lift producing components $u_V = u_F - (u_F \cdot e_z)$ for the vertical and the horizontal fin $u_H = u_F - (u_F \cdot e_y)$. With the angles $\sin \alpha_V = (u_V \cdot e_y)/\|u_V\|$ and $\sin \alpha_H = (u_H \cdot e_z)/\|u_H\|$ between the fin and the lift producing components of the incident flow an estimate of the lift on the vertical and horizontal fin, acting on the quarter-cord point ρ_Q is given by

$$l_V = \tilde{A}_F \sin \alpha_V |u_V| \begin{pmatrix} -u_{Vy} \\ u_{Vx} \\ 0 \end{pmatrix} \quad \text{and} \quad l_H = \tilde{A}_F \sin \alpha_H |u_H| \begin{pmatrix} -u_{Hz} \\ 0 \\ u_{Hx} \end{pmatrix} \quad (20.17)$$

with $\tilde{A}_F = -\pi \varrho a_F b_F$. These results are applicable within an range of $\alpha = -10\ldots10°$ (Newman 1977, p. 20 ff), provided the plate has a smooth surface. The breakdown of the lift at higher angles of attack due to stall is accounted for by an additional factor of $\Theta(\alpha_0 - \alpha)\,\Theta(\alpha_0 + \alpha)$ for the respective force components with Θ being the Heaviside step function and α_0 the critical angle. The lift forces on the quarter-cord point have the force and torque resultants

$$f_L = l_V + l_H \quad \text{and} \quad m_L = (l_V + l_H) \times \rho_Q \qquad (20.18)$$

about O, which may be combined for a compact notation to the 6D force vector $\mathfrak{f}_L = (f_L, m_L)^T$. As mentioned previously, the effective lift of hydrofoils or wings of finite length is reduced in 3D contrary to the 2D results due to the flow over the tip of the wing. The lift therefore enters the equations of motion in the dynamical model with an additional safety margin 1/2 ensure that the stabilising effects of the fins are not overestimated.

Viscosity induced pressure drag on the fins in x direction is modelled due to the absence of better alternatives by the drag coefficient $C_P = 1.28$ on a flat plate perpendicular to the incident flow $F_P = \frac{1}{2} C_P \varrho \hat{A}_F u_F^2$, corrected by the net frontal area $\hat{A}_F = a_F b_F \sin(\alpha)$ exposed to the incident flow, which is also taken as a rough estimate for the lift-induced drag for $|\alpha| < 10°$.

Correct estimates of the drag on the fins due to sideward motion or rotations are challenging. Since lift and drag scale with u_F^2, an error in the estimates of these forces does not cause large effects on the dynamics, since in general the rotational velocity of the robot is small. Drag coefficients of similar shapes, e.g. a cube, a cube at an angle of 45° and a circular cylinder, are in the range of 0.8–1.3 for Reynolds numbers $\text{Re} \approx 1,000$. A safe estimate of the lower bound of the drag forces of the fins at sideward or vertical motion is therefore given by the drag coefficient $C_F \approx 1$ of a cylinder with the net frontal area of the fin A_F with mathematically convenient independence of the roll angle. The drag of the fin acts upon the geometric centre of the fin located at ρ_F. The incident flow u_F is decomposed in the x component and the components perpendicular to the fins $u_{Fy} = e_y \cdot u_F$ and $u_{Fz} = e_z \cdot u_F$ and the drag on the fins is given by

$$d_F = -\frac{1}{2} \varrho C_F A_F \sqrt{u_{Fy}^2 + u_{Fz}^2} \begin{pmatrix} 0 \\ u_{Fy} \\ u_{Fz} \end{pmatrix}. \qquad (20.19)$$

The resulting force and torque $\mathfrak{f}_F = (f_F, m_F)^T$ on the origin O of the BFS are

$$f_F = F_P e_x + d_F \quad \text{and} \quad m_F = f_F \times \rho_F. \qquad (20.20)$$

The fins stabilise the forward motion. Their disadvantages are increased added masses constraining the manoeuvrability. The added masses of the perpendicular arrangement of the vertical and horizontal fin are approximated by the 2D result of a cross-shaped section composed of plates of zero thickness. Both plates have a length of $a_F = 10.4$ cm and a width of $b_F = 25$ cm. The added masses per unit length cross section (Newman 1977, p. 144 ff) are given by $\lambda_{11} = 0$, $\lambda_{22} = \lambda_{33} = \pi\varrho(l/2)^2$, $\lambda_{44} = 2/\pi\varrho(h/2)^4$, and $\lambda_{55} = \lambda_{66} = 1/8\pi\varrho(l/2)^4$, which results in the added masses

$$m_F = \lambda_{22} = \lambda_{33} = \varrho\frac{\pi a_F b_F^2}{4}, \quad \iota_{F1} = \lambda_{44} = \varrho\frac{b_F^4 a_F}{8\pi},$$

$$\iota_{F2} = \lambda_{55} = \lambda_{66} = \varrho\frac{\pi a_F^4 b_F}{128} \tag{20.21}$$

of the cross-shaped fins computed about their common geometric centre. Their numerical values are $m_F = 5.105$ kg, $\iota_{F1} = 16.16 \times 10^{-3}$ kg m^2 and $\iota_{F2} = 0.718\times 10^{-3}$ kg m^2.

Vertical motion remains unstable since the motion in z direction cannot be stabilised by fixed fins without affecting motions in x direction. During submerging and descending Munk's moment appears due to both, the round shape of the bow and the large angle of an attack at the tip of the fin. Since the vertical speed is usually small, the thrust needed for balancing is also small.

20.3.2.8 Thrusters

The thrusters are neglected in the computation of the added masses, the drag and the lift. They provide the robot with acceleration in forward/backward and vertical direction, and also angular momentum about pitch, roll and yaw. The thrusters enter the equations of motions via the force generated by the propellers.

The four horizontal thrusters are placed at $\boldsymbol{\rho}_{Ti} = (l_H, \pm r_H/\sqrt{2}, \pm r_H/\sqrt{2})^T$ with $i \in \{1, 2, 5, 6\}$ symmetrically in a plane $l_H = 22.0$ cm behind the geometric centre of the hull parallel to the xy plane in $\boldsymbol{e}_x$ direction at a distance of $r_H = 18.5$ cm to the longitudinal axis of the hull; see Fig. 20.2. The two vertical thrusters are located at $\boldsymbol{\rho}_{T3} = (s_{Vx}, r_V, 0)^T$ and $\boldsymbol{\rho}_{T4} = (s_{Vx}, -r_V, 0)^T$ symmetrically in a plane parallel to the xy plane at $\boldsymbol{\rho}_V = (s_{Vx} = -6.2, \pm r_V = \pm 19.0, 0)^T$ cm in z direction. The added masses of the hull and the fins depend of the direction of motion, therefore the total centre of mass also depends on the direction on motion; see Sect. 20.3.2.9. Any other arrangement of the vertical thrusters would effectively cause angular momentum about the y axis at pure vertical motion.

So far, a fairly simple thruster model is used. Thrust, torque and efficiency not only depend on the geometry of the propeller, but also on the relative speed u_T of the incident flow and the rotation number v of the axle, usually expressed in terms of the advance ratio $J = u_T/vd$ with diameter d of the propeller. The thrust $T = K_T(J)\, \varrho v^2 d^4$ and the torque $Q = K_Q(J)\, \varrho v^2 d^5$ of the propeller are determined by

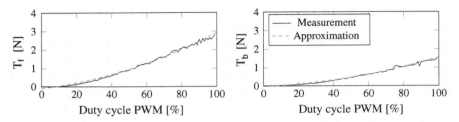

Fig. 20.3 *Solid line*: measured forward (*left*) and reverse (*right*) thrust. *Dashed line*: second-order polynomial approximation

non-dimensional parameters, the thrust coefficient K_T and the torque coefficient K_Q, both functions of J (Newman 1977, p. 24 ff; Breslin and Andersen 2008, p. 162 ff), representing the specific properties of the propeller. The thrust and torque coefficient have their maximum and the smallest slope at zero advance ratio. The focus of the robot is put on slow motion and high acceleration. Accordingly, the coefficients were taken to be constant. It was further assumed, that a unique relations between pulse width and thrust force in the regime of 0–2 m/s incident flow velocity exists. This is acceptable for small advance ratios.

Figure 20.3 shows the measured non-linear characteristic line of thrust force in both directions for a thruster with a three-blade 50 mm diameter propeller. Measurements were carried out with a thruster mounted on a JR3 six axis force-torque sensor. It automatically sweeps through the PWM of the motor driver, which results approximately in a current control of the motor for low advance ratios. Since the motor control is open-loop and friction of the seals, inertia and viscosity keep the motor on hold for small PWM signals, a dead zone with no thrust exists, which can be clipped. The resulting characteristic line for forward and backward thrust is fitted with a second-order polynomial function. The forward T_f and backward T_b thrust are given by

$$T_f = 2 \times 10^{-4}x^2 + 1.11 \times 10^{-2}x - 1.186 \times 10^{-1} \text{ and} \tag{20.22}$$

$$T_b = 10^{-4}x^2 + 3.4 \times 10^{-3}x - 3.5 \times 10^{-2} \tag{20.23}$$

with $x \in [0,100]$ for the PWM duty cycle. Since the prediction made by the simple motion controller on how the pulse width will affect the state of motion of the robot is approximated to lowest order, this approach conies with sustaining demands for control activity and therefore increased power supply.

The external forces and torques acting on the total centre of mass given the current thrust T_i of the thrusters $i = 1 \ldots 6$ are given by

$$f_T = \sum_{i \in \{1,2,5,6\}} T_i e_x - \sum_{i \in \{3,4\}} T_i e_z \quad \text{and}$$

$$m_T = \sum_{i \in \{1,2,5,6\}} T_i e_x \times \rho_{Ti} - \sum_{i \in \{3,4\}} T_i e_z \times \rho_{Ti}. \tag{20.24}$$

The respective 6D force vector is denoted by $\mathfrak{f}_T = (f_T, m_T)$.

20.3.2.9 Combination of Mass, Moment of Inertia and Added Masses

Snookie is buoyancy neutral, meaning that its mass is equal to the equivalent volume of water. But the mass is not distributed homogeneously. Counterweights are mounted below the longitudinal axis so that the robot is balanced about the pitch axis including fins and thrusters and stable about the roll axis. The symmetry with respect to the vertical plane is preserved. The mass of *Snookie* including balancing weights and the flooded bow and stern is equal to its water displacement, $m = 32.324$ kg. The actual centre of mass is shifted below the origin O of the BFS by the vector $\boldsymbol{\rho}_M = (0, 0, \rho_M)^T$. Integration over the mass distribution carried out by the computer-aided design (CAD) programme SolidWorks leads to a centre of mass $\boldsymbol{\rho}_M = (0, 0, 1.5)^T$ cm and moments of inertia $\iota_{11} = 0.23, \iota_{22} = 1.68, \iota_{33} = 1.70$, and $\iota_{ij} = 0$ for $i \neq j$ in units of kg m^2 computed in body-fixed coordinates about the centre of mass. Accordingly, the axis of rotation must be shifted by $-\boldsymbol{\rho}_M$,

$$\lambda'_M = \begin{pmatrix} m\mathbb{1} & -m\boldsymbol{\rho}_M \times \\ m\boldsymbol{\rho}_M \times & \boldsymbol{\iota} - m\boldsymbol{\rho}_M \times \boldsymbol{\rho}_M \times \end{pmatrix}, \tag{20.25}$$

which was already incorporated in the inertia of the rigid body in Sect. 20.3.2.2.

The added mass matrices have been determined with respect to the geometric centre of each shape independently and now have to be transformed and combined according to their location in the BFS relative to O; see Sect. 20.3.2.5. The added mass matrix of the hull (20.15a) is already computed about the origin of the BFS and does not need any further treatment. The coordinate system of the added mass matrix of the fin (20.21) has to be shifted by $-\boldsymbol{\rho}_F$. The resulting total mass matrix is then given by

$$\lambda = \lambda'_M + \lambda'_H + \lambda'_F \tag{20.26}$$

with the entries $\lambda_{11} = m + m_{H1}$, $\lambda_{22} = m + m_{H2}$, $\lambda_{33} = \lambda_{22}$, $\lambda_{15} = \lambda_{51} = -\lambda_{24} = -\lambda_{42} = -m\rho_M$, $\lambda_{26} = \lambda_{62} = -\lambda_{35} = -\lambda_{53} = -m_F\rho_F$, $\lambda_{44} = \iota_{11} + m\rho_M^2 + \iota_{F1}$, $\lambda_{66} = \iota_{22} + m\rho_M^2 + m\rho_F^2 + \iota_H + \iota_{F2}$, $\lambda_{55} = \iota_{33} + m_F\rho_F^2 + \iota_H + \iota_{F2}$. The remaining entries of $\boldsymbol{\lambda}$ vanish.

A closer look at the combined added masses reveals, that for example the volume displaced by the hull, and therefore also the added mass (20.15a), depends on the direction of motion. The net frontal area for forward motion is much smaller than the net frontal area for sideward motion, and accordingly the displaced fluid. The centre of total mass for pure forward motion for example is given by

$$s_F = \left(0, 0, \frac{m\rho_M}{m + m_{H1}}\right)^T, \tag{20.27}$$

the centre of total mass for pure sideward or vertical motion by

$$s_V = (s_{Vx}, 0, s_{Vz})^T = \left(\frac{m_F \rho_F}{m + m_{H2} + m_F}, 0, \frac{m \rho_M}{m + m_{H2} + m_F} \right)^T. \qquad (20.28)$$

Obviously, the centre of total mass has its own dynamics coupled to the motion of the robot. Strictly speaking, in the presence of added masses *Snookie* is not a rigid body. Although the position of each mass contribution—the added masses of the hull, the added mass of the fin and the body mass fixed in the BFS—the centre of total mass moves since the quantity of the added masses—their relative weights in the barycentre—change with the direction of motion. With the origin of the BFS fixed in an arbitrary point on the rigid body, e.g. *O*, the location of the masses is fixed, and the equations of motion of the rigid body can be used, keeping in mind that the total mass depends on the direction of motion.

20.3.2.10 Rigid-Body Motion with Drag, Lift and Thrust

The only thing left to do now is to balance the inertial forces of the rigid body (20.8) complemented by the added masses (20.26) with the origin of the BFS at the geometric centre of hull

$$\frac{\mathfrak{D}\lambda \mathfrak{u}}{\mathfrak{D}t} = \mathcal{Q} \qquad (20.29)$$

with the external forces

$$\mathcal{Q} = (\mathfrak{f}_L + \mathfrak{f}_F + \mathfrak{f}_H + \mathfrak{f}_T)^T, \qquad (20.30)$$

consisting of lift, drag and thrust, which results in the standard equations of submarine motion (Newman 1977, p. 135 ff; Feldman 1979; Fossen 1994). These equations can be solved easily by numerical integration in real-time on recent hardware (Sosnowski et al. 2010). The transformation of all dynamical quantities to the FOR is given in Sect. 20.3.2.1.

20.3.2.11 Validity and Benefit of the Model of the Dynamics

Numerous assumptions and simplifications have been made to arrive at Eq. (20.29). The most important ones shall be briefly reviewed and discussed. The assumptions and approximation were necessary to obtain a treatable model of the dynamics of the robot. The model is far from being perfect since there is no easy solution for the forces exerted on the robot by the surrounding fluid. However, it provides reasonable estimates of the dynamics at low computational efforts.

As briefly discussed in the introduction to Sect. 20.3.2, the model is restricted to the motion of the robot in an unbounded inviscid irrotational fluid at rest. No location dependent fluid-mechanical forces like additional pressure forces due to the presence of a wall (Korotkin 2010) on the robot exist. The state of motion is fully determined by the translational and rotational velocity. In case additional boundaries were present, the added masses of *Snookie*, if computable at all, had to be adapted by an expression dependent on the current position of the robot relative to all surrounding boundaries (Korotkin 2010, Chaps. 4 and 5). The effects of stationary walls and a free surface are negligible at sufficiently large distances, typically larger than 5 or 10 times the size of the robot.

While the inertia due to the physical mass of the robot are obtained directly from CAD, the estimates of the added masses are composed from the fluid-mechanical inertia of simple shapes resembling the shape of the robot. Each element is treated independently of the others including the thrusters, which means that in the model disturbances of the fluid caused by each element do not interfere. In future experiments, it is planned to determine the coefficients of the equations of motion experimentally. The conditions for the non-viscous estimates of the inertial forces are strictly only met at Reynolds numbers up to 10 or 15 or during the early stages of rapid acceleration from rest (Newman 1977, p. 34 ff).

In particular, it is assumed that the disturbance of the flow field of the robot due to the action of the thrusters is negligible. This assumption is justified for small thrust values and large distances between the thruster and the hull, which is not fulfilled very well for *Snookie*. If one wanted to take the interactions of the thrusters with the hull and the fin into account, one had to deal with added masses depending on the state of all six thrusters in the equations of motions. The added masses could be estimated by 3D boundary-element method (BEM) simulations or tow car experiments, with the robot attached to a force metre as a function of the six instantaneous thrust values.

Nevertheless, as soon as additional boundaries like a free surface or a solid wall are present, possibly significant errors are made in the estimation of the added masses as well as in the estimates of forces generated by the thrusters and forces that result from changing added masses. The order of magnitude of the thrust forces required to stop the robot due to the appearance of an obstacle in the range of the ALL, however, remains the same. The added masses are indeed increased in the vicinity of a stationary obstacle. But, without further external forces, the total kinetic energy of the robot remains constant, since the added masses are an effect of the motion in an inviscid irrotational and inviscid fluid. The robot and the fluid moving with the robot are decelerated to the same extent to which the add mass is increased while approaching the stationary object. The power necessary to reduce the kinetic energy within a certain distance remains unchanged, no matter if the robot moves close to a wall or in open water. Due to the presence of a wall not only decelerating forces, but also torque might be exerted onto the robot. Therefore, it should be taken care of sufficient thrust reserve.

Viscosity is accounted for by quasi-stationary semi-empirical drag coefficients. This approximation breaks down at high accelerations. Furthermore, the assumption that viscous drag forces just add linearly to inertial forces and that viscous forces can be decomposed into forward and sideward forces is only an approximation. The decomposition is correct in the special cases that the vehicle moves forward, sideward, upward or downward.

Even if the model of the dynamics of the robot would be perfect, due to the a priori unknown environment of the robot, the motion controller must be flexible enough, react fast enough and have enough power available to compensate for external effects such as changing boundary conditions like the presence of a wall.

20.3.3 Motor Control

The equations of motion (20.29) are the basis to set up a control strategy for the robot. The equations are independent of the position and the orientation of the robot, but non-linearly couple the velocity in every degree of freedom. For conventional submarines with a main propeller and steering fins, the equations of motion are usually modelled as decoupled in longitudinal, lateral and angular motion (Fossen 1994). For the control of *Snookie* a similar approximation is used. Each degree of freedom is treated independently, since by design the state of motion for basic operation can be reduced to a much lower number of degrees of freedom. The decoupling of the degrees of freedom is achieved simply by restricting the motion and by adding passive stabilising forces due to the lift of the fins and to shift the centre of body mass below the centre of buoyancy.

Snookie is supposed to always maintain a horizontal orientation, which means that pitch and roll and the respective angular velocities are small. A change in depth is supposed to happen in pure vertical motion. *Snookie* shall move mainly forward. The yaw angle during forward motion is kept small except for turns in place.

The centre of mass ρ_M is shifted below the centre of buoyancy. A deflection in roll angle ϕ leads to a restoring force and a slowly damped oscillation about ϕ_0. This oscillation must be damped by a proportional-derivative (PD) controller in ϕ. The lowered centre of mass also leads to a self-stabilisation with a small stability margin about the pitch angle θ. Pitch θ and yaw are stabilised by the fins counteracting Munk's moment—see Sect. 20.3.2.7—to reduce thrust forces necessary to maintain the orientation with the PD controllers. The angular velocity ω is implicitly given by the derivative part of the PD controllers for θ and ψ.

At pure forward motion the robot is self-stabilising in the horizontal plane aiding the desired horizontal orientation. The shift of the centre of mass below the geometric centre would lead to a roll motion induced for an acceleration $a_y \neq 0$ in y direction and a change in pitch for $a_x \neq 0$, which is counteracted by the fins, as described in Sect. 20.3.2.7. With the robot being kept horizontal, a_z is decoupled for $\phi \approx 0$ and $\theta \approx 0$, since depth change happens solely through vertical motion. Depth control is achieved by a proportional-derivative (PD) controller in z. Finally, forward velocity v_x is controlled by a proportional-integral (PI) controller to counteract a steady-state error.

The control of decoupled linearised equations of motion with a helicopter-like thruster layout has been previously demonstrated in AMOUR V (Vasilescu et al.

2010). Currently, more sophisticated control methods are investigated to account for the non-linear dynamics.

20.4 Flow Sensing in Water

As described in Sect. 20.2 several groups have already used different types of sensor concepts to realise an ALL. At present none of the sensors is commercially available yet. Flow sensors available on the market, which would promise acceptable accuracy and stability, can hardly be integrated to an ALL and mounted on a robot. For the ALL of *Snookie* a very conservative design decision was made in favour of hot thermistor velocimetry for several reasons.

There is plenty of theory and experience with a very similar sensor concept, the hot wire. Hot wires were shown to in principle provide the necessary accuracy and temporal resolution (Coombs et al. 1989; Franosch et al. 2010). The electronics and the sensors are relatively easy to develop and fabricate. The energy dissipation of the smallest commercially available thermistors allows high integration densities and low energy consumption. A thermistor promises a better signal-to-noise ratio for small relative signal changes due to its steeper resistance curve compared to a hot wire. And finally, a small thermistor can be embedded in solid material providing the robustness necessary for operation on a moving robot.

20.4.1 Physics of Hot Thermistor Velocimetry

The temperature of the heated element is given by $T = T_\infty + T_\theta$ with T_∞ being the ambient temperature of the fluid and T the over-temperature. The heat dissipation in a fluid from a small element $P \approx (A + Bv^n)T_\theta$ is a function of the relative velocity v of the fluid, where $n \approx 0.5$ and the constants A and B depend on the geometry and the properties of fluid (Middlebrook and Piret 1950; Felix 1962; Strickert 1974; Emsmann and Lehmann 1975; Perry 1982; Itsweire and Helland 1983; Lomas 1986; Eser 1990; Bruun 1996). For a sphere with diameter d (Emsmann and Lehmann 1975; Eser 1990) the dissipated power can be approximated by

$$P = \left[2 + 0.55\left(\frac{vc_p\rho}{k}\right)^{0.33}\left(\frac{vd}{v}\right)^{0.5}\right]4\pi\left(\frac{d}{2}\right)^2\frac{k}{d}T_\theta \qquad (20.31)$$

with specific heat capacity c_p, heat conductivity k, and kinematic viscosity v. Constant temperature anemometer sense the velocity of a fluid or gas by measuring the power P necessary to keep a heated element at an over-temperature T_θ.

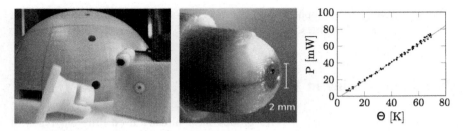

Fig. 20.4 *Left*: Hull integration. *Middle*: Close-up view of the artificial lateral-line sensor. *Right*: Power dissipated by a thermistor ($R_0 = 1{,}523$ at $T_0 \approx 293$ K, mounted on a PCB board and coated) in water versus over-temperature T_θ. *Black dots*: Measurement of dissipated power P. *Red line*: Linear fit. The relation between energy dissipation and over-temperature is perfectly linear (1.8 mW/K) as predicted by theory

20.4.2 The Artificial Lateral-Line System of Snookie

Glass-coated thermistors with a diameter of 0.36 mm from the Honeywell 111 series are used as heated elements for the artificial lateral-line sensors of *Snookie*. Thermistors are semiconductors with a non-linear negative dependency of electrical resistance upon the temperature. The resistance $R_\vartheta \approx 240\ \Omega$ of a thermistor at working temperature $T \approx 80$ °C with an over-temperature T_θ of approximately 60 °C is

$$R_\vartheta = R_0 e^{\beta_\vartheta(1/T_0 - 1/T)}, \tag{20.32}$$

given the resistance $1{,}400\ \Omega < R_0 < 2.4$ kΩ at room temperature $T_0 = 20$°C and the constant $2{,}000$ °K $< \beta_\vartheta < 5{,}000$ °K. To sustain a constant thermistor temperature the supplied electrical power $P = P_{el} = UI = U^2/R$ must equal the dissipated energy, if all energy is converted to heat and no leakage currents, e.g. due to deficient isolation appear.

The following rough estimates show that it is entirely legitimate to treat the thermistor adiabatically in the sense that it immediately adapts its temperature and thereby its resistance to changes in the transport of heat from it as it has been implicitly assumed in the thermistor model. The voltage necessary to maintain a stable resistance about 240 Ω of the thermistor in water at rest is approximately 1.5 V, though the properties of the individual sensors are scattered. This results in a dissipated power of approximately 9 mW. For comparison, the total heat stored in a sphere of the size of the thermistor with over-temperature 60 °C made of silicon or glass is only less than a factor of three larger than the heat dissipated per second. Within a temperature range of 60° the thermistor changes its resistance by approximately a factor of five. A change of heat transport due to changing flow conditions must be therefore immediately compensated by a change in the voltage supplied to the thermistor to hold a constant temperature. The voltage

$$U^2 \approx R_\vartheta(A + Bv^n)T_\theta \tag{20.33}$$

over the sensor is therefore an adiabatic measure for the fluid velocity.

The voltage to keep the thermistor at a constant resistance—and thus at constant temperature—is provided by specially designed boards, which incorporate a Wheatstone bridge and a two-stage amplifier. For easy mounting and maintenance in the snout of *Snookie*, the thermistor is embedded in a bullet-shaped packaging; see Fig. 20.4. The thermistor is placed at the tip of the bullet.

20.4.3 Flow Sensor Calibration

The measured power dissipation of a thermistor (Franosch et al. 2010) confirmed the power law (20.33) with $n = 0.34$, $A = 1.03$ mW/K and $B = 0.74$ mW/[K(m/s)^n]; see Fig. 20.4. The actual sensor calibration was done with a tow car. A linear axis (thrust tube, motor 2504 from Copley Controls, encoder resolution 1 μm) pulled the sensor through an aquarium of length 1 and 0.5 m in width and height. With constant acceleration from rest all velocities from 0 to 10 cm/s were present in one measurement. The flow velocity at the tip of the sensors was approximately given by the velocity of the linear axis. The voltage applied to the sensors was sampled with 10 kHz and filtered with a digital low-pass filter with a cut-off frequency of 5 Hz, since higher frequencies are generally not expected to occur in the hydrodynamic image of stationary objects. The filtered thermistor voltage at given velocity v_A of the linear axis was least-square fitted with

$$v_A = \sqrt[n]{\tilde{A}U^2 + \tilde{B}}, \qquad (20.34)$$

which can be easily derived from (20.33) with the substitutions $\tilde{A} = R_\vartheta A(T - T_0)$ and $\tilde{B} = R_\vartheta B(T - T_0)$, to obtain $\tilde{A}, \tilde{B}$ and n.

Figure 20.5 shows a sensor calibration run. For $v_A > 0.7$ cm/s the shape of the curve resembles very strongly the model described by (20.34). But for $v_A < 0.7$ cm/s essentially no change in the sensor readings could be observed. This effect was present at all sensors and calibration runs conducted. Additional complications arise from the fact that the parameters of the model (20.34) vary not only strongly between the sensors, but also for one sensor in different trials. It was found that this is partially caused by air bubbles generated by the sensors, which locally heat up the water. Gases dissolved in water and released due to heating form small bubbles and cling to the surface of the thermistor. The bubbles partly isolate the sensor. This results in a drift of the initial sensor offset and also in a decreased sensitivity. After removal of the bubbles, the sensor behavior returns to its initial state, but the generation of the bubbles continues.

The motion of the sensors always started from a longer period at rest with the sensors heating up the surrounding fluid and probably slowly driving thermal convection. As the sensors move away from their resting position at the beginning of the experiment, presumably a transition to the operating regime as described by

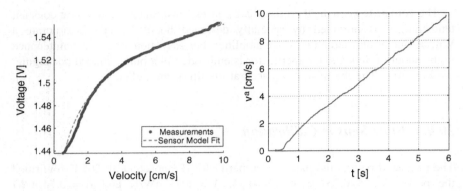

Fig. 20.5 *Left*: Sensor voltages plotted against velocity of the linear axis. *Right*: Time course of the linear axis velocity of an exemplary calibration run

the theory (20.34) takes place. It is planned to investigate the reason for the lower bound of velocity by a change of the settings of the experiments. Above 10 cm/s signals from the sensors quite often become ambiguous which probably attributes to instabilities in the flow over the tip of the sensor.

20.4.4 Object Detection

For object detection experiments the sensors are placed on a test sphere with a diameter of 15 cm and 15 sensor mountings as shown in Fig. 20.6b, which resembles the snout of *Snookie*. Of the 15 available slots, 7 are occupied with sensors as indicated in Fig. 20.6c. The linear axis is accelerated up to the velocity $v_A = 0.05$ m/s. This velocity is then kept constant till the end of the linear axis is reached. For the analysis only the constant velocity part is considered, which can be easily compared to a fluid-dynamics simulation of the experiment. A description of the simulation method can be found in Sect. 20.5.2.

An obstacle with rectangular cross section and rounded edges is placed in the aquarium. The whole scenario is depicted in Fig. 20.6a. At the closest distance between sphere and obstacle, the sensor is 3 mm away from the obstacle. Figure 20.6 shows the result of the experiment. The simulated velocity at the sensor with the obstacle present is shown for each sensor in the upper plot. The lower plots display the actual measured sensor voltages of both, the experimental run with and without the obstacle for comparison.

The quantitative conclusions that can be drawn from this experiment are limited in a number of ways. The circumstances in the laboratory available at the time the experiments were carried out allowed only a very limited volume of water. Water waves caused by the motion of the spherical mounting and vibrations of the floor

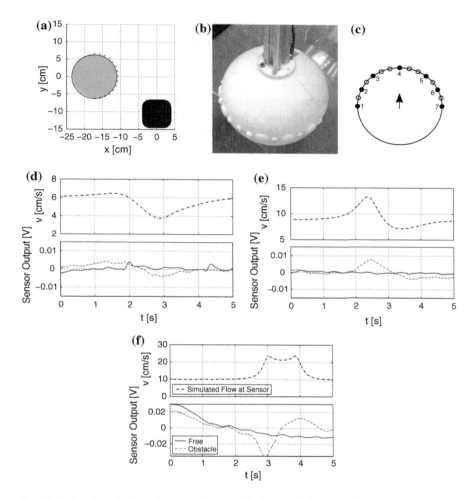

Fig. 20.6 An obstacle detection experiment with simulated flow velocity at the sensors and actual measurements. **a** Schematic of obstacle in relation to test sphere at the beginning of the experiment. **b** Submerged sphere. **c** Schematic of sphere with sensor distribution. Positions with *filled circles* contain a sensor, **d, e, f** Simulated flow velocity at the sensor (*upper plot*) and actual measurements (*lower plot*). The measurements were carried out with and without obstacle for comparison. The mean of each measurement was subtracted to allow a better comparison between the runs. For an explanation of the large deviations between simulation and experiment see Sect. 20.4.4

contaminated the measurements. The BEM simulation only considered the presence of the spherical mounting and the obstacle, and not the walls and the free surface of the aquarium. For an accurate analysis the distance of the walls of the aquarium and the free surface to the obstacle and the sphere should be at least 5–10 times the diameter of the sphere. The simulation was reduced to 2D, which is obviously not possible given the experimental situation. The large deviations between simulation and experiment shown in Fig. 20.6f demonstrate such effects. The sensors, in particular the formation of a gas bubbles, were not accurately monitored. An initial state with the fluid at rest in the small aquarium could not be guaranteed under the given conditions in the laboratory. In any case, our experiments show that the sensors are mechanically and electronically stable over months and that the sensors deliver signals as expected by theory (20.34) in the range from 7 mm/s to 10 cm/s. Most importantly the hydrodynamic image of the obstacle shows up in the sensory data. By improving the experimental situation and a careful calibration, acceptable signal-to-noise ratios suitable for the arrangement of the sensors to a useful artificial lateral line can be expected.

20.5 Flow-Field Reconstruction

Once the flow velocities measured by the ALL are available for further analysis, the question arises how to proceed.

Simple approximative fluid-mechanical considerations like the investigation of the flow field of a sphere approaching a solid wall (Franosch et al. 2010) by means of the mirror charge method inspire simple heuristics to conclude to the presence of a solid object in close neighbourhood of the ALL. The presence of a solid object would for example increase the mean flow velocity measured by a subset of the sensors. If the object was placed asymmetrically with respect to the longitudinal axis of the robot, it would slightly shift the stagnation point at the stern in dependence of its relative position. By storing the flow patterns of a set of known objects (Fernandez et al. 2007, 2009, 2011) and comparing these flow patterns (templates) with the actual flow velocities, shape and distance of a known object could be identified. But, those heuristics somewhat look at side effects or particular aspects of the hydrodynamic mapping of the environment onto the flow sensory system. Accordingly, these heuristics are not universally applicable and will suffer from ambiguities. The question then is first, if it is possible to formulate the hydrodynamic mapping in an as general as possible way so that its properties can be analysed, and second, what this mapping tells about the environment without application of prior knowledge or strong assumptions.

The most universal solution to the problem of extracting information from the hydrodynamic image would be the inversion of the mapping. To the knowledge of the authors not much has been published yet on inverse problems in fluid-mechanics (Derou et al. 1995; Murray and Ukeiley 2003; Suzuki and Colonius 2003). In the following the inversion of a 2D hydrodynamic image is carried out from the flow velocities measured on a circle and on a fish-like shape given incompressible inviscid and irrotational flow. In a real fluid, these conditions are found around the front of the blind Mexican cave fish or *Snookie*, where the lateral line or the ALL is placed, moving through the fluid at rest. The vorticity produced at the surface of the moving body is convected to the rear with the incident flow.

This is usually expressed by a high Reynolds number in front of the moving observer, resulting in a inviscid and irrotational flow well described by a velocity potential Φ for the incompressible Euler equations (Lamb 1945; Panton 2005).

The fluid domain shall be bounded by stationary solid walls of arbitrary geometry and number.

20.5.1 Properties of the Hydrodynamic Image

For the sake of simplicity the following analysis is restricted to two spatial dimensions. The body is supposed to move with velocity U. As already mentioned in Sect. 20.3.2.3, for the aforementioned conditions the flow field around the front of the moving body is well described by a velocity potential $V = \nabla \Phi(x, y)$ that suffices the Laplace Equation

$$\Delta \Phi = 0 \qquad (20.35)$$

on the fluid domain $\mathcal{D}$. The domain is bounded by the surface of the moving body $\mathcal{S}$ and eventually by other solid walls with surfaces $\mathcal{W}$. Again, the conventions of Sect. 20.3.2.1 are used to distinguish between quantities in the FOR and in the BFS. The closed surface of the moving body is composed of the disjoint surfaces $\mathcal{S}_S$, on which the flow sensory system measures the tangential flow velocity $v_\parallel = \partial \Phi / \partial t$ with tangent t of the surface, and the rest of the body $\mathcal{S}_B$ where nothing except the no-penetration condition $v_\perp = \partial \Phi / \partial n = 0$ with surface normal n is known about the flow field. Of course, the no-penetration condition is also given on the surface $\mathcal{S}_S$ of the sensory system.

The tangential velocity $v_\parallel$ measured by the lateral line may be integrated to obtain the potential $\Phi|_{\mathcal{S}_S}$ on the lateral-line system up to an irrelevant constant. Then, together with the Neumann boundary condition given by the no-penetration condition on the surface $\mathcal{S}_S$ with mounted flow sensors, Cauchy boundary conditions are obtained.

Expressed in coordinates of the FOR, which requires the knowledge of the velocity U, the problem is to determine the reconstructed $\hat{\Phi}$ on $\mathcal{D}$ given

$$\Phi|_{\mathcal{S}_S} \quad \text{and} \quad \left.\frac{\partial \Phi}{\partial N}\right|_{\mathcal{S}_S} = U \cdot N \qquad (20.36)$$

on $\mathcal{S}_S$. The choice of the coordinate system—FOR or BFS—does not affect the reconstruction. The transformation of the reconstructed flow field between the FOR $\hat{V}$ and the BFS $\hat{v}$ is identical to the directives described in Sect. 20.3.2.1.

The Cauchy-Kowalevski theorem guarantees the existence and uniqueness of the solution of the cauchy problem in some neighbourhood of $\mathcal{S}_S$. The solution of the potential problem (20.35) is a harmonic and analytic function on $\mathcal{D}$. Therefore, the analytic continuation of the potential to the whole fluid domain on $\mathcal{D}$ (Courant and Hilbert 1989, p. 505) and even beyond is possible. Given the exact knowledge of the potential and the normal derivative of the potential on S_S the reconstruction of the potential on $\mathcal{D}$ exists and it is unique. But, similar to many other inverse problems, the problem is ill-posed in the sense that any small error in determining the boundary values on S_S is amplified exponentially with the distance to S_S (Hadamard 1902; Isakov 1998) and therefore needs regularisation. Given a finite number of sensors and in the presence of noise, it is, however, unlikely to obtain valid boundary values an S_S so that an inverse exists. And if it exist, it was always possible to find an additive potential with zero velocity at the sensors, which destroys uniqueness. Properly choosen regularisation also restricts possible solutions so that existence and uniqueness are preserved. In summary, it is possible to construct a unique regularized inversion of the hydrodynamic image.

A polar coordinate system is the most suitable system for the problem raised in 2D as it is closest to the geometry of horizontal branch of the ALL mounted on the spherical snout of *Snookie*. The flow-field reconstruction is extensible to further 2D geometries by the application of conformal mapping—see Sect. 20.5.4. The general solution of Laplace's equation is known and must be adapted to the boundary conditions on the sensory system S_S to obtain the reconstructed flow field $\hat{V}$ on $\mathcal{D}$.

20.5.2 Boundary-Element Method

Before the solution of the inverse problem is presented, the computation of the forward problem, i.e. the hydrodynamic image on the lateral-line system, is described. The solution of the potential flow on $\mathcal{D}$ can be expressed as surface integral over a distribution of monopole and/or dipole sources of a priori unknown strength distributed over S_S, S_B and $\mathcal{W}$ (Liu 2009). In a simple version of the so-called boundary-element method (BEM)—as implemented here—pure monopole line sources with constant strength over the line segment are assigned to each element of the discretised boundary. The source strengths are determined such that the boundary conditions on all boundary elements are fulfilled. To be precise, to avoid further complications due to the computation of singular integrals and to precisely fulfil the boundary conditions, the monopole line sources were placed on a second surface in the interior of the solid bodies at close distance to the actual surface. This construction does not affect the validity of the solution (Lamb 1945). The resulting linear system of equations for the monopole strengths was solved directly. Then, the flow field was evaluated at any point on $\mathcal{D}$ by the resulting surface integral over the boundary elements and monopoles.

This method has several advantages. It does not require a mesh of the fluid domain, just the boundaries need to be meshed. Independent meshes can simply be moved against each other. The quality of the simulation can be easily assessed by checking the boundary conditions. A major disadvantage is its limitation to

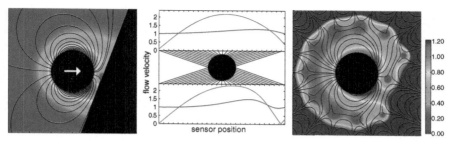

Fig. 20.7 Reconstruction of the flow field around a circle. The flow velocity is colour coded in units of the speed of the sphere, lengths are scaled by the size of the sphere. *Left*: A circle with 1,000 sensors equally distributed on its surface moves towards a wall at a 20° angle. The flow field is plotted with respect to the FOR. *Middle*: The tangential velocities on the surface of the circle are used to determine the coefficients A_α. The measured flow velocities are shown in the BFS (*pink line*) and the FOR (*blue line*). The middle part shows the unrolling of the surface of the circle onto the *x*-axis of the velocity plot. *Right*: In the reconstructed flow field the wall can be deduced from the parallel streamlines in front of the circle

potential flow. In a simple version as used here the BEM suffers from high memory consumption.

20.5.3 The Inversion of the Hydrodynamic Image

The origin of a polar coordinate system (r, ϕ) shall be placed in the centre of the circular flow sensory system with radius r_0; see Fig. 20.7. The surface of the moving circle is a streamline and the flow velocity component normal to the circle is zero. Then, the Laplace equation (20.35) on $\mathcal{D}$ is solved by the Ansatz

$$\Phi(r, \phi) = \sum_\alpha \left(A_\alpha \frac{r^\alpha}{r_0^{\alpha-1}} + B_\alpha \frac{r^{-\alpha}}{r_0^{-\alpha-1}} \right) e^{i\alpha\phi}, \tag{20.37}$$

and $i = \sqrt{-1}$. In the BFS the no-penetration condition (20.36) on the surface of the moving circle requires $A_\alpha = B_\alpha$, and the radial $\hat{v}_r$ and angular $\hat{v}_\phi$ velocities of the flow field on D are given by

$$\hat{v}_r(r, \phi) = \frac{\partial \hat{\Phi}}{\partial r} = \sum_\alpha A_\alpha \alpha \left[\left(\frac{r}{r_0} \right)^{\alpha-1} - \left(\frac{r}{r_0} \right)^{-\alpha-1} \right] e^{i\alpha\phi} \tag{20.38a}$$

$$\hat{v}_\phi(r, \phi) = \frac{1}{r} \frac{\partial \hat{\Phi}}{\partial \phi} = i \sum_\alpha A_\alpha \alpha \left[\left(\frac{r}{r_0} \right)^{\alpha-1} + \left(\frac{r}{r_0} \right)^{-\alpha-1} \right] e^{i\alpha\phi}. \tag{20.38b}$$

The coefficients A_α have to be determined from the measured tangential velocities

$$v_{\parallel}(r_0, \phi) = 2i \sum_{\alpha} \alpha A_{\alpha} e^{i\alpha\phi} \tag{20.39}$$

on the surface of the circle by the Fourier transform

$$A_a = \frac{1}{4\pi i\alpha} \int_0^{2\pi} v_{\parallel}(r_0, \phi) e^{-i\alpha\phi} d\phi. \tag{20.40}$$

Figure 20.7 shows the flow field around a circle moving towards a wall under an angle of 20°. For a finite number of flow sensors the Fourier coefficients are computed by a Fourier series. The frequency regularisation is implicitly carried out by the finite number of sensors considered to obtain A_{α} by (20.40) and a finite number of non-zero Fourier coefficients matched to the sensor readings.

20.5.4 Flow-Field Reconstruction from a Fish-Like Shape

While the circular shape is directly applicable to the artificial lateral system of *Snookie*, the Joukowski transformation allows to apply the reconstruction to the inspiring role model, the blind Mexican cave fish. In the complex plane $z = x + iy$ the circle and the surrounding flow field can be transformed into a fish-like shape and the corresponding flow field by application of a Joukowski transformation (Hassan 1985)

$$Z(z) = z + \frac{c^2}{z+s} + s \tag{20.41}$$

with the shape parameters c and s of the resulting aerofoil by the following procedure: measure the velocities on surface of fish; apply the inverse of the Joukowski transformation (20.41) to obtain the corresponding flow velocities on surface of the unit circle; reconstruct the flow field around the circle using Eqs. (20.40) and (20.38a); finally, apply the Joukowski transformation (20.41) on the reconstructed flow field to obtain the flow field around fish; see Fig. 20.8.

20.6 Discussion and Outlook

As soon as speed and shape of surrounding objects are available, the lateral line can contribute to more complex tasks involving more than just the detection of the pure presence and classification of an object. For the biological counterparts, some behavioral experiments about map formation of the environment and self localisation with the lateral-line system are already available (Teyke 1985; Burt de Perera 2004; Sharma et al. 2009; Patton et al. 2010). Using a lateral-line sensor

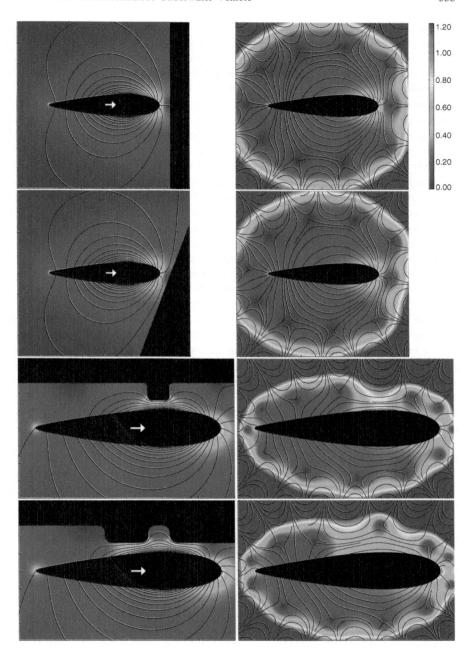

Fig. 20.8 *Left column*: Flow field and streamlines around a fish. *Black*: Solid objects and walls, respectively, in the vicinity of the moving fish. The flow velocity is colour coded in units of the speed of the fish, lengths are scaled by the size of the fish. *Right column*: Reconstructed flow field and streamlines in the FOR. The fish can deduce the shape of the wall from the streamlines. The lateral-line organ, which samples the flow field on the surface of the fish, was modelled by 1,000 equally spaced sensors

processing technique as the one described in the previous sections, similar capabilities can be implemented on *Snookie*. *Snookie* is equipped with an inertia sensor system (see Sect. 20.3.1) enabling the robot to estimate the current acceleration and by integration the current speed and the relative position. One challenge is that the quality of a map generated during motion depends on the accuracy of the position estimate, which is subject to drift. To counteract this drift, usually external references such as GPS are used, which is highly attenuated underwater. A method without external signals is called simultaneous localisation and mapping (SLAM) (Thrun et al. 2005), building up a map on the fly, which can then be used for matches to verify a position. A method to extract walls from the reconstructed flow field and a theoretical study of SLAM with inertial data and the extracted walls will be published somewhere else (Lenz et al. 2014). Although the reconstruction of the environment around the robot as described in Sect. 20.5.3 is, due to its mathematical nature, restricted to a close range, the additional matching of the reconstructed wall against the existing continuously refined map allows to compensate for the drifts of the inertial sensory system.

However, for a widely applicable flow sensory system useful in arbitrary environmental condition, several problems have to be solved. First, the reconstruction due to the nature of the problem is very sensitive against small errors in the flow velocity measurement. Any small disturbance increases exponentially with the distance to the sensory system. Therefore, the flow reconstruction requires strong regularisation. The easiest regularisation one can perform is frequency regularisation, as it was carried out for Figs. 20.7 and 20.8 by a finite number of sensors used to compute the Fourier Series of the flow velocity on the surface of the sensory system. Simply by omitting the higher spatial frequencies from the Fourier transform of the measured velocities along the flow sensory system and matching the remaining coefficients to the measurements (20.40), the reconstruction is stabilised for the prize of resolving less details. A follow-up publication with a Thikonov-regularised flow-field reconstruction (TFR) that allows to regularise in dependence of the distance to the sensory system, is in preparation (Urban et al. 2014). While frequency regularisation cuts the fine details globally, the TFR preserves details at least at small distances.

The next step towards a real-world application is to formulate the reconstruction of potential flow in three spatial dimensions. While this seems to be easily doable since it is quite clear how to proceed mathematically, the step towards more realistic flow conditions with vorticity might be challenging.

Fish are certainly interested in the fluid motion surrounding them for wake tracking or improved motion control. The fluid may be also considered as the medium through which boundaries and the motion of boundaries are mapped onto the lateral-line system. It is possible to extract the boundaries (Lenz et al. 2014) from the reconstructed flow field. However, one could also think of methods to directly conclude the shape and location of boundaries given the hydrodynamic image on the sensory system without preceding reconstruction of the flow field.

From the more practical side, the next steps in realising an ALL that can actually be mounted on a swimming robot are improvements in calibration and the experimental test conditions. For a convincing comparison with simulations a 3D BEM is desirable. For the reliable and enduring operation on an autonomous underwater robot the formation of bubbles on the thermistor remains an issue to be solved. Experiments with hydrophilic coatings so far did not lead to convincing results. It has already been shown experimentally by several groups that in principle detection (Martiny et al. 2009; Yang et al. 2011) and classification (Fernandez et al. 2011) based on an artificial lateral with different sensor concepts is possible. The flow-field reconstruction requires an albeit small but carefully calibrated set of flow sensors.

The discussion of the dynamics of the robot shows that additional forces appear in the close vicinity of boundaries such as a free surface or solid walls. The control of the thrusters, the correct treatment and computation of the added masses, and incorporating corrections to the added masses in the presence of a wall offer large space for the improvements of the dynamics of a highly manoeuvrable robot. This also requires extending the design of the hardware, e.g. a separate controller for each thruster surveying rotation speed of the propeller and electric current. Of course, the flow-field reconstruction by the ALL may also be used to improve the dynamical model used for control online during operation.

Acknowledgments This work was supported in part by the DFG excellence initiative research cluster Cognition for Technical Systems (CoTeSys)—see www.cotesys.org—and the Bernstein Center for Computational Neuroscience (BCCN) Munich—see www.bccn-munich.de. We thank all students, who have participated in this project.

References

Akanyeti O, Venturelli R, Visentin F, Chambers L, Megill WM, Fiorini P (2011) What information do Kármán streets offer to flow sensing? Bioinspir Biomim 6(3):036001

Bleckmann H (2008) Peripheral and central processing of lateral line information. J Comp Physiol A 194(2):145–158

Bleckmann H, Zelick R (2009) Lateral line system of fish. Integr Zool 4(1):13–25

Bleckmann H, Breithaupt T, Blickhan R, Tautz J (1991) The time course and frequency content of hydrodynamic events caused by moving fish, frogs, and Crustaceans. J Comp Physiol A 168(6):749–757

Boufianais R, Weymouth GD, Yue DKP (2010) Hydrodynamic object recognition using pressure sensing. Proc Royal Soc A: Math Phys Eng Sci 467(2125):19–38

Brennen CE (1982) A review of added mass and fluid inertial forces. Technical report

Breslin JP, Andersen P (2008) Hydrodynamics of ship propellers. Cambridge University Press, Cambridge

Bruun HH (1996) Hot wire anemometry. Oxford University Press, Oxford

Burt de Perera T (2004) Spatial parameters encoded in the spatial map of the blind Mexican cave fish, *Astyanax fasciatus*. Anim Behav 68(2):291–295

von Campenhausen C, Riess I, Weissert R (1981) Detection of stationary objects by the blind cave fish *Anoptichthys jordani* (*Characidae*). J Comp Physiol A 143:369–374

Chagnaud BP, Bleckmann H, Engelmann J (2006) Neural responses of goldfish lateral line afferents to vortex motions. J Exp Biol 209(2):327–342

Chen J, Engel J, Chen N, Pandya SD, Coombs S, Liu C (2006) Artificial lateral line and hydrodynamic object tracking. In: 19th IEEE international conference on micro electro mechanical systems, IEEE, pp 694–697

Conley RA, Coombs S (1998) Dipole source localization by mottled sculpin III. Orientation after site-specific, unilateral denervation of the lateral line system. J Comp Physiol A 183(3):335–344

Coombs S (1994) Nearfield detection of dipole sources by the goldfish (*Carassius auratus*) and the mottled sculpin (*Cottus bairdi*). J Exp Biol 190(1):109–129

Coombs S, Conley RA (1997a) Dipole source localization by mottled sculpin I. Approach strategies. J Comp Physiol A 180(4):387–399

Coombs S, Conley RA (1997b) Dipole source localization by the mottled sculpin II. The role of lateral line excitation patterns. J Comp Physiol A 180(4):401–415

Coombs S, van Netten SM (2005) The hydrodynamics and structural mechanics of the lateral line system. Fish Physiol 23:103–139

Coombs S, Patton P (2009) Lateral line stimulation patterns and prey orienting behavior in the Lake Michigan mottled sculpin (*Cottus bairdi*). J Comp Physiol A 195(3):279–297

Coombs S, Fay RR, Janssen J (1989) Hot-film anemometry for measuring lateral line stimuli. J Acoust Soc Am 85(5):2185–2193

Coombs S, Hastings M, Finneran J (1996) Modeling and measuring lateral line excitation patterns to changing dipole source locations. J Comp Physiol A 178(3):359–371

Coombs S, Braun CB, Donovan B (2001) The orienting response of Lake Michigan mottled sculpin is mediated by canal neuromasts. J Exp Biol 204(2):337–348

Courant R, Hilbert D (1989) Methods of mathematical physics, vol 1. Wiley-VCH, Germany

Curcić-Blake B, van Netten SM (2005) Rapid responses of the cupula in the lateral line of ruffe (*Gymnocephalus cernuus*). J Comp Physiol A 191(4):393–401

Curcić-Blake B, van Netten SM (2006) Source location encoding in the fish lateral line canal. J Exp Biol 209(8):1548–1559

Dagamseh A, Wiegerink R, Lammerink T, Krijnen G (2013) Imaging dipole flow sources using an artificial lateral-line system made of biomimetic hair flow sensors. J Roy Soc Interface 10(83):162

Dehnhardt G, Mauck B, Hanke W, Bleckmann H (2001) Hydrodynamic trail-following in harbor seals (*Phoca vitulina*). Science 293(5527):102–104

Derou D, Dinten J, Herault L, Niez J (1995) Physical-model based reconstruction of the global instantaneous velocity field from velocity measurement at a few points. In: Proceedings of the workshop on physics-based modeling in computer vision, IEEE Computer Society Press, p 63

Dijkgraaf S (1933) Untersuchungen über die Funktion der Seitenorgane an Fischen. Zeitschrift für Vergleichende Physiologic 20(1–2):162–214

Dijkgraaf S (1963) The functioning and significance of the lateral-line organs. Biol Rev 38(1):51–105

Emsmann S, Lehmann A (1975) Entwicklung eines Thermistoranemometers zur Messung instationärer Wassergeschwindigkeiten. Fortschrittsberichte der VDI Zeitschriften 18(8)

Engelmann J, Bleckmann H (2004) Coding of lateral line stimuli in the goldfish midbrain in still and running water. Zoology 107(2):135–151

Engelmann J, Hanke W, Bleckmann H (2002) Lateral line reception in still-and running water. J Comp Physiol A 188(7):513–526

Eser U (1990) Thermisches Anemometer mit Kugelsonde zur Bestimmung kleiner Geschwindigkeitsvektoren. Ph.D. thesis, Universität Essen

Essen Feldman J (1979) DTNSRDC revised standard submarine equations of motion. Technical report, DTNSRDC

Felix W (1962) Strömungsmessung mit Thermistoren. Naunyn-Schmiedberg's Arch exp Path u Pharmak 244:254–269

Fernandez VI, Hou SM, Hover F, Lang JH, Triantafyllou M (2007) Lateralline inspired MEMS-array pressure sensing for passive underwater navigation. Technical report, MIT Sea Grant

Fernandez VI, Hou SM, Hover FS (2009) Development and application of distributed MEMS pressure sensor array for AUV object avoidance. Technical report, MIT Sea Grant

Fernandez VI, Maertens A, Yaul FM, Dahl J, Lang JH, Triantafyllou MS (2011) Lateral-line-inspired sensor arrays for navigation and object identification. Mar Technol Soc J 45(4):130–146

Fossen TI (1994) Guidance and control of ocean vehicles. Wiley, New York

Franosch JMP, Hagedorn H, Goulet J, Engelmann J, van Hemmen JL (2009) Wake tracking and the detection of vortex rings by the canal lateral line of fish. Phys Rev Lett 103(7):078102

Franosch JMP, Sichert AB, Suttner MD, van Hemmen JL (2005) Estimating position and velocity of a submerged moving object by the clawed frog *Xenopus* and by fish—a cybernetic approach. Biol Cybern 93(4):231–238

Franosch JMP, Sosnowski S, Chami NK, Kühlenck Hirche S (2010) Biomimetic lateral-line system for underwater vehicles. In: 9th IEEE sensors conference, pp 2212–2217

Geer J (1975) Uniform asymptotic solutions for potential flow about a slender body of revolution. J Fluid Mech 67(04):817–827

Goulet J, Engelmann J, Chagnaud BP, Franosch JMP, Suttner MD, van Hemmen JL (2008) Object localization through the lateral line system of fish: theory and experiment. J Comp Physiol A 194(1):1–17

Hadamard J (1902) Sur les problemes aux dérivés partielles et leur signification physique. Princeton Univ Bull 13:49–52

Handelsman RA, Keller JB (1967) Axially symmetric potential flow around a slender body. J Fluid Mech 28(01):131–147

Hanke W (2004) The hydrodynamic trails of *Lepomis gibbosus* (*Centrarchidae*), *Colomesus psittacus* (*Tetraodontidae*) and *Thysochromis ansorgii* (*Cichlidae*) investigated with scanning particle image velocimetry. J Exp Biol 207(9):1585–1596

Hanke W, Brucker C, Bleckmann H (2000) The ageing of the low-frequency water disturbances caused by swimming goldfish and its possible relevance to prey detection. J Exp Biol 203(7):1193–1200

Hassan ES (1985) Mathematical analysis of the stimulus for the lateral line organ. Biol Cybern 52(1):23–36

Hassan ES (1986) On the discrimination of spatial intervals by the blind cave fish (*Anoptichthys jordani*). J Comp Physiol A 159(5):701–710

Hassan ES (1989) Hydrodynamic imaging of the surroundings by the lateral line of the blind cave fish *Anoptichthys jordani*. In: The mechanosensory lateral line. Springer, New York, pp 217–227

Hassan ES (1992a) Mathematical description of the stimuli to the lateral line system of fish derived from a three-dimensional flow field analysis I. The cases of moving in open water and of gliding towards a plane surface. Biol Cybern 66(5):453–461

Hassan ES (1992b) Mathematical description of the stimuli to the lateral line system of fish derived from a three-dimensional flow field analysis II. The case of gliding alongside or above a plane surface. Biol Cybern 66(5):443–452

Hassan ES (1993) Mathematical description of the stimuli to the lateral line system of fish, derived from a three-dimensional flow field analysis III. The case of an oscillating sphere near the fish. Biol Cybern 69(5–6):525–538

Howe MS (2006) Hydrodynamics and sound. Cambridge University Press, Cambridge

Hsieh TY, Huang SW, Mu LJ, Chen E, Guo J (2011) Artificial lateral line design for robotic fish. In: Underwater technology (UT), 2011 IEEE symposium on and 2011 workshop on scientific use of submarine cables and related technologies (SSC), pp 1–6

Isakov V (1998) Inverse problems for partial differential equations, Applied Mathematical Sciences. Springer, New York

Itsweire IC, Helland KN (1983) A high-performance low-cost constant-temperature hot-wire anemometer. J Phys E-Scientific Instrum 16(6):549–553

Izadi N, de Boer MJ, Berenschot JW, Krijnen GJM (2010) Fabrication of superficial neuromast inspired capacitive flow sensors. J Micromech Microeng 20(8):85041

Kinsey JC, Eustice RM, Whitcomb LL (2006) A survey of underwater vehicle navigation: recent advances and new challenges. In: IFAC conference of manoeuvering and control of marine craft, Lisbon, Portugal

Kirchhoff GR (1870) Über die Bewegung eines Rotationskörpers in einer Flüssigkeit. J für die reine und angewandte Mathematik 71:237–262

Klein A, Bleckmann H (2011) Determination of object position, vortex shedding frequency and flow velocity using artificial lateral line canals. Beilstein J Nanotechnol 2:276–283

Klein A, Herzog H, Bleckmann H (2011) Lateral line canal morphology and signal to noise ratio. In: Proc. SPIE 7975, Bioinspiration, Biomimetics, and Bioreplication, p 797507

Klein A, Münz H, Bleckmann H (2013) The functional significance of lateral line canal morphology on the trunk of the marine teleost *Xiphister atropurpureus* (*Stichaeidae*). J Comp Physiol A 199(9):735–749

Korotkin AI (2010) Added masses of ship structures. Fluid mechanics and its applications. Springer, New York

Kröther S, Mogdans J, Bleckmann H (2002) Brainstem lateral line responses to sinusoidal wave stimuli in still and running water. J Exp Biol 205(10):1471–1484

Künzel S, Bleckmann H, Mogdans J (2011) Responses of brainstem lateral line units to different stimulus source locations and vibration directions. J Comp Physiol A 197(7):773–787

Lamb H (1945) Hydrodynamics, 6th edn. Dover publications, New York

Lenz D, Sosnowski S, Vollmayr AN, van Hemmen JL, Hirche S (2014) SLAM with an artificial lateral-line system. In preparation

Lewandowski EM (2003) The dynamics of marine craft: maneuvering and seakeeping. World Scientific Pub Co Inc, Singapore

Li F, Liu W, Stefanini C, Fu X, Dario P (2010) A novel bioinspired PVDF micro/nano hair receptor for a robot sensing system. Sensors 10(1):994–1011

Liu C (2007) Micromachined biomimetic artificial haircell sensors. Bioinspir Biomim 2(4):162–169

Liu Y (2009) Fast multipole boundary element method: theory and applications in engineering, 1st edn. Cambridge University Press, Cambridge

Lomas CG (1986) Fundamentals of hot wire anemometry. Cambridge University Press, Cambridge

Martiny N, Sosnowski S, Hirche S, Nie Y, Franosch JMP (2009) Design of a lateral-line sensor for an autonomous underwater vehicle. In: 8th IFAC international conference on manoeuvring and control of marine craft, Guaruja, Brazil, pp 292–297

McHenry MJ, Strother JA, van Netten SM (2008) Mechanical filtering by the boundary layer and fluid-structure interaction in the superficial neuromast of the fish lateral line system. J Comp Physiol A 194(9):795–810

Meirovitch L (2004) Methods of analytical dynamics. Dover civil and mechanical engineering. Dover Publications, New York

Meyer G, Klein A, Mogdans J, Bleckmann H (2012) Toral lateral line units of goldfish, *Carassius auratus*, are sensitive to the position and vibration direction of a vibrating sphere. J Comp Physiol A 198(9):639–653

Middlebrook GB, Piret EL (1950) Hot wire anemometry-solution of some difficulties in measurement of low water velocities. Ind Eng Chem 42(8):1511–1513

Mogdans J, Bleckmann H (2012) Coping with flow: behavior, neurophysiology and modeling of the fish lateral line system. Biol Cybern 106(11–12):627–642

Motamed M, Yan J (2005) A review of biological, biomimetic and miniature force sensing for microflight. In: 2005 IEEERSJ international conference on intelligent robots and systems, pp 2630–2637

Murray NE, Ukeiley LS (2003) Estimation of the flow field from surface pressure measurements in an open cavity. AIAA J 41(5):969–972

van Netten SM (2006) Hydrodynamic detection by cupulae in a lateral line canal: functional relations between physics and physiology. Biol Cybern 94(1):67–85

Newman JN (1977) Marine hydrodynamics. MIT Press, Cambridge

Nguyen N, Jones DL, Yang Y, Liu C (2011) Flow vision for autonomous underwater vehicles via an artificial lateral line. EURASIP J Adv Signal Process 2011(1):406, 806

Nicholson JW, Healey AJ (2008) The present state of autonomous underwater vehicle (AUV) applications and technologies. Marine Technol Soc J 42(1):8

Obasaju ED, Bearman PW, Graham JMR (1988) A study of forces, circulation and vortex patterns around a circular cylinder in oscillating flow. J Fluid Mech 196(1):467–494

Oertel H, Mayes K (eds) (2004) Prandtl's essentials of fluid mechanics. Applied mathematical sciences, 2nd edn. Springer, New York

Pandya S, Yang Y, Liu C, Jones DL (2007) Biomimetic imaging of flow phenomna. In: IEEE international conference on acoustics, speech and signal processing, vol 2, 2007 ICASSP 2007, pp 933–936

Pandya SD, Yang Y, Jones DL, Engel J, Liu C (2006) Multisensor processing algorithms for underwater dipole localization and tracking using MEMS artificial lateral-line sensors. EURASIP J Adv Signal Process 1–8

Panton R (2005) Incompressible flow, 3rd edn. Wiley, New York

Partridge BL, Pitcher TJ (1980) The sensory basis of fish schools: relative roles of lateral line and vision. J Comp Physiol A 135(4):315–325

Patton P, Windsor S, Coombs S (2010) Active wall following by Mexican blind cavefish (*Astyanax mexicanus*). J Comp Physiol A 196(11):853–867

Peleshanko S, Julian MD, Ornatska M, McConney ME, LeMieux MC, Chen N, Tucker C, Yang Y, Liu C, Humphrey JAC, Tsukruk VV (2007) Hydrogel-encapsulated microfabricated haircells mimicking fish cupula neuromast. Adv Mater 19(19):2903–2909

Perry AE (1982) Hot-wire anemometry. Clarendon Press, Oxford

Plachta DTT (2003) A hydrodynamic topographic map in the midbrain of goldfish *Carassius auratus*. J Exp Biol 206(19):3479–3486

Pohlmann K, Grasso FW, Breithaupt T (2001) Tracking wakes: the nocturnal predatory strategy of piscivorous catfish. Proc Natl Acad Sci USA 98(13):7371–7374

Pohlmann K, Atema J, Breithaupt T (2004) The importance of the lateral line in nocturnal predation of piscivorous catfish. J Exp Biol 207(17):2971–2978

Qualtieri A, Rizzi F, Todaro MT, Passaseo A, Cingolani R, De Vittorio M (2011) Stress-driven AlN cantilever-based flow sensor for fish lateral line system. Microelectron Eng 88(8):2376–2378

Sharma S, Coombs S, Patton P, Burt de Perera T (2009) The function of wall-following behaviors in the Mexican blind cavefish and a sighted relative, the *Mexican tetra* (*Astyanax*). J Comp Physiol A 195(3):225–240

Sichert A, Bamler R, van Hemmen JL (2009) Hydrodynamic object recognition: when multipoles count. Phys Rev Lett 102(5):058104

Sosnowski S, Franosch JMP, Zhang L, Nie Y, Hirche S (2010) Simulation of the underwater vehicle "Snookie": navigating like a fish. In: 1st international conference on applied bionics and biomechanics

Strickert H (1974) Hitzdraht-und Hitzfilmanemometrie. VEB Verlag Technik, Berlin

Suzuki T, Colonius T (2003) Inverse-imaging method for detection of a vortex in a channel. AIAA J 41(9):1743–1751

Teyke T (1985) Collision with and avoidance of obstacles by blind cave fish *Anoptichthys jordani* (*Characidae*). J Comp Physiol A 157(6):837–843

Thrun S, Burgard W, Fox D (2005) Probabilistic robotics (intelligent robotics and autonomous agents series). MIT Press, Cambridge

Urban S, Vollmayr AN, Sosnowski S, Hirche S, van Hemmen JL (2014) Hydrodynamic imaging on a 1-dimensional manifold and its inversion in 2-dimensional potential flow. In preparation

Van Trump WJ, McHenry MJ (2008) The morphology and mechanical sensitivity of lateral line receptors in zebrafish larvae (*Danio rerio*). J Exp Biol 211(13):2105–2115

Vasilescu I, Detweiler C, Doniec M, Gurdan D, Sosnowski S, Stumpf J, Rus D (2010) AMOUR V: a hovering energy efficient underwater robot capable of dynamic payloads. Int J Robot Res 29:547–570

Vogel D, Bleckmann H (2001) Behavioral discrimination of water motions caused by moving objects. J Comp Physiol A 186(12):1107–1117

Weissert R, von Campenhausen C (1981) Discrimination between stationary objects by the blind cave fish *Anoptichthys jordani* (*Characidae*). J Comp Physiol A 143(3):375–381

Wieskotten S, Mauck B, Miersch L, Dehnhardt G, Hanke W (2011) Hydrodynamic discrimination of wakes caused by objects of different size or shape in a harbour seal (*Phoca vitulina*). J Exp Biol 214(11):1922–1930

Windsor SP, Tan D, Montgomery JC (2008) Swimming kinematics and hydrodynamic imaging in the blind Mexican cave fish (*Astyanax fasciatus*). J Exp Biol 211(18):2950–2959

Windsor SP, Norris SE, Cameron SM, Mallinson GD, Montgomery JC (2010a) The flow fields involved in hydrodynamic imaging by blind Mexican cave fish (*Astyanax fasciatus*). Part I: open water and heading towards a wall. J Exp Biol 213(22):3819–3831

Windsor SP, Norris SE, Cameron SM, Mallinson GD, Montgomery JC (2010b) The flow fields involved in hydrodynamic imaging by blind Mexican cave fish (*Astyanax fasciatus*). Part II: gliding parallel to a wall. J Exp Biol 213(22):3832–3842

Yang Y, Chen J, Engel J, Pandya S, Chen N, Tucker C, Coombs S, Jones DL, Liu C (2006) Distant touch hydrodynamic imaging with an artificial lateral line. Proc Natl Acad Sci USA 103(50):18891–18895

Yang Y, Nguyen N, Chen N, Lockwood M, Tucker C, Hu H, Bleckmann H, Liu C, Jones DL (2010) Artificial lateral line with biomimetic neuromasts to emulate fish sensing. Bioinspiration Biomimetics 5(1):16001

Yang Y, Klein A, Bleckmann H, Liu C (2011) Artificial lateral line canal for hydrodynamic detection. Appl Phys Lett 99(2):023701